The Facts On File

DICTIONARY OF MILITARY SCIENCE

The Facts On File

DICTIONARY OF MILITARY SCIENCE

JAY M. SHAFRITZ

TODD J. A. SHAFRITZ

DAVID B. ROBERTSON

Facts On File

New York • Oxford

The Facts On File Dictionary of Military Science

Facts On File, Inc.
460 Park Avenue South
New York, NY 10016
USA

or

Facts On File Limited
Collins Street
Oxford OX4 1XJ
United Kingdom

Library of Congress Cataloging-in-Publication Data

Shafritz, Jay M.
The Facts On File dictionary of military science / by Jay M. Shafritz, Todd J. A. Shafritz, David B. Robertson.
p. cm.
Bibliography: p.
ISBN 0-8160-1823-5
1. Military art and science—Dictionaries. I. Shafritz, Todd J. A. II. Robertson, David B. III. Facts On File, Inc. IV. Title. V. Title: Dictionary of military science.
U24.S47 1989 88-28648
355'.003'21—dc19 CIP

British CIP data available on request.

Jacket Design by Linda Kosarin
Composition by Vail-Ballou
Manufactured by Maple-Vail
Printed in the United States of America

10 9 8 7 6 5 4 3 2 1

This book is printed on acid-free paper.

Preface

We live in an era of perpetual garrison states, in which preparation for total war is an everyday, as opposed to a sometime, fact of life. Thus the concerns of military science are far broader than in an earlier, more innocent time when war (and its preparations) confined itself to an occasional outburst and did not dominate the daily life of a nation. Because millitary affairs are now so thoroughly integrated with all other aspects of national life, it is almost impossible to distinguish accurately concerns that are purely military from those that are purely industrial, educational, scientific or of some other nature. While the U. S. Department of Defense has about two million men and women under arms (in uniform), it also employs a million more civilians. Defense contractors employ millions more. This reality presents major problems for defining military science.

When one considers the many traditional noncombatants involved in modern military concerns, one must immediately conclude, as the authors did, that the coverage of the subject must go beyond any immediate battlefield. But how far do we delve into foreign policy, logistical support, nuclear physics, industrial mobilization, law, and the deepest cauldron of all—military history, which, lamentably, in itself is almost the whole history of humankind. Having established that the legitimate concerns of military science—the ways and means as well as the hows and whys of military affairs—are as big as the great outdoors, how could we cut it down to a reasonable size for a single volume? We immediately developed a rule of thumb for guidance: We decided that this dictionary should focus on the strategy, tactics, and technology of fighting; on the terms and techniques of warfare that have been the essence of Western combat since the age of Napoleon. Thus this is a dictionary that on the whole concentrates on the means used to kill people in large numbers; and equally important, to defend people from an aggressor's intentions. We hope it will be as useful to the professional military, whose job it is to prepare for war, as to concerned citizens, who ultimately determine whether a nation will start or sustain a war.

Nevertheless, while we have a specific focus in this book, we will not ignore peripheral yet critical concerns such as logistics, law, administration, and history. We constantly had to make judgment calls about how far to go into such related areas. In the end we wanted a dictionary that would be essential for any commissioned or noncommissioned officer—one that contained definitions of the vocabulary of his or her profession and brief discussions of all of its major strategic doctrines and concerns. The length of each entry reflects a judgment of how important each topic was considered. The entries also reflect a bias toward the military concerns of the United States and its North Atlantic Treaty Organization (NATO) allies. This is what you might expect from two Americans and a Briton. The authors have an international focus; but it is unavoidably the perspective of the West.

The references at the end of entries serve both as examples of the usage of the entry, as well as sources for further information. A word or phrase that appears in SMALL CAPITAL LETTERS is defined in another part of the dictionary. Perhaps the greatest single source of information in our preparation of the book was the United States Government. Extensive use was made of materials in the public domain, printed by a variety of federal agencies (especially the U. S. Department of Defense and the U. S. Army), and we acknowledge the legions of anonymous public servants who over the years have produced the various manuals, glossaries, directories, and guidelines that were so useful.

We thank the following individuals for reviewing the manuscript and making literally thousands of helpful suggestions for improvement: J. Michael Kelly of the Department of the Air Force of the United States Department of Defense; Frederick C. Thayer of the University of Pittsburgh (Colonel, USAF, ret.); John McAllister (Colonel, United States Army Reserve); Donald Goldstein of the University of Pittsburgh (Lt. Col., USAF, ret.); the Facts On File editors (Kate Kelly, Neil Maillet, Kenneth Lane, and Joe Reilly); Albert C. Hyde, Tim Tieperman, Gregory Scott, and Chad Pfeffer of the University of Pittsburgh; Laurence J. Korb of the Brookings Institution; Harry A. Bailey, Jr. of Temple University; Sir Michael Howard, Regius Professor of History in Oxford University; Liz Oberle-Robertson of Jones, Day, Reavis and Pogue, New York; John Mroz, President of the Institute for East-West Security Studies; Stephen Larrabee, Vice-President of the Institute for East-West Security Studies; Rachael Tricket, Principal of St. Hugh's College, Oxford; and A. I. Marsh of St. Edmund Hall, Oxford.

While we had considerable assistance in producing this book, all mistakes, omissions, or other flaws found in it are solely our responsibility. Yet we remain hopeful that as the years go by, this work will warrant a subsequent edition. We would therefore encourage those of you who might care enough to help us do it better the next time around to communicate with us.

Jay M. Shafritz
Todd J. A. Shafritz
David B. Robertson

A

A Attack, when used to designate a type of aircraft, as in the U.S. Navy's A-6 or A-7.

AA (or AAA) Antiaircraft artillery.

AAGS See ARMY AIR-GROUND SYSTEM.

AAM See AIR-TO-AIR MISSILE.

abatis 1. A road OBSTACLE constructed by felling trees on both sides of a roadway in such a manner that they fall, interlocked, at a 45-degree angle to the roadway. 2. Wooden stakes, usually trees that have been felled and sharpened to a point at one end, placed to make it difficult to storm defensive positions.

ABC warfare The use of atomic, biological, and chemical weapons. See entries for ATOMIC BOMB, BIOLOGICAL WARFARE, and CHEMICAL WARFARE. Compare to NBC WARFARE.

ablative shield A shield that evaporates when heated, absorbing energy and protecting the object behind it from heat damage.

ablative shock The generation of a mechanical SHOCK WAVE at the surface of an object exposed to intense, pulsed ELECTROMAGNETIC RADIATION. A thin layer of the object's surface rapidly boils off; the resulting vapor exerts pressure against the surface, generating a mechanical shock wave. This wave then propagates deeper into the object and can cause melting, vaporization, and spallation of surface material, and structural failure of the object.

ABM See ANTIBALLISTIC MISSILE.

A-bomb See ATOMIC BOMB.

abort 1. To fail to accomplish a mission for any reason other than enemy action. It may occur at any point from the initiation of an operation to its completion. 2. To discontinue aircraft takeoff or launch.

about face A command to make a 180-degree turn, usually given to soldiers standing at attention.

absent without leave (AWOL) Absence without proper authority from the properly appointed place of duty, or from a unit, organization, or other place of duty at which one is required to be at the time prescribed. Commonly pronounced "A-wall," this may be a court-martial offense. In peacetime one can be AWOL for up to 30 days, after which AWOL is usually considered desertion.

absolute altitude The height of something, usually an aircraft, with respect to the ground below it, as opposed to sea level.

absolute deviation The shortest distance between the center of the TARGET and the point where a PROJECTILE hits or bursts.

absolute dud A nuclear weapon that when fired at or dropped on a target fails to explode. Compare to DUD.

absolute error 1. The shortest distance between the CENTER OF IMPACT of a group of shots and the point of impact or burst of a single shot within the group. 2. The error of a sight, consisting of its error in relation to the reference sight with which it is tested, including the known error of the reference sight. Relative error, which is a part of absolute error, includes only the error of a sight in relation to a reference sight.

absolute war 1. A TOTAL WAR. 2. All-out NUCLEAR WAR.

absolute weapon 1. A weapon that would cause so much destruction that activating it can never be an alternative. This is a purely theoretical concept because humankind (being an imperfect species) can never make the "perfect" weapon. 2. Any NUCLEAR WEAPON that functions as a deterrent. See also BERNARD BRODIE.

absorbed dose The amount of ionizing radiation, measured in RADS or REMS, that is absorbed into the body of a particular human.

ACA See AIRSPACE COORDINATION AREA.

acceptable casualties 1. The number of killed and wounded one side in a military engagement is willing to endure before making a decision to withdraw and accept a defeat. 2. A strategic planning concept for the number (in the tens of millions) of civilian deaths one side might be willing to accept to "win" a nuclear war.

acceptance trial The testing of a new weapon or item of equipment, carried out by nominated representatives of its eventual military users, to determine whether the specified performance and characteristics have been met; also called operational test and evaluation (OT&E).

accidental attack An unintended attack that occurs without deliberate national design as a direct result of a random event, such as a mechanical failure, a human error, or an unauthorized action by a subordinate individual.

accidental war A war started not by the formal leader of either side of a conflict, but by a technical malfunction or an act of insubordination. The nuclear age has brought about a pervasive fear of a total nuclear war triggered by accident. In the past, a mistaken belief by one party that it was under attack from an enemy could be relatively easily verified. But accidental war is now seen to be much more likely, and still more disastrous, since automatic alert systems could theoretically launch missiles that cannot be recalled. Should either superpower launch a major attack on the other, retaliation would have to be ordered so rapidly that there would be virtually no time for the two opponents to assess the situation and agree that neither side intended aggression. See Daniel Frei, and Christian Catrina, *Risks of Unintentional Nuclear War* (London: Taylor and Francis, 1982).

accompanied tour A tour of duty on which dependents (spouses or children) are allowed to accompany the service member, as opposed to an unaccompanied tour, on which dependents are not allowed.

accompanying supplies 1. The basic load of AMMUNITION and other supplies that are packaged and loaded by a unit or carried by individual soldiers. 2. All supplies carried by the ASSAULT and FOLLOW-UP ECHELONS in an airborne or amphibious operation.

accordion effect The tendency of COLUMNS of soldiers or vehicles to bunch up or stretch out as they move alternately across broken and flat terrain.

accountable strength An administrative term referring to all personnel assigned to a unit regardless of DUTY STATUS. It includes those who are present for duty, absent from duty, in transit incoming, and in transit outgoing.

accuracy life The estimated average number of ROUNDS that a particular weapon can fire before its tube becomes so worn that its accuracy tolerance is exceeded.

accuracy of fire 1. The precision of FIRE expressed by the closeness of a grouping of shots at and around the center of the target. 2. The measure of the deviation of fire from the point of aim, expressed in terms of the distance between the point of aim and the mean of bursts.

accuser In MILITARY LAW, a person who signs and swears to charges, a person who directs that charges should nominally be signed and sworn to by any other person who has an interest, other than an official interest, in the prosecution of the accused.

ace An unofficial term for a fighter pilot who has shot down at least five enemy planes.

ACE Allied Command Europe, a command organization within NATO.

ack ack 1. Antiaircraft artillery. 2. The fire from antiaircraft artillery.

acknowledgment A message from an addressee informing the originator that a communication has been received and is understood.

acoustic Having to do with sound and its associated vibrations.

acoustic jamming The deliberate radiation or reradiation of mechanical or electro-acoustic signals with the objective of obliterating or obscuring signals that the enemy is attempting to receive and of deterring enemy weapon systems.

acoustic mine A MINE with an acoustic circuit that responds to the sound or vibration field of a ship or minesweeping operations.

acoustic mine hunting The use of SONAR to detect mines or mine-like objects that may be on or protruding from the seabed, or buried.

acoustic warfare The use of underwater acoustic devices to determine, exploit, reduce, or prevent hostile use of the underwater acoustic spectrum.

acoustical surveillance The employment of electronic devices, including sound-recording, -receiving, or -transmitting equipment, for the collection of information.

ACP See AIR CONTROL POINT.

acquire To obtain the exact position of a target visually, electronically, or by other means so that FIRE can be directed upon it.

acquisition The detection of a potential target by the SENSORS of a WEAPONS SYSTEM.

action 1. Any military engagement between opposing forces. Marshal Foch (1851–1929) of France insisted that "action is the governing rule of war." CLAUSEWITZ elaborated: "A fundamental principle is never to remain completely passive, but to attack the enemy frontally and from the flanks, even while he is attacking us." 2. A command ordering weapon crews to prepare to fire in any direction designated by the leader. 3. Responsibility, as when an officer is assigned an action on a particular matter.

action, direct See DIRECT ACTION.

action, economic See ECONOMIC ACTION.

action agent An INTELLIGENCE operative who has access to and performs actions against a target.

action deferred A tactical action on a specific track that is temporarily delayed for better tactical advantage.

action front (rear) (right) (left) A command ordering small units and weapon crews to prepare to fire to the front (rear) (right) (left).

action left See ACTION FRONT.

action link, permissive See PERMISSIVE ACTION LINK.

action mine, delayed See DELAYED ACTION MINE.

action mission, direct See DIRECT ACTION MISSION.

action officer (AO) 1. The officer responsible for an assigned task. 2. The key person in an operation.

action-reaction 1. A phenomenon often used to explain behavior during international crises. It implies that each actor responds to the behavior of others with pre-planned moves, and that the sequence of actions as a crisis unfolds is not necessarily directed by long-term goals or genuine motivation. The classic example remains the events of 1914, where the initial response to the assassination of Archduke Ferdinand of Austria set off a chain reaction: mobilization by Russia sparked mobilization by Germany, which in turn led to mobilization by others. In each case one participant was reacting in a more or less automatic way to another, rather than making a deliberately thought-out move. 2. In an arms race context, the action-reaction pattern occurs when the deployment of a new weapons system by one side automatically leads to the development of an equivalent or superior system by the opposition. Its automaticity is evidenced by the fact that the first actor most certainly did not deploy the system simply to have its opponent do likewise, and that the opponent had no intention, originally, of developing (or purchasing) such weaponry. See Alex Mintz, "Arms Exports as an Action-Reaction Process," *Jerusalem Journal of International Relations* 8, 1 (March 1986).

action rear See ACTION FRONT.

action right See ACTION FRONT.

action station 1. An assigned duty in anticipation of or in actual combat. 2. An assigned position to be taken by an individual in case of air attack.

activate 1. To put into existence by official order a unit, post, camp, station, base, or shore activity that has previously been constituted and designated by name or number, or both, so that it can be organized to function in its assigned capacity. 2. To prepare for active service a naval ship or craft that has been in an inactive or reserve status. See also COMMISSION; CONSTITUTE. 3. To turn on equipment.

activated mine A MINE having a secondary FUZE that will cause detonation when the mine is moved or otherwise interfered with. The device may be attached either to the mine itself or to a second mine, or to an auxiliary charge beside or beneath the mine.

active 1. An adjective describing full-time military service. 2. In SURVEILLANCE, an adjective applied to actions or equipment that emit energy capable of being detected.

active aircraft Aircraft currently and actively engaged in supporting flying missions, either through direct assignment to operational units or in the preparation for such assignment or reassignment through any of the logistic processes of supply, maintenance, and modification.

active air defense Direct defensive action taken to destroy attacking enemy aircraft or missiles or to nullify or reduce the effectiveness of such attack. It includes such measures as the use of AIRCRAFT, interceptor missiles, air defense artillery, non-air defense weapons in an air-defense role, and ELECTRONIC COUNTERMEASURES and COUNTER-COUNTERMEASURES.

active ballistic missile defense Direct defensive action taken to intercept and destroy, or reduce the effectiveness of, an enemy attack by BALLISTIC MISSILES. It includes, but is not limited to, such measures as the use of ANTIBALLISTIC MISSILES (ABMs) and ELECTRONIC COUNTERMEASURES. See also BALLISTIC MISSILE DEFENSE.

active defense 1. The employment of limited offensive action and counterattacks to deny a contested area or position to the enemy. 2. A flexible and elastic defense used by mechanized and armored forces to defeat the attacker by confronting him aggressively and continually, with strong combined arms teams fighting from mutually supporting battle positions organized in-depth.

active duty Full-time duty in a military service without regard to duration or purpose. This is in contrast to part-time reserve duty.

active homing guidance A system of HOMING GUIDANCE wherein both the source for illuminating the target and the receiver for detecting the energy reflected from the target as the result of illuminating the target are carried within the missile.

active list 1. A list of all officers currently in service. 2. A list of all active units assigned to a command.

active material Material, such as plutonium and certain isotopes of uranium, that is capable of supporting a nuclear-fission chain reaction.

active measures Operations by the Soviet Union designed to influence policies of other nations that are distinct from normal diplomacy and traditional espionage. Active measures include the use of communist parties and front organizations abroad; the dissemination of false rumors (i.e., disinformation); forgery of documents; manipulation of the press; and personal and economic blackmail. The KGB has primary responsibility for developing and implementing active measures. Examples of active measures directed against the United States are campaigns "sponsored" by local communist parties to have U. S. military bases removed in Greece and Spain. See Lawrence S. Eagleburger, "Unacceptable Intervention: Soviet Active Measures," *NATO Review* 31, 1 (April 1983); John J. Dziak, "Soviet 'Active Measures,' " *Problems of Communism* 33, 5 (November–December 1984).

active mine A MINE to be actuated by the reflection from a target of a signal emitted by the mine.

active satellite defense Direct defensive action taken to destroy or reduce the effectiveness of enemy satellite capabilities; for example, ANTISATELLITE WEAPONS, ELECTRONIC COUNTERMEASURES, SATELLITE DEFENSE.

active sector An area where combat is occurring.

active sensor A sensor that illuminates a target, producing return secondary radiation that is then detected in order to track or identify the target, or both.

activity 1. A unit, organization, or installation performing a function or mission, e.g., reception center, redistribution center, naval station, or naval shipyard. 2. A function or mission, e.g., recruiting, schooling.

actual ground zero See GROUND ZERO.

actual range In bombing, the horizontal distance a bomb travels from the instant of its release until the time of its impact.

actuated mine A MINE whose detecting element has been operated and either electrically signaled a control station or caused the mine to explode.

acute radiation dose The total dose of ionizing radiation received over a period so short that biological recovery cannot occur.

ACV See AIR CUSHION VEHICLE.

adaption kit Those items that must be associated with a WARHEAD to install it in a delivery vehicle, thereby permitting the mating of the warhead and its carrier.

ADC 1. An AIDE-DE-CAMP. 2. An assistant division commander. 3. Aerospace Defense Command; formerly Air Defense Command.

add In ARTILLERY AND NAVAL GUNFIRE SUPPORT, a correction used by an observer/spotter to indicate that an increase in range is desired along a SPOTTING LINE.

additive A number (or series of numbers) or alphabetical intervals added to code, cipher, or plain text to encipher it. It is often referred to as the key.

ADGE See AIR DEFENSE GROUND ENVIRONMENT.

adjust To align a weapon, especially ARTILLERY.

adjusted elevation Elevation based on firing and computed to place the CENTER OF IMPACT of a projectile on the target.

adjusted range 1. The range corresponding to the ADJUSTED ELEVATION. 2. The range setting, based on firing, computed to place the CENTER OF IMPACT on the target.

adjust fire In ARTILLERY AND NAVAL GUNFIRE SUPPORT, an order or request to initiate an ADJUSTMENT OF FIRE.

adjusting point A distinctive terrain feature or some portion of a target, at or near the center of the area on which the observer wishes to place FIRE.

adjustment of fire The process used in artillery and naval gunfire to obtain the correct bearing, range, and height of a burst (if TIME FUZES are used) when engaging a target by observed fire. See also BRACKETING.

adjutant An officer acting as the chief administrative assistant to a unit commander. Usually only units of the battalion or regimental level have adjutants. In the British Army, the adjutant is also responsible for discipline among junior officers. In other European armies the title is used for senior non-commissioned officers.

adjutant general 1. The ADJUTANT of a unit having a general staff. 2. The chief administrative officer of an army. 3. An administrative officer in the U.S. Air Force. 4. The head of a National Guard unit.

adjutant general corps An army branch that specializes in administrative and personnel-management matters.

adjutant's call A bugle call announcing that the ADJUTANT is about to form the guard, battalion, or regiment for a ceremony.

administration 1. The management and execution of all military matters not included in TACTICS and STRATEGY; primarily in the fields of logistics and personnel management. 2. The internal management of military units. ALFRED THAYER MAHAN wrote in his 1903 *Naval Administration and Warfare:* "There has been a constant struggle on the part of the military element to keep the end—fighting, or readiness to fight—superior to mere administrative considerations . . . The military man, having to do the fighting, considers that the chief necessity; the administrator equally naturally tends to think the smooth running of the machine the most admirable quality." But it remained for novelist Leo Rosten to observe in his World War II novel, *Captain Newman, M.D.*, that "It is difficult to know whether a man is a good administrator because he is so busy, or a bad one for the same reason."

administrative chain of command See OPERATIONAL CHAIN OF COMMAND.

administrative control The direction or exercise of authority over subordinate or other organizations with respect to administrative matters such as personnel management, supply, services, and other matters not included in the operational missions of the subordinate or other organizations. See also CONTROL; OPERATIONAL COMMAND; OPERATIONAL CONTROL.

administrative estimate A survey made by a command or staff of the necessary arrangements for the supply, transportation, evacuation, and other administrative matters relating to a military force. An administrative estimate is used as the basis for an ADMINISTRATIVE PLAN.

administrative landing An unopposed landing involving debarkation from vehicles that have been administratively loaded. See also ADMINISTRATIVE LOADING; ADMINISTRATIVE MOVEMENT.

administrative leadtime The time interval between the initiation of a procurement action and the letting of contract or placing of an order. See also PROCUREMENT LEADTIME.

administrative loading A loading system that gives primary consideration to achieving the maximum utilization of troop and cargo space without regard to tactical considerations. Equipment and supplies must be unloaded and sorted before they can be used. See also LOADING.

administrative map A map on which information is graphically recorded that pertains to administrative matters, such as supply and evacuation installations, personnel installations, and medical facilities.

administrative movement The movement of troops, vehicles, or both, arranged to conserve time and energy when no enemy interference, except by air, is anticipated. Also called an administrative march.

administrative order 1. An order covering traffic, supply, maintenance, evacuation, personnel, and other administrative details. 2. Any order not related to combat.

administrative plan A plan, normally relating to and complementing the OPERATION PLAN or ORDER, that provides information and instructions covering the logistic and administrative support of an operation. It must be based on a survey of the situation, called the ADMINISTRATIVE ESTIMATE, and is put into effect by the ADMINISTRATIVE ORDER.

administrative restriction A form of command restraint that a commanding officer may, within his discretion and without imposing arrest, use to restrict a person to specified areas of a military command with the further provision that the person so restrained will participate in all military duties and activities of his organization while under such restraint. Administrative restrictions are only imposed for minor infractions and are usually of short duration.

administrative shipping Support shipping that is capable of transporting troops and cargo from a point of origin to a destination, but which cannot be loaded or unloaded without non-combat personnel, equipment, or both.

administrative unit 1. A unit organically able to do its own interior management. It may be both administrative and tactical. 2. A unit organized and used for purposes of administration.

admiral The highest ranking or a FLAG OFFICER of a navy; comparable to a general in an army. It was Voltaire in his 1759 novel *Candide* who is usually first credited with observing that "an admiral has to be put to death now and then to encourage the others." He was referring to the presumed policy of the British Royal Navy.

admiralty 1. The naval agency in the United Kingdom roughly parallel to the U.S. Department of the Navy. 2. Maritime law; the general laws of the sea as modified by Congress and applicable on both the high seas and internal navigable waters. Article III, Section 2, of the U.S. Constitution provides that the federal courts will have exclusive jurisdiction in all cases of admiralty law. 3. A court that handles maritime cases, such as collisions at sea. 4. The specific office or jurisdiction of an admiral.

Adolphus, Gustavus See GUSTAVUS ADOLPHUS.

advance by bounds To move forward in a series of separate advances, usually from cover to cover or from one point of observation to the next.

advance by echelon Advance by separate elements of a command moving at different times.

advanced base A base located in or near a THEATER OF WAR, whose primary mission is to support military operations in the field.

advance detachment The leading element of an ADVANCE GUARD. It is set out from the advance guard. See also ADVANCE PARTY, POINT MAN.

advanced fleet anchorage A secure anchorage for a large number of naval ships, mobile support units, and auxiliaries located in or near a theater of operations.

advanced individual training Training given to enlisted personnel subsequent to completion of basic training, so as to render them qualified for the award of a military occupational specialty.

advanced unit training The final stages of unit training, in which small (company-size) units are trained together in rehearsal of their role in the mission of the parent organization.

advance force A temporary organization within an AMPHIBIOUS TASK FORCE that precedes the main body to an OBJECTIVE AREA. Its function is to participate in preparing the objective for the main assault by conducting such operations as RECONNAISSANCE, seizure of supporting positions, MINESWEEPING, PRELIMINARY BOMBARDMENT, underwater demolitions, and air support.

advance guard A detachment sent ahead of the main force to insure its uninterrupted advance; to protect the main body against surprise; to facilitate the advance by removing obstacles and repairing roads and bridges; and to COVER the deployment of the main body if it is committed to action. It may or may not be preceded by a separate COVERING FORCE. In the absence of a covering force, it seeks to find and exploit gaps in the enemy's defensive system; to prevent the main body of the advancing force from running blindly into enemy positions; and to clear away minor opposition. While NAPOLEON wrote that "the duty of an advance guard does not consist in advancing or retiring, but in maneuvering," British Field Marshall William Slim (1891–1970) maintained that nevertheless, "the first duty of an advance guard is to advance." See also ADVANCE PARTY.

advance-guard reserve The second of the two main parts of an ADVANCE GUARD, the other being the ADVANCE-GUARD SUPPORT. It protects the main force and is itself protected by the advance guard support. Small advance guards do not have reserves.

advance-guard support The first of the two main parts of an ADVANCE GUARD, the other being the ADVANCE-GUARD RESERVE. It is made up of three smaller elements, in order from front to rear, the advance guard point, the advance guard party, and the support proper. The advance guard support protects the advance guard reserve.

advance officer An officer designated to precede a column in order to reconnoiter the route of march and to select alternate routes or detours if required, to place guides and route markers where appropriate, and to take other necessary action. This officer may also command the advance party.

advance party 1. A security element of an ADVANCE GUARD. It is sent out from, and precedes, the ADVANCE-GUARD SUPPORT on a march. It sends forward and is preceded by the advance-guard point. 2. A group of unit representatives dispatched to a port area in advance of the main body of the unit for the purpose of identifying and supervising the shipside delivery of the unit's equipment. 3. A group of unit representatives dispatched to a probable new site of operations (or base camp) in advance of the main body of the unit for the purpose of arranging for the unit and its equipment upon arrival. See also ADVANCE DETACHMENT.

advance rate The speed with which a military unit advances, often expressed as miles or kilometers per day.

advance to contact An offensive operation designed to gain or re-establish contact with the enemy. See also APPROACH MARCH.

adventurism The implementation of ill-considered, provocative, or dangerous international policies; the term implies a major disregard for the normal standards of international behavior. The word comes from the Russian *avantyurizm,* and was often used by the Soviets to denounce U.S. foreign policies during the Cold War. See Christopher Coker, "Adventurism and Pragmatism: The Soviet Union, COMECON, and Relations with African States," *International Affairs (Great Britain)* 57, 4 (Autumn 1981).

advisors 1. Military forces or civilian military experts from one country sent to "advise" the military forces of another country. Sometimes such advisors are limited to teaching tactics and weapons usage. At other times such advisors become more like "lead workers" and show their "students" how to take action against an actual enemy in the field. 2. A euphemism for the military forces of one country sent to assist the military forces of another country. For example, the first U.S. forces in the Vietnam War were called "advisors." Compare to MAAG.

AEC Atomic Energy Commission. See NUCLEAR REGULATORY COMMISSION.

Aegis A totally integrated shipboard WEAPON SYSTEM that combines computers, radar, and missiles to provide an area defense umbrella for surface shipping. The system is capable of automatically detecting, tracking, and destroying airborne, seaborne, and land-launched weapons.

aerial port An airfield that has been designated for the sustained air movement of PERSONNEL and MATERIEL, and to serve as an authorized port for entrance into or departure from the country in which it is located.

aerial reconnaissance Operations conducted from aircraft to obtain, by visual observation or other detection methods, information about the activities and resources of an enemy or potential enemy; or to secure data concerning the meteorological, hydrographic, or geographic characteristics of a particular area.

aerodynamic missile A missile that uses aerodynamic forces to maintain its flight path generally employing propulsion guidance. See also BALLISTIC MISSILE; GUIDED MISSILE.

aeromedical evacuation The movement of patients to and between medical treatment facilities by air transportation. It is a major factor in the decreasing mortality rate of combat-related injuries since World War II. The evacuation by air from the battlefield itself became feasible during the Korean War with the advent of helicopters. In the Vietnam conflict, the vast majority of American casualties were transported from the battlefield by air ambulance helicopters. See also DUSTOFF.

aeronautical chart A representation of a portion of the Earth, its CULTURE and relief, specifically designed to meet the requirements of air navigation.

affirmative 1. Response signifying agreement with a message or communication. 2. Response confirming that a message has been received.

aft The rearmost part of a boat or ship.

afterburner A device within or attached to a jet engine exhaust pipe for the afterburning of exhaust gases.

afterburning 1. The process of fuel injection and combustion in the exhaust of a jet engine which produces additional thrust and speed. 2. The characteristic of some ROCKET motors to burn irregularly for some time after the main burning and thrust has ceased.

afterwinds Wind currents set up in the vicinity of a nuclear explosion directed toward the burst center, resulting from the updraft accompanying the rise of the FIREBALL.

AFV See ARMORED FIGHTING VEHICLE.

agency 1. In intelligence usage, an organization or individual engaged in collecting and/or processing information. See also AGENT; INTELLIGENCE CYCLE; SOURCE. 2. An organization under the direct supervision of a headquarters; it can be functionally described as having either a staff-support or field-operating mission. 3. A unit or organization that has primary responsibility for performing duties or functions as representative of, and within the assigned authority of, the headquarters to which it is subordinate. 4. A slang term for the Central Intelligence Agency.

agent In INTELLIGENCE usage, a person who is recruited, trained, controlled, and employed to obtain and report information.

agent, action See ACTION AGENT.

agent, antimaterial See ANTIMATERIAL AGENT.

agent authentication The technical support task of providing an INTELLIGENCE agent with personal documents, accouterments, and equipment that have the appearance of authenticity as to the agent's claimed origin, and which support and are consistent with the agent's cover story.

agent net An organization for clandestine purposes which operates under the direction of a principal agent.

agent orange The predominant chemical agent used by the United States during the Vietnam War to defoliate dense jungle cover and destroy enemy food supplies. While the U.S. Department of Defense maintained that the chemical was "not harmful," it was later held responsible for deformed (or stillborn) children of the Vietnamese and the U.S. Vietnam veterans as well as for high cancer rates among these groups. In 1984 the makers of agent orange, after extended litigation, agreed to pay $180 million for the medical costs of U.S. Vietnam veterans and their families. See James B. Jacobs and Dennis McNamara, "Vietnam Veterans and the Agent Orange Controversy," *Armed Forces and Society* 13, 1 (Fall 1986).

agent provocateur A person hired by an organization or country to associate himself with opposing rival organizations by feigning sympathy with their aims so that they can be induced to take actions contrary to their interests.

aggression The unprovoked attack by one state on another. What one state may see as unprovoked, another may view as justifiable retaliation. Utlimately, each nation defines aggression in its own interest. The Charter of the United Nations does not define aggression, although Article 39 holds that the "Security Council shall determine the existence of any threat to the peace, breach of the peace, or act of aggression . . ." But this provision did not stop the UN General Assembly from defining agression in a 1974 resolution: "Aggression is the use of armed force by a State against the sovereignty, territorial integrity or political independence of another State, or in any other manner inconsistent with the Charter of the United Naitons . . ." According to this General Assembly definition, "The first use of armed force by a State in contravention of the Charter shall constitute prima facie evidence of an act of aggression, although the Security Council may, in conformity with the Charter, conclude that a determination that an act of aggression has been committed would not be justified in the light of other relevant circumstances, including the fact that the act concerned or [its] consequences are not of sufficient gravity." But in the end this resolution backs down from its assertion that "no consideration of whatever nature, whether political, economic, military or otherwise, may serve as a justification for aggression," when it sanctions specified categories of aggression by asserting that "nothing in the definition in any way prejudices the right to self-determination, freedom and independence of peoples forcibly deprived of that right . . . particularly peoples under colonial and racist regimes or other forms of alien domination; nor the right of these peoples to struggle to that end and to seek and receive support." 2. ECONOMIC WARFARE. See Vernon Cassin, Whitney Debevoise, Howard Kailes, and Terence W. Thompson, "The Definition of Aggression," *Harvard International Law Journal* 16, 3 (Summer 1975); Julius Stone, "Hopes and Loopholes in the 1974 Definition of Aggression," *American Journal of International Law* 71, 2 (April 1977); Jack I. Garvey, "The UN Definition of 'Aggression': Law and Illusion in the Context of Collective Security," *Virginia Journal of International Law* 17, 2 (Winter 1977).

aggressor forces Those forces engaged in aggressive military action. In the context of training exercises, the "enemy" created to add realism in training maneuvers and exercises.

AGOC See AIR GROUND OPERATIONS CENTER.

AGOS See AIR-GROUND OPERATIONS SYSTEM.

aid, landing See LANDING AID.

aid, mutual See MUTUAL AID.

aide-de-camp (ADC) An officer serving as a personal assistant to a general.

aide-memoire A French term for an aid to the memory; an informal summary of a diplomatic event such as an interview, a conversation at a social gathering, or any other matter worth retaining for military files or sharing with colleagues.

aid station A medical treatment facility in or near a combat zone where medical care, limited health service, and the sorting and the disposition of sick, injured, and wounded personnel are accomplished under the technical supervision of a medical officer.

aiguillette The ornamental rope worn around the shoulder by a military ATTACHÉ, a general's aide, whole military units, or honorary societies such as the Arnold Air Society.

aimed fire FIRE that is intentionally placed on targets.

aiming circle An instrument for measuring horizontal and vertical angles. It is equipped with a magnetic needle so that magnetic azimuths can also be set off or read. An aiming circle is used in surveying and for similar work in connection with artillery or machine gun fire.

aiming point The common target point toward which ARTILLERY sights are adjusted. If no obvious naturally occurring point is available, aiming posts (poles painted with contrasting colors) are put in front of the guns and the sights adjust via angular movements relative to the aiming posts.

airborne 1. Personnel and equipment transported by air; e.g. airborne infantry. Airborne troops should not be confused with AIRPORTABLE troops, which simply can be moved great distances by air with all their equipment. Various and subtle gradations of airborne troops exist. A rough distinction is made between those troops that can be dropped by parachute directly over an AIRHEAD and those that can only be delivered by helicopter near a battle zone. The first category, equivalent to paratroops of the Second World War, is not very important in modern tactical doctrine (but could be critically important in LOW-INTENSITY CONFLICT or as an ADVANCE GUARD). Even during the Second World War there were very few completely successful paratroop drops. John Slessor wrote in his 1954 *Strategy for the West:* "It would probably astonish the reader were I able . . . to state the cost in manpower and materiel of the airborned forces of the late war, complete with all the aircraft and manpower devoted to training and carrying them . . . compared with their impact upon the enemy. They would certainly find no place in the early stages of another great war. They are too vulnerable." The second category, AIRMOBILE troops flown by helicopter to a point near or in a combat area, with long-range support from ground-based artillery of TACTICAL AIR FORCES, is much more important. Such success as the United States had in Vietnam was very much dependent on this form of troop transport. 2. Materials being or designed to be transported by aircraft, as distinguished from weapons and equipment installed in and remaining a part of the aircraft. 3. The state of an aircraft in flight. 4. A qualified PARATROOPER. 5. A response from airborne troops meaning "affirmative" or "can do;" normally said in a very loud voice. See Fletcher K. Ware, "The Airborne Division and a Strategic Concept," *Military Review* (March 1976); Mark L. Urban, "The Strategic Role of Soviet Airborne Troops," *Jane's Defense Weekly* (July 14, 1984); James F. Holcomb, "Soviet Airborne Forces and the Central Region," *Military Review,* 17, 11 (November 1987).

airborne alert A state of aircraft readiness wherein combat-equipped aircraft and airborne forces are ready for immediate action. It is designed to reduce reaction time and to increase survivability.

airborne assault weapon An unarmored, mobile, FULL-TRACKed gun providing a mobile antitank capability for airborne troops. It can be airdropped.

airborne beacon An infrared-light transmitter used to assist in the reorganization of forces at night; also used to mark DROP and LANDING ZONES.

airborne command post 1. A suitably equipped aircraft used by a commander for the control of his forces. 2. The always airborne command and control center of the SAC of the United States; known as "Looking Glass."

airborne early warning The warning transmitted to friendly units upon the detection of enemy air or surface units by radar or other equipment carried in an airborne vehicle.

airborne lift The total capacities expressed in terms of personnel and cargo that are, or can be, carried by available aircraft in one trip.

airborne operation The delivery by air of combat forces into an area for tactical or strategic purposes, and their LOGISTIC SUPPORT. Airborne operations are executed in four phases: MOUNTING PHASE, AIR MOVEMENT PHASE, ASSAULT PHASE, and SUBSEQUENT OPERATIONS PHASE.

airborne optical adjunct A set of SENSORS designed to detect, track and discriminate an incoming WARHEAD. The sensors are typically optical or infrared devices flown in an aircraft stationed above clouds.

air-breathing missile A MISSILE with an engine requiring the intake of air for combustion of its fuel, as is the case with jet engines. To be contrasted with the rocket missile, which carries its own oxidizer and can operate beyond the atmosphere.

airborne warning and control (AWAC) Air surveillance and weapons control provided by airborne early-warning vehicles equipped with search and height-finding radar and communications equipment. See also AIR PICKET, AWACS.

airburst An explosion of a BOMB or PROJECTILE above the surface of the earth, as distinguished from an explosion on contact with the surface or after penetration. An airburst causes destruction over a wider area, as opposed to a GROUND BURST, which in the context of nuclear weapons would only be used against HARD TARGETS.

airburst, high See HIGH AIRBURST.

airburst, low See LOW AIRBURST.

airburst, nuclear See NUCLEAR AIRBURST.

air cavalry Helicopter-transported infantry units with rapid response and support capabilities.

air command, tactical See TACTICAL AIR COMMAND.

air commodore An officer in some air forces (the United Kingdom, for example) equivalent to a BRIGADIER GENERAL.

air control point (ACP) An easily identifiable point on the terrain or an electronic navigational aid used to provide necessary control during air movement. Air control points are generally designated at each point at which the route of a flight makes a definite change in direction, and at any other point deemed necessary for timing or control of the operation.

air controller 1. An individual especially trained for and assigned the duty of traffic control (by use of radio, radar, and other means) of such aircraft as may be allotted for operation within an area. 2. A GCA (ground control approach) operator who is responsible for landings at air fields. 3. A GCI (ground control interceptor) who directs friendly aircraft to enemy targets.

air controller, forward See FORWARD AIR CONTROLLER.

air coordinator, tactical See TACTICAL AIR COORDINATOR.

air corridor A restricted air route of travel specified for use by friendly aircraft and created to prevent friendly aircraft from being fired on by friendly forces.

air cover 1. Airborne protection of ground troops from the enemy. 2. The specific aircraft or total forces providing this protection.

aircraft, active See ACTIVE AIRCRAFT.

aircraft, assault See ASSAULT AIRCRAFT.

aircraft carrier, attack See ATTACK AIRCRAFT CARRIER.

aircraft, lead See LEAD AIRCRAFT.

aircraft cross-servicing That servicing performed on an aircraft by an organization other than that to which the aircraft is assigned, and for which the assigned organization may be charged.

aircraft landing mat A prefabricated steel portable mat so designed that any number of planks (sections) may be rapidly fastened together to form surfacing for emergency runways; commonly referred to as pierced steel planking (PSP).

aircraft scrambling The immediate take-off of aircraft from a ground-alert condition of readiness. See SCRAMBLE.

air cushion vehicle (ACV) Any vehicle that rides on land or water on an air cushion created by powerful downward thrusting fans. Traditional air propellers then move the ACV forward.

air defense All defensive measures designed to destroy attacking enemy aircraft or missiles or to nullify or reduce the effectiveness of such attack. See Jim Bussert, "Soviet Air Defense Systems Show Increasing Sophistication," *Defense Electronics* (May 1984); David F. Bond, "The Return of Air Defense," *Air Force* (October 1987).

air defense, active See ACTIVE AIR DEFENSE.

air defense, passive See PASSIVE AIR DEFENSE.

air defense controller The person charged with the specific responsibility of controlling (by radio, radar, and other means) aircraft used in air defense; known as the ground control interceptor (GCI).

air defense early-warning station An installation located and equipped to detect and report the approach of hostile aircraft or missiles.

air defense ground environment (ADGE) The overall organization for air security, incorporating air-traffic control and air defenses over a region. For example, the system for air security over the whole of the United Kingdom is known as the UK ADGE.

air defense restricted area 1. An airspace in which there are special restrictive measures employed to prevent or minimize interference between friendly forces. 2. Airspace from which both civilian and military (unless specifically authorized) aircraft are excluded for reasons of safety or national security.

air defense warning conditions A degree of air raid probability according to the following code:

1. *Air defense warning yellow*—Attack by hostile aircraft, missiles, or both is probable. This means that hostile aircraft or missiles are en route toward an air defense division/sector, or unknown aircraft or missiles suspected to be hostile are en route.
2. *Air defense warning red*—Attack by hostile aircraft, missiles, or both is imminent or is in progress. This means that hostile aircraft or missiles are within an air defense sector or are in the immediate vicinity.
3. *Air defense warning white*—Attack by hostile aircraft, missiles, or both is improbable.

air defense weapons control status Three degrees of weapons fire control used by a commander to indicate the current rules concerning the fire of air defense weapons. See: WEAPONS FREE, WEAPONS TIGHT, and WEAPONS HOLD.

air doctrine, tactical See TACTICAL AIR DOCTRINE.

airdrop The unloading and delivery of personnel or materiel from aircraft in flight. It is usually done by parachute.

air force 1. Air power; whatever destructive capability can be brought to the enemy by air. 2. A military service that focuses on air and space operations regardless of whether it is independent (such as the U.S. Air Force) or part of another service (such as naval aviation). 3. An organization unit within a larger air command; such as the 15th Air Force in SAC or the 9th Air Force within TAC. GUILIO DOUHET wrote in his 1921 *The Command of Air:* "I have mathematical certainty that the future will confirm my assertion that aerial warfare will be the most important element in future wars, and that in consequence not only will the importance of the Independent Air Force rapidly increase, but the importance of the army and the navy will decrease in proportion."

air force, tactical See TACTICAL AIR FORCE.

airfield, dispersal See DISPERSAL AIRFIELD.

airfield, recovery See RECOVERY AIRFIELD.

airframe 1. The structural components of an airplane, including the framework and skin of such parts as the fuselage, empennage, wings, landing gear (minus tires), and engine mounts. 2. The framework, envelope, and cabin of an airship. 3. The assembled principal structural components, less the propulsion system, control, electronic equipment, and payload, of a MISSILE.

air-ground liaison code A set of symbols for a limited number of words, phrases, and sentences used for communications between air and ground forces. These symbols can be given by radio, telephone, blinker, or strips of cloth called ground-liaison panels. Also called air-ground code.

air-ground operations center A unit that coordinates air operations in the field.

air-ground operations system (AGOS) A system that provides a ground commander with the means for receiving, processing, and forwarding the requests of subordinate ground commanders for air-support missions and for the rapid dissemination of information and INTELLIGENCE.

air group, carrier See CARRIER AIR GROUP.

airhead 1. A designated area in a hostile or threatened territory which, when seized and held, insures the continuous air landing of troops and MATERIEL and provides the maneuvering space necessary for projected operations. Normally it is the area seized in the ASSAULT PHASE of an AIRBORNE OPERATION and is considered a DROP ZONE until it is secured. This would most probably be achieved by means of a paratroop landing, although in principle an AIRHEAD could also be gained by helicopter-delivered AIRMOBILE troops. The need is to capture, clear, and secure an area into which the heaviest equipment and less immediately combat-ready forces can be dropped. Once this activity has ceased the airhead becomes a base from which to advance in pursuit of military objectives. A slightly different example involves an airborne assault by paratroopers to secure an airfield to provide a simple route for the delivery of major forces of an AIRPORTABLE nature behind the enemy's lines. 2. A designated location in an area of operations used as a base for supply and evacuation by air. See also BEACHHEAD; BRIDGEHEAD. 3. Military slang for someone who isn't considered too intellectually able. See Michael B. Duncan, "Defending the Division Airhead," *Army Logistician* 15, 1 (January–February 1983).

airhead line A line described or portrayed in an OPERATION ORDER which marks the outside limit of that part of an AIRHEAD to be denied to the enemy.

air intercept control, common A tactical air-to-ground radio frequency, monitored by all AIR INTERCEPTION control facilities within an area, which is used as a backup for other discrete tactical control frequencies and for ground control intercept missions.

air interception To effect visual or electronic contact by a friendly aircraft with another aircraft. Normally, the air intercept of hostile aircraft is conducted in the following five phases:

1. *Climb phase*—Airborne to cruising altitude.
2. *Maneuver phase*—Receipt of initial vector to target until beginning transition to attack speed and altitude.
3. *Transition phase*—Increase or decrease of speed and altitude required for the attack.
4. *Attack phase*—Turn to attack heading, acquire target, complete attack, and turn to breakaway heading.
5. *Recovery phase*—Breakaway to landing.

air interception, broadcast-controlled See BROADCAST-CONTROLLED AIR INTERCEPTION.

air interdiction Air operations conducted to destroy, neutralize, or delay the enemy's military potential before it can be brought to bear effectively against friendly forces; this is a primary mission of a tactical air command. Done at such a distance from friendly forces that detailed integration of each air mission with the fire and movement of friendly forces is not required. See Edmund Dews, *Air Interdiction: Lessons from Past Campaigns* (Santa Monica, CA: Rand Corporation, 1981).

air interdiction battlefield (BAI) See BATTLEFIELD AIR INTERDICTION.

airland To disembark or unload troops and materiel after an aircraft has landed or while a helicopter is hovering.

airland battle concept The U.S. Army doctrinal concept that integrates TACTICAL AIR FORCES with ground troops into one concerted TACTICAL PLAN for an entire theater of war. In the European context this involves using airpower, and perhaps long-range army weapons, for an INTERDICTION attack on the second ECHELON Soviet forces while the ground troops hold and try to defeat the first echelon (see ECHELONED ATTACK). One hazard inherent to the air-land battle concept is its call for an integration of nuclear, chemical, and conventional weapons which might, in fact, lead to an early escalation to nuclear war. See Wayne M. Hall, "A Theoretical Perspective of Airland Battle Doctrine," *Military Review* 66, 3 (March 1986); John B. Rogers, "Synchronizing the Airland Battle," *Military Review* 66, 4 (April 1986); W. Baxter, *Soviet Airland Battle Tactics* (Novato, CA: Presidio Press, 1986). See also DEEP STRIKE and FOLLOW ON FORCES ATTACK.

airland operation An operation involving air movement, in which personnel and supplies land from the air at a destination for further deployment of units and personnel and further distribution of supplies.

air-launched cruise missile (ALCM) One of three variants of CRUISE MISSILE developed by the United States (as well as the Soviet Union) since the mid-1970s. Its main purpose is to maintain the utility of the U.S. Air Force bomber fleet, which is now highly vulnerable to Russian antiaircraft defenses. Air-launched cruise missiles can be carried to within range of Soviet targets and launched while the aircraft are still at a safe distance from antiaircraft defenses. Because ALCMs have a range of about 2,500 kilometers, bombers approaching the Soviet Union from one compass direction or another could target virtually any spot in Soviet territory. American nuclear strategy (like the French, but unlike the British) has always held that missiles launched from piloted aircraft form an indispensable part of the deterrent force, along with ICBMS and SLBMS. (The three types of missile systems together are known as the deterrent TRIAD).

air liaison officer An officer (aviator/pilot) attached to a ground unit who functions as the primary advisor to the ground commander on air operation matters.

airlift The carrying of troops and equipment over large distances by air to bring them into crisis areas rapidly. Because wide-bodied jets have made tourist travel so easy, it is often thought that military airlifts must be relatively easy. In fact the numbers of troops and the weight of equipment needed for any serious military activity are so enormous, compared with the still very restricted ability of aircraft to lift them, that no major power is even close to having sufficient airlift capacity for its possible strategic needs. With even a moderate amount of luggage, a big civilian aircraft, such as the Boeing 747, can carry only some 400 passengers. This is equivalent to about two companies, or only half a battalion, of troops. In fact, with the amount of weapons, equipment, ammunition, and supplies needed to keep a soldier active for even a few days, one company is probably the most that could be carried on a single aircraft. See Lloyd K. Mosemann II, "The Air Force's Solution to the Airlift Shortfall," *Defense Management* 19, 2 (1983); Thomas D. Pitsch, "The Airlift Master Plan: Evolution and Implementation," *Defense Management* 20, 4 (1984); David Wragg, *Airlift: The History of Military Air Transport* (Novato, CA: Presidio Press, 1987).

airlift, tactical See TACTICAL AIRLIFT.

airman The enlisted rank in the U.S. Air Force that is equivalent to a PRIVATE in the Army and a SEAMAN in the Navy; any enlisted rank below sergeant.

air maneuver forces AIR CAVALRY and attack helicopter units that operate in the ground environment. Their operations are similar to ground combat operations, and are integrated into the TACTICAL PLAN of the ground-force commander. They can dominate terrain by denying the enemy its use by direct aerial fire for limited periods.

air marshall A rank equivalent to general in some air forces. (This term is not used by United States forces.)

air mine A MINE dropped from aircraft, with or without a parachute, and designed for use against water targets, but sometimes used against land targets.

airmobile A term describing a unit that can be delivered directly into a battle zone by helicopters. Inevitably the unit must be lightly armed, but will usually be fighting within the range of support of artillery and ground-attack aircraft. Such units are to be distinguished from AIRBORNE troops in general, which include the traditional paratroops of World War II, and AIRPORTABLE troops, which are simply those that can be carried long distances by air. The first extensive use of AIRMOBILE fighting units was by the United States during the Vietnam War, in which whole regiments of infantry frequently were landed deep within enemy territory on SEARCH AND DESTROY MISSIONS.

airmobile field artillery A FIELD-ARTILLERY unit that is transported by helicopters in a tactical configuration to accomplish a field-artillery mission without using the transporting aircraft as a firing platform.

airmobile operation An operation in which combat forces and their equipment maneuver about the battlefield in helicopters under the control of a ground-force commander to engage in ground combat. Airmobile operations are conducted in four phases: the loading phase, the AIR-MOVEMENT PHASE, the landing phase, and the ground-operations phase.

airmobile support party An organization formed for employment in a landing zone to facilitate the assault landing and interim logistical support of elements in the LANDING ZONE.

airmobile task force A grouping of aviation and ground units under one commander for the purpose of carrying out a specific operation.

airmobile task force commander The senior ground commander responsible for the overall conduct of an AIRMOBILE OPERATION.

airmobility The capability of AIRMOBILE forces which permits them to move by air while retaining the ability to engage in ground combat.

air movement The air transport of units, personnel, supplies, and equipment, including AIRDROPS and air landings.

air movement phase The second phase of an AIRBORNE OPERATION or AIRMOBILE OPERATION, which begins with the takeoff of loaded aircraft from departure areas and ends with the delivery of units to their drop or landing zones.

air movement plan An AIRBORNE plan prepared jointly by ground and airlift units and covering that phase of an AIRBORNE OPERATION or AIRMOBILE OPERATION from the time that units have been loaded onto aircraft until they arrive in the objective area. It indicates unit-loading times at specific departure airfields or pickup zones, and includes the takeoff time, flight routes, order of flight, and arrival time over drop zones or landing zones. It is usually published as an annex to the operations plan.

air operations, tactical See TACTICAL AIR OPERATIONS.

air patrol, barrier combat See BARRIER COMBAT AIR PATROL

air patrol, combat See COMBAT AIR PATROL.

air picket An AIRBORNE early-warning aircraft positioned primarily to detect, report and track approaching enemy aircraft or missiles and to control intercepts. See also AIRBORNE WARNING AND CONTROL.

air platform Any weight-carrying vehicle capable of navigating under its own power above the earth's surface.

air policing The use of interceptor aircraft, in peacetime, to preserve the integrity of a specified airspace.

airportable A term referring to forces that can be carried considerable distances by aircraft, along with all their equipment. They are typically light forces, either purely infantry or at most accompanied with only light armor and artillery. They can be brought close to a battle area as long as there are suitable nearby air bases, but cannot be inserted directly into the battlefield. Thus they are distinguished from AIRBORNE and AIRMOBILE forces. In theory, any military unit is airportable if given enough time and aircraft to become so.

air power 1. A nation's air and space military capabilities. This would include destructive (bombers and missiles) power, transport capacities, and air defense. 2. The fire power or destructive potential of an air unit.

air raid A vague term meaning any attack from the air using any combination of weapons.

airship A blimp; any aircraft that is lighter than air.

air signal A signal from an aircraft to communicate between air and ground forces when radio cannot be used. An air signal may be made by dipping the wing, dropping a flame, or firing fireworks (such as a flare) that give off an intense colored light. Also called aircraft signal.

airspace coordination area (ACA) 1. A three-dimensional block of airspace in a target area in which friendly aircraft are reasonably safe from friendly surface fire. It may be imposed by division or higher level commanders, and constitutes a restrictive FIRE-CONTROL measure when in effect. 2. In FIRE-SUPPORT operations, a safety measure that establishes a three-dimensional area that is reasonably safe from friendly, surface-delivered, non-nuclear fire.

airspace management The coordination, integration, and regulation of airspace use in a defined area.

airspace reservation (or restriction) 1. The airspace located above an area on the surface of the land or water, designated and set apart by Executive Order of the president or by a state, commonwealth, or territory, over which the flight of aircraft is prohibited or restricted for the purpose of national defense or for other governmental purposes. 2. An agreement with an air-traffic control agency for the special use of airspace at a particular time and point; usually for a limited duration for a specific purpose, such as aerial refueling.

airspeed The speed of an aircraft relative to its surrounding air mass. The unqualified term "airspeed" can mean any one of the following:

1. *Indicated Airspeed:* the airspeed shown by an airspeed indicator.
2. *Calibrated Airspeed:* indicated airspeed corrected for instrumental installation error.
3. *Equivalent Airspeed:* calibrated airspeed corrected for compressibility error.
4. *True Airspeed:* equivalent airspeed corrected for error due to air density (altitude and temperature).

air strike An attack on specific objectives by fighter, bomber, or attack aircraft on an offensive mission. It may comprise several air organizations under a single command in the air. The 1925 President's Board to Study Development of Aircraft for the National Defense was prophetic when it analyzed the utility of air strikes and found that "new weapons operating in an element hitherto unavailable to mankind will not necessarily change the ultimate character of war. The next war may well start in the air but in all probability it will wind up, as did the last war, in the mud."

air superiority That degree of dominance in battle by the air forces of one entity over another that permits the conduct of operations by the former and its related land, sea, and air forces at a given time and place without prohibitive interference by the opposing air force. Air superiority consists of two elements: 1. being able to prevent enemy aircraft, especially bombers and reconnaissance planes, from operating over one's own lines; this requires a sizeable interceptor or FIGHTER force; and 2. being able to fly missions over the enemy's lines, attacking its troop concentrations and supply network, and observing military movements.

air support Military aircraft operations primarily designed to assist the land or sea operations of other units.

air support, close See CLOSE AIR SUPPORT.

air support, general See GENERAL AIR SUPPORT.

air support, immediate See IMMEDIATE AIR SUPPORT.

air support, indirect See INDIRECT AIR SUPPORT.

air support, offensive See OFFENSIVE AIR SUPPORT.

air supremacy That degree of air superiority wherein the opposing air force is incapable of effective interference with the operations of friendly ground and/or naval forces.

air surveillance The systematic observation of airspace by electronic, visual, or other means, primarily for the purpose of identifying and determining the movements of aircraft and missiles, friendly and enemy, in the airspace under observation.

air-to-air missile (AAM) A missile launched from an airborne vehicle at an air target; the Sidewinder is an example.

air-to-surface missile (ASM) A missile launched from an airborne vehicle at a surface target; a smart bomb is an example.

air transport, strategic See STRATEGIC AIR TRANSPORT.

air transportable unit A unit, other than AIRBORNE, whose equipment is adaptable for air movement.

air warfare, strategic See STRATEGIC AIR WARFARE.

air wing See WING.

AK-47 The standard assault rifle of Soviet bloc countries since the 1950s. On full automatic, it can fire at the rate of 600 rounds per minute.

ALCM An air-launched cruise missile; see CRUISE MISSILE.

alert 1. Ready for action, defense or protection. 2. A warning signal of a real or threatened danger, such as an air attack. 3. The period of time during which troops stand by in response to an alarm. 4. To forewarn; to prepare for action. 5. A warning received by a unit or a headquarters which forewarns of an impending operational mission. See also AIR DEFENSE WARNING CONDITIONS; WARNING ORDER.

alert, airborne See AIRBORNE ALERT.

alert, ground See GROUND ALERT.

alert force A specified force maintained in a special degree of readiness.

alert station The position taken up by defensive aircraft between expected enemy aircraft and the objective to be defended.

alien, enemy See ENEMY ALIEN.

all-clear signal A prearranged signal to indicate that danger from attack by aircraft, forces, or ships has passed.

all hands 1. A ship's company. 2. All of the individuals in a naval unit. 3. The total personnel of a navy.

alliance A formal agreement between two or more nations for mutual assistance in case of war (or to fight a war). The main reason for peacetime alliances such as the North Atlantic Treaty Organization (NATO) is to deter war. By that criterion, the "Western Alliance," as NATO is often called, has been a resounding success given that there has been no European war since 1945. See Erich Weede, "Extended Deterrence by Superpower Alliance," *Journal of Conflict Resolution* Vol. 27, No. 2 (June 1983); Stephen M. Walt, "Alliance Formation and the Balance of World Power," *International Security* Vol. 9, No. 4 (Spring 1985).

allied commander The head of a military command composed of elements of two or more allied nations working together; also known as the combined commander.

allied staff A staff or headquarters composed of two or more allied nations working together.

allies 1. Nations that are formally bound to each other by a mutual defense treaty. Thus the North Atlantic Treaty Organization (NATO) countries refer to each other as allies. 2. "The Allies" refers to the association of nations led by the United States, the United Kingdom, and the Soviet Union in the fight against the Axis powers in World War II. During World War I this term also referred to the winning coalition of the United States, the United Kingdom (and its dominions), France, and Italy. John Slessor, in his 1954 *Strategy for the West,* expressed the feelings of many when he observed: "War without allies is bad enough—with allies it is hell!"

allocation 1. The apportionment of specific numbers and types of NUCLEAR WEAPONS to a commander for a stated period for use in the development of war plans. 2. The designation, by an air-force commander, of specific numbers and types of SORTIES available for TACTICAL AIR SUPPORT, AIR INTERDICTION, and COUNTERAIR missions during a specific period. Upon notification of the allocation, the command officer suballocates a number of the available tactical air-support sorties to each of several subordinate commands. 3. The designation of specific units and other resources to subordinate commands to carry out a given tactical scheme.

allotment 1. A portion of the pay of military personnel that is voluntarily authorized to be paid to another person, bank or to an institution. 2. The specific authorization of numbers of personnel to a command organization or unit.

allowance 1. Money or some equivalent furnished in addition to the prescribed rate of pay for military personnel. An allowance is given to provide for expenses for which a soldier's pay is considered inadequate, such as a travel allowance, quarters allowance, clothing allowance, or subsistence allowance. 2. The prescribed number or portion of items of supply or equipment provided for an individual or organization.

all-volunteer force (AVF) A term used to describe the United States military after the termination of the DRAFT, when it became dependent upon voluntary enlistments to fill its ranks. The last period of CONSCRIPTION in American history ended in 1973. Men turning 18 years old are currently required to register with the SELECTIVE SERVICE ADMINISTRATION, but there is no actual draft now in practice.

all-weather fighter A fighter aircraft with RADAR devices and other special equipment which enable it to intercept its target in darkness or in daylight weather conditions that do not permit visual interception.

alphabet, phonetic See PHONETIC ALPHABET.

alphabet-code flag A signal flag that represents a letter of the alphabet or a numeral.

alternate airfield An airfield specified in a flight plan to which a flight may proceed when a landing at the intended destination becomes inadvisable. An alternate airfield may be the airfield of departure.

alternate command post A location designated by a commander to assume the functions of a COMMAND POST in the event that the primary command post becomes inoperative. It may be partially or fully equipped and manned, or it may be the command post of a subordinate unit.

alternate headquarters An existing headquarters of a component or subordinate command that is predesignated to assume the responsibilities and functions of another headquarters under prescribed emergency conditions.

alternate position A place near the primary position from which a weapon, unit, or individual can perform a designated task if the primary position becomes untenable or unsuitable.

alternate traversing fire A method of covering with fire a TARGET that has both width and depth, involving the firing of a succession of traversing groups of fire whose normal range dispersion will provide for distribution in depth.

altitude, absolute See ABSOLUTE ALTITUDE.

altitude, coordinating See COORDINATING ALTITUDE.

altitude, indicated See INDICATED ALTITUDE.

altitude azimuth An AZIMUTH determined by solution of the navigational triangle (using the vector sum of an aircraft's air speed and wind velocity to determine its true ground speed) with altitude, delineation, and latitude given.

ambush A surprise attack from concealed positions on a moving or temporarily halted enemy. VEGETIUS wrote in his *De Re Militari* of 378 A.D. that: "an ambuscade, if discovered and promptly surrounded, will repay the intended mischief with interest."

American Legion A membership organization for veterans of the U.S. armed forces. Created in 1919, it has long been an effective lobby for veterans' interests.

ammo (plus, minus, zero) In air interception, a code designating the amount of ammunition that a warplane has left (the type may be specified.) For example:

> *Ammo plus*—I have more than half my ammunition left.
> *Ammo minus*—I have less than half my ammunition left.
> *Ammo zero*—I have no ammunition left.

ammunition A device such as a ROUND, charged with EXPLOSIVES, PROPELLANTS, PYROTECHNIC materials, an initiating composition, or nuclear, biological, or chemical material for use in connection with military defense or offense, including demolition. Certain ammunition can be used for training, ceremonial, or nonoperational purposes. Because of its critical importance, ammunition has often been associated with prayers. Oliver Cromwell (1599–1658) was the first commander to tell his troops to "put your trust in God, my boys, and keep your powder dry." And American forces were told they would win World War II if they remembered to "praise the Lord and pass the ammunition." This was first said by Navy Chaplain H. M. Forgy on board the cruiser *New Orleans,* during the December 7, 1941 attack on Pearl Harbor.

ammunition, antimateriel See ANTIMATERIEL AMMUNITION.

ammunition, armed See ARMED AMMUNITION.

ammunition, ball See BALL AMMUNITION.

ammunition, biological See BIOLOGICAL AMMUNITION.

ammunition, chemical See CHEMICAL AMMUNITION.

ammunition, drill See DRILL AMMUNITION.

ammunition, estimated expenditure of See ESTIMATED EXPENDITURE OF AMMUNITION.

ammunition, fixed See FIXED AMMUNITION.

ammunition, inert See INERT AMMUNITION.

ammunition, live See LIVE AMMUNITION.

ammunition, practice See PRACTICE AMMUNITION.

ammunition, semi-fixed See SEMI-FIXED AMMUNITION.

ammunition, separated See SEPARATED AMMUNITION.

ammunition, service See SERVICE AMMUNITION.

ammunition, subcaliber See SUBCALIBER AMMUNITION.

ammunition barricade A structure consisting essentially of concrete, earth, metal, or wood, so constructed as to reduce or confine the blast effect or fragmentation of an explosive.

ammunition belt 1. A fabric or metal band with loops for CARTRIDGES that are fed from it into a MACHINE GUN or other automatic weapon. In this meaning, usually called a feed belt. 2. A belt with loops or pockets for carrying cartridges or clips of cartridges. In this meaning, usually called a cartridge belt.

ammunition clip A device that holds a number of rounds of ammunition for loading into certain types of automatic, semiautomatic, and bolt-action rifles.

ammunition-controlled supply rate In U.S. Army usage, the amount of ammunition estimated to be available to sustain operations of a designated force for a specified time if expenditures are controlled at that rate. It is expressed in terms of ROUNDS per weapon per day for ammunition items fired by weapons, and in terms of units of measure per organization per day for bulk-allotment ammunition items.

ammunition day of supply The estimated quantity of conventional ammunition required per day to sustain operations in an active theater of combat. It is expressed in terms of ROUNDS per weapon per day for ammunition items fired by weapons, and in terms of other units of measure for bulk-allotment ammunition items. See John R. Drebus, "Ammunition Loads—Defining the Problem," *Army Logistician* 13, 6 (November–December 1981).

ammunition handler 1. One whose primary duty is the handling and servicing of AMMUNITION. 2. A soldier who prepares ammunition for firing and who, as a member of a weapons crew, assists in the final delivery of ammunition to the loader.

ammunition pit A hole or trench dug in the ground where ammunition is stored temporarily. An ammunition pit is usually near the weapon from which the ammunition is to be fired.

ammunition required supply rate The amount of AMMUNITION, expressed in terms of ROUNDS per weapon per day for ammunition items fired by weapons, and in terms of other units of measure per day for bulk-allotment ammunition items, estimated to be required to sustain the operations of any designated force without restriction for a specified time. Tactical commanders use this rate to state their requirements for ammunition to support planned tactical operations at specific intervals. It is submitted through command channels and is consolidated at each echelon.

ammunition supply point (ASP) A point at which ammunition is available for distribution to military units on an area basis. It is normally operated by an ORGANIC ordnance unit.

ammunition supply point, special See SPECIAL AMMUNITION SUPPLY POINT.

ammunition train An organization, consisting of personnel and equipment, whose main function is the transportation of ammunition.

amphibious control group Personnel, ships, and craft designed to control the waterborne ship-to-shore movement in an amphibious operation.

amphibious demonstration A type of amphibious operation conducted for the purpose of deceiving the enemy by a show of force with the expectation of deluding the enemy into a course of action unfavorable to it.

amphibious force A naval force and landing force (usually marines), together with supporting forces that are trained, organized, and equipped for amphibious operations.

amphibious group A command within the amphibious force, consisting of the commander and his staff, designed to exercise operational command of assigned units in executing all phases of a division-size amphibious operation.

amphibious lift The total capacity of assault shipping utilized in an amphibious operation, expressed in terms of personnel, vehicles, and measurement or weight tons of supplies.

amphibious objective area A geographical area, delineated for purposes of command and control within which is located the objective(s) to be secured by the amphibious task force. This area must be of sufficient size to ensure accomplishment of the amphibious task force's mission, and must provide sufficient area for conducting necessary sea, air, and land operations.

amphibious operation An attack launched from the sea by naval and landing forces embarked in ships or craft, and involving a landing on a hostile shore. As an entity, the amphibious operation includes the following phases:

1. *Planning*—the period extending from issuance of the initiating directive to embarkation.
2. *Embarkation*—the period during which the forces, with their equipment and supplies, are embarked in the assigned shipping.
3. *Rehearsal*—the period during which the prospective operation is rehearsed for the purpose of: testing the adequacy of plans, timing detailed operations, and determining the combat-readiness of participating forces; insuring that all echelons are familiar with plans; and testing communications.
4. *Movement*—the period during which various components of the AMPHIBIOUS TASK FORCE move from points of embarkation to the objective area.
5. *Assault*—the period between the arrival of the major assault forces of the amphibious task force in the objective area and the accomplishment of the amphibious task-force mission.

See Roger M. Jaroch, "Amphibious Forces: Theirs and Ours," *U.S. Naval Institute Proceedings* (November 1982); Milan Vego, "Soviet Amphibious Forces," *Navy International* (May 1983); Dov S. Zakheim, "The Role of Amphibious Operations in National Military Strategy," *Marine Corps Gazette* (March 1984); John F. Brosnan, Jr., "An Amphibious Landing? With Civilian Ships?" *Naval War College Review* 39, 2 (March–April 1986).

amphibious raid A type of limited AMPHIBIOUS OPERATION; a landing from the sea on a hostile shore involving swift incursion into, or a temporary occupation of, an objective, followed by a planned withdrawal. The cross-channel Dieppe raid of 1942 is a classic example.

amphibious reconnaissance An amphibious landing conducted by minor ELEMENTS, normally involving stealth rather than force of arms, for the purpose of securing information, and usually followed by a planned withdrawal.

amphibious squadron A tactical and administrative organization composed of amphibious-assault shipping to transport troops and their equipment for an amphibious-assault operation.

amphibious task force The task organization formed for the purpose of conducting an amphibious operation. An amphibious task force always includes naval forces and a landing force, with their organic aviation. The military historian B. H. LIDDELL HART observed that "amphibious flexibility is the greatest strategic asset that a sea power possesses." See also AMPHIBIOUS TASK GROUP; ATTACK GROUP.

amphibious transport dock (LPD) A ship designed to transport and land troops, equipment, and supplies by means of embarked landing craft, amphibious vehicles, and helicopters. It is known as an LPD, which stands for "landing platform, dock."

amphibious vehicle A wheeled or tracked vehicle capable of operating on both land and water.

amphibious task group A subordinate force that may be formed within an AMPHIBIOUS TASK FORCE. It is comprised of a naval ATTACK GROUP and a LANDING GROUP.

AMRAAM The advanced medium range air-to-air missile with a launch and leave capability that allows pilots to race for safety immediately after firing at an enemy.

anchorage, advanced fleet See ADVANCED FLEET ANCHORAGE.

anchorage, emergency See EMERGENCY ANCHORAGE.

anchorage, holding See HOLDING ANCHORAGE.

anchored 1. Stationary ships, usually in a harbor. 2. In AIR INTERCEPTION, a code meaning, "Am orbiting a visible orbit point."

angels In AIR INTERCEPTION and CLOSE AIR SUPPORT, a code meaning aircraft altitude (in thousands of feet).

angle, dropping See DROPPING ANGLE.

angle of approach The angle, relative to the horizontal, formed by a plane tangent to the front tires of a wheeled land vehicle and touching the lowest part of the forward position of the vehicle. See also ANGLE OF DEPARTURE.

angle of arrival The vertical angle between the horizontal and the tangent of the trajectory at any point.

angle of attack The angle between a longitudinal reference line on an aircraft and the velocity VECTOR relative to the ambient undisturbed air.

angle of defense The angle formed where the FLANK meets a front line of defense.

angle of departure 1. The angle, relative to the horizontal, formed by a plane tangent to the rear tires of a wheeled vehicle and touching the lowest part of the rear portion of the vehicle. See also ANGLE OF APPROACH. 2. In artillery, the vertical angle between the tangent to the trajectory at the origin and the horizontal or base of the trajectory.

angle of depression Vertical angle between the horizontal and the axis of the bore of a gun when the gun is pointed below the horizontal.

angle of fall The vertical angle between the tangent to the trajectory of a projectile at the level point and the base of the trajectory.

angle of jump See VELOCITY JUMP.

angle of repose 1. The steepest slope at which a heap of material, such as earth, will stand without sliding.

angle of safety The minimal permissible angular clearance, at the gun barrel, of the path of a projectile above friendly troops. It is the angle of clearance corrected to insure the safety of the troops.

angle of site The vertical angle between the level base of the trajectory of a projectile (the horizontal) and the LINE OF SITE.

angle of site, complementary See COMPLEMENTARY ANGLE OF SITE.

angle of traverse 1. The horizontal angle through which a gun can be turned on its mount. 2. The angle between the lines from a gun to the right and left limits of the front that is covered by its fire; that is, the angle through which it is traversed.

angle T In artillery and naval gunfire support, the angle formed by the intersection of the GUN-TARGET LINE and the OBSERVER-TARGET LINE.

angular travel The angular distance covered by a moving target in a given time. It is equal to the ANGULAR VELOCITY of the target multiplied by its time of travel.

angular velocity The speed of a moving target, measured at the observing point in terms of the rate of change of the angular position of the target in direction and elevation.

angular travel method A method of calculating FIRING DATA based on the rate of ANGULAR TRAVEL of the target in terms of its direction and elevation.

animus belligerendi Latin meaning the intention to wage war. It is sometimes contended that a state of war cannot exist unless warlike acts are committed with *animus belligerendi;* with intent. This is a legal figleaf under which nations may seek to hide acts of war by calling them something else, thus proclaiming there was no intent.

annexation The formal extension of sovereignty over new territory. For example, the United States annexed the Republic of Texas in 1845.

Antarctic Treaty The 1959 agreement that internationalizes and outlaws the militarization of the continent of Antarctica. All signatories, including the United States and the Soviet Union, have the right to inspect the facilities of the others to make sure the continent is used solely for peaceful purposes.

antenna mine In naval MINE warfare, a contact mine fitted with antennae which, when touched by a steel ship, explodes.

antiair warfare Actions required to destroy or reduce to an acceptable level an enemy air or missile threat.

antiballistic missile (ABM) A missile designed to shoot down incoming missiles. In 1972, the United States signed an ABM treaty with the Soviet Union limiting ABM deployments. A 1974 protocol to the treaty restricts both sides to one ABM site each. The Soviets deployed an ABM system around Moscow although it was never brought to full strength. The United States presently has no deployed ABM system. While the ABM treaty specifically limits "space-based" systems, the Reagan Administration has contended that research and testing of exotic weapons is allowed by the treaty. This has proved controversial because many members of the U.S. Senate feel that the treaty is unambiguous when it states that: "Each party undertakes not to develop, test or deploy ABM systems or components which are sea-based, air-based, space-based, or mobile land based." It is generally agreed that research of some form is legitimate, if only because of the impossibility of preventing theoretical and small-scale laboratory research. It is also generally accepted that actual deployment of any device that is a clear development of the ABM systems envisioned in the early 1970s, based on interceptor rockets that would be fired into the path of an incoming ballistic missile, would be within the bounds of the treaty. The problems with the treaty arise from an ambiguous area: Does the treaty preclude the testing of new methods of destroying ballistic missiles?

The Reagan Administration began, in 1986, to insist that the correct interpretation of the treaty (but one never thought of for the first 14 years of its life) actually allowed systems based on physical principles or "new concepts" that had not been in operation in 1972. The new United States interpretation is in part a response to claims that the Soviet Union is itself in breach of the treaty, from having built a very large PHASED ARRAY RADAR system of the type best suited for handling a space-based ABM system. The Soviet Union has indeed built such a radar station, at Krasnoyarsk. It could run an ABM system, but it could also have other more traditional uses—and the Soviet Union insists that it does. Does the ABM Treaty prohibit the construction of systems with ABM potential, or only systems actually intended for such use? There is no clear answer. See Kevin C. Kennedy, "Treaty Interpretation by the Executive Branch: The ABM Treaty and 'Star Wars' Testing and Development," *American Journal of International Law* 80 (October 1986); Alan B. Sherr, "Sound Legal Reasoning or Policy Expedient? The 'New Interpretation' of the ABM Treaty," *International Security* 11 (Winter 1986–87).

antidisturbance fuze A FUZE designed to be in a sensitive condition after it is armed, so that any further movement or disturbance will result in its detonation. See ANTIWITHDRAWAL DEVICE.

antimaterial agent A living organism or chemical used to cause the deterioration of, or damage to, a selected material.

antimateriel ammunition A type of ammunition specifically designed to defeat MATERIEL targets, such as light armored vehicles.

antipersonnel mine A mine designed to cause maximum casualties to personnel as opposed to equipment damage.

antipersonnel weapon Any destructive device whose prime purpose is to injure enemy soldiers as opposed to damaging their equipment.

antisatellite weapon (ASAT) A weapon designed to destroy SATELLITES in space. Such weapons are seen by many sources as extremely dangerous because they threaten to destabilize the nuclear balance between the superpowers. Satellites are now crucial to the military command and control functions, intelligence gathering, navigation, and targeting operations of both superpowers (See C^3I). A clear and unrivalled capacity for one superpower quickly to destroy the other's orbiting satellites would give it a considerable advantage, and would produce a very real fear on the part of the vulnerable power that it could be subject to a FIRST-STRIKE attack. If one of the superpowers suddenly lost a large portion of its satellites, it would be unable to detect signs of an impending NUCLEAR attack, and might not even know that a strike had been launched until a few minutes before it suffered the resulting damage. In a non-nuclear-warfare scenario, loss of satellites would seriously impair both communications with one's own forces and the tracking of the enemy's, resulting in a major strategic, and probably tactical, advantage to the enemy. See K. L. Eichelberger, "A New Duel: Antisatellite Combat in Space," *Naval War College Review* 35, 3 (May–June 1982); Jeffrey Boutwell, Donald Hafner, and Franklin A. Long, eds., *Weapons in Space: The Politics and Technology of Ballistic Missile Defense and Anti-Satellite Weapons* (New York: W. W. Norton, 1985); The Aspen Strategy Group, *Anti-Satellite Weapons and U.S. Military Space Policy* (Lanham, MD: University Press of America, 1986).

antisimulation Deceiving adversary SENSORS by making a strategic target look like a decoy.

antisubmarine barrier A line formed by a series of static devices or mobile units arranged for the purpose of detecting, denying passage to, or destroying hostile submarines.

antisubmarine carrier group A formed group of ships consisting of one or more antisubmarine carriers and a number of escort vessels whose primary mission is to detect and destroy submarines. Such groups may be employed in convoy support or HUNTER/KILLER roles.

antisubmarine minefield A MINEFIELD laid specifically against submarines. It may be laid at a shallow depth and be unsafe for all craft, including submarines, or laid deep with the aim of allowing the safe passage of surface ships.

antisubmarine warfare (ASW) Operations designed to detect and destroy enemy submarines. Submarines have long been serious threats to surface fleets; but with the deployment of strategic missiles in them, they have become important nuclear-weapon carriers to both superpowers, all the more so because they are relatively safe from a FIRST STRIKE. Each superpower seeks a method of detecting and monitoring the other's strategic-missile-carrying submarines, but it is generally accepted that neither side yet has this capacity, or is anywhere near to developing it. Although it is obviously very difficult to be sure, most commentators agree that the United States and its allies, particularly the United Kingdom, retain a considerable lead over the Soviet Union in ASW, in part because the Soviet Union has only relatively recently acquired a BLUE-WATER NAVY, even though its submarine fleet is large and advanced. See R. W. Atkins, "ASW: Where is the Inner Screen?" *Naval War College Review* 35, 1 (January–February 1982); James M. McConnel, "New Soviet Methods for Antisubmarine Warfare?" *Naval War College Review* 38, 4 (July–August 1985); Louis Gerkin, *ASW Versus Submarine Technology Battle* (Chula Vista, CA: American Scientific Corporation, 1987).

antitank mine A mine designed to immobilize or destroy a tank.

antitank weapons Guns or missiles (such as the TOW) whose main purpose is to destroy tanks and other armored vehicles; they may be hand held or mounted on armored vehicles. Hand-held antitank weapons that were originally designed to protect infantry are now considered a great threat to them because the relatively lightly ARMORED FIGHTING VEHICLES in which infantry often ride are excellent targets for antitank weapons.

antiwithdrawal device A device used in bombs, made integrally with the FUZE, that will set off the fuze and subsequently set off the bomb upon any attempt at withdrawal of the fuze. See also ANTIDISTURBANCE FUZE.

anvil A rigid metal part in a PRIMER assembly. When a blow from the firing pin forces in the primer cap, the charge is compressed against the anvil and set off.

ANZUS Pact The mutual defense treaty between Australia, New Zealand, and the United States signed in 1951. In 1986 this treaty of mutual defense became less mutual when New Zealand decided that it could not tolerate American warships (because they might be carrying nuclear weapons) using New Zealand ports. Both countries therefore agreed that in the future they would both defend Australia, but not each other. See F. A. Mediansky, "ANZUS in Crisis," *Australian Quarterly* Vol. 57. No. 1–2 (Autumn-Winter 1985); Peter Samuel and F. P. Serong, "The Troubled Waters of ANZUS," *Strategic Review* 14, 1 (Winter 1986).

AO 1. ACTION OFFICER. 2. AREA OF OPERATIONS. 3. Air officer: an officer of the day for aviation operations.

AP Armor piercing. As used in conjunction with ammunition, AP means that the ammunition has some degree of armor-piercing ability.

APA See ATTACK TRANSPORT.

APC See ARMORED PERSONNEL CARRIER.

aperiodic compass A magnetic compass in which the pointer comes to rest in the final position in one direct movement and without oscillating.

apex angle 1. The horizontal angle at a target between a line from the target to the gun aimed at the target and a line from the target to the observation post. 2. The angle of a triangle opposite the base.

APO See ARMY POST OFFICE.

apparatchik A Russian word meaning "bureaucrat," now used colloquially to refer to any administrative functionary. The word as used in English seems to have no political connotations; it merely implies that the individual referred to mindlessly follows orders.

appeasement Giving in to the demands of others who make explicit or implied threats. The term is most associated with England's and France's permitting Germany to occupy the Sudentenland in Czechoslovakia in 1938. Thereupon the British Prime Minister, Neville Chamberlin, said there would be "peace in our time," and the German Chancellor, Adolph Hitler, declared that he had no further territorial ambitions in Europe. Because the policy of appeasement only encouraged Hitler's aggression, which led to World War II, the word "appeasement" (which was once merely descriptive of a policy of acceding) has taken on a decidedly negative connotation. When Chamberlain returned from appeasing Hitler in Munich, Winston Churchill told him "Prime Minister, you had the choice between war and dishonor. You have chosen dishonor, and you will get war." See C. A. MacDonald, *The United States, Britain, and Appeasement, 1936–1939* (New York: St. Martin's Press, 1981); Paul M. Kennedy, "The Study of Appeasement: Methodological Crossroads or Meeting Place?" *British Journal of International Studies* Vol. 6, No. 3 (October 1980).

apportionment The decision by a JOINT-FORCE commander on the division of tactical air resources among the combat functions of COUNTERAIR, AIR INTERDICTION, and TACTICAL AIR SUPPORT. It is expressed in terms of percentages of assets to be devoted to different functions, and is based, in part, on the recommendations of the component commanders.

appreciations Assumptions, estimates, facts, and analyses about an opponent's intentions and military capabilities, used in planning and decision making.

approach 1. The area immediately in front of an enemy. 2. The fortified positions built by an attacker while laying siege to an enemy position. 3. The first segment of an aircraft landing procedure.

approach march The advance of a combat unit when direct contact with the enemy is imminent. Troops may be fully or partially deployed during the march. The approach march ends when ground contact with the enemy is made or when the attack position is occupied. The approach march hasn't changed much in 2,000 years. Archidamus II, King of Sparta, told his troops in 431 B.C.: "When invading an enemy's territory, men should always be confident in spirit, but they should fear, too, and take measures of precaution; and thus they will be at once most valorous in attack and impregnable in defense." In this century John W. Thomason, Jr., would write in his 1926

Fix Bayonets that "There is no sight in all the pageant of war like young, trained men going up to battle. The columns look solid and business-like . . . There is no singing—veterans know, and they do not sing much—and there is no excitement at all; they are schooled craftsmen, going up to impose their will, with the tools of their trade, on another lot of fellows and there is nothing to make a fuss about." See also ADVANCE TO CONTACT.

approach route A route that joins a port to a coastal or transit route.

approach schedule The schedule that indicates, for each scheduled WAVE, the time of departure from the rendezvous area, from the LINE OF DEPARTURE, and from other control points, and the time of arrival at a beach or other destination.

approach sequence The order in which two or more aircraft are cleared for an approach to landing.

approach time The time at which an aircraft commences its final approach preparatory to landing.

approximate contour A line on a map representing an imaginary line on the ground that passes through points of estimated equal elevation.

apron A defined area, on an airfield, intended to accommodate aircraft for purposes of loading or unloading passengers or cargo, refueling, parking, or maintenance.

arc of fire The segment of a circle through which FIRE may be directed from a position or weapon.

area 1. Any clearly defined space. 2. All of the buildings and grounds assigned to a military unit. 3. An intangible aspect of responsibility or knowledge.

area assessment In UNCONVENTIONAL WARFARE, the prescribed collection of specific information by the commander of a force, which commences immediately after infiltration and is a continuous operation. It confirms, corrects, refutes, or adds to previous intelligence of the area.

area bombing The bombing of a target that is in effect a general area rather than a small or PINPOINT TARGET.

area command 1. A command composed of those organized elements of one or more of the armed services, designated to operate in a specific geographical area, and functioning under a single commander, e.g., the commander of a unified command, or an area commander. 2. In unconventional warfare, the organizational structure established within an UNCONVENTIONAL WARFARE operational area to command and control resistance forces. Usually it will integrate the SPECIAL FORCES operational detachment and the RESISTANCE FORCE. See also COMMAND.

area damage control Measure taken before or after hostile actions, or natural or manmade disasters, to reduce the probability of damage and minimize its effects.

area defense 1. A defense organized to protect a general geographical area as opposed to specific points or property within it. 2. An ANTI-BALLISTIC MISSILE defense covering a large area, usually implying the capability to protect "soft" (i.e. not HARDENED missile silos or bunkers) targets.

area fire Fire delivered on a prescribed area. The term is applicable regardless of the tactical purpose of the fire, but area fire is generally NEUTRALIZATION FIRE.

area of influence A geographical area wherein a commander is directly capable of influencing operations, by maneuver or FIRE-SUPPORT systems normally under his command or control.

area of interest That area of concern to a commander, including the AREA OF INFLUENCE, areas adjacent thereto, and areas extending into enemy territory to the objectives of current or planned operations. This also includes areas occupied by enemy forces who could jeopardize the accomplishment of a mission.

area of operations (AO) 1. That portion of an area of war necessary for military operations and for the administration of such operations. 2. In naval usage, operations conducted in a geographical area and not related to the protection of a specific force.

area-oriented A term applied to personnel or units whose organization, mission, training, and equipping are based on projected operational deployment to a specific geographical area.

area reconnaissance A directed effort to obtain detailed information about the terrain or enemy activity within a prescribed area such as a town, ridge line, woods, or other feature critical to military operations.

area of responsibility 1. A defined area of land in which responsibility is specifically assigned to a commander for the development and maintenance of installations and control of the movement and conduct of tactical operations involving troops under his control, along with parallel authority to exercise these functions. 2. In naval usage, a predefined area of enemy terrain for which supporting ships are responsible for covering known targets or targets of opportunity with fire and by observation.

area study In UNCONVENTIONAL WARFARE, the prescribed collection of specific information pertaining to a given geographical area and developed from sources available prior to entering the area.

area target A target consisting of an area rather than a single point.

areal feature 1. A topographic feature, such as sand, a swamp, or vegetation, which extends over an area. It is represented on a published map or chart by a solid or screened color, by a prepared pattern of symbols, or by a delimiting line. 2. Any area enclosed by a delimiting line that has any unique characteristic (e.g., a forest, residential area, or industrial area).

arm 1. A weapon for use in war. In this meaning usually called arms. 2. The supplying of military personnel with prescribed stores of ammunition, bombs, and other armament items in order to make them ready for combat. 3. A branch of an army primarily concerned with combat and COMBAT SUPPORT missions. 4. To make ammunition ready for detonation, as by removing safety devices or aligning the explosive elements in the EXPLOSIVE TRAIN of the FUZE.

armageddon 1. A great last battle between good and evil that the Bible (Apocalypse 16:14–16) predicts will precede the end of the world. 2. A decisive major battle. 3. A full nuclear exchange between the superpowers that, it is believed, will destroy all life and civilization. See Roy A. Werner, "Down the Road to Armageddon?" *Atlantic Community Quarterly* 13, 4 (Winter 1975–76); Michael S. Sherry, *The Rise of American Air Power: The Creation of Armageddon* (New Haven, CT: Yale University Press, 1987).

armament 1. The offensive weapons and defensive equipment of a military unit, whether a single tank or an entire army. 2. The FIREPOWER of a ship.

armament error The dispersion of shots from a particular gun; the deviation (the error) of any shot from the center of impact of a series of shots from a gun after all human and sight adjustment errors have been accounted for.

armed ammunition An explosive device ready for actuation.

armed conflict 1. War; a battle. 2. Any fight in which the combatants use weapons other than parts of their bodies.

armed forces, all volunteer See ALL VOLUNTEER FORCE.

Armed Forces of the United States A collective phrase for all military components of the U.S. Army, Navy, Air Force, Marine Corps, and Coast Guard. The term includes not only active forces but reserve and NATIONAL GUARD units as well.

armed reconnaissance An air mission with the primary purpose of locating and attacking TARGETS OF OPPORTUNITY (i.e., enemy materiel, personnel, and facilities) in assigned general areas or along assigned ground communications routes, and not for the purpose of attacking specific targets.

Armed Services Committees The permanent standing committees in the Senate and the House of Representatives of the United States Congress where most debate, analysis, and amendment of legislation concerning military policy takes place. Their primary function is to review the military budget and, in particular, procurement plans. More recently they have also used their power to influence military doctrine and strategy. See MILITARY REFORM CAUCUS. The membership of both committees has always been heavily unrepresentative of their respective legislative houses. They tend to be dominated by members who, whether Democrat or Republican, are notably more conservative than other members of Congress. Therefore the committees have had a general ideological inclination to be sympathetic toward Pentagon requests for defense expenditures. Furthermore, the members typically have major military bases or defense industry plants in their districts. Thus there is a tendency to accept Pentagon requests benefiting those areas.

arming As applied to explosives, weapons, and ammunition, the changing from a safe condition to a state of readiness for initiation.

arming device, intermittent See INTERMITTENT ARMING DEVICE.

arming pin A safety device that is inserted into a FUZE to prevent the arming cycle from starting until its removal.

arming range In artillery, the range at which a FUZE arms.

arming vane A small, rotating propeller attached to the FUZE mechanism of a bomb that puts the bomb into a condition to explode when it hits.

arming wire 1. A cable, wire, or lanyard attached to an aircraft and routed to a WEAPONS SYSTEM (i.e., a fuze fin, parachute pack, etc.) to prevent arming initiation prior to weapon release. Also called "safety wire"; "arming lanyard"; "safety lanyard." 2. A wire attached to the fuze mechanism of bombs.

armistice The cessation of hostilities pending a formal peace agreement between warring parties. A capitulary armistice leads to a surrender. See H. W. Lewis, "The Nature and Scope of the Armistice Agreement," *American Journal of International Law,* 50 (1956).

Armistice Day November 11, 1918, the day World War I ended on the Western front. Once a holiday celebrating the end of the "war to end all wars," it has since been converted to a day to commemorate the veterans of all wars.

armor 1. Protective covering, especially metal plates used on ships, tanks, motor vehicles, aircraft, and other military equipment. 2. A fighting COMBINED ARMS TEAM consisting of tanks and ARMORED CAVALRY reconnaissance/security units, supported on the battlefield by army aviation, a flexible and rapid communications network, and a mobile logistics system, all trained and equipped for mounted ground combat. 3. A branch of an army equipped with tanks or armored cars, analogous to infantry or artillery branches; the metallic descendent of the cavalry. See Simon Dunston, *Vietnam Tracks: Armor in Battle 1945–75* (Novato, CA: Presidio Press, 1983); John W. Mountcastle, "On the Move: Command and Control of Armor Units in Combat," *Military Review* 65, 11 (November 1985); A. J. Bacevich and Robert R. Ivany, "Deployable Armor Today," *Military Review* 67, 4 (April 1987).

armor, Chobham See CHOBHAM ARMOR.

armor, composite See COMPOSITE ARMOR.

armor, reactive See REACTIVE ARMOR.

armor group A FIELD ARMY unit designed to exercise command control and supervision of one or more separate tank, ARMORED INFANTRY, and ARMORED CAVALRY battalions, assigned to a corps or field army.

armor sweep A RAID or other limited attack without terrain objective by a rapidly moving armor unit through or across enemy controlled territory. An armor sweep may be conducted for reconnaissance in force, destruction or capture of personnel or materiel, or to harass or disrupt enemy plans and operations.

armored artillery 1. Self-propelled artillery weapons that are completely or partially armored. 2. Artillery units equipped with armored artillery weapons and appropriate armored auxiliary vehicles, and organized primarily to function with armored units.

armored car A wheeled, as opposed to tracked, ARMORED FIGHTING VEHICLE.

armored cavalry Combat units characterized by a high degree of mobility, firepower, SHOCK-ACTION capacity and multiple communications. The units are especially designed to execute reconnaissance, security, combat, or ECONOMY OF FORCE operations utilizing organic surface and air modes of transport. See Field Marshall Lord Carver, *The Apostles of Mobility: The Theory and Practice of Armoured Warfare* (New York: Holmes & Meier, 1979).

armored fighting vehicle (AFV) An armored vehicle used for transporting infantry into combat. It is somewhat like a tank, being both armored to protect the occupants from machine-gun and shell fire, and moving on tracks rather than wheels so that it can traverse rough ground. However, it is usually only lightly armed with a small-caliber cannon or machine-guns, and its armor plating is not strong enough to protect it against armor-piercing weapons. The advent of such vehicles has transformed the nature of infantry combat, as well as considerably increasing the pace of battle by removing the need for infantry to move by walking. The AFV is a later generation, more heavily armed and armored APC.

armored infantry A FIELD ARMY unit designed to close with and destroy the enemy by fire and maneuver, to repel hostile assault in close combat, and to provide support for tanks. Unlike ordinary infantry, armored infantry is equipped with ARMORED FIGHTING VEHICLES.

armored personnel carrier (APC) A lightly armored, highly mobile, full-tracked vehicle, sometimes amphibious or AIRPORTABLE, used primarily for transporting personnel and their individual equipment during tactical operations. Production modifications or the application of special kits permit the use of APCs as mortar carriers, COMMAND POSTS, FLAME THROWERS, or antiaircraft artillery chassis.

armored vehicle A wheeled or tracked vehicle with an armored hull or body and with or without major armament, used for combat, security, or cargo carrying.

armorer One who services and makes repairs on SMALL ARMS and performs the duties necessary to keep small arms ready for use.

armory 1. An ARSENAL. 2. A place where part-time reserve units train and store their weapons and equipment.

arms 1. Hand-held weapons. 2. The total weapons of an army or a nation. 3. The total military power of a nation. Niccolo Machiavelli wrote in *The Prince* (1513) that "There cannot be good laws where there are not good arms." And President John F. Kennedy announced in his 1961 inaugural address that: "Only when our arms are sufficient beyond doubt can we be certain that they will never be employed."

arms control 1. Any intentional agreement between countries which limits the numbers, types, and performance characteristics of WEAPONS SYSTEMS or armed forces. Arms control can usefully be divided into two types: the first is the control over existing weapons systems, the second is the attempted prevention of the original deployment of some new or potential weapon, sometimes called pre-emptive arms control. 2. A general reference to any measures taken to reduce international military instability. 3. Any measures taken by potential adversaries to reduce the likelihood or scope of a future war. See Richard Dean Burns, ed. *Arms Control and Disarmament: A Bibliography* (Santa Barbara, California: ABC-Clio Press, 1977); Bruce D. Berkowitz, *Calculated Risks: A Century of Arms Control, Why It Has Failed, and How It Can Be Made to Work* (New York: Simon & Schuster, 1988).

Arms Control and Disarmament Act of 1961 The law that created the Arms Control and Disarmament Agency to conduct research in and to aid in ARMS CONTROL and DISARMAMENT negotiations, and to provide public information on this highly technical field. The Act specifically states that "adequate verification of compliance should be an indispensable part" of any arms-control treaties.

arms control measures, unilateral See UNILATERAL ARMS CONTROL MEASURES.

arms, inspection See INSPECTION ARMS.

arms race A process by which potential enemies gear their arms procurement to each other's military development, with the intention of gaining a specific level of comparative military strength. Perhaps the first important modern arms race was the competition between Britain and Germany at the turn of the century to build bigger and better battleships, the "Dreadnoughts." In current usage the arms race refers to the competition between the United States and the Soviet Union to build up more powerful nuclear weaponry, especially INTERCONTINENTAL BALLISTIC MISSILES, in the hope of achieving a FIRST STRIKE CAPABILITY over the enemy. An arms race is tied to the idea of a BALANCE OF POWER: any technological advance by one state threatens such a balance by giving it a preponderance of power over another, who then tries to build even better weapons, forcing the first mover to improve his weapons, and so on. See Hugh G. Mosley, *The Arms Race: Economic and Social Consequences* (Lexington, MA: Lexington Books, 1985).

army 1. An inclusive term meaning the land military forces of a nation; an army in the true sense of the word is a permanent and bureaucratically organized standing force rather than a temporary and amateur force assembled only during emergencies. In a nation relying on temporary, amateur troops, the army cannot be a threat to other social and political institutions. But as soon as an army in the bureaucratic sense comes into being, and with its own legitimacy and power base, it becomes a potential contender to control the state. An army must constantly be trained if it is to be effective in time of war. Edward Gibbon, in his 1776 *Decline and Fall of the Roman Empire*, observed: "So sensible were the Romans of the imperfections of valor without skill and practice that, in their language, the name of an Army was borrowed from the word which signified exercise." 2. The largest ADMINISTRATIVE and TACTICAL UNIT of military forces, consisting of two or more army corps and supporting troops; a field army. 3. A FORMATION larger than an army corps but smaller than an army group. 4. All the armed forces of a nation. 5. When capitalized, the U.S. Army. For the best one volume history of the U.S. Army, see: Russell F. Weigley, *History of the United States Army* (New York: Macmillan, 1967).

army, field See FIELD ARMY.

army, regular See REGULAR ARMY.

army, standing See STANDING ARMY.

army air-ground system (AAGS) The U.S. Army system that provides the interface between the Army and TACTICAL AIR SUPPORT agencies of other services in the planning, evaluating, processing, and coordinating of air-support requirements and operations.

army attaché The army officer who serves in a dual capacity as the senior representative of his nation's military in a foreign country, and as a member of the official staff of the nation's ambassador or minister posted there. He serves as a military OBSERVER and reports to his government on the military plans and developments of the country where he is stationed. In most countries he is referred to as the military attaché. Parallel positions may be occupied by naval and air attachés.

army aviation Aircraft, allied aircraft equipment, and associated personnel organically assigned to army organizations.

army base A base or group of installations for which a local commander is responsible, consisting of facilities necessary for the support of army activities, including security, internal lines of communication, utilities, plants and systems, and real property for which the army has operating responsibility.

army corps A tactical unit larger than a DIVISION and smaller than a FIELD ARMY. A corps usually consists of two or more divisions, together with auxiliary arms and services.

army group The largest formation of military land forces, normally comprising two or more armies or ARMY CORPS under a designated commander.

army in the field All types of military personnel and units utilized in, or intended for utilization in, a THEATER OF WAR.

army landing force The army component of an AMPHIBIOUS TASK FORCE; a task organization comprising all army units assigned for participation in an amphibious operation. The commander of an army component of the amphibious task force is the army landing force commander.

army of excellence A morale-building phrase used by the U. S. Army to describe what it strives to be.

army of occupation An army in effective control of enemy territory for the purpose of maintaining law and order within the area or to insure the carrying out of ARMISTICE or surrender terms.

Army Post Office A U.S. address through which military mail is transferred to theaters of operations. For example, all mail going to U.S. forces in the Pacific goes to APO San Francisco.

army service area The territory between the corps rear boundary and the COMBAT ZONE rear boundary. Most of an army's administrative establishment and service troops are usually located in this area. See also REAR AREA.

Aron, Raymond (1905–1983) The French sociologist and political commentator who has been the foremost European intellectual supporter of the American leadership of the Western World. His *The Imperial Republic: The United States and the World, 1945–1973,* translated by Frank Jellineck (Cambridge, MA: Winthrop Publishers, 1974) continues to serve as a response to critics of an American foreign policy that protects Western Europe from the Soviet empire. See Richard M. Swain, "Clausewitz for the 20th Century: The Interpretation of Raymond Aron," *Military Review* 66, 4 (April 1986).

arrival date, latest See LATEST ARRIVAL DATE.

arrogance of power A phrase that called into question the premises of United States foreign policy during the 1960s. It was most associated with former U.S. Senator J. William Fulbright of Arkansas who, as chairman of the Senate Foreign Relations Committee, was a severe critic of United States intervention in Vietnam and the Dominican Republic. See William J. Fulbright, *The Arrogance of Power* (New York: Random House, 1966).

arsenal 1. A place where WEAPONS and AMMUNITION are stored. 2. A factory where an army manufactures its weapons. 3. The totality of weapons belonging to an individual or a nation. 4. An armory.

arsenal of democracy President Franklin D. Roosevelt's phrase, first used in a 1940 speech, to describe the United States role in supplying arms to the nations opposed to the Axis powers in World War II. When the United States entered the war a year later, the phrase became even more significant and literal.

art of war 1. A scientific approach to military actions implying that appropriate skill and learning can solve any military problem. 2. A mystical approach to military actions implying that an intangible aptitude for command, all things being equal, will carry the day. Two classics on the art of war are Sun Tzu's *The Art of War* (400 B.C.) and Henri Jomini's *Precis on the Art of War* (1836). Jomini in his book insisted that: "War in its ensemble is not a science but an art." Also see Harriet Fast Scott and William F. Scott, eds. *The Soviet Art of War* (Boulder, Colo.: Westview Press, 1982); Michael A. Rogalla, "The 'Art' of War," *Military Review* 66, 9 (September 1986); Archer Jones, *The Art of Warfare in the Western World,* (Champaign, IL: University of Illinois Press, 1987).

Articles of War The laws under which the United States military governed itself from 1775 until 1950, when they were superseded by the UNIFORM CODE OF MILITARY JUSTICE.

artillery 1. Any war machine capable of shooting PROJECTILES other than small arms. 2. All gunpowder-based weapons too large to be carried by hand. 3. That branch of an army equipped with heavy ORDNANCE. NAPOLEON maintained that "artillery, like the other arms must be collected in mass if one wishes to attain a decisive result," and that "the best generals are those who have served in the artillery."

artillery, armored See ARMORED ARTILLERY.

artillery, direct support See DIRECT SUPPORT ARTILLERY.

artillery, division See DIVISION ARTILLERY.

artillery, field See FIELD ARTILLERY.

artillery, general support See GENERAL SUPPORT ARTILLERY.

artillery, nuclear Any cannon or self-propelled gun containing or employing nuclear force.

artillery, supporting See SUPPORTING ARTILLERY.

artillery and naval gunfire support The fire support that ground units can call upon from their own unit's artillery or from offshore naval units.

artillery battalion group A tactical grouping of two or more artillery battalions for a specific mission, commanded by one of the battalion commanders.

artillery cannon calibration, field See FIELD ARTILLERY CANNON CALIBRATION.

artillery ears A loss of hearing caused by exposure to loud noises.

artillery fire plan table A presentation of PLANNED TARGETS giving data for engagement. Scheduled targets are fired on in a definite time sequence. The starting time may be on call, prearranged, or at the occurrence of a specific event.

artillery observer, field See FIELD ARTILLERY OBSERVER.

artillery preparation Artillery fire delivered before an attack to "soften up" the enemy, to disrupt his communications, and to disorganize his defenses.

artillery raid The airlifting of artillery BATTERIES into enemy territory where they could fire at preselected targets and then be removed (again by air) before the enemy could locate them.

artillery survey, field See FIELD ARTILLERY SURVEY.

artillery tactical operations center, field See FIELD ARTILLERY TACTICAL OPERATIONS CENTER.

ASAT See ANTISATELLITE WEAPON.

ASM See AIR-TO-SURFACE MISSILE.

ASP See AMMUNITION SUPPLY POINT.

asparagus bed A ground obstacle made of railway rails or similar items stuck vertically into the ground to inhibit the movement of opposition forces; so called because they resemble a field of growing asparagus. According to General James M. Gavin in *On To Berlin* (1978), "*Rommelspargel* (Rommel's asparagus) were poles about six to twelve inches in diameter and eight to twelve feet long. They were sunk a foot or two into the ground and stood seventy-five to a hundred feet apart. They were obviously intended to deny the use of good landing zones to the Allied airborne troops."

assailable flank The flank of a unit that is vulnerable to attack by virtue of terrain, unit positioning, or adjacent unit location. An assailable flank may be created by fire or maneuver.

assault 1. The climax of an attack; closing with the enemy in hand-to-hand combat. CLAUSEWITZ wrote: "On no account should we overlook the moral effect of a rapid, running assault. It hardens the advancing soldier against danger, while the stationary soldier loses his presence of mind." 2. In an AMPHIBIOUS OPERATION, the period of time between the arrival of the major assault forces of the AMPHIBIOUS TASK FORCE in the objective area and the accomplishment of the amphibious task-force mission. 3. To make a short, violent, but well-ordered attack against a local objective, such as a gun emplacement, a fort, or a machine-gun nest. 4. A phase of an AIRBORNE OPERATION beginning with delivery by air of the ASSAULT ECHELON of the force into the objective area and continuing through attack on the assault objectives and consolidation of the initial AIRHEAD. 5. In river crossings, the period from the launching of the first crossing effort until the BRIDGEHEAD has been secured and responsibility passed to a CROSSING-AREA COMMANDER. See also ASSAULT PHASE; LANDING ATTACK.

assault aircraft 1. Aircraft, including helicopters, that move assault troops and cargo into an OBJECTIVE AREA and provide for their resupply. 2. Aircraft used for direct fire support.

assault breach The BREACH of an obstacle or obstacles, characterized by the rapid employment of breaching techniques to take advantage of a situation. It is normally conducted by combat units without engineering aid or assistance.

assault course An area of ground used for training soldiers in attacking an enemy in CLOSE COMBAT.

assault craft A LANDING CRAFT or AMPHIBIOUS VEHICLE primarily employed for landing troops and equipment in the assault waves of an AMPHIBIOUS OPERATION.

assault echelon 1. The element of a force that is scheduled for initial assault on an objective area. 2. Those forces required in the initial stages of an airborne or river-crossing operation to secure the assault objectives.

assault fire 1. That FIRE delivered by attacking troops as they close with the enemy. 2. In artillery, extremely accurate, short-range destructive fire at POINT TARGETS.

assault follow-on echelon In an AMPHIBIOUS OPERATION, that echelon of the assault troops, vehicles, aircraft, equipment and supplies, which though not needed to initiate the assault, is required to support and sustain the assault.

assault force 1. In an AMPHIBIOUS or AIRBORNE OPERATION, those units assigned to seize a LODGEMENT area. 2. In an offensive river crossing, the major subordinate units conducting the assault to, across, and beyond the water obstacle to the final objective.

assault gun Any of various sizes and types of guns that are self-propelled or mounted on tanks, and are used for direct fire from close range against POINT TARGETS.

assault objectives Key terrain features whose immediate seizure facilitates the overall accomplishment of an AIRBORNE OPERATION.

assault phase 1. That phase of an airborne, airmobile, amphibious, or river crossing operation that begins with the delivery of the assault forces into the OBJECTIVE AREA and ends when all assault objectives have been seized. 2. In an AMPHIBIOUS OPERATION, the period of time between the arrival of the major assault forces of the AMPHIBIOUS TASK FORCE in the objective area and the accomplishment of their mission. 3. In an AIRBORNE OPERATION, a phase beginning with delivery by air of the ASSAULT ECHELON of the force into the objective area and extending through attack on the assault objectives and consolidation of the initial AIRHEAD. See also ASSAULT.

assault position That position between the LINE OF DEPARTURE and the objective in an attack from which forces assault the objective. Ideally, it is the last covered and concealed position before the objective (primarily used by DISMOUNTED infantry).

assault wire A very light field-telephone wire, wound on reels small enough for one man to carry over difficult terrain under front-line conditions.

assemblage A collection of items designed to accomplish one general function, and identified and issued as a single item. It may be made up of items included in more than one class of SUPPLIES, and may include items for which logistic responsibilities are assigned to more than one agency (e.g., a pontoon bridge, baking outfit, FIRE-CONTROL EQUIPMENT, etc.).

assemble anchorage An anchorage intended for the assembly and onward routing of ships.

assembly 1. An item forming a portion of a piece of equipment, that can be provisioned and replaced as an entity and which normally incorporates replaceable parts or groups of parts. 2. A signal given by drum or bugle for units of troops to gather or come together, usually in CLOSE ORDER formation. 3. Groupings of units, usually in close order formation. 4. A point in the air to which all elements of a massed formation of aircraft are assigned.

Assembly, General See GENERAL ASSEMBLY.

assembly area An area in which a force prepares or regroups for further action.

assembly area, boat See BOAT ASSEMBLY AREA.

assessment 1. Analysis of the security, effectiveness, and potential of an existing or planned INTELLIGENCE activity. 2. Judgment of the motives, qualifications, and characteristics of present or prospective employees or "agents." 3. Financial contributions made by a government (such as the United States) to the regular budget of an international organization (such as the United Nations) to which it belongs.

assessments, scale of The formula for assessing membership dues in an international organization.

asset Any resource, whether a person, group, relationship, instrument, installation, or supply, at the disposition of an INTELLIGENCE organization for use in an operational or support role. Often used with a qualifying term such as agent asset or propaganda asset.

assign To relatively permanently place units or personnel in a military organization; such organizations control and administer their assigned units or personnel according to the primary function of the organization.

assigned forces Forces in being that have been placed under the operational command or operational control of a commander. See also FORCE(s).

assignment, duty See DUTY ASSIGNMENT.

astimatizer A device attached to a RANGE FINDER for observing small lights at night.

ASW See ANTISUBMARINE WARFARE.

as you were 1. A command cancelling an immediately previous command. 2. A command from a newly arrived senior officer that all of the persons under his command should continue with their normal duties. 3. An order that cancels a previous order of "attention"; in effect, "as you were" allows soldiers to resume their duties.

at ease 1. A command releasing subordinates from the obligation to stand at ATTENTION. 2. A command to be quiet and listen for announcements or new orders. 3. An instruction to calm down.

at my command In ARTILLERY AND NAVAL-GUNFIRE SUPPORT, the command used when it is desired to control the exact time of delivery of fire.

at priority call A first obligation task for an artillery unit to provide fire to a formation or unit on a guaranteed basis. Normally, observer, communications, and liaison units are not provided. An artillery unit in "direct support" or "in support" may simultaneously be placed "at priority call'' to another unit or agency for a particular task or for a specific period.

Atlantic Alliance An informal phrase for the NORTH ATLANTIC TREATY ORGANIZATION (NATO). See Colin Gordon, *The Atlantic Alliance: A Bibliography* (London: Francis Pinter, 1978); Jeane J. Kirkpatrick, "The Atlantic Alliance and the American National Interest," *World Affairs* Vol. 147, No. 2 (Fall 1984).

Atlantic Charter The statement of general principles for the world following World War II, issued by President Franklin D. Roosevelt of the United States and British Prime Minister Winston Churchill after they met on a warship in the Atlantic Ocean off Newfoundland on August 14, 1941.

Atlantic Community 1. The west in general. 2. THE NORTH ATLANTIC TREATY ORGANIZATION. 3. The United States, Great Britain, and Canada.

Atlanticist Someone who in the debate over an appropriate U. S. world strategy would argue that the U. S. commitment to NATO is America's paramount military obligation, that the defense of Western Europe is integral to the security of the United States itself. In contrast, "maritime"' theorists believe that the United States should be much more independent of Western Europe, that the United States must be able to protect its national interests anywhere in the world, and must be able to rely on its own power to do so.

atomic bomb The type of NUCLEAR WEAPON used twice by the United States in August 1945, on the Japanese cities of Hiroshima and Nagasaki. Until the early 1950s it was the only form of nuclear weapon, but has since been replaced for most uses by HYDROGEN- or THERMONUCLEAR BOMBS. The atomic bomb is the product of nuclear fission rather than nuclear fusion, which is the source of power in the hydrogen bomb. The first atomic bombs were built in the United States by the Manhattan Project during the 1940s. See Richard Rhodes, *The Making of the Atomic Bomb* (New York: Simon & Schuster, 1987).

Atomic Energy Commission See NUCLEAR REGULATORY COMMISSION.

atomic-demolition munition A NUCLEAR device designed to be detonated on or below ground or underwater as a demolition munition against material-type targets (such as bridges) to block, deny, or CANALIZE the enemy. They are essentially very powerful mines. Now regarded obsolete, the North Atlantic Treaty Organization is removing them from Europe.

atomic-demolition munition, medium See MEDIUM ATOMIC DEMOLITION MUNITION.

atomic-demolition munition, special See SPECIAL ATOMIC DEMOLITION MUNITION.

Atoms for Peace The Eisenhower Administration's phrase for the civilian use of nuclear power. Unlike the bomb-making aspects of atomic power, this provided a vast field of co-operative research and development with the United States' allies as well as nonaligned countries. Unfortunately, it also gave those countries much of the technical capability they would need to make nuclear weapons. See Gerald E. Marsh, "If 'Atoms for Peace' Are Used for War," *Bulletin of the Atomic Scientists* Vol. 38, No. 2 (February 1982); Joseph F. Pilat, Robert E. Pendley and Charles K. Ebinger, *Atoms for Peace: An Analysis After Thirty Years* (Boulder, CO: Westview Press, 1985).

atropine A chemical substance prepared from plants that is used to counteract the effects of NERVE-AGENT poisoning.

attach 1. To place units or personnel in an organization on a relatively temporary basis. Subject to limitations imposed in the attachment order, the commander of the formation, unit, or organization receiving an attachment will exercise the same degree of command and control over it as he does over the units and persons organic to his command. However, the responsibility for transfer and promotion of personnel in the attachment will normally be retained by its parent formation, unit, or organization. See also CROSS ATTACHMENT. 2. To detail individuals to specific functions that are secondary or relatively temporary (i.e., attach for quarters and rations, attach for flying duty).

attaché A French term meaning "one assigned to." It usually refers to a technical specialist (military, economic, cultural, etc.) who is assigned to a diplomatic mission abroad.

attaché, army See ARMY ATTACHÉ.

attached strength A term that applies to personnel assigned to an organization, other than the one to which they are formally assigned, and who perform their duties in an attached status.

attack An offensive action characterized by weapons fire and maneuvering, and culminating in a violent assault or, in an attack by fire, in the delivery of intensive direct fire from an advantageous position. Its purpose is to direct a decisive blow at the enemy to hold him, destroy him in place, or force him to capitulate. Frederick the Great of Prussia told his generals in 1747: "I approve of all methods of attacking provided they are directed at the point where the enemy's army is weakest and where the terrain favors them the least." It is often asserted that during the American Civil War, Robert E. Lee explained a decision to attack thusly: "I was too weak to defend, so I attacked." See Erwin Rommel, ed. by Lee Allen, tr. by J. R. Driscoll, *Attacks,* (Vienna, VA: Athena Press Inc., 1979).

attack, accidental See ACCIDENTAL ATTACK.

attack, casualty See CASUALTY ATTACK.

attack, catalytic See CATALYTIC ATTACK.

attack, cloud See CLOUD ATTACK.

attack, coordinated See COORDINATED ATTACK.

attack, deep See DEEP ATTACK.

attack, deliberate See DELIBERATE ATTACK.

attack, direction of See DIRECTION OF ATTACK.

attack, echeloned See ECHELONED ATTACK.

attack, flanking See FLANKING ATTACK.

attack, frontal See FRONTAL ATTACK.

attack, harassing See HARASSING ATTACK.

attack, hasty See HASTY ATTACK.

attack, holding See HOLDING ATTACK.

attack, internal See INTERNAL ATTACK.

attack, landing See LANDING ATTACK.

attack, main See MAIN ATTACK.

attack, piecemeal See PIECEMEAL ATTACK.

attack, preemptive See PREEMPTIVE ATTACK.

attack, spoiling See SPOILING ATTACK.

attack, structured See STRUCTURED ATTACK.

attack, supporting See SUPPORTING ATTACK.

attack, surprise See SURPRISE ATTACK.

attack, vectored See VECTORED ATTACK.

attack aircraft carrier A warship designed to support and operate aircraft, engage in attacks on targets afloat or ashore, and engage in sustained operations in support of other forces. The aircraft carrier first became strategically significant during World War II. The Japanese used carriers to attack Pearl Harbor in 1941. In 1942 United States carriers fought Japanese carrier forces at the Battle of the Coral Sea and the Battle of Midway, the first major sea battles in which the combating ships never came within sight of each other. Since then all surface fleets of the United States have been organized around carriers.

attack altitude, minimum See MINIMUM ATTACK ALTITUDE.

attack cargo ship A naval ship designed to or converted so as to be able to transport combat-loaded cargo in an assault landing. Its capabilities for carrying LANDING CRAFT, its speed and armament, and the size of its hatches and booms are greater than those of comparable cargo ship types.

attack carrier striking force A naval force whose primary offensive weapon is carrier-based aircraft. Often referred to as a carrier battle group. The U.S. Navy aims to have 16 of these groups by the end of the twentieth century. Ships other than aircraft carriers act primarily to support and screen against submarine and air threats and secondarily against surface threats to an attack carrier striking force.

attack group A subordinate task organization of the navy forces of an AMPHIBIOUS TASK FORCE. It is composed of assault shipping and supporting naval units designated to transport, protect, land, and initially support a LANDING GROUP.

attack heading 1. The air intercept heading during the attack phase that will lead to contact with the enemy. 2. The assigned magnetic compass heading to be flown by aircraft during the delivery phase of an AIR STRIKE.

attack origin 1. The location or source from which an attack is initiated. 2. The nation initiating an attack.

attack position The last position occupied by an ASSAULT ECHELON before it crosses the LINE OF DEPARTURE. See also FORMING-UP PLACE.

attack transport A naval ship designed for COMBAT LOADING a BATTALION LANDING TEAM with its equipment and supplies, and having the facilities, including LANDING CRAFT, for landing them on a hostile beach.

attack unit, search See SEARCH ATTACK UNIT.

attention 1. A prescribed erect body posture of readiness and alertness for military personnel, with eyes straight ahead, hands at the sides, heels together, and toes turned out at an angle of 45 degrees. Positions of attention also are prescribed for mounted individuals, persons carrying weapons, and others. Complete silence and immobility are required. 2. The command to take this position.

attention to orders A command given by the ADJUTANT or other officer of a military organization to announce that he is about to issue orders.

attenuation 1. In the context of electronics, a decrease in intensity of a signal, beam, or wave as a result of absorption of energy and scattering out of the path of a detector. 2. In MINE warfare, the reduction in intensity of an influence as distance from the source increases.

attrition The reduction of the effectiveness of a force caused by loss of PERSONNEL and MATERIEL.

attrition rate A factor, normally expressed as a percentage, reflecting the degree of loss in PERSONNEL or MATERIEL due to various causes within a specified period.

attrition reserve aircraft Aircraft procured for the specific purpose of replacing the aircraft anticipated to be lost through peacetime or wartime attrition.

attrition sweeping The continuous sweeping of MINEFIELDS to keep the risk of mines to all ships as low as possible.

augmentation The reinforcement of UNIFIED or SPECIFIED COMMANDS through the deployment or redeployment of forces assigned to other commands.

autarchy 1. An autocracy. 2. A policy of national (or regional) self-sufficiency that prevents a nation from being dependent for critical materials on any nondomestic source. This meaning of autarchy is also spelled autarky.

authentication 1. A security measure designed to protect a communications system against a fraudulent or simulated transmission by establishing the validity of a transmission, message, or originator. 2. A means of identifying individuals and verifying their eligibility to receive specific categories of information. 3. Evidence by proper signature or seal that a document is genuine and official.

authentication, agent See AGENT AUTHENTICATION.

authentication, challenge and reply See CHALLENGE AND REPLY AUTHENTICATION.

authenticator A symbol or group of symbols, or a series of bits, selected or derived in a prearranged manner and usually inserted at a predetermined point within a message or transmission for the purpose of attesting to the validity of the message or transmission.

authoritarianism Rule by an individual whose claim to sole power is supported by subordinates who sustain control of the political system by carrying out the ruler's orders, and by a public that is unwilling or unable to rebel against that control. The ruler's personality may be a significant element in maintaining the necessary balance of loyalty and fear for maintaining rule. An authoritarian regime differs from a totalitarian one only in that the latter may have a specific ideology that rationalizes it, although it may require a leader who embodies that ideology to sustain public support.

authoritative source A term applied to a public official whose particular position implies a special closeness to information even though neither the individual nor the position can be identified publicly. The information thus gets to the public even though the source may be "off the record."

authority, coordinating See COORDINATING AUTHORITY.

authority, establishing See ESTABLISHING AUTHORITY.

authorized strength The total of the personnel spaces contained in current personnel authorization vouchers issued by a higher headquarters to a subordinate element of a military command.

autofrettage An operation whereby the BORE of a cannon TUBE is stressed to a predetermined value in excess of the yield strength of the tube, thus introducing a permanent prestress into the tube. This helps resist the pressure generated during firing, and inhibits the growth of fatigue cracks.

automatic direction finder A radio instrument in an aircraft that continues to point toward any radio station to which it is tuned. It is used to aid in the navigation of an aircraft.

Automatic Secure Voice Communications Network The Defense Department's worldwide telephone network. Also called AUTOSEVOCOM.

automatic supply A system by which certain required supplies are automatically shipped or issued for a predetermined period without requisition by the using unit. It is based upon estimated or experience-usage factors.

Automatic Voice Network The principal long-haul, unsecure telephone network within the Department of Defense. Also called AUTOVON.

automatic weapon 1. A firearm that will fire continuously with one squeeze of the trigger. All machine-guns are automatic. In contrast, a semi-automatic weapon will reload itself after firing but the trigger must be pulled before it will fire again. 2. A semi-automatic weapon; because of its automatic loading feature, it is commonly, but incorrectly, called an automatic.

autonomous operation A mode of operation of a military unit in which the unit commander assumes full responsibility for the control of weapons and ENGAGEMENT of hostile targets. This mode may be either directed by a higher authority or result from a loss of all means of communication with a higher authority.

autorotation The process through which a helicopter is brought to a safe landing in the event of mechanical failure.

auxiliary 1. A branch of a military service, often recruiting only women, that operates as a non-combatant force. 2. A reserve branch of a military force; for example, the Royal Auxilliary Air Force in the United Kingdom. 3. In UNCONVENTIONAL WARFARE, that element of a RESISTANCE FORCE established to provide the organized civilian support of the resistance movement. 4. Military units available to the state but not fully under the command of the regular military forces of the state; for example, in Nazi Germany, the SS, which grew to be an army within an army not under the control of otherwise normal command structures.

auxiliary target 1. In the context of artillery, a point at a known distance from a target; a registration target. An auxiliary or registration target is used as an adjusting point before firing on the actual target. Fire is delivered and adjusted on the auxiliary target. When the adjustment is complete, the necessary correction is made on each gun to shift its fire to the actual target. Auxiliary targets are used when fire on actual targets is intended to surprise the enemy. 2. In the context of a bomber raid, a back-up target that will be bombed if the primary target cannot be reached.

avenue of approach An air or ground route taken by an attacking force and leading to its objective or to key terrain in its path.

aviation, army See ARMY AVIATION.

avionics The electronic systems that control navigation, targeting, and communications in aircraft and missiles.

AWACS Airborne Warning and Control System, the E-3 Sentry, a modified Boeing 707 commercial airframe with a rotating radar dome. The 30-foot diameter dome contains a radar system that permits surveillance from the Earth's surface up into the stratosphere. It can look down to detect, identify, and track enemy and friendly low-flying aircraft by eliminating ground clutter returns that confuse other radar systems. The E-3 Sentry can gather and present detailed battlefield information, including position and tracking information on enemy aircraft and ships, and the location and status of friendly aircraft, naval vessels, and ground troops. The information can be sent to major command and control centers in rear areas and aboard ships, or to the National Command Authorities in the United States. In its tactical role, the E-3 Sentry can provide information needed for interdiction, reconnaissance, airlift, and close-air support for friendly ground forces. As an air defense system, it can detect, identify, and track airborne enemy forces far from the boundaries of the United States or NATO countries. The E-3 Sentry can fly more than 11 hours without refueling. The first E-3 Sentry was delivered to NATO in January 1982.

AWOL See ABSENT WITHOUT LEAVE.

AWOL bag A small gym bag only large enough to hold toilet articles and one change of clothing.

axial route A route running through the REAR AREA and into the FORWARD AREA of a war zone. See also ROUTE.

axis 1. Any route used for control purposes. 2. The World War II coalition of Germany, Italy, and Japan, describing the Rome-Berlin Axis. 3. Any political or military alliance between two powers.

axis of advance 1. A line of advance assigned for purposes of control; often a road or a group of roads, or a designated series of locations, extending in the direction of the enemy. 2. A general route of advance extending in the direction of the enemy, which is assigned to ground forces for purposes of control. It follows terrain suitable for the size of the force assigned to follow the axis, and is often a road, a group of roads, or a designated series of locations. A commander may maneuver his forces and supporting fire to either side of an axis of advance, provided they remain oriented on the axis and the objective. A deviation from an assigned axis of advance should not interfere with the maneuvering of adjacent units without prior approval of the route commander's superior. Enemy forces that do not threaten security or jeopardize the accomplishment of a mission may be bypassed.

axis of communications The route used for front-to-rear communications during a battle.

axis of trunnions The axis about which a gun is rotated in elevation to increase or decrease the range of its fire.

azimuth A direction expressed as a horizontal angle, usually in degrees, and measured clockwise from a reference datum; thus an azimuth is always relative depending upon which reference datum is used.

azimuth, altitude See ALTITUDE AZIMUTH.

azimuth, back See BACK AZIMUTH.

azimuth, compass See COMPASS AZIMUTH.

azimuth, corrected See CORRECTED AZIMUTH.

azimuth adjustment slide rule A circular slide rule by which a known angular correction for fire at one elevation can be changed to the proper correction for fire at any other elevation.

aximuthal equidistant projection An azimuthal map projection on which straight lines radiating from the center or pole of the projection represent great circles in their true AZIMUTHS from the center; the lengths along these lines are of exact scale.

azimuthal projection A projection in which the bearings from the center of a map to all other points on the map are correctly shown.

azimuth circle An instrument for measuring AZIMUTHS. It is a graduated circle on a sight, GUN CARRIAGE, searchlight, or other device.

azimuth deviation The angular difference in AZIMUTH between the lines from a gun to its target and from the gun to the point at which a PROJECTILE strikes or bursts.

azimuth difference The apparent difference in the position of an object viewed from two different points, especially from a gun position and DIRECTING POINT. Also called parallax.

azimuth indicator 1. An electrical or mechanical device that shows the AZIMUTH or deflection to be used in aiming guns or other weapons. 2. A mechanical or electrical device to measure TRAVERSE.

azimuts, tous See TOUS AZIMUTS.

B

B The designation for a bomber, such as the B-29 of World War II or today's B-52.

back azimuth A reciprocal bearing. The reverse or backward direction of an AZIMUTH; that is, the azimuth plus or minus 180 degrees.

back azimuth method A method of locating an observer's position by sighting directions from it to three or more points of known position.

back blast The rearward blast of gases and debris from recoilless weapons, rocket launchers, and certain ANTIPERSONNEL MINES when initiated.

back channel 1. A term referring to any use of informal methods for government-to-government communications, in place of normal or routine methods. 2. Unofficial communications; usually discouraged in military organizations.

backsight method Adjusting the sights of two pieces of equipment, such as guns or tanks, directly at each other in order to orient and synchronize one with the other in AZIMUTH and ELEVATION.

back tell The transfer of information from a higher to a lower ECHELON of command. See also TRACK TELLING.

bad-conduct discharge (BCD) A formal, punitive separation of an enlisted person from the military service under conditions one degree above those producing a DISHONORABLE DISCHARGE; and a grade below an undesirable discharge.

baggage train An older and declining term for the logistical support needed by a military force in the field. The "train" can move by any means (mule, truck, railroad, etc.).

BAI BATTLEFIELD AIR INTERDICTION; a tactical air mission.

Bailey bridge A prefabricated portable bridge that can be carried to and assembled over a river in rapid order. It was named after its inventor, Donald Coleman Bailey, and first used in World War II.

balance 1. A condition in which the armed forces and equipment of one nation do not have an advantage or disadvantage over the military posture of another nation. 2. The internal military adjustments that a state may make to cope with remaining threats to its security after an ARMS-CONTROL agreement is implemented. 3. Appropriate numbers of each component. A military unit is in balance if it has the appropriate numbers of infantry to armor elements. Supplies are in balance if they have the correct proportions of food, clothing, and medical supplies.

balance of power 1. The international-relations policy of rival states whose goal is to prevent any one state or alliance of states from gaining a preponderance of power in relation to its rival state or a rival alliance, thus maintaining an approximate military balance. 2. A principle of international relations which asserts that when any state seeks to increase its military potential, neighboring or rival states will take similar actions in order to maintain the military equilibrium. Consideration of the balance of power might also limit a military build-up on the grounds that one state's increased capacity would only provoke a similar increase by its potential enemy, leading at best to a stabilization at a higher, more dangerous, and more expensive level of armaments. The concept of OVERKILL suggests that one nation might have fewer nuclear weapons than its rival, but that the system would still be in balance as long as the weaker nation retained a guaranteed SECOND-STRIKE CAPABILITY.

balance of terror Winston Churchill's (1874–1965) phrase for the nuclear stalemate between the United States and the Soviet Union. Because the terror lies in the balance of their military capability, neither nation would in theory risk nuclear war because neither could win. See also MUTUAL ASSURED DESTRUCTION.

balisage The marking of a military route by a system of dim beacon lights, enabling vehicles to be driven at near daytime speed under BLACKOUT conditions.

ball ammunition SMALL-ARMS cartridges with a general-purpose, solid-core bullet intended for use against personnel and material targets not otherwise requiring armor-piercing or other special ammunition.

ballistic conditions The conditions that affect the motion of a PROJECTILE in the BORE of a weapon and through the atmosphere, including MUZZLE VELOCITY, weight of projectile, size and shape of projectile, rotation of the earth, density of the air, and elasticity of the air and wind.

ballistic conditions, standard See STANDARD BALLISTIC CONDITIONS.

ballistic correction An adjustment in FIRING DATA that is based on conditions, such as wind and temperature, affecting the flight of a PROJECTILE. It does not include adjustment based on observation of fire.

ballistic curve The actual path or trajectory of a bullet or shell as influenced by wind, and other factors.

ballistic density The computed constant air density that would have the same total effect on a PROJECTILE during its flight as the varying densities actually encountered.

ballistic director A combined observing and predicting instrument that computes FIRING DATA for the future position of a moving target.

ballistic efficiency The ability of a PROJECTILE to overcome the resistance of air. Ballistic efficiency depends chiefly on the weight, diameter, and shape of the projectile.

ballistic match A condition in which projectiles have identical ballistic characteristics. Projectiles are ballistically matched when they are fired under common conditions and their burst-point distributions have the same mean and equal standard deviations.

ballistic missile Any missile that does not rely on aerodynamic surfaces to produce lift, and consequently follows a BALLISTIC TRAJECTORY when its thrust is terminated. The missile in question behaves essentially like an ARTILLERY shell or MORTAR bomb, launched by a powerful explosion and then coasting for most of its trajectory. A ballistic missile burns its engines very quickly for a relatively short time, giving it enough velocity to complete the remainder of its journey without further power. Most ballistic missiles, in fact, briefly leave the Earth's atmosphere and accumulate gravitational acceleration on re-entry. As with a bullet or shell, and unlike an aircraft or other continually-powered flying vehicle, the entire track of the missile and its impact point are determined by its initial position, thrust, and angle. See also AERODYNAMIC MISSILE; GUIDED MISSILE.

ballistic missile, intercontinental See INTERCONTINENTAL BALLISTIC MISSILE.

ballistic missile, intermediate-range See INTERMEDIATE-RANGE BALLISTIC MISSILE.

ballistic missile, medium-range See MEDIUM-RANGE BALLISTIC MISSILE.

ballistic missile, sea-launched See SEA-LAUNCHED BALLISTIC MISSILE.

ballistic missile, short-range See SHORT-RANGE BALLISTIC MISSILE.

ballistic missile defense (BMD) All measures designed to nullify or reduce the effectiveness of an attack by BALLISTIC MISSILES after they are launched; usually conceived as having several independent layers of defense. The problems, involved in achieving any fully effective BMD system are enormous, not only on the engineering side but, perhaps even more so, in the computing and software aspects. Some estimates of the required amount of programming suggest that a full ballistic missile defense, such as is planned under the STRATEGIC DEFENSE INITIATIVE, might run to 10-million lines of computer code, none of which, critics contend, could ever be tested in realistic conditions. Ballistic missile defenses such as ABMs can be constructed to protect a large area, for example a city or region, or as a POINT DEFENSE protection around a missile SILO. Naturally the more precise the target to be defended, the more possible the task. But many analysts fear that any BMD system can be overcome because it will always be possible for the opponent to build enough offensive weapons to saturate the defense. There is a powerful and vocal school of thought which feels that the attempt to build a BMD shield is destabilizing and inevitably leads to an ARMS RACE. Their argument is that international stability rests on MUTUAL ASSURED DESTRUCTION; anything that lets one nation feel less vulnerable automatically makes the other more fearful and behave less predictably. See also ACTIVE BALLISTIC MISSILE DEFENSE; PASSIVE AIR DEFENSE. Nevertheless, other equally credible sources, agree that an effective ballistic missile defense, such as SDI, is both technically feasible and strategically desirable. See William E. Burrows, "Ballistic Missile Defense: The Illusion of Security," *Foreign Affairs,* 62, 4 (Spring 1984); Hans A. Bethe, et al, "Space-Based Ballistic Missile Defense," *Scientific American* 251:4 (October 1984); Ashton B. Carter and David N. Schwartz, eds., *Ballistic Missile Defense* (Washington, D.C.: The Brookings Institution, 1984); Alexander Flax, "Ballistic Missile Defense: Concepts and History," *Daedalus* Vol. 114, No. 2 (Spring 1985).

ballistic missile defense, active See ACTIVE BALLISTIC MISSILE DEFENSE.

ballistic missile early-warning system An electronic system for providing detection and early warning of an attack by enemy INTERCONTINENTAL BALLISTIC MISSILES. The main ballistic missile early-warning system for the United States is known as the Distant Early Warning (DEW) Line, and draws on three enormous RADAR stations, two in Canada and one in England. These radar stations can give about 20 minutes warning of an incoming missile attack over the polar route from the Soviet Union. Several other subsidiary and interlinking chains of radar stations exist for the North American continent, including systems set up to spot missiles coming in from the west, for which the DEW Line is useless. Increasingly, however, these ground-based systems are of secondary importance to SATELLITE observation.

ballistic missile submarine, fleet See FLEET BALLISTIC MISSILE SUBMARINE.

ballistics The science or art that deals with the motion, behavior, appearance, and modification of MISSILES or other vehicles acted upon by propellants, wind, gravity, temperature, or any other modifying substance, condition, or force.

ballistics, exterior See EXTERIOR BALLISTICS.

ballistics, terminal See TERMINAL BALLISTICS.

ballistics of penetration The science that treats the motion of a PROJECTILE as it forces its way into targets of solid or semi-solid construction such as earth, concrete, or steel.

ballistic table A compilation of ballistic data from which TRAJECTORY elements such as the angle of fall, range to vertex, time of flight, and ordinate at any time can be obtained. See also EXTERIOR BALLISTIC TABLE.

ballistic temperature A computed constant temperature that would have the same total effect on a PROJECTILE traveling from a gun to a target as the varying temperatures the projectile actually encounters.

ballistic trajectory The TRAJECTORY a projectile traces after the propulsive force is terminated and the body is acted upon only by gravity and aerodynamic drag.

ballistic wave An audible disturbance or wave caused by the compression of air ahead of a projectile in flight. See also SHELL WAVE.

ballistite A smokeless gunpowder used as a PROPELLING CHARGE in small arms and mortar ammunition.

balloon reflector In ELECTRONIC WARFARE, a balloon-supported CONFUSION REFLECTOR to produce fraudulent echoes.

band 1. Two or more lines of wire entanglements or other OBSTACLES arranged one behind the other. Each line of obstacles is called a belt. 2. A particular range of wavelengths in radio broadcasting. 3. A unit organized for military music. 4. A grouping of radio frequencies; for example, the frequency-modulated (FM) band or the short-wave band.

band of fire GRAZING FIRE, usually from one or more AUTOMATIC WEAPONS, that has a CONE OF DISPERSION so dense that a man trying to cross the LINE OF FIRE would probably be hit. A final protection line uses a band of fire. Sometimes referred to as the "BEATEN ZONE." These bands or zones can be set up so that fire can be applied under conditions of bad visibility or at night.

Bangalore torpedo A metal tube or pipe packed with a HIGH-EXPLOSIVE charge. Chiefly used to clear a path through barbed wire or MINEFIELDS; first used by the British in 1799 at Bangalore, India.

banquette A step or small wall within a TRENCH or parapet that soldiers stand on when firing over the crest.

barbed wire Metal wire with sharpened barbs; used in defensive positions to slow the advance of an attack.

barbette Any support that raises a gun; it can be dirt piled up on a specially built platform. When guns are elevated enough to fire over their protective parapets or armor, they are said to be "in barbette."

bargaining chip 1. Anything one might be willing to trade in a negotiation. 2. Any military force, WEAPONS SYSTEM, or other resource, present or projected, which a country expresses willingness to downgrade or discard in return for a concession by a particular military rival. The phrase was first used in this context during the STRATEGIC ARMS LIMITATION TALKS (1969–1979) between the United States and the Soviet Union. Because the Nixon Administration believed that the Soviet Union only signed SALT I because the United States had approved the development of the anti-ballistic missile (ABM), the POSEIDON submarine, and the MINUTEMAN III missile systems, it asked Congress to approve, as a bargaining chip for SALT II, the development of the TRIDENT submarine, the B-1 bomber, and the CRUISE MISSILE. The bargaining-chip strategy has been attacked on two fronts: first, because it is too expensive to develop weapons systems simply to trade them away at the negotiating table; and second, because it is inflammatory to the ARMS RACE to develop weapons systems merely as a hedge in negotiations. See Robert J. Bresler and Robert C. Gray, "The Bargaining Chip and SALT," *Political Science Quarterly* Vol. 92, No. 1 (Spring 1977); Robert C. Gray and Robert J. Bresler, "Why Weapons Make Poor Bargaining Chips," *Bulletin of the Atomic Scientists* Vol. 33, No. 7 (September 1977). See also SALT.

bargaining strength The relative power that each of two or more parties holds during a negotiating process. The final settlement often reflects the bargaining power of each side.

barracks 1. Building(s) for lodging troops or other military personnel. 2. The quarters of enlisted personnel as opposed to officers' quarters.

barrage 1. A prearranged barrier of FIRE, not including SMALL-ARMS fire, designed to protect friendly troops and installations by impeding enemy movements across defensive lines or areas. 2. A protective screen of balloons that are moored to the ground and maintained at given heights to prevent or hinder operations by enemy aircraft. Also called balloon barrage. 3. A method of fire employed against a fast-opening or -closing target, whereby a gun RANGE or FUZE setting is used that will place the initial shots ahead of the target in the direction of its anticipated advance.

barrage, creeping See CREEPING BARRAGE.

barrage, rolling See ROLLING BARRAGE.

barrage balloons A series of blimps anchored by metal wires to an area in order to inhibit low flying aircraft attacks; widely used in World War II.

barrage fire Fire that is designed to fill a volume of space or area rather than being aimed specifically at a given target. See also FIRE.

barrage jamming An ELECTRONIC COUNTERMEASURE that produces simultaneous JAMMING over a broad band of frequencies. See also ELECTRONIC WARFARE.

barrel 1. A metal or plastic tube through which AMMUNITION is fired and which controls the initial direction of a projectile. 2. A standard unit of measurement of liquids in petroleum pipeline and storage operations, 42 United States standard gallons.

barrel erosion Wearing away of the surface of the BORE of a gun due to combined effects of gas washing, scoring, and mechanical abrasion. Barrel erosion causes a reduction in MUZZLE VELOCITY.

barrel reflector A device used for inspecting the BORE and chamber of a gun or rifle barrel. A barrel reflector consists of a mirror mounted in a frame, and a tube that is inserted into the chamber of the gun and gives a view of the bore.

barrier A coordinated series of OBSTACLES designed or employed to CANALIZE, direct, restrict, delay, or stop the movement of an opposing force, and to impose additional losses in personnel, time, and equipment on the opposing force.

barrier combat air patrol One or more divisions or elements of FIGHTER aircraft employed between a force and an OBJECTIVE AREA as a barrier across the probable direction of an enemy attack. It is used as far from the force as control conditions permit, giving added protection against RAIDS that would use the most direct routes of approach.

barrier forces Air, surface, and submarine units and their supporting systems positioned across the likely courses of expected enemy transit for early detection, blocking, and destruction of the enemy.

barrier line An artificial traffic control line over land or water; ships and vehicles can be ordered to move in relation to it. For example, in a search for a missing plane, search aircraft may be ordered not to search beyond a specially established barrier line.

barrier plan That part of an OPERATION PLAN (or order) that is concerned with the employment of OBSTACLES to CANALIZE, direct, restrict, delay, or stop the movement of an opposing force, and with the infliction of additional losses in personnel and equipment upon that opposing force.

barrier study An analysis of the terrain in an area designed to determine the most effective use of existing and reinforced obstacles and the area's potential for COMBAT operations. See also TERRAIN ANALYSIS.

barrier system A coordinated series of related BARRIERS located in depth and designed to CANALIZE and disorganize enemy forces, to delay or stop enemy movement, and to aid in the accomplishment of a unit mission.

barrier tactics Tactics based on the use of fortified lines, both natural and artificial, supported by fire.

bar sight The rear sight of a FIREARM, consisting of a movable bar, usually with an open notch.

base 1. A locality from which military operations are projected or supported. 2. An area or locality containing INSTALLATIONS that provide logistic or other support. 3. A home airfield or a home aircraft carrier. 4. The foundation or part upon which an object or instrument rests, such as a gun base. 5. A unit or other organization around which a MANEUVER is planned and performed, usually called a BASE UNIT. 6. Part of a PROJECTILE below the ROTATING BAND. 7. A line used in mapping, surveying, or FIRE CONTROL as a reference from which distances and angles are measured.

base, advanced See ADVANCED BASE.

base, army See ARMY BASE.

base command An area containing a military base or group of such bases organized under one commander. See also COMMAND.

base defense The local military measures, both normal and emergency, required to nullify or reduce the effectiveness of enemy attacks on, or sabotage of, a base.

base end station An observing point at the end of a BASELINE from which angles are measured for determining FIRING DATA, or for surveying, or both.

baseline 1. A surveyed line established with more than usual care, to which surveys are referred for coordination and correlation. 2. The line between the principal points of two consecutive vertical air photographs. It is usually measured on one photograph after the principal point of the other has been transferred. 3. The shorter arc of the great circle joining two radio transmitting stations of a navigation system. 4. The side of one of a series of coordinated triangles the length of which is measured with prescribed accuracy and precision and from which lengths of the other triangle sides are obtained by computation.

base mortar A mortar for which initial FIRING DATA are computed and with reference to which the data for other mortars in a unit are computed.

base of fire Fire placed on an enemy force or position to reduce or eliminate its capability to interfere with the movement of friendly elements; it may be provided by a single weapon or a grouping of WEAPONS SYSTEMS. A base of fire supports the advance of other military units and serves as the base around which attack operations are carried out.

base of operations An area or facility from which a military force begins its offensive operations, to which it falls back in case of reverses, and in which supply facilities are organized.

base of trajectory A straight horizontal line from the center of the muzzle of a weapon to the point in the downward curve of the path of a projectile fired from the weapon that is level with the muzzle.

base operations support Administrative and logistical services for a permanent military installation.

base piece 1. The GUN or HOWITZER in a battery for which the initial FIRING DATA may be calculated, and with reference to which the firing data for other guns or howitzers may be computed. 2. The gun or howitzer nearest the BATTERY CENTER.

base plate 1. A plate or support used to distribute the weight of a heavy structure or apparatus so as to prevent it from sinking or collapsing under direct thrust. 2. A metal plate with a socket into which the base of the barrel of a MORTAR is seated.

base point An aiming point for an ARTILLERY unit, usually something easily identifiable by sight or on a map; all fire is related or adjusted to it.

base ring A circular metal track in the concrete platform of a fixed CANNON. A base ring is used to support the cannon and to enable it to be fired in any direction.

base spray Fragments of a bursting shell that are thrown rearward into the line of flight of the PROJECTILE of which the shell is a part. See also NOSE SPRAY; SIDE SPRAY.

base surge A cloud that rolls out from the bottom of the column produced by a NUCLEAR WEAPON bursting beneath the earth's surface. For underwater bursts, the surge is, in effect, a cloud of liquid droplets which flows almost as if it were a homogeneous fluid. For subsurface land bursts the surge is made up of small solid particles that also behave like a fluid.

base unit The unit of organization in a tactical operation around which a movement or MANEUVER is planned and performed.

basic branch The military branch to which an OFFICER is assigned upon being commissioned (or upon branch transfer). For example: INFANTRY or ARTILLERY.

basic combat training Training in basic military subjects and fundamentals of basic infantry combat, given to newly inducted and enlisted personnel without prior military service.

basic communication An original letter, report, or other document. Added materials, such as endorsements and enclosures, are not part of a basic communication.

basic data Essential facts needed to place FIRE on a TARGET. The location of the target relative to the battery must be known in terms of direction or DEFLECTION; distance or RANGE; and difference in altitude or site; all of which are basic data.

basic encyclopedia A compilation of identified INSTALLATIONS and physical areas of potential significance as OBJECTIVES for attack.

basic intelligence Fundamental INTELLIGENCE concerning the general situation, resources, capabilities, and vulnerabilities of foreign countries or areas, which may be used as reference material in the planning of OPERATIONS at any level and in evaluating subsequent information relating to the same subject.

basic load 1. The quantity of supplies required to be on land within, and which can be moved by, a UNIT or FORMATION. It is expressed according to the wartime organization of the unit or formation and maintained at the prescribed levels. 2. That quantity of non-nuclear ammunition that is authorized and required to be on hand within a unit or formation at all times.

basic military route network Axial, lateral, and connecting routes designated in peacetime by a host nation to meet the anticipated military movements and transport requirements, both allied and national. See also TRANSPORT NETWORK.

basic military training Training in military subjects given a soldier during the first phase of BASIC TRAINING.

basic pay The salary (other than ALLOWANCES or hazardous duty pay) of a member of a military unit based on grade and length of service.

basic stopping power The probability, expressed as a percentage, of a single vehicle being stopped by MINES while attempting to cross a minefield.

basic tactical organization The conventional organization of LANDING-FORCE units for combat, involving combinations of infantry, supporting ground arms, and aviation for the accomplishment of missions ashore. This organizational form is employed as soon as possible following the landing of the various assault components of the landing force.

basic tactical unit A fundamental UNIT capable of carrying out an independent tactical mission, such as RIFLE COMPANY in the INFANTRY, or a battery in the ARTILLERY.

basic training The first physical and professional training that soldiers receive at the beginning of their military service. According to Gwynne Dyer in *War* (Chicago: Dorsey, 1985), basic training is "a feat of psychological manipulation on the grand scale which has been so consistently successful and so universal that we fail to notice it as remarkable. In countries where the army must extract its recruits in their late teens, whether voluntarily or by conscription, from a civilian environment that does not share the military values, basic training involves a brief but intense period of indoctrination whose purpose is not really to teach the recruits basic military skills but rather to change their values and their loyalties."

basic undertakings The essential actions, expressed in broad terms, that must be taken in order to successfully implement a commander's concept. These may include military, diplomatic, economic, psychological, and other measures. See also STRATEGIC CONCEPT.

basic unit training Applicatory training given during the final phase of BASIC TRAINING when soldiers assigned to units are first assembled and trained together in rehearsal of their role as an officially designated military organization.

basing, deceptive See DECEPTIVE BASING.

basing mode The way in which INTERCONTINENTAL BALLISTIC MISSILES are deployed or stored; for example, in hardened underground SILOS, in submarines, or on movable vehicles.

bastion 1. Any fortified position. 2. A triangular fortification that extends from a main element or wall to give its defenders two additional walls from which to fire on an attacking enemy.

batman A British term for an enlisted man who functions as a servant to an officer.

battalion A unit composed of a headquarters and two or more COMPANIES or BATTERIES; anywhere from 300 to 1,000 soldiers. It is usually commanded by a LIEUTENANT COLONEL and may be part of a REGIMENT (or BRIGADE), and be charged with only tactical functions, or it may be a separate unit and be charged with both administrative and tactical functions. In the British army, only infantry is organized into battalions, units of the same size in other areas being referred to as regiments. Voltaire (1694–1778) is often credited with first observing that "God is always for the big battalions." In this century Joseph Stalin (1879–1953) made a different analysis and came to a similar conclusion when he was asked to accommodate the Pope of the Roman Catholic Church: "The Pope! How many divisions has he got?"

battalion group, artillery See ARTILLERY BATTALION GROUP.

battalion landing team (BLT) In an AMPHIBIOUS OPERATION, an infantry battalion normally reinforced by necessary combat and service elements; the basic UNIT for planning an ASSAULT landing.

battalion task force A combined force of TANK and mechanized infantry units under a single battalion commander. A battalion task force may be TANK-heavy, MECHANIZED-INFANTRY-heavy, or balanced, depending on the concept and plan of operation.

battery 1. A tactical and administrative ARTILLERY unit or subunit corresponding to a COMPANY or similar unit in other branches of the Army. 2. All guns, TORPEDO tubes, searchlights or missile launchers of the same size or CALIBER or used for the same purpose, either installed in one ship or otherwise operating as an entity. 3. A group of weapons or artillery support equipment such as MORTARS, ARTILLERY pieces, launchers, searchlights, or target-acquisition equipment under one tactical commander in a certain area. 4. A gun in firing position is said to be "in battery." A gun out of battery is a gun not in firing position, i.e., not ready to fire. 5. A tube (BARREL) in battery is a gun tube fully returned from recoil upon its cradle. A tube out of battery is a tube not fully returned from recoil. 6. Military power in general. See also FIRING BATTERY.

battery, firing See FIRING BATTERY.

battery center A point on the ground, the coordinates of which are used as a reference indicating the location of a battery in the production of FIRING DATA. Also called the chart location of the battery.

battery front The lateral distance between the flank guns of a BATTERY.

battery ground pattern The shape and dimensions of the pattern made by the location of the guns of a BATTERY emplaced for firing.

battery left (or right) A method of FIRE in which weapons are discharged from the left (or right) one after the other, at five-second intervals, unless otherwise specified.

battery operations center In FIELD-ARTILLERY operations, a facility established to serve as an alternate FIRE-DIREC-

TION CENTER and as the battery COMMAND POST.

battery right See BATTERY LEFT.

battle Major hostilities between two warring sides, in which each seeks to destroy or neutralize (by rendering ineffective) the military forces of the other. Modern battles are made up of many ENGAGEMENTS, and can last days or even months. A battle ends when one side wins or both sides simply withdraw. Winston Churchill wrote in his 1923 history of World War I, *The World Crisis* that: "Battles are won by slaughter and maneuver. The greater the general, the more he contributes in maneuver, the less he demands in slaughter." This was confirmed by General Douglas MacArthur (1880–1964), when he told President Franklin D. Roosevelt: "The days of the frontal attack should be over. Modern infantry weapons are too deadly, and frontal assault is only for mediocre commanders. Good commanders do not turn in heavy losses." See John Keegan, *The Face of Battle* (New York: Viking Press, 1976); Trevor Nevitt Dupuy, *Numbers, Predictions, and War: Using History to Evaluate Combat Factors and Predict the Outcome of Battles,* (Fairfax, Va.: Hero Books, 1985); Hans Henning von Sandrart, "Considerations of the Battle in Depth," *Military Review,* 17, 10 (October 1987).

battle, pitched See PITCHED BATTLE.

battle, set-piece See SET-PIECE BATTLE.

Battle Act of 1951 The Mutual Defense Assistance Control Act of 1951, which calls for the automatic embargo of military and strategic materials to states that the U.S. government declares to be a threat. This law also forbids foreign aid both to designated "threat" states and their military suppliers.

battle area, main See MAIN BATTLE AREA.

battle casualty Any CASUALTY incurred in action. "In action" characterizes the casualty as having been the direct result of hostile action, sustained in combat or relating thereto, or sustained going to or returning from a combat mission, provided that the casualty was directly related to hostile action. Included are persons killed or wounded mistakenly or accidentally by FRIENDLY fire directed at a hostile force or what is thought to be a hostile force. However, not to be considered as sustained in action and thereby not to be interpreted as battle casualties are injuries due to the elements, self-inflicted wounds, and, except in unusual cases, wounds or death inflicted by a friendly force while the killed or injured individual is in ABSENT-WITHOUT-LEAVE (AWOL) or dropped-from rolls status or is voluntarily absent from a place of duty.

battle cry The yelling of attacking troops used to demoralize the enemy and to encourage their own efforts to close with the opposition.

battle dress uniform A field uniform with minimal ornamentation that is generally made of camouflaged material

battle fatigue A World War II term for mental exhaustion brought on by extended combat. Another term for the same phenomenon is COMBAT-HAPPY. Also called battle stress.

battlefield The location of a specific military ENGAGEMENT. The battlefield in conventional warfare has been relatively small in area, and has usually been linear in composition: two lines of troops fight over an area only a few miles long, with death and destruction restricted to a narrow strip, perhaps not much more than a mile wide. A projection from presumed WARSAW PACT and NORTH ATLANTIC TREATY ORGANIZATION (NATO) strategy and tactics in the event of war would suggest that their battlefield would be very different from the traditional model. First, it is probable that the entire length of the central front would be assaulted at once.

Second, NATO tactics are designed to deepen the battlefield by attacking second-echelon Warsaw Pact forces as far as 500 kilometers behind the front line. New doctrines anticipate a much more fluid and non-linear battle, rather than a firmly held, but thin, line.

battlefield air interdiction (BAI) Air action against hostile ground targets that threaten friendly forces. Battlefield air interdiction missions require joint planning, but may not require continuous coordination during execution. Compare to INTERDICTION.

battlefield nuclear weapons Tactical NUCLEAR WEAPONS used within the context of a battle, for direct and immediate tactical ends. "Battlefield" obviously suggests short-range weapons and, as such, they have to be weapons of low YIELD, to prevent endangering the user's own troops. Typical of such weapons are nuclear shells for firing from field guns, with ranges of approximately 15 to 25 kilometers, or MISSILES having a range of about 110 kilometers. Such weapons will have YIELDS from as little as one KILOTON to perhaps 10 kilotons. More modern designs of battlefield weapons in fact have what is called a "dial-a-yield" capacity, so that the field commander can select the detonation power to suit the exact tactical needs. The NORTH ATLANTIC TREATY ORGANIZATION (NATO) stocks of battlefield nuclear weapons are plentiful; there are more than 6,000 WARHEADS in this category, the majority of them in the form of artillery shells. See James H. Polk, "The Realities of Tactical Nuclear Warfare," *Orbis* 17, 2 (Summer 1973); David L. Nichols, "Who Needs Nuclear Tacair?" *Air University Review* (March–April 1976); Manfred Worner, "NATO Defenses and Tactical Nuclear Weapons," *Strategic Review,* 5, 4 (Fall 1977); Peter D. Zimmerman and G. Allen Greb, "The Bottom Rung of the Ladder: Battlefield Nuclear Weapons in Europe," *Naval War College Review* 35, 6 (November-December 1982); Ilana Kass and Michael J. Dean, "The Role of Nuclear Weapons in the Modern Theater Battlefield: The Current Soviet View," *Comparative Strategy* 4, 3 (1984).

battle force A standing operational naval TASK FORCE organization of aircraft carriers, surface combatants, and submarines assigned to numbered fleets. A battle force is subdivided into BATTLE GROUPS.

battle group A standing naval task group consisting of an aircraft carrier, surface combatants, and submarines as assigned in direct support, operating in mutual support with the task of destroying hostile submarine, surface, and air forces within the group's assigned area of responsibility. The battle group is a larger unit than an ATTACK CARRIER STRIKING FORCE.

battle honor An award to a military unit or individual denoting participation in a CAMPAIGN; a streamer or band attached to the staff of the FLAG, COLOR, standard, or GUIDON of a unit denoting battle participation or the award of a unit DECORATION.

battle management 1. The command, control and communications function (see C^3I). 2. The set of instructions and rules and the corresponding hardware controlling the operation of a BALLISTIC MISSILE DEFENSE system. SENSORS and interceptors are allocated by the system, and the updated battle results are presented to the (human) command for analysis and possible intervention.

battle position (BP) A defensive location on the ground, selected on the basis of TERRAIN and available weapons, from which units can defend or attack.

battleship The largest and most powerful of traditional warships. New battleships have not been built since World War II when the aircraft carrier replaced the battleship as the core of a modern fleet.

battle sight A predetermined sight setting that, carried on a WEAPON, will enable the firer to engage targets effectively at battle ranges when conditions do not permit exact sight settings.

battle stations The various duty positions that all members of a ship's company assume to ensure complete combat readiness, or to proceed into battle. See GENERAL QUARTERS.

battlewagon A modern battleship; only the U. S. Navy currently operates traditional World War II-vintage battleships such as the USS *Iowa* or USS *New Jersey*.

bay 1. A section of a floating bridge extending from the center of one pontoon to the center of the next. 2. A designated area within a section of a warehouse or depot shop, usually outlined or bounded by posts, pillars, or columns. 3. A straight section of a TRENCH between two bends. 4. A body of water partially enclosed by land, but which has a direct outlet to the sea; for example, Hudson Bay in Canada.

Bay of Pigs 1. The landing site in 1961 of the American-sponsored invasion of Cuba by expatriate Cubans who were trained by the CENTRAL INTELLIGENCE AGENCY (CIA) to overthrow the government of Fidel Castro. The action was a total failure and major embarrassment to the Kennedy Administration. 2. Any fiasco or major flop. Just as Napoleon had his Waterloo, Kennedy had his Bay of Pigs. But where Napoleon made an honest try and nearly won, Kennedy's effort was especially embarrassing because of the incompetence in its execution. The invasion was based on grossly inaccurate intelligence, was poorly planned and led, and lacked adequate AIR COVER. If you have a Bay of Pigs, you haven't merely lost, you've disgraced yourself as well. See Peter Wyden, *Bay of Pigs: The Untold Story* (New York: Simon & Schuster, 1979); Trumbell Higgins, *The Perfect Failure: Kennedy, Eisenhower and the CIA at the Bay of Pigs* (New York: Norton, 1987).

bayonet A knife that can be attached to the muzzle of a rifle, especially useful in close-quarter fighting, when silence is essential, or when all ammunition has been used. It was Confederate General Thomas Jonathan "Stonewall" Jackson who wrote in 1862 that: "Under Divine blessing, we must rely on the bayonet when firearms cannot be furnished."

bazooka The shoulder-fired, direct, line-of-sight antitank rocket of World War II. By analogy, any antitank or antiaircraft rocket that can be fired by only one individual.

BCD See BAD-CONDUCT DISCHARGE.

beach 1. The area extending from the shoreline of a land mass inland to a marked change in physiographic form or material, or to the line of permanent vegetation (coastline). 2. In AMPHIBIOUS OPERATIONS, that portion of the shoreline designated for the landing of a TACTICAL UNIT.

beach capacity An estimate, expressed in terms of measurement tons, or weight tons, of cargo that may be unloaded over a designated strip of shore per day.

beach diagram A diagram showing the landing areas to be used by various landing teams. Such landing areas are usually named by colors, as beach red, yellow, blue, and so forth.

beach dump 1. An initial and temporary supply point established on a BEACHHEAD for the receipt and issue of supplies. 2. An area adjacent to a beach utilized by a SHORE PARTY for the temporary storage of supplies.

beach flag A FLAG with a colored background and vertical stripes. The background indicates the color designating the beach. It is usually used on control and other craft that serve a designated beach area.

beach group, naval See NAVAL BEACH GROUP.

beachhead A designated area on a hostile shore which, when seized and held, insures the continuous landing of troops and material, and provides maneuver space requisite for subsequent projected operations ashore. It is the physical objective of an AMPHIBIOUS OPERATION. See also AIRHEAD; BRIDGEHEAD.

beachhead line The line that fixes the inshore limits of a BEACHHEAD.

beachmaster The naval officer in command of the BEACHMASTER UNIT of a NAVAL BEACH GROUP

beachmaster unit A unit of the NAVAL BEACH GROUP, designed to provide to a SHORE PARTY a naval component known as a BEACH PARTY which provides support for the amphibious landing of a DIVISION.

beach matting Fabricated material placed on soft ground or sand surfaces to improve the traction of vehicles.

beach obstacle An artificial obstacle placed on possible landing beaches between the high-water line and the shoreline vegetation, intended for use against personnel or vehicles.

beach party The naval component of a shore party. See also BEACHMASTER UNIT; SHORE PARTY.

beach party commander The naval officer in command of the naval component of a SHORE PARTY.

beach-support area In AMPHIBIOUS OPERATIONS, the area to the rear of a LANDING FORCE or elements thereof, established and operated by SHORE-PARTY units, which contains the facilities for the unloading of troops and materiel and the support of forces ashore; it includes facilities for the evacuation of wounded, prisoners of war, and captured materiel.

beam rider A MISSILE guided by RADAR or a radio beam.

beam weapon, electronic See ELECTRONIC BEAM WEAPON.

bear 1. The national symbol of the Soviet Union. 2. A series of Soviet bombers. 3. To aim a weapon. 4. An indication of direction, as in "bears north."

beaten zone The area of ground on which a CONE OF FIRE falls. Compare to BAND OF FIRE.

beaten zone, effective See EFFECTIVE BEATEN ZONE.

beehive ammunition An informal term for anti-personnel artillery shells, loaded with small steel FLECHETTEs, and designed for fire at close range against massed infantry attack.

beleaguered Describes any organizational element that has been surrounded by a hostile force to preclude its escape.

belligerent 1. A nation at war with another. 2. The fighting armed forces of a nation at war, as opposed to its ordinary citizens. GIULIO DOUHET wrote in his *The Command of the Air* (1921): "Any distinction between belligerents and non belligerents is no longer admissable today either in fact or theory . . . When nations are at war, everyone takes part in it; the soldier carrying his gun, the woman loading shells at a factory, the farmer growing wheat, the scientist experimenting in his laboratory . . . It begins to look now as if the safest place may be the trenches."

besieged A term describing any organized element that has been surrounded by a hostile force to compel it to surrender.

bessel method A method of locating an observer's position on a map or chart by sighting along points on the map that represent visible terrain features.

B-52 The 1950s-vintage heavy bomber operated by the United States Air Force's STRATEGIC AIR COMMAND. Compare to B-1B.

bias One of two very different measures of a MISSILE's accuracy, the other being CIRCULAR ERROR PROBABLE (CEP), the most frequently cited measure of accuracy. CEP refers to the size of the area within which it can be predicted that 50 percent of missile WARHEADS will hit, and does not reflect the ordinary meaning of accuracy. Bias refers to the probable average distance a warhead will be from the actual aiming point, or designated GROUND ZERO, when it detonates. Clearly the degree of bias in a missile's FLIGHT PATH is crucial if the objective is to destroy a HARD TARGET, but it is much more difficult to measure than CEP.

big stick 1. Military power. 2. President Theodore Roosevelt's foreign policy, derived from the adage "speak softly and carry a big stick." The best example of his "big stick" policy was the separation of the Isthmus of Panama from Colombia in order to create a government that would be more cooperative in the American effort to build a ship canal across Central America. When Roosevelt met with his cabinet to report what had happened, he asked Attorney General Philander C. Knox (1853–1921) to construct a legal defense. The Attorney General is reported to have remarked, "Oh, Mr. President, do not let so great an achievement suffer from any taint of legality." Later, when Roosevelt sought to defend his heavily criticized actions to the cabinet, he made a lengthy statement and asked, "Have I defended myself?" The Secretary of War, Elihu Root (1845–1937), replied, "You certainly have. You have shown that you were accused of seduction and you have conclusively proved that you were guilty of rape." See David McCullough, *The Path Between the Seas: The Creation of the Panama Canal: 1870–1914* (New York: Simon & Schuster, 1977).

bilateral infrastructure An infrastructure that concerns only two North Atlantic Treaty Organization (NATO) members and is financed by mutual agreement between them (e.g., facilities required for the use of forces of one North Atlantic Treaty Organization (NATO) member in the territory of another). See also INFRASTRUCTURE.

bilateralism 1. Joint economic policies between states; specifically, the agreement to extend to each other privileges (usually relating to trade) that are not available to others. 2. Joint security policies between states; specifically treaties of alliance in the event of war. 3. Joint diplomatic postures or actions by states, whether or not in the form of a formal alliance. This is in contrast to unilaterialism, in which each state goes its own way without necessarily regarding the interests of the others.

billet 1. Shelter for troops. 2. To quarter troops. 3. A PERSONNEL position or assignment that may be filled by one person. 4. To assign troops to live in civilian housing. The Third Amendment to the Constitution of the United States holds that: "No Soldier shall, in time of peace, be quartered in any house, without the consent of the Owner, nor in time of war, but in a manner to be prescribed by law."

binary explosive A mixture containing two HIGH EXPLOSIVES.

binary weapon A SHELL or BOMB containing two substances that remain inert while kept separate, but which will mix upon impact to form a new substance that is harmful in some way. Normally this term pertains to CHEMICAL WEAPONS. Such weapons are designed to be safer to store and transport than previous models whose poisonous chemicals were already combined.

bingo 1. A term that when originated by an aircraft pilot, means, "I have reached minimal fuel for safe return to base or designated alternate." 2. When originated by air control authorities, means, "Proceed to alternate airfield or aircraft carrier as specified." 2. An expression of pride in achieving something, meaning, "We did it."

bingo field An alternate airfield.

biological agent A micro-organism that causes disease in humans, plants, or animals, or causes the deterioration of material. See also BIOLOGICAL OPERATION; BIOLOGICAL WEAPON; CHEMICAL AGENT.

biological ammunition A type of ammunition, the filler of which is primarily a BIOLOGICAL AGENT.

biological operation 1. The employment of BIOLOGICAL AGENTS to produce casualties in humans or animals and damage to plants or MATERIEL. 2. Defense against such employment.

biological warfare Warfare waged with BIOLOGICAL (or bacteriological) AGENTS. The essence of biological warfare is simple. It involves taking a highly infectious virus or bacteria that occurs naturally, and developing it so that its lethality is enormously increased. This would make it possible to infect large populations or large territorial areas with very small amounts of such a preparation. Thus a war would be fought by deliberately infecting the enemy's armies or civilian population with deadly diseases, some of which, in their scientifically enhanced form, could kill in minutes. The ideal biological agent is one with a very fast infection and killing rate, but which also becomes inert very rapidly, allowing troops to quickly occupy no-longer-defended areas. Although extremely virulent diseases can easily be bred in laboratories, and even deployed in weapons, the latter requirement, fortunately, seems harder to guarantee. See Susan Wright and Robert L. Sinsheimer, "Recombinant DNA and Biological Warfare," *Bulletin of the Atomic Scientists* 39, 9 (November 1983); Erhard Geissler, *Biological and Toxin Weapons Today* (New York: Oxford University Press, 1986).

bird colonel See COLONEL.

birth-to-death tracking The ability to track a MISSILE and its PAYLOAD from launch until it is intercepted or reaches its target.

bistatic radar A RADAR system in which the RECEIVER and transmitter are separated.

biting angle The smallest angle of impact at which a PROJECTILE will penetrate or pierce ARMOR.

bivouac 1. A temporary military encampment. 2. A rear area rest site for military units.

black In INTELLIGENCE handling, a term used in certain phrases (e.g., "living black," "black border crossing") to indicate reliance on illegal concealment.

black-bag job 1. A Federal Bureau of Investigation (FBI) term for illegal searches to gather INTELLIGENCE. 2. Bribing someone to obtain information. 3. A non-military covert operation of the Central Intelligence Agency (CIA).

black box 1. Any complicated technical, mechanical, or electronic element that is installed or replaced as a unit. 2. A device that records aircraft instrument readings and which is designed to survive in the event of a crash.

black forces A term used in the reporting of intelligence on WARSAW PACT exercises, to denote those units actually representing Warsaw Pact forces.

black list 1. Any list of people to be treated badly in some way. 2. An official

COUNTERINTELLIGENCE listing of actual or potential enemy collaborators, sympathizers, intelligence suspects, and other persons whose presence menaces the security of FRIENDLY forces.

blackout 1. A state in which lights or fires are put out to avoid enemy detection. 2. An electrical-utility failure. 3. The disabling of RADAR by means of a nuclear explosion. The intense electromagnetic energy released generates a large background that obscures signals and renders many types of radar useless for minutes or longer.

black powder An unstable, sensitive, and easily ignitable LOW-EXPLOSIVE charge used as a component of IGNITERS, IGNITING PRIMERS, FUZES, and blank fire charges.

black propaganda Propaganda that purports to originate from a source other than the true one.

black rain Rainwater which appears black due to its high content of radioactive debris; it may cause the same harmful effects as radiation from other sources.

blank file A position in a DRILL formation that has not been filled.

blast The brief and rapid movement of air, vapor, or fluid away from a center of outward pressure, as in an explosion or in the combustion of rocket fuel; the pressure accompanying this movement. This term is commonly used for "explosion," but the two terms may be distinguished.

blast effect The destruction of or damage to structures and PERSONNEL by the force of an explosion on or above the surface of the ground. A blast effect may be contrasted with the cratering and ground-shock effects of a PROJECTILE or CHARGE that goes off beneath the ground surface.

blasting cap Small cylindrical case with a thin wall that contains sensitive explosives used to set off another explosive charge, such as demolition blocks or dynamite.

blasting fuze A time FUZE with BLASTING CAPS, used for setting off an EXPLOSIVE CHARGE.

blasting machine A small, hand-operated generator for electrically firing one or more detonators or SQUIBS to explode or ignite MUNITIONS or series of explosive charges.

blast line A horizontal radial line on the surface of the Earth originating at GROUND ZERO, on which measurements of the blast from an explosion are taken.

blast wave A sharply defined wave of increased pressure rapidly propagated through a surrounding medium from a center of detonation of an explosive, or a similar disturbance.

bleaching material A white powder with a chlorine-like odor that is the most commonly used decontaminant for BLISTER AGENTS and G-SERIES CHEMICAL AGENTS.

blind 1. A CAMOUFLAGE cover for combat forces. 2. In COVERT OPERATIONS an activity or organization which conceals its true purposes.

blind bombing The bombing by air of a target that can be located only by navigational estimates.

blind circuit A circuit in which communication is possible in only one direction.

blind transmission A transmission made without the expectation of a receipt or reply.

blister agent A CHEMICAL AGENT that injures the eyes and lungs, and burns or blisters the skin. The MUSTARD H gas of World War I is one example. Also called a "vesicant agent."

blitz Concentrated bombing on a target. The original blitz was the World War II bombing of London by the German air force.

blitzkrieg From the German: "lightning war;" the tactical method used by the German army in the invasion of Poland, France, and the Soviet Union during World War II. It stresses speed and maneuverability. A blitzkrieg campaign involves a series of army COLUMNS searching for weak spots in the enemy LINE. By exploiting these weaknesses, UNITS pass behind the enemy, destroying its lines of communication and disrupting its plans. The emphasis is on avoiding the opposition's strength rather than trying to meet it head on, and winning not so much by destroying the enemy's TROOPS and MATERIEL as by paralyzing them with multiple unexpected attacks that destroy military coordination. If properly executed, a blitzkrieg offers a weaker army the chance to overcome one it could not defeat in PITCHED BATTLE. Winston Churchill described the classic German Blitzkrieg in his 1948 history of World War II, *The Gathering Storm,* thusly; "We had seen a perfect specimen of the modern Blitzkrieg; the close interaction on the battlefield of army and air force; the violent bombardment of all communications and of any town that seemed an attractive target; the arming of an active Fifth Column; the free use of spies and parachutists: and above all, the irresistible forward thrust of great masses of armor." See Paul Tiberi, "German versus Soviet Blitzkrieg," *Military Review* 65, 9 (September 1985); Stephen J. Cimbala, "Soviet 'Blitzkrieg' in Europe: The Abiding Nuclear Dimension," *Strategic Review* 14, 3 (Summer 1986); Noyes B. Livingston III, "Blitzkrieg in Europe: Is It Still Possible?" *Military Review* 66, 6 (June 1986).

block 1. A group of explosive units fastened together to initiate all at once. 2. An obstacle that prevents or hinders the advance of enemy troops. 3. To hinder the movement of ground troops by placing obstacles across their routes of advance. 4. To interfere with enemy radio broadcasts by transmitting on the same frequency. 5. A self-supporting, regular stack of supplies, two or more units wide, two or more deep, and two or more high. A block may be either rectangular or pyramidal.

blockade A military action in which one country attempts to prevent another by force from importing either some or all of its goods. Traditionally, a blockade involves the exercise of sea power, where one navy patrols the coastline of another, stopping shipping from entering, and preventing the enemy's merchant and naval ships from leaving harbor. The aim of a blockade is to starve the enemy into surrender by depriving him of food, raw materials, and military supplies, and by impeding its international trade, thereby squeezing its financial resources.

blockade runner Any ship, whether merchant or naval, that attempts to pass through a formally declared naval blockade. See SHOT ACROSS THE BOW.

blockhouse A defensive structure, usually built of wood or concrete, with small windows for observation and fire. Unless they are part of a fort, blockhouses are frequently partially underground or otherwise camouflaged.

blocking position A DEFENSIVE POSITION so situated as to deny the enemy access to a given area, or to prevent his advance in a given direction.

blocks, burster See BURSTER BLOCKS.

block shipment A method of shipping supplies to overseas areas to provide balanced (appropriate amounts of food, clothing, ammunition, etc.) stocks for an arbitrary balanced force for a specific number of days, e.g., the shipment of a 30 days' supply of materials for an average force of 10,000 individuals.

block-stowage loading A method of loading whereby cargo for a specific destination is stowed together. The purpose is to facilitate rapid offloading at the destination, with the least possible disturbance of cargo intended for other points.

blood agent A chemical compound, including compounds of the cyanide group, that affects bodily functions by preventing the normal transfer of oxygen from the blood to body tissues. Also called "cyanogen agent."

blood chit A small cloth chart depicting the American flag and a statement in several languages that anyone assisting the bearer to safety will be rewarded.

blouse A DRESS UNIFORM coat.

bloused Pertaining to an item of clothing tucked or placed inside or beneath other clothing or items, as in bloused trousers (those tucked into the tops of the boots).

blowback 1. The escape, to the rear and under pressure, of gases formed during the firing of a WEAPON. Blowback may be caused by a defective BREECH mechanism, a ruptured CARTRIDGE CASE, or a faulty PRIMER. 2. A type of weapon function in which the force of expanding gases, acting to the rear against the face of the bolt, furnishes all the energy required to initiate the complete cycle of operation of the weapon. A weapon that employs this method of operation is characterized by the absence of any breech-lock or bolt-lock mechanism.

blue berets/blue helmets Any of the various United Nations PEACEKEEPING FORCES who wear pale blue headgear on top of their national uniforms.

BLT See BATTALION LANDING TEAM.

blue water The open sea, as opposed to coastal areas or white water.

blue-water navy A naval force with the ability to patrol and fight effectively anywhere in the world; in contrast to coastal protection forces, which many countries maintain instead of investing in long-range naval capacity. The main focus of interest in blue-water navies since the 1960s has been the development of such a capacity by the Soviet Union.

blunting mission 1. Any military operation intended to destroy enemy MATERIEL and not necessarily enemy TROOPS. 2. In early war plans the main mission of nuclear forces like the STRATEGIC AIR COMMAND to destroy the Soviet Union's industry, depriving its armies of support.

BMD See BALLISTIC MISSILE DEFENSE.

board A body of persons, either military, civilian, or both, appointed to act as a fact-finding agency or as an advisory body to the appointing authority. A board may be authorized either to recommend or to take final action on such matters, such as promotions, as may be placed before it.

boat-assembly area 1. A designated area for the assembly of empty LANDING CRAFT prior to their being called alongside a ship or to a shore EMBARKATION point for loading. 2. A designated area in which loaded landing craft or AMPHIBIOUS VEHICLES are assembled in formation for an overwater movement.

boat group The basic organization of LANDING CRAFT in an AMPHIBIOUS OPERATION. One boat group is organized for each BATTALION LANDING TEAM (or equivalent) to be landed in the first wave of landing craft or amphibious vehicles in an assault.

boat-rendezvous area A designated area in which loaded LANDING CRAFT or AMPHIBIOUS VEHICLES assemble in assault-landing formation for movement to the LINE OF DEPARTURE and thence to the shore in an amphibious assault.

boattail The conical section of a ballistic body that progressively decreases in diameter toward the tail to reduce overall aerodynamic drag.

boat team The troops assigned to one landing craft or amphibious vehicle for the ship-to-shore movement in an AMPHIBIOUS OPERATION.

bog The bow gunner of a TANK.

bogey An air contact that is unidentified but is assumed to be the enemy. (Not to be confused with an unknown contact.) See also FRIENDLY; HOSTILE.

bolt 1. The sliding part of a breech-loading rifle that pushes the cartridge into firing position and holds it there. Also called the breechlock. 2. To suddenly run away. 3. The small arrow used in a crossbow.

bolt action Any rifle that is manually operated by means of a bolt.

bomb 1. An explosive device dropped from an aerial platform. 2. To drop, from the air, explosives on a TARGET. In 1923, a Hague convention held: "Bombardment from the air is legitimate only when directed at a military objective, the destruction of which could constitute a distinct military disadvantage to the belligerent." In the light of subsequent experience this seems quaint and naive. 3. Any attack with bombs. 4. The terrorist act of planting a bomb among innocent, unsuspecting civilians in order to gain some political or psychological advantage.

bomb, atomic See ATOMIC BOMB.

bomb, delayed action See DELAYED-ACTION BOMB.

bomb, depth See DEPTH BOMB.

bomb, dumb See DUMB BOMB.

bomb, fire See FIRE BOMB.

bomb, fragmentation See FRAGMENTATION BOMB.

bomb, glide See GLIDE BOMB.

bomb, gravity See GRAVITY BOMB.

bomb, hung See HUNG BOMB.

bomb, hydrogen See HYDROGEN BOMB.

bomb, ideal See IDEAL BOMB.

bomb, leaflet See LEAFLET BOMB.

bomb, neutron See ENHANCED RADIATION WEAPONS.

bomb, scatter See SCATTER BOMB.

bomb, standoff See STANDOFF BOMB.

bombardier 1. The member of a bomber aircraft crew who operates the bombsight and the bomb-release mechanism. 2. An artilleryman.

bombardment 1. An attack by artillery or naval gunfire. 2. Any bomb attack.

bombardment, preliminary See PRELIMINARY BOMBARDMENT.

bombardment, strategic See STRATEGIC BOMBARDMENT.

bombardment line, shore See SHORE BOMBARDMENT LINE.

bomb-release line An imaginary line around a defended area or objective, over which an aircraft should release its bomb in order to obtain a hit or hits on the area or objective.

bomb-release line, final See FINAL BOMB RELEASE LINE.

bomb-release point The point in space at which bombs must be released to reach the desired point of detonation.

bomb unit, cluster See CLUSTER BOMB UNIT.

bomber See INTERMEDIATE-RANGE BOMBER AIRCRAFT; LONG-RANGE BOMBER AIRCRAFT; MEDIUM-RANGE BOMBER AIRCRAFT.

bomber aircraft, intermediate-range See INTERMEDIATE-RANGE BOMBER AIRCRAFT.

bomber aircraft, long-range See LONG-RANGE BOMBER AIRCRAFT.

bomber aircraft, medium-range See MEDIUM-RANGE BOMBER AIRCRAFT.

Bomber, Stealth See STEALTH BOMBER.

bombing, area See AREA BOMBING.

bombing, blind See BLIND BOMBING.

bombing, carpet See CARPET BOMBING.

bombing, dive See DIVE BOMBING.

bombing, laydown See LAYDOWN BOMBING.

bombing, loft See LOFT BOMBING.

bombing, low-altitude See LOW-ALTITUDE BOMBING system.

bombing, medium-altitude See MEDIUM-ALTITUDE BOMBING.

bombing, minimum-altitude See MINIMUM-ALTITUDE BOMBING.

bombing, offset See OFFSET BOMBING.

bombing, over-the-shoulder See OVER-THE-SHOULDER BOMBING.

bombing, pattern See PATTERN BOMBING.

bombing, precision See PRECISION BOMBING.

bombing, shuttle See SHUTTLE BOMBING.

bombing, skip See SKIP BOMBING.

bombing, toss See TOSS BOMBING.

bombing, volley See VOLLEY BOMBING.

bombing height In air operations, the height above ground level at which an aircraft is flying at the moment it releases its ORDNANCE. Bombing heights are classified as follows:

Very low —Below 100 feet
Low —From 100 to 2,000 feet
Medium —From 2,000 to 10,000 feet
High —From 10,000 to 50,000 feet
Very high—50,000 feet and above

bombing run In air bombing, that part of the flight that normally begins at an INITIAL POINT, includes the approach to the target, and target ACQUISITION, and usually ends at the weapon-release point.

bombing system, low-altitude See LOW-ALTITUDE BOMBING SYSTEM.

bombs, smart See SMART BOMBS.

B-1B The United States Air Force's long-range strategic bomber originally designed to replace the B-52 as the "manned bomber" leg of the TRIAD. Because it took more than a decade to produce, engineers were able to include some aspects of STEALTH BOMBER technology in it.

booby trap An explosive or non-explosive device or other harmful material, usually hidden and deliberately placed to cause CASUALTIES when an apparently harmless object is touched.

booby-trapped mine A MINE laid with an antilift device so that it will explode if lifted.

booster 1. A HIGH-EXPLOSIVE element sensitive enough to be actuated by small explosive elements in a FUZE or PRIMER, and powerful enough to cause detonation of the main explosive filling. 2. An auxiliary or initial propulsion system that travels with a MISSILE or aircraft and which may or may not separate from the parent craft when its impulse has been delivered.

booster, fast-burn See FAST-BURN BOOSTER.

boost phase That portion of the flight of a BALLISTIC MISSILE or space vehicle during which the BOOSTER and sustainer engines operate. The boost phase lasts for the first few minutes of flight before the missile leaves the earth's atmosphere and starts to coast. The missile is easily visible to orbiting SATELLITES during this phase, not only by RADAR but by INFRARED detectors, and would be highly vulnerable. The boost phase is the ideal time for a BALLISTIC MISSILE DEFENSE system to operate against a missile. By destroying the missile at this stage, before the front end, or BUS, has separated and any MULTIPLE RE-ENTRY VEHICLES have been released, the problems of DECOYS and multiple targets are avoided. See also MIDCOURSE PHASE; RE-ENTRY VEHICLE; TERMINAL PHASE.

boot camp Where Naval or Marine recruits receive BASIC TRAINING.

boots 1. Heavy footgear. 2. Naval or Marine recruits.

boots and saddle The classic cavalry command to mount up. Strangely enough it has nothing to do with boots; the phrase comes from the French *boute selle* meaning to "put on the saddle."

booty Goods legally taken from an enemy during combat operations, or goods informally (and illegally) taken from the enemy or citizens of OCCUPIED TERRITORY.

BOQ Bachelor Officer's Quarters; where unmarried officers or officers without accompanying dependents often reside on a military base.

bore 1. The cylindrical interior of a gun BARREL. 2. The diameter, or CALIBER, of the interior of a gun barrel.

boresafe fuze Type of FUZE having an interrupter in the EXPLOSIVE TRAIN that prevents a PROJECTILE from exploding until after it has cleared the muzzle of a weapon. See also FUZE.

boresight 1. A device used to align the axis of the BORE of a gun with an aiming point. 2. To so align a weapon.

boresight, breech See BREECH BORESIGHT.

boresight, muzzle See also MUZZLE BORESIGHT.

boresight line An optical reference line used in the HARMONIZATION of guns, rockets, or other weapon launchers.

boring from within A tactic that calls for a military group to secretly place agents in the midst of an enemy in order to weaken and eventually destroy them. This is a classic technique of Communist infiltration into non-Communist organizations.

Bouncing Betty An antipersonnel land mine with two charges: the first to "bounce" it up to waist level; the second to explode the mine.

bound 1. A single movement, usually from one covered and concealed position to another, by DISMOUNTED troops or COMBAT VEHICLES. 2. The distance covered in one movement by a unit that is advancing by bounds.

boundary 1. An international border. 2. A control measure drawn along identifiable terrain features and used to delineate areas of tactical responsibility for subordinate units of a military force. Within their boundaries, units may FIRE and MANEUVER in accordance with the overall plan without close coordination with neighboring units, unless otherwise restricted.

boundary (de facto) An international or administrative boundary whose existence and legality are not recognized, but which is a practical division between separate national and provincial administrative authorities.

boundary (de jure) An international or administrative boundary whose existence and legality is recognized.

boundary disclaimer A statement on a map or chart that the status or alignment of international or administrative boundaries is not necessarily recognized by the government of the publishing nation.

bound barrel A gun BARREL that is touching parts of the STOCK in such a manner that expansion due to heat from firing causes the barrel to bind and bend, resulting in inaccurate fire.

bounding mine A type of ANTIPERSONNEL MINE, usually buried just below the surface of the ground. It has a small charge that throws the entire case into the air after which it explodes at a height of three or four feet and throws SHRAPNEL or fragments in all directions.

bounding overwatch See MOVEMENT TECHNIQUES.

bounty 1. A bonus paid upon re-enlistment in an armed force. 2. A payment to an individual or a group if they capture someone or something.

bouquet mine In naval MINE WARFARE, a mine in which a number of buoyant mine cases are attached to the same sinker, so that when the mooring of one mine case is cut, another mine rises from the sinker to its set depth.

bourrelet A finely machined band or ring of metal just behind the OGIVE of a PROJECTILE, and designed to support the front portion of the projectile, as it travels through the BORE of a gun.

bow The forward end of an aircraft, ship, or boat.

box magazine A boxlike device that holds ammunition and feeds it into the receiver mechanism of certain types of AUTOMATIC WEAPONS.

BP See BATTLE POSITION.

brace A command to assume an exaggerated position of attention.

bracket 1. The distance between two STRIKES or series of strikes, one of which is over a TARGET and the other short of it, or one of which is to the right and the other to the left of the target. 2. To deliver fire that places a bracket on a target.

bracketing A method of adjusting ARTILLERY and MORTAR fire in which a bracket is established by obtaining an OVER and a SHORT, with respect to the observer, then successively splitting the bracket in half until a target hit is obtained or the smallest practicable RANGE change has been made. See ADJUST FIRE.

bracketing salvo A group of shots in which the number of shots going over the target equals the number falling short of it. Also called bracketing volley.

Bradley Fighting Vehicle A full-track, lightly ARMORED FIGHTING VEHICLE for screening, reconnaissance and other missions that carries a nine-man squad consisting of a commander, gunner, driver, and six infantrymen.

brassard A band worn on the upper arm to indicate a special function or organizational affiliation. The word is derived from plate armor that in medieval times covered the upper arms.

breach A break achieved by any means through an enemy DEFENSE, MINEFIELD, or FORTIFICATION.

breaching force A military GROUP or a UNIT engaged in breaching operations.

break ranks A command for troops to leave a military formation.

breakout An OFFENSIVE operation conducted by an encircled force. A breakout normally consists of an attack by a PENETRATION force to open a gap through the enemy for the remainder of the force to pass.

breakthrough An offensive action designed to rupture the enemy's FORWARD DEFENSES to permit the passage of an EXPLOITATION force.

breakup 1. In detection by RADAR, the separation of one solid return signal into a number of individual returns which correspond to various objects or structure groupings. This separation is contingent upon a number of factors, including RANGE, object size, and the distance between objects. 2. In IMAGERY INTERPRETATION, the result of magnification or enlargement, which causes the imaged item to lose its identity and the resultant presentation to become a random series of tonal impressions.

breakup point An AIR-CONTROL POINT at which helicopters returning from a LANDING ZONE break formation and are released to return to base, or are dispatched for other employment.

breastwork An earthwork structure that gives protection for defenders in a standing position, firing over its crest. Breastworks are constructed wholly or partly above the surface of the ground.

breech The rear of a gun, where the SHELL is inserted.

breechblock A movable steel block that closes the BREECH of a cannon.

breechblock carrier A hinged member of a breech mechanism which supports the rotating breechblock of a cannon.

breechblock tray A traylike support for the BREECHBLOCK, hinged to the BREECH of a large cannon and which supports the breechblock when it is withdrawn, and permits it to be swung clear of the breech.

breech boresight A disk with a small opening at its center that fits snugly in the BREECH chamber of a gun. The breech boresight is commonly used with a MUZZLE BORESIGHT to BORESIGHT a gun.

breech lock See BOLT.

breech recess The opening at the rear of a gun into which the BREECHBLOCK is inserted.

breech ring BREECHBLOCK housing, screwed or shrunk onto the rear of a cannon, in which the breechblock engages.

brevity code A CODE that provides no security but which has as its sole purpose the shortening of messages rather than the concealment of their content.

Brezhnev Doctrine The concept that the Soviet Union has the right to protect Communist regimes, particularly in Eastern Europe, even if this requires the use of force. It was first put forth by Soviet Premier Leonid Brezhnev in a speech on November 12, 1968, just a few months after the Soviet Union's invasion of Czechoslovakia. In that it asserted the status quo in a sphere of influence, it resembled the Monroe Doctrine. See R. Judson Mitchell, "The Brezhnev Doctrine and Communist Ideology," *Review of Politics,* Vol. 34, No. 2 (April 1972).

bridge, Bailey See BAILEY BRIDGE.

bridgehead 1. An area of ground held or to be gained on the enemy's side of an OBSTACLE that must be held or at least controlled to permit the continuous embarkation, landing, or crossing of troops and materiel. 2. In river-crossing operations, an area on the enemy's side of the water obstacle that is large enough to accommodate the majority of the crossing force, has adequate terrain to permit defense of the CROSSING SITES, and provides a base for continuing the attack. At the very least, ground must be secured that would permit INDIRECT FIRE on the crossing site to be observed. See also AIRHEAD; BEACHHEAD.

bridgehead line In offensive river-crossing operations, the limit of the OBJECTIVE AREA in the development of the BRIDGEHEAD.

brigade A unit usually smaller than a DIVISION, to which are attached groups or BATTALIONS and smaller UNITS tailored to meet anticipated requirements. They are less administrative and more operational than regiments and divisions. Though in the past brigades were the basic constituents of divisions, they are rarely used now as organizing units.

brigade, retraining See RETRAINING BRIGADE.

brigade landing team An assault landing team. It is a balanced TASK ORGANIZATION composed of a brigade headquarters, two or more BATTALION-level COMBAT UNITS, and the reinforcing combat and service elements required for combat and interim LOGISTIC SUPPORT during the period it conducts independent tactical operations.

brigade support area A location for brigade trains and brigade headquarters company elements not required in the main COMMAND POST.

brigadier The initial level of GENERAL OFFICER; formally a brigadier general, whose insignia is a single silver star.

brinkmanship 1. Taking very large risks in negotiations in order to force the other side to back down; this tactic is always reckless and sometimes a bluff. According to Thomas C. Schelling in *The Strategy of Conflict* (New York: Oxford University Press, 1963), brinkmanship is the "deliberate creation of a recognizable risk of war, a risk that one does not completely control. It is the tactic of deliberately letting the situation get somewhat out of hand, just because its being out of hand may be intolerable to the other party and force his accommodation." 2. A critical description of the foreign policies of President Eisenhower's Secretary of State John Foster Dulles, who advocated going to the "brink of war" as a negotiating tactic. In a famous *Life* magazine interview (January 16, 1956), he asserted that "The ability to get to the verge without getting into the war is the necessary art. If you cannot master it, you inevitably get into war. If you try to run away from it, if you are scared to go to the brink, you are lost." See Richard Ned Lebow, "Soviet Incentives for Brinkmanship?" *Bulletin of the Atomic Scientists* Vol. 37, No. 5 (May 1981).

broadcast-controlled air interception An AIR-INTERCEPTION technique in which the interceptor is given a continuous broadcast of information concerning an enemy RAID, and effects interception without further control.

Brodie, Bernard (1910–1978) The first major academic theorist of NUCLEAR WEAPONS. In *The Absolute Weapon* (1946), Brodie argued that the nuclear age presented the United States with the dilemma of either pre-emptively striking before a potential enemy could develop nuclear weapons, or accepting DETER-

RENCE and all that it implies as the only sane nuclear strategy. See Bernard Brodie, *Strategy in the Missile Age,* (Princeton, N.J.: Princeton University Press, 1959); Bernard Brodie, *Escalation and the Nuclear Option* (Princeton New Jersey: Princeton University, 1966).

B-2 The STEALTH BOMBER.

Buchan, Alastair (1918–1976) A major influence on the post-World War II development of strategic studies as an academic specialty, Buchan wrote widely on the causes, probabilities, and effects of wars. He founded the London-based Institute for Strategic Studies in 1958, whose annual handbook, *The Military Balance,* quickly became required reading by both NATO and Warsaw Pact analysts and spawned many imitations. See Ian Smart, "Alastair Buchan as Strategist," *International Journal* 31, 4 (Autumn 1976); David Curtis Skaggs, "Between the Hawks and the Doves: Alastair Buchan and the Institute for Strategic Studies," *Conflict* 7, 1 (1987).

buck 1. Responsibility for something, as in "the buck stops here." 2. The lowest of a series of grades, such as a buck private or buck sergeant. 3. To work hard to achieve a goal, as in "he's bucking for sergeant."

buddy system A system that requires two or more persons to work and remain near each other in certain areas and on certain missions so that they can give each other mutual protection and assistance. Often seen as the basis for development of UNIT morale.

buffer 1. A small state situated between two larger powers that functions to reduce the possibility of conflict between them; for example, Poland between Germany and Russia. See Michael Greenfield Partem, "The Buffer System in International Relations," *Journal of Conflict Resolution* Vol. 27, No. 1 (March 1983); 2. Organizational procedures or structures that absorb disruptive inputs and thus protect the continuity or equilibrium of some core group. For example, people in positions near the boundaries of organizations, such as receptionists, often absorb a wide variety of messages and demands. These inputs are filtered, processed, and passed to the "technical core" of the organization in a sequential and routine form. Because the inputs have been buffered, the central work processes of the organization are not disrupted. 3. In automatic data processing, a routine or storage device used to compensate for a difference in rate of flow of data, or time of occurrence of events, when transmitting data from one device to another.

buffer distance (nuclear) The horizontal distance added to the RADIUS OF SAFETY around GROUND ZERO that serves as a boundary outside of which the specified degree of risk to friendly troops is not exceeded. The buffer distance is normally expressed quantitatively in multiples of the DELIVERY ERROR.

build-down Reductions in nuclear arsenals by destroying more old WARHEADS than new ones are built. A build-down does not necessarily change strategic relationships because the fewer new weapons may be more accurate and powerful than the more numerous older ones. See Alton Frye, "Strategic Build-Down: A Contest for Restraint," *Foreign Affairs* Vol. 62, No. 2 (Winter 1983–84); Jack N. Barkenbus and Alvin M. Weinberg, "Defense-Protected Build-Down," *Bulletin of the Atomic Scientists* (October 1984).

buildup 1. The process of attaining the prescribed strength of units and prescribed levels of vehicles, equipment, stores, and supplies. It may also be applied to the means of accomplishing this process. 2. An ARMS RACE.

built-up area Any concentration of structures, facilities, and population in a geographic area; any urban environment.

bullet 1. The shaped projectile fired from a small arm. 2. A complete round of ammunition which includes the bullet (its tip that is fired) plus the cartridge case, gun powder, and primer.

bullet, frangible See FRANGIBLE.

bunker 1. A fortified structure for the protection of PERSONNEL, a defended gun position, or a DEFENSIVE POSITION. 2. A reinforced concrete COMMAND POST built below ground level.

buoy, dan See DAN BUOY.

burden sharing In the context of the NORTH ATLANTIC TREATY ORGANIZATION (NATO) this refers to the periodically asked question of which allies should bear what expenses for the alliance. It often refers to the relative share of total NATO defense expenditures paid by each member nation. This has become an ever increasingly contentious issue between the United States and its European allies (and increasingly, Japan). Almost from the beginning American voices have complained that European NATO has not been paying enough for its defense. Now that Europe has recovered from the devastation of World War II and on the whole is as rich as the United States, these few voices have turned into a large chorus. But assessing an appropriate "burden" for each NATO member is a difficult task. While all sides agree that each should pay a fair share, consensus breaks down thereafter. More than half of all U.S. defense spending goes for defending its European allies. How can this be fair? The answer depends on just how dangerous and how much of a threat a Soviet-controlled Western Europe would be to the United States. If Americans could live in perfect contentment with the whole of Europe under Soviet control, then they are indeed spending the money for other countries. However, the position adopted by those Americans who defend U.S. spending on NATO is that the defense of Western Europe is integral to the defense of the United States—that Western defense in general is a seamless web. See Klaus Knorr, "Burden-Sharing in NATO: Aspects of U.S. Policy," *Orbis* 29, 3 (Fall 1985); Jack A. LeCuyer, "Burden Sharing: Has the Term Outlived Its Usefulness?" *Atlantic Community Quarterly* 24, 1 (Spring 1986).

burn 1. To deliberately expose the true status of a person under COVER. 2. To legitimately destroy and burn classified (secret) material, usually accomplished by the custodian of the material as prescribed in regulations.

burned A term used to indicate that a clandestine agent has been exposed to the opposition (especially in a SURVEILLANCE operation), or that his reliability as a source of information has been compromised.

burn notice An official statement by one INTELLIGENCE agency to other agencies, domestic or foreign, that an individual or group is unreliable for any of a variety of reasons.

burst 1. A series of shots fired by a single pull on the trigger of an automatic weapon. 2. The explosion of a PROJECTILE or BOMB in the air or when it strikes the ground or target. 3. A series of shots from multiple independently fired weapons as in a VOLLEY.

burster An EXPLOSIVE CHARGE within a projectile used to break it open at a predetermined time and spread its contents of smaller projectiles such as bombs or mines.

burster blocks Prefabricated, reinforced concrete blocks so designed that they can be wired together to form a BURSTER COURSE.

burster course A fortification of BURSTER BLOCKS constructed to detonate projectiles before they can penetrate it deeply enough to cause great destruction. Also called a detonating slab.

bursting charge A BURSTER.

bursting layer A layer of hard material used in the roofs of DUGOUTS or cave shelters. It sets off incoming shells fuzed for short delay or immediate detonation before the explosive shells can enter deeply enough to cause great destruction.

burst range The horizontal distance from an artillery piece to the point of burst of its shell.

burst wave A wave of compressed air caused by a bursting projectile or bomb that may cause extensive local damage; also called a detonation wave.

bus The final stage of a multi-stage ROCKET. It is the bus that continues the trajectory path once the drive stages have burned out and been jettisoned (see BOOST PHASE). The bus contains the WARHEADS (more properly called re-entry vehicles), guidance computers, DECOYS, and other counter-ABM defense technology. The bus and its contents are also referred to as the front end.

bus-deployment phase The portion of a MISSILE flight during which multiple WARHEADS are deployed on different paths to different targets (also referred to as the POST-BOOST PHASE).

butt 1. The retaining wall at a target range, backed up with dirt on the side next to the firing point, with a target pit on the opposite side. Also called a target butt. 2. The rear end of the stock of a rifle or other SMALL ARM.

button up To close all the hatches on ARMORED VEHICLES.

butt plate A metal or rubber piece covering the end of the stock on SMALL ARMS, particularly rifles.

butt stroke A blow with the butt end of a rifle. A butt stroke is used in close combat, especially in BAYONET fighting and bayonet drill.

bypass Maneuvering around an obstacle, position, or enemy force in order to maintain the momentum of advance.

by the numbers 1. A preparatory command given in a CLOSE-ORDER DRILL to signify that the movement ordered is to be carried out step by step, at the command of the DRILL SERGEANT. 2. To do anything in a prescribed way with no room for individuality.

C

C The United States Armed Forces letter designation for transport or cargo aircraft, such as the C-5 or C-130.

C² See COMMAND AND CONTROL

C³CM See COMMAND, CONTROL, AND COMMUNICATIONS COUNTERMEASURES.

C³I See COMMAND, CONTROL, COMMUNICATIONS AND INTELLIGENCE ASSETS.

cable block A road obstruction made by stretching a cable diagonally across a road so as to ditch a vehicle that hits it.

cadence 1. The timing of a DRILL or march movement. 2. A command to count in unison while marching; for example 1, 2, 3, 4 continuously repeated.

cadet An individual in training to be a military OFFICER; one in training to be a naval officer is a midshipman.

cadre A detachment capable of being the training nucleus about which a new, larger organization can be built. Cadres of officers (or potential officers) are militarily and politically very significant. The German Wehrmacht was able to expand rapidly under the Nazi Party in the 1930s because it kept its cadre of officers intact during the 1920s (when the Treaty of Versailles severely restricted its size). The potential power of a cadre is suggested by the Katyn Massacre of 1940, in which thousands of imprisoned Polish officers were killed by the Russians (according to the Germans) or the Germans (according to the Russians) in order to prevent a dismembered Poland from reviving resistance. The killers would seem to have acted on NAPOLEON's observation that "it is very difficult for a nation to create an army when it has not already a body of officers and non-commissioned officers to serve as a nucleus, and a system of military organization."

cadre division See DIVISION.

caisson 1. A horse-drawn vehicle once used to transport ARTILLERY ammunition; now used only at military funerals to transport coffins. 2. Any large AMMUNITION box.

caliber 1. The diameter of the BORE of a GUN. In rifled gun barrels the caliber is obtained by measuring between opposite lands. A caliber .45 revolver has a barrel with a LAND DIAMETER 45/100 of an inch. 2. The diameter of a PROJECTILE. 3. Unit of measure expressing the length of the bore of a WEAPON. The caliber is determined by dividing the length of the bore of the weapon, from the BREECH face of the barrel to the muzzle, by the diameter of its bore. A gun barrel that is 40 feet (480 inches) long and 12 inches in diameter is said to be 40 calibers long.

calibration 1. Comparing a measuring instrument against a standard and making necessary adjustments to assure accurate readings. 2. Determining the correct range settings for artillery by observing where fire lands. See FIELD ARTILLERY CANNON CALIBRATION.

call 1. A request for the delivery of SUPPLIES covered by credits or allocations; also called a draft. 2. A signal on a bugle, DRUM, or whistle. 3. That part of a message containing the CALL SIGNS of the station calling and the station called. 4. A request from a port transportation officer for shipments, to include the time and place at which they are desired to reach the port. 5. A request by a port commander for movement to the port. It is forwarded to the agency responsible for issuing MOVEMENT ORDERS, and includes the date and time the unit and equipment accompanying or shipped by the unit will arrive at the port.

call fire Fire delivered on a specific target in response to a request from the unit being supported.

call for fire A request for FIRE, containing data necessary for obtaining the required fire on a TARGET.

call mission A type of air support mission that is not requested sufficiently in advance of the desired time of execution to permit detailed planning and briefing of pilots prior to take-off. Aircraft scheduled for this type of mission are on air, ground, or carrier alert, and are armed with a prescribed load of weapons.

call off A command to the members of unit to call out their titles or numbers in order. Also called "SOUND OFF."

call sign Any combination of characters or pronounceable words that identifies a communication facility, a COMMAND, an authority, an activity, or a UNIT; used primarily for establishing and maintaining COMMUNICATIONS.

call up 1. A set of signals used by a radio station to establish contact with another station. 2. The process of summoning conscripts or RESERVES to ACTIVE DUTY.

camisado A night attack. The word comes from the Spanish word for a light colored shirt that soldiers would wear over their armor so that they could identify themselves in the dark. Also spelled *camisade.*

cammies Camouflage uniforms.

camouflage The use of CONCEALMENT and disguise to minimize the possibility of the detection or identification of TROOPS, MATERIEL, equipment and INSTALLATIONS. It includes taking advantage of the natural environment as well as the application of natural and artificial materials. Camouflage is an ancient technique. In Shakespeare's *Macbeth,* Macduff tells his troops:

Let every soldier hew him down a bough
And bear't before him: thereby shall we shadow
The numbers of our host, and make discovery
Err in report of us.

See also CONCEALMENT, COVER and DISRUPTIVE PATTERN. See Byron Lester, "Mass Camouflage Painting," *Army Logistician* 15, 5 (September–October 1983).

camouflage, electromagnetic See ELECTROMAGNETIC CAMOUFLAGE.

camouflage, radar See RADAR CAMOUFLAGE.

camouflet 1. A MINE designed to collapse a tunnel. 2. The underground cave caused by an underground explosion.

camp A group of tents, huts, or other shelter set up temporarily for troops, and more permanent than a BIVOUAC. A military post, temporary or permanent, may be called a camp.

campaign A series of continuous field operations that is a part of a war; the period of time that begins when a military force leaves its home base to engage the enemy and ends when it returns home—victorious or defeated. A campaign often has a single strategic objective and is often located in a single region.

campaign, naval See NAVAL CAMPAIGN.

campaign hat The wide, flat-brim, big crown, "smokey-the-bear" hat worn by U.S. Army DRILL SERGEANTS.

campaign plan A plan for a series of related military operations to accomplish a common objective, normally within a given time and space.

campaign ribbon 1. A colored or patterned strip of ribbon worn on a uniform to indicate the military campaigns in which the wearer participated. 2. A STREAMER.

canalize To restrict enemy operations and movement to a narrow zone by the use of existing or reinforcing OBSTACLES or by fire or bombing.

canister 1. A special, short-range antipersonnel PROJECTILE consisting of a casing of light metal, loaded with preformed submissiles such as FLECHETTES or steel balls. The casing is designed to open just beyond the muzzle of the weapon that fires the canister, dispersing the submissiles. 2. A component of a canister-type PROTECTIVE MASK containing a mechanical filter and chemical filling to filter, neutralize, or absorb toxic CHEMICAL, BIOLOGICAL, and RADIOLOGICAL AGENTS. 3. A component of a projectile containing colored or screening smoke or a riot agent.

cannae The maneuver named after the classic 216 B.C. battle in which the Carthaginian general Hannibal used a DOUBLE ENVELOPMENT to annihilate the Roman forces. Today, any attack on both flanks of an enemy when that enemy is attacking forward is a "cannae" when the result is complete encirclement and destruction of the enemy.

cannelure 1. A ringlike groove in the JACKET of a BULLET which provides a means of securely crimping the CARTRIDGE CASE to the bullet. 2. A ringlike groove for locking the jacket of an armor-piercing bullet to the rest of the bullet. 3. A ringlike groove in the ROTATING BAND of a gun projectile to lessen its resistance to the gun RIFLING. 4. A groove around the base of a cartridge case. 5. The ring-like groove cut into the outside surface of a water-cooled MACHINE-GUN barrel, into which packing is placed to prevent the escape of water. Also called a cannelure cut.

cannibalization 1. The removal of serviceable parts and assemblies from unserviceable equipment to repair similar items. 2. Using personnel of one or more units to complete the authorized strength of another unit. 3. The use of human flesh as emergency rations. See Warren W. Fisher and J. J. Brennan, "The Performance of Cannibalization Policies in a Maintenance System with Spares, Repair, and Resource Constraints," *Naval Research Logistics Quarterly*, 33, 1 (February 1986).

cannon 1. A complete assembly, consisting of an artillery tube (the barrel) and a BREECH mechanism, firing mechanism or base cap, which is a component of a gun, howitzer or mortar. It may include MUZZLE appendages. 2. The *ultima ratio regum*, a Latin phrase meaning the final argument of Kings, the motto inscribed on French cannons by order of Louis XIV (1638–1715).

cannoneer 1. A member of a FIELD ARTILLERY GUN or HOWITZER crew whose primary duty is servicing the piece. In the United Kingdom the title for such a person is "gunner." 2. Any member of an artillery team.

cannon primer An assembly containing the PRIMER and IGNITER, to initiate burning of the propelling charge.

canteen 1. A flask for holding liquid, usually water, small enough to be comfortably carried by a soldier in the field. 2. A snack bar or social club operated at a military base. 3. An officer's chest containing culinary utensils and supplies.

cantonment 1. A military post. 2. Temporary military housing. 3. The quartering of troops in detached housing.

capital satellite A highly valued or costly SATELLITE, as distinct from an inexpensive DECOY satellite.

capital ship The largest and most powerful warship of any given era in military history.

capitulation 1. Surrender. 2. The articles of surrender. Napoleon took a dim view of capitulation, writing: "No sovereign can be secure, if individual units are permitted to capitulate in the field and lay down their arms by virtue of an agreement favorable to themselves and to the troops under their command, but opposed to the interests of the remainder of the army. To withdraw from peril themselves, and thus render the position of their comrades more dangerous, is manifestly an act of baseness. Such conduct ought to be proscribed, pronounced infamous and punishable with death."

captain 1. In most armies and air forces the senior company-grade OFFICER above a FIRST LIEUTENANT and below a MAJOR. (The equivalent U.S. Navy rank is lieutenant.) The INSIGNIA of a captain is two silver bars. 2. In most navies the officer just below flag rank, comparable to a full COLONEL in the army. 3. The pilot in charge of a civilian aircraft. (The pilot in charge of a military aircraft is called the aircraft commander.) 4. A now dated way to refer to the head of a large military force; for example, it has often been said that Hannibal and Napoleon were great captains. 5. Any commanding officer of a ship no matter what the officer's formal rank.

capture The taking into custody of hostile forces or equipment as a result of military operations.

carbine 1. A short-barrelled rifle originally designed to be used by mounted troops. 2. Any light semi-automatic rifle.

careerism An attitude on the part of officers that the troops they command and the responsibilities they are given are basically the means to advance their personal careers rather than an opportunity for profitless self-sacrifice. According to Richard A. Gabriel and Paul L. Savage in *Crisis in Command* (New York: Hill and Wang, 1978), in the private sector "careerism and entrepreneurialism are accepted and considered to be desirable. Military life, on the other hand, is unique in that it clearly levels upon the officer, or any other member for that matter, responsibilities which transcend his career or material self-interest. The problem has been, however, a failure to realize this and to regard the military life as the same as working at any other occupation. This equation is false, misleading, and ultimately dangerous, for it does not recognize that at some point an officer may be called upon to do his duty and 'be faithful unto death.' That alone, the burden of expectation, is sufficient in itself to distinguish the military ways from the business way."

In many respects, the distinction between careerism and professionalism has little operational meaning. In a professional group, the evaluation of one's performance by superiors will determine when and if he or she is promoted ahead of contemporaries. Careerism, then, is often little more than an accusation made against others by those whose career advancement has been slow. There is often no way to determine if those promoted to higher rank are those who should have been promoted. See William L. Hauser, "Careerism versus Professionalism in the Military," *Armed Forces and Society* Vol. 10, No. 3 (Spring 1984).

carpet bombing The progressive distribution of a mass BOMB load on an area defined by designated boundaries, in such manner as to inflict damage to all portions thereof. Also called saturation bombing.

carrier air group Two or more aircraft squadrons formed under one commander for administrative and tactical control of operations from a carrier.

carrier, breechblock See BREECHBLOCK CARRIER.

carrier striking force A naval TASK FORCE composed of aircraft carriers and supporting combatant ships capable of conducting STRIKE operations.

carry 1. A prescribed position for holding and carrying the COLOR, GUIDON, or other marker of a UNIT in a military formation. 2. To hold a color, guidon, or other marker in this prescribed position.

carry light A searchlight used to keep a target that has been spotted by a pickup light constantly illuminated, so that the target can be tracked and fired upon.

cartel 1. An agreement between two hostile forces for a mutual exchange of prisoners. 2. Any formal agreement between two warring powers for non-hostile interactions. 3. A group organized to obtain a monopolistic advantage such as the Organization of Petroleum Exporting Countries (OPEC).

cartel ship An unarmed ship sailing under a guarantee of freedom from attack or capture in time of war. A cartel ship usually carries prisoners to be exchanged.

Carter Doctrine The policy announced by President Jimmy Carter in his State of the Union address to the U.S. Congress on January 23, 1980, maintaining: "An attempt by any outside forces to gain control of the Persian Gulf region will be regarded as an assault on the vital interests of the United States of America, and such an assault will be repelled by any means necessary, including military force." The press has labeled the statement "The Carter Doctrine," and has characterized it as a reversal of the NIXON DOCTRINE. See, for example, Leslie Gelb, "Beyond the Carter Doctrine," *New York Times Magazine,* February 10, 1980.

cartridge 1. AMMUNITION for a WEAPON which contains in a unit assembly all of the components required to make the weapon function as intended, and which is loaded into the weapon in one operation. 2. An explosive item designed to produce gaseous pressure for performing a mechanical operation (such as starting an engine) other than the common one of expelling a projectile. An object similar to a blank cartridge (one without a projectile) that is utilized by a device known as a CARTRIDGE ACTUATED DEVICE. 3. A container for explosives loaded separately from the PROJECTILE in certain types of naval guns.

cartridge-actuated device (CAD) A device that utilizes the gases produced by explosives to initiate or operate a mechanical device such as a catapult, canopy remover, or aircraft-ejection seat.

cartridge case A container that holds the PRIMER and propellant of a projectile, and to which the projectile may be affixed.

case 1. To fold the COLOR, or any FLAG, and cover it with a case. 2. A particular instance of disease or injury.

cashier To dismiss an officer dishonorably from a military service.

castrametation The art of encampment; the building of a military CAMP.

casual detachment A military unit consisting of officers or soldiers separated from their own units or awaiting assignment.

casualties, mass See MASS CASUALTIES.

casualty Any person who is lost to a military organization by reason of having been declared dead, wounded, injured, diseased, interned, captured, retained, missing, missing in action, beleaguered, besieged or detained. See also BATTLE CASUALTY; NON-BATTLE CASUALTY.

casualty, battle See BATTLE CASUALTY.

casualty agent A CHEMICAL AGENT that is capable of producing serious injury or death when used in lethal concentrations.

casualty attack A surprise attack in which a high concentration of a toxic CHEMICAL AGENT is built up in a short time (2 minutes or less) on an area occupied by enemy PERSONNEL, to obtain a maximum number of casualties.

casualty drill A method or procedure used to promote the successful continuation of a MISSION or ENGAGEMENT when the mission crew or team is reduced by casualties.

casus belli From the Latin: "cause of war." An event used to justify a formal declaration of war.

catalytic attack An attack designed to bring about a war between major powers through the machinations of a third power who may attack one major power in the expectations that others will be drawn into the war.

catalytic war 1. A war that results from a CATALYTIC ATTACK. 2. A cataclysm between the superpowers triggered by a small nuclear engagement, or the use of NUCLEAR WEAPONS by a minor nuclear power. During the early development of nuclear strategic theory, this possibility was a topic of common concern. Strategists feared that a major war could start by accident due to the fallibility of detection methods and the impossibility of controlling escalation.

cavalier 1. A traditional cavalryman. 2. A high fortification; something built to create a high command of fire. 3. A rude, distainful manner. 4. A soldier with great panache and style.

cavalry 1. Military forces that travel and fight on horseback. Even before the American Civil War, cavalry FRONTAL AT-

TACKS were tactically foolish against prepared troops and artillery. Pierre Bosquet, a French divisional commander during the Crimean War of 1853 to 1856, witnessed the famous charge of the British Light Brigade and remarked: "C'est magnifique, mais ce n'est pas la guerre" (It's magnificent, but it's not war). 2. In a modern context, troops that travel or fight from tanks, armed fighting vehicles, or helicopters.

C-day The unnamed day on which a DEPLOYMENT operation commences or is to commence. The deployment may be a movement of troops, cargo, WEAPONS SYSTEMS, or a combination of these elements utilizing any or all types of transport.

cease engagement 1. A command to disengage weapons from firing on a particular target or targets, and to prepare to engage another target. Missiles already in flight will continue to intercept their targets. 2. To call a truce between two warring parties. 3. An order indicating that the enemy has surrendered. When Admiral William F. Halsey learned of the Japanese intention to surrender on August 15, 1945, he issued the following order: "Cease firing, but if any enemy planes appear, shoot them down in a friendly fashion."

cease fire 1. The general order to stop firing during an ENGAGEMENT of any military unit. 2. A command given to AIR-DEFENSE artillery units to refrain from firing on, but to continue to track, an airborne object.

cease loading In artillery and naval gunfire support, the command used during the firing of two or more rounds to cease inserting ROUNDS OF AMMUNITION into a weapon.

celestial sphere An imaginary spherical shell of infinite extent, the center of which is a given observer's position on the earth.

cellular unit A unit composed of teams, each of which includes personnel and equipment required for the performance of a specific function. A group of one or more teams may be selected to form a military unit to meet a special requirement not currently provided for in fixed or flexible tables of organization and equipment. Teams also may be used to augment units organized under fixed or flexible tables where increments of less than company size are required.

center line A line indicated on the ground and representing the center of traverse of a PIECE of artillery. It is used to facilitate the emplacement of heavy FIELD ARTILLERY so as to avoid subsequent shifting of the trails (the rearmost part of a cannon that rests on the ground).

center of dispersion The theoretical center of hits or bursts that would have been made in artillery fire if an unlimited number of shots had been fired with the same FIRING DATA.

center of impact The center of the dispersion pattern of impact bursts from an artillery piece. Considered from the viewpoint of RANGE only, it is the range center; from the viewpoint of direction, the direction center.

central command See RAPID DEPLOYMENT FORCE.

central control officer The naval officer designated by an AMPHIBIOUS TASK FORCE commander for the overall coordination of the waterborne ship-to-shore movement. He is embarked in the central control ship.

central front The border between East and West Germany, along which are arrayed the NORTH ATLANTIC TREATY ORGANIZATION (NATO) and WARSAW PACT armies. This is the area with the highest concentration of modern weapons and combat troops in the world, and traditionally is seen as the primary theater of war in a future third world war. Apart from the northern extension of the central front in Arctic Norway, it is the only place in the world where American and Soviet troops face each other. Nevertheless, aside from a brief period at the close of World War I, U.S. and Soviet forces have never fought each other. See Colin S. Gray, "Targeting Problems for Central War," *Naval War College Review* 33, 1 (January–February 1980).

Central Intelligence Agency (CIA) The federal agency created by the NATIONAL SECURITY ACT of 1947 to coordinate the various INTELLIGENCE activities of the United States. The Director of Central Intelligence is a member of the president's cabinet, and is the principal spokesperson for the American INTELLIGENCE COMMUNITY. Both the director and deputy director of the CIA are appointed by the president by and with the advice and consent of the Senate. Under the direction of the president or the NATIONAL SECURITY COUNCIL, the CIA:

1. Correlates and evaluates intelligence relating to national security, and provides for its appropriate dissemination.
2. Collects, produces, and disseminates foreign intelligence and counterintelligence. The collection of foreign intelligence or counterintelligence within the United States must be coordinated with the Federal Bureau of Investigation (FBI).
3. Collects, produces, and disseminates intelligence on foreign aspects of narcotics production and trafficking.
4. Conducts counterintelligence activities outside the United States and, without assuming or performing any internal security functions, conducts counterintelligence activities within the United States in coordination with the FBI.
5. Conducts special activities approved by the president (see COVERT operations).

The CIA has no police, subpoena, or law-enforcement powers, and has no INTERNAL SECURITY functions. See William M. Leary, editor *The Central Intelligence Agency: History and Documents* (University, AL: University of Alabama Press, 1984); John Ranelagh, *The Agency: The Rise and Decline of the CIA* (New York: Simon & Schuster, 1986).

central strategic warfare A full-scale nuclear war—the exchange of long-range BALLISTIC MISSILES between the United States and the Soviet Union. Even central strategic war can come in a variety of forms, given the endless distinctions between COUNTERFORCE and COUNTERVALUE strikes. Nevertheless, it is widely accepted in strategic thought that the transition to central strategic war, whatever precedes it, would be the biggest single step in any escalation ladder. To the extent that the only real purpose of American and Soviet possession of ballistic forces is to deter them from attacking one another's respective homelands, the onset of central strategic war would coincide with the breakdown of DETERRENCE.

central war A war between the superpowers.

CEP See CIRCULAR ERROR PROBABLE.

certification, security See SECURITY CERTIFICATION.

CFE See MUTUAL AND BALANCED FORCE REDUCTION.

C4 composition See COMPOSITION, C4.

CHAFF RADAR CONFUSION REFLECTORS, which consist of thin, narrow metallic strips of various lengths and frequency responses, used to reflect echoes for confusion purposes. See also ROPE; ROPE-CHAFF; WINDOW.

chain of command The succession of commanding officers, from a superior to a subordinate, through which the command of a military force is exercised. Also called command channel. See also OPERATIONAL CHAIN OF COMMAND.

chain of evacuation 1. A series of PRISONER-OF-WAR collecting points and cages, and routes by which prisoners of war, RETAINED ENEMY PERSONNEL, and CIVILIAN INTERNEES are collected and evacuated from a COMBAT ZONE to REAR AREAS. 2. A series of medical-treatment stations and facilities and the evacuation routes along which they are positioned, where medical evacuation and treatment functions are performed. 3. A series of points or INSTALLATIONS indicating the direction for evacuating disabled or salvaged MATERIEL.

challenge 1. Any process carried out by one military unit or person with the object of ascertaining the friendly or hostile character or identity of another. See also COUNTERSIGN; PASSWORD. 2. An individual invitation to fight a duel. This is usually illegal. 3. A legal objection to a member of a COURT-MARTIAL. 4. The traditional challenge of a sentinel: "Who goes there?"

challenge-and-reply authentication A prearranged procedure whereby one communicator requests authentication of another communicator and the latter establishes his or its validity by a proper reply.

challenge inspection The inspection by one nation of another's nuclear stockpiles and facilities, to determine whether or not ARMS-CONTROL or other agreements have been followed.

channel 1. The route of official COMMUNICATION between the HEADQUARTERS or commanders of military units. "To go through channels" is to follow the regularly established means for getting things done. 2. A facility for telecommunications on a system or circuit. The number of independent channels on a system or circuit (derived by frequency or time division) is measured by the number of separate communication facilities that it can provide.

channel, back See BACK CHANNEL.

chaplain 1. A clergyman attached to a military unit. Military chaplains come from all major religious denominations. While they conduct religious services for troops on bases and in the field, they also function as de facto social workers and counselors. "Tell it to the chaplain" is a typical response to a soldierly complaint.

chaplain's flag A blue FLAG, 2 feet by 3 feet, with religious symbols in white, used in the field to indicate a place of worship or the presence of a chaplain.

charge 1. The amount of propellant required as part of fixed, SEMI-FIXED, or SEPARATE-LOADING AMMUNITION. It may also refer to the quantity of explosive filling contained in a BOMB, MINE, or the like. 2. In combat engineering, a quantity of explosive, prepared for demolition purposes. 3. An enthusiastic ATTACK. General Thomas Jonathan "STONEWALL" JACKSON said in 1863: "My idea is that the best mode of fighting is to reserve your fire till the enemy gets you—or you get them—to close quarters. Then deliver one deadly, deliberate volley—and charge!" 4. An indictment by a COURT-MARTIAL. 5. A ROUND of SMALL-ARMS ammunition. 6. Assigned responsibility for a task. 7. An allegation. For example, the Confederate Civil War General John S. Mosby wrote in his 1887 *War Reminiscences:* "In one sense the charge that I did not fight fair is true. I fought for success and not for display. There was no man in the Confederate army who had less of the spirit of knight-errantry in him, or who took a more practical view of war than I did."

Charlie A slang word for the enemy; commonly used by U.S. forces in Vietnam.

chart, aeronautical See AERONAUTICAL CHART.

chauvinism An excessive, unreasoning, and unreasonable patriotism. The word comes from Nicholas Chauvin, a fanatically uncritical supporter of NAPOLEON.

check firing In artillery and naval gunfire support, a command to temporarily halt firing.

checkout A sequence of functional, operational, and calibrational tests to determine the condition and status of a WEAPONS SYSTEM or element thereof.

checkpoint 1. A predetermined point on the surface of the earth used as a means of controlling movement; a REGISTRATION target for fire adjustment, or reference for location. 2. A CENTER OF IMPACT; a burst center of an exploding shell. 3. A geographical location on land or water above which the position of an aircraft in flight may be determined by observation or by electrical means. 4. A place where MILITARY POLICE check vehicular or pedestrian traffic in order to enforce circulation control measures and other laws, orders, and regulations.

chemical agent A chemical substance that is intended for use in military operations to kill, seriously injure, or incapacitate through its physiological effects. Excluded from consideration are riot-control agents, herbicides, smoke, and flame. See also BIOLOGICAL AGENT, BLISTER AGENT, CASUALTY AGENT and CHOKING AGENT.

chemical ammunition A type of ammunition, the filler of which is primarily a CHEMICAL AGENT.

chemical defense The methods, plans, and procedures involved in establishing and executing defensive measures against an attack through the use of chemical agents. See also NBC DEFENSE.

chemical deterrence The prevention of an enemy from using CHEMICAL WEAPONS by having the known ability to escalate to CHEMICAL WARFARE as a counteraction if he does so first. World War II is the best example of chemical deterrence. All sides had easy access to chemical weapons and the means (via air strike) to deliver them on the enemy. In consequence and to the great surprise of many, they were never used against military targets. The Germans, however, made extensive use of chemicals (poison gas) to kill millions of innocent civilians (men, women, and children) in concentration camps. See John Tower, "The Politics of Chemical Deterrence," *Washington Quarterly*, 5, 2 (Spring 1982).

chemical hand grenade A burning or bursting type GRENADE which depending on the filler, can be used for CASUALTY, incendiary, training, screening, signaling, or riot-control purposes.

chemical mine A mine containing a CHEMICAL AGENT designed to kill, injure, or incapacitate PERSONNEL or to contaminate MATERIEL or terrain.

chemical projectile A BOMB, GRENADE, ROCKET, or SHELL containing a CHEMICAL AGENT.

chemical warfare 1. All aspects of military operations involving the employment of lethal and incapacitating chemical MUNITIONS or agents (and the warning and protective measures associated with such offensive operations). "Chemical warfare" is the modern way of referring to the use of gas as a weapon. Its most infamous example is the use of poison gas in World War I. Perhaps the one clear example of WEAPONS SYSTEMS deterrence between two hostile alliances is the non-use of gas or chemical weapons in World War II. The objective of a chemical attack is to kill, to incapacitate, and to DEGRADE (people wearing chemical protective clothing don't do whatever it is they are supposed to do wearing the protective gear nearly as well as they can do it when they're not wearing the gear). Riot-control agents and herbicides generally are not considered to be chemical warfare agents. 2. The employment of chemical agents to deny or hinder the use of areas, facilities, or materiel. See Joseph D. Douglass, Jr., "The Expanding Threat of Chemical Biological Warfare: A Case of U.S. Tunnel-Vision," *Strategic Review* 14, 4 (Fall 1986); W. Andrew Terrill, Jr., "Chemical Weapons in the Gulf War," *Strategic Review* 14, 2 (Spring 1986); George G. Weickhardt and James M. Finberg, "New Push for Chemical Weapons," *Bulletin of the Atomic Scientists* 42, 9 (November 1986).

chemical weapon An item or material that projects, disperses, or disseminates a CHEMICAL AGENT.

cheval-de-frise 1. From the French: "horse of Frieseland." A defense consisting of obstacles from which spikes or stakes protrude often strung with barbed wire. Named after the Dutch location where such defensive devices were first used against CAVALRY in the sixteenth century. 2. A line of spikes atop a fortification or wall (usually called chevaux-de-frise).

chevron A cloth device of varying design denoting a military grade, battle wound, enlisted service, or overseas service.

chicken 1. Cowardly. 2. A short version of "chicken shit," meaning petty or demanding; a soldier who "wants out of this chicken outfit" wants to avoid its "spit-and-polish" and is not implying that it is cowardly. Shortly before D-DAY in 1944, General George S. Patton told American troops: "All through your army careers, you've been bitching about what you call 'chicken-shit drill'. That, like everything else in the army, has a definite purpose. That purpose is instant obedience to orders and to create and maintain constant alertness! This must be bred into every soldier. A man must be alert all the time if he expects to stay alive." 3. Young soldiers that older soldiers may look upon with homosexual intent. 4. The "game" of nuclear deterrence, nuclear bluff, and nuclear diplomacy; each side waiting to see who will "chicken out" and back down first.

chicken colonel See COLONEL.

chicken hawk A public figure, whether a congressman or movie star, who legally avoided military service during the Vietnam War despite his eligibility, and who now advocates a "hard line" foreign policy that might lead to American troops being sent into combat.

chicks Friendly fighter aircraft.

chief of section 1. The NONCOMMISSIONED OFFICER in charge of a small unit. 2. The noncommissioned officer in charge of a gun crew.

Chief of Staff 1. The military title for the officer who supervises the work of all of the other officers on a commander's staff. German Field Marshal Gebhardt von Blucher (1742–1819) once said of Gneisenau, his chief of staff, "Gneisenau makes the pills which I administer." 2. The top aide to the president of the United States. 3. The highest ranking officer in the U. S. Army and Air Force. In the U. S. Navy the comparable title is Chief of Naval Operations; in the U. S. Marine Corps, it is Commandant. Compare to: JOINT CHIEFS OF STAFF.

Chiefs of Staff, Joint See JOINT CHIEFS OF STAFF.

China card A phrase that refers to a policy of strengthening the relationship between the United States and the People's Republic of China relationship as a means of influencing Soviet policy and the development of United States–Soviet relations.

China syndrome A slang expression for the meltdown of the fuel element in a NUCLEAR REACTOR; the resulting molten mass could theoretically sink through the entire earth beneath it, so as to emerge on the other side of the world, perhaps in China.

Chobham armor Layered ceramic armor developed to protect tanks against ever more powerful weapons. The traditional method of protection had been simply to thicken the armor, but the resulting weight increase soon began to produce a much slower vehicle. Furthermore, the heavier a tank is the more difficult it becomes to traverse soft ground. The solution has been to build much more sophisticated armor, relying on the strength obtained by combining layers of different sorts of material, particularly ceramics, between the usual layers of steel. The technique is often referred to as "Chobham" armor, after the site of the British Army's experimental station where it was pioneered. All modern armies now use some form of Chobham armor.

choke points 1. Places where the transport of military materiel and personnel is concentrated and slowed down. A typical choke point would be one of a small number of bridges across a major railway junction through which supplies for a front line had to pass. Choke points provide important targets for INTERDICTION strikes by an enemy, who can cause serious delay to troop movements and reinforcements with a small number of attacks. In this way a military power can win a battle in which it would face certain defeat were the opponents able to bring their full force forward. 2. In geopolitics certain vital sea lanes such as the Strait of Gibraltar or the Strait of Dardanelles. 3. A targeting doctrine concerned with destroying an enemy's economic capability for war.

choke ring A metal ring used in the reaction chambers of certain recoilless weapons to control gas escape. The same function is carried out by the throat rings, throat blocks, and restricting plugs in other types of recoilless weapons.

choking agent A CASUALTY AGENT that causes irritation and inflammation to the bronchial tubes and lungs. Phosgene is an example of this type of agent.

chronic radiation dose A dose of ionizing radiation received either continuously or intermittently over a prolonged period. A chronic radiation dose may be high enough to cause radiation sickness and death but, if received at low dose rate, a significant portion of the acute cellular damage can be repaired. See also ACUTE RADIATION DOSE; RADIATION DOSE RATE.

cipher Any CRYPTOGRAPHIC system for using symbols to represent text that can only be deciphered by another party privy to the code used.

cipher machine A mechanical or electrical apparatus for enciphering and deciphering coded material.

cipher system A CRYPTOSYSTEM in which the cryptographic treatment is applied to plain text elements of equal length.

cipher text Unintelligible text or signals produced through the use of cipher systems.

ciphony 1. The CRYPTOGRAPHY of telephonic communications. 2. Enciphered speech signals.

circuit, blind See BLIND CIRCUIT.

circular distribution 90 (CD90) The radius of a circle around the MEAN POINT OF IMPACT within which a single artillery round has a 90 percent probability of impacting, or within which 90 percent of the rounds fired will impact.

circular error probable (CEP) An indicator of the delivery accuracy of a WEAPONS SYSTEM, used as a factor in determining probable damage to a target. It is the radius of a circle within which half of all WARHEADS fired at the same target will fall. Thus, if 10 single-warhead rockets, each with a CEP of 100 meters, are fired at the same point, at least half of them should fall within a CLUSTER where they are not more than 200 meters apart. (The CEP is measured by the radius not the diameter of the hypothetical circle.) This measure is used to gauge the suitability of a MISSILE for a particular task. Against a city, for example, high CEPs can be accepted because the force of a nuclear explosion will cause enough damage, even if the missile lands some distance from its actual target. See also BIAS; DELIVERY ERROR; DISPERSION ERROR; HORIZONTAL ERROR.

circumvallation A wall, TRENCH, or other works built by besiegers around a BESIEGED area but facing outward to protect them from attacks from outside the besieged location.

citadel 1. A major FORTRESS in a strategic location. 2. A small FORT constructed within a larger one and intended as a last refuge for the GARRISON manning the fort. This is where the battle can be prolonged after the larger fort has fallen. 3. The central, most heavily protected part of a warship. 4. A military academy in South Carolina.

civic action, military See MILITARY CIVIC ACTION.

civil affairs 1. In MILITARY GOVERNMENT, the administrative processes by which an occupying power exercises executive, legislative, and judicial authority over occupied territory. 2. A general term for all those matters concerning the relationship between military forces and the surrounding civil authorities.

civil affairs agreement An agreement that governs the relationship between allied armed forces located in a friendly country and the civil authorities and people of that country.

Civil Air Patrol A volunteer auxiliary of the U.S. Air Force. Its goals include helping in national or regional emergencies and making youth more mindful of aerospace concerns and education. Its headquarters are at Maxwell Air Force Base in Alabama.

civil defense 1. The mobilization, organization, and direction of the civilian population of a nation that is designed to minimize, by passive measures, the effects of enemy action against all aspects of civilian life. There are several schools of thought on the efficacy, and even the desirability, of organizing civil-defense measures in the face of nuclear war. In some circles, civil defense is thought to be not only useless because of the inescapable horrors of a nuclear war, but actually dangerous. A belief that adequate civil-defense measures had been taken could encourage governments to risk nuclear war, or persuade populations to support nuclear armament policies. There is, in the 1980s, little general emphasis on civil defense in the West, because most evidence suggests that the investment would be wasted. For example, the evacuation of major cities could well lead to more deaths, not only through panic and possible rioting, but by trapping people in the open during an attack, when many of them would have been marginally better protected indoors, following an ordinary daily routine. Furthermore, some of the comparative statistics of injuries in nuclear war compared with health facilities are so stark as to strengthen the case for taking no action. A few 200-kiloton warheads exploding over the center of London or New York would produce more serious burn injuries than there are beds in burn units in the whole of the respective countries—probably by a factor of 10. Thus the provision of nuclear air-raid shelters in the United States has been allowed to lapse, and no Western European government has had a policy of providing sophisticated shelters for the general populace. In the United States, the Federal Emergency Management Agency has responsibility for overall civil defense. 2. The emergency repairs to, or the restoration of, vital utilities and facilities destroyed or damaged by enemy action. See John M. Weinstein, "Soviet Civil Defense and the U.S. Deterrent," *Parameters* (March 1982); Allan M. Winkler, "A Forty-Year History of Civil Defense," *Bulletin of the Atomic Scientists* Vol. 40, No. 6 (June–July 1984); Jonathan Mostow, "An Issue for the People: Civil Defense in the Nuclear Age," *Fletcher Forum* Vol. 8, No. 1 (Winter 1984).

civilian internee 1. A civilian who is interned during an ARMED CONFLICT or occupation for security reasons or for protection, or because he has committed an offense against the detaining power. 2. A term used to refer to persons interned and protected in accordance with the GENEVA CONVENTIONS relative to the Protection of Civilian Persons in Time of War (Geneva Convention 12 August 1949). See also PRISONER OF WAR.

civilian-type items Those items, including demilitarized items, that have a commercial equivalent or civilian market.

civil-military relations The dynamics between a nation's military leadership and its civilian society. For two classics on this, see: Samuel P. Huntington, *The Soldier and the State: The Theory and Politics of Civil-Military Relations* (Cambridge, Mass.: The Belknap Press of Harvard University Press, 1967); Bernard Brodie, *War and Politics* (New York: MacMillan, 1973).

civil war 1. An armed conflict between military units of the same nation or political entity. Edmund Burke wrote in a 1777 letter to the sheriffs of Bristol: "Civil wars strike deepest of all into the manners of the people. They vitiate their politics: they corrupt their morals; they pervert even the natural taste and relish of equity and justice. By teaching us to consider our fellow-citizens in a hostile light, the whole body of our nation becomes gradually less dear to us." Most of organized warfare since World War II has been civil war. Examples include the Korean Conflict, the Vietnam War, Nicaragua in the 1980s, and Northern Ireland. 2. The American Civil War between the North and the South, fought from 1861 to 1865 initially over the issue of secession; later over the issue of slavery. 3. The English Civil War of the 17th century between the forces of Oliver Cromwell and King Charles I.

civision 1. The CRYPTOGRAPHY of television signals. 2. Enciphered television signals.

clandestine operations See COVERT OPERATIONS.

classes of supplies The grouping of military SUPPLIES, by type, into ten categories to facilitate supply management and planning. The U.S. Army uses the following classes:

Class I —Rations and the gratuitous issue of health, morale, and welfare items.
Class II —Clothing, individual equipment, tentage, tool sets, and administrative and housekeeping supplies and equipment.
Class III —Petroleum, oil, and lubricants.
Class IV —Construction materials.
Class V —Ammunition.
Class VI —Personal-demand items sold through post exchanges.
Class VII —Major end items such as tanks, armored personnel carriers, and attack helicopters.
Class VIII—Medical.
Class IX —Repair parts and components for equipment maintenance.
Class X —Nonstandard items to support nonmilitary programs such as agriculture and economic development.

classified information 1. Secrets, usually military. 2. Any matter in any form that requires protection against disclosure in the interests of NATIONAL SECURITY.

classify 1. To group bureaucratic positions according to their duties and responsibilities, and assign them a class title. 2. To make secret; to determine that official information requires, in the interests of NATIONAL SECURITY, a high level of protection against unauthorized disclosure.

Clausewitz, Karl Maria von (1780–1831) The Prussian general who wrote the classic analysis of military strategy and tactics, *On War* (1832), which is most famous for asserting that "war is the continuation of diplomacy by other means." By this he meant to stress that all strategy, all military activity, must be subordinated to clear political motivation and aims. Clausewitz believed that the only difference between military conflicts and conflicts in the social realm was the fact that the former were resolved by bloodshed. His philosophy unequivocally subordinates the military instrument of power to the objectives of the state; war is nothing but the politics of violence, and the nation able to bring the greatest violence to bear should always win. Thus the proper aim of military force is to destroy the enemy in the field—any consideration of moderation is irrational. Clausewitz's position on the political nature of military force forms the basis for Communist military doctrine. According to Lenin: "Politics is the guiding force and war is only a tool, not vice versa. Consequently, it remains only to subordinate the military point of view to the political." According to Mao: "Politics is war without bloodshed, and war is politics with bloodshed."

A second very famous concept of Clausewitz concerns what he called "the friction of war." He argued, effectively, that everything that can go wrong will, and that all plans, however elegant or simple, will not quite work out properly. Considered the father of modern strategic thought, he upheld as a basic principle that war had to be absolute or total—violence at its utmost limits. He believed that the key to success lay in victorious battles, regardless of how bloody they might be. He held that the basic objective of war was to break the will of the enemy so that he will conform to your will. See Julian Lider, "War and Politics: Clausewitz Today," *Cooperation and Conflict* Vol. 12, No. 3 (1977); Raymond B. Furlong, "Clausewitz and Modern War Gaming," *Air University Review* 35, 5 (July–August 1984); Raymond Aron, *Clausewitz, Philosopher of War* (Englewood Cliffs, NJ: Prentice-Hall, 1985); Gertmann Sude, "Clausewitz in U.S. and German Doctrine," *Military Review* 66, 6 (June 1986).

claymore The name given to a type of ANTIPERSONNEL MINE designed to produce a directionalized, fan-shaped FRAGMENTATION pattern. 2. Originally, a large Scottish double-edged sword.

clear 1. The text of a SIGNAL that has not been encoded. 2. To authorize or gain the authorization for something. 3. To authorize an aircraft to take-off, land, or perform some other action. 4. To remove the enemy from an area.

clear enemy in zone An order to eliminate organized resistance from an assigned zone by destroying, capturing, or forcing the withdrawal of the enemy's forces.

clearing block A wooden block placed between the bolt and the rear of the BARREL of an AUTOMATIC WEAPON to prevent closing of the action and to show that the gun is unloaded.

clearing station An operating field-medical facility which provides emergency or resuscitative treatment for patients until they are evacuated, and definitive treatment for patients with minor illnesses, wounds, or injuries.

clinometer An instrument for measuring a degree of incline; for determining the angles of inclination or slope.

clinometer rest A device placed in the BORE of a gun to support a clinometer. Also called a bore rest.

clip, ammunition See AMMUNITION CLIP.

cloak-and-dagger A melodramatic phrase for the covert operations of INTELLIGENCE agents. Compare to: COVERT OPERATIONS. See David Wise, "Cloak and Dagger Operations: An Overview," *Society* Vol. 12, No. 3 (March–April 1975).

clock-code position The position of a target in relation to an aircraft or ship, with a dead-ahead position considered as 12 o'clock.

clock method A method of describing how shots hit a target by reference to the figures on an imaginary clock dial assumed to have the target at its center. Thus, a shot directly above the target is at 12 o'clock. Also called clock face method, clock system.

clock system, horizontal See HORIZONTAL CLOCK SYSTEM.

close 1. A preparatory command used to bring men marching at a normal interval to CLOSE INTERVAL. 2. An order to decrease the distances between vehicles or units in a march column, or to bring the tail of a column into an area. 3. An order to reduce the angle of divergence between ARTILLERY PIECES of a BATTERY to form a narrower SHEAF. 4. A term used in a firing message to indicate that the target is near friendly forward elements. 5. An order to discontinue operations in preparation for movement to another site.

close air support 1. Air action against hostile targets that are in close proximity to friendly forces and which require detailed integration of each air mission with the fire and movement of these forces. 2. The mission of a TACTICAL AIR COMMAND. See also AIR INTERDICTION; AIR SUPPORT; IMMEDIATE MISSION REQUEST; PREPLANNED MISSION REQUEST.

close combat Fighting at close quarters with the enemy, utilizing SMALL ARMS, BAYONETS, and other hand weapons.

close confinement 1. A mass formation in a CLOSE-ORDER DRILL in which the COMPANIES are arranged in columns of platoons at reduced distances, each platoon remaining in line. Also called CLOSE COLUMN. 2. A motor column in which the vehicles are closed up to safe driving distance behind each preceding vehicle. 3. The confinement of prisoners apart from the main prisoner group in quarters especially designated by the commanding officer of a unit for that purpose, and under constant supervision.

close defensive fire Firing on an attacking force with the intent to destroy the integrity of his assault, to disrupt the enemy's COMMAND, cover ATTACK POSITIONS, neutralize observation, and weaken SUPPORTING FIRE.

close in security The employment of cover CAMOUFLAGE, OBSTACLES, antitank weapons, sentinels, and PATROLS for protection for a unit against attack at close range.

close interval 1. A space between soldiers standing abreast, measured by placing the palm of the hand on the hip so that the elbow touches the arm of the next soldier in line. 2. The smallest prescribed interval between two units standing abreast in a FORMATION.

close march The command to take a CLOSE INTERVAL in marching.

close order The positioning of troops within the minimum space possible, usually to maximize FIREPOWER within a small area.

close-order drill A drill formation and drill movements that are done at NORMAL INTERVAL or at CLOSE INTERVAL. The formations and movements are those usually performed in drill marching, parades, and reviews, and those involving the manuals (the prescribed drills) of various hand weapons. See also DRILL and COMBAT DRILL.

close ranks 1. To lessen the distance between rows of men; to bring a unit from OPEN RANKS formation to NORMAL INTERVAL. 2. A preparatory command to close ranks.

close station A command dismissing all military personnel engaged in a DRILL, practice, or action at a given gun or communications station.

close stick Bombs released at the same time.

close-support mission A MISSION with the primary purpose of closely supporting friendly ground forces in the accomplishment of their immediate task or for the prevention of front-line enemy forces from accomplishing their missions. Close coordination of air, naval, and ground activities is required prior to and during the mission.

close supporting fire Fire placed on enemy troops, weapons, or positions which, because of their proximity, present the most immediate and serious threat to the supported unit. See also SUPPORTING FIRE.

closing plug A device used to close openings of various components in a ROUND OF AMMUNITION, such as the nose of an unfuzed projectile.

closure minefield In naval mine warfare, a MINEFIELD that is planned to present such a threat that waterborne shipping is prevented from moving.

closure time The time at which the last element of a unit has arrived at a specific location.

cloud attack An attack made by means of a toxic CHEMICAL AGENT or aerosol cloud for harassment or to produce casualties.

cluster 1. A fireworks signal in which various pyrotechnics burn at the same time. 2. A group of bombs released together. A cluster usually consists of fragmentation or incendiary bombs. 3. Two or more parachutes for dropping light or heavy loads. 4. In LAND-MINE WARFARE, a component of a pattern-laid minefield. It may be an ANTITANK, ANTIPERSONNEL, or mixed MINE. It consists of one to five mines and no more than one antitank mine. 5. Two or more engines coupled together so as to function as one power unit. 6. In naval mine warfare, a number of mines laid in close proximity to each other as a pattern or coherent unit. They may be of mixed types. 7. In MINEHUNTING, a term that designates a group of minelike contacts. See Michael Krepon, "Weapons Potentially Inhumane: The Case of Cluster Bombs," *Foreign Affairs* 52, 3 (April 1974).

cluster bomb unit An aircraft store composed of a dispenser (a container) and SUBMUNITIONS.

CN solutions Irritant liquid CHEMICAL AGENTS that cause lacrimation or tearing; popularly known as tear gas.

CO 1. Commanding officer. 2. Conscientious objector. 3. Carbon monoxide.

coalition defense The basic long-term security strategy of the United States. It recognizes the United States' inability to defend without assistance all areas of the world in which it has vital interests. The North Atlantic Treaty Organization (NATO) is the linchpin, though not the whole, of the United States' coalition defense strategy. The strategy assumes that Western European powers are not able to defend their own interests independently, and that the United States has an interest in a free Western Europe, but that it could not afford to defend this interest independently either—only a combination of both European and American forces can guarantee Western European security. Other major alliances are justified in a similar way.

A drawback to coalition defense is that while two or more nations may have generally compatible and overlapping interests in defending a certain area, the interests are not likely to be identical, and hence strain will occur. Furthermore, there is a natural tendency for weaker powers to rely rather more than is in some sense 'fair' on the fact that the United States will, for its own reasons, defend them anyway, thus giving rise to arguments over BURDEN SHARING. The United States, for its part, naturally expects to be the leader in these coalitions, and to define the specific targets and methods of coalition policy.

coarse setting The preliminary adjustment of the sight in laying a gun. A coarse setting is first made on the main scale; then a FINE SETTING is made on the associated scale of smaller gradations. See also LAY.

coarse sighting Adjustment of the sight of a gun so that a part of the front sight is seen through the notch in the rear sight.

coastal zone All of the navigable waters adjacent to a seacoast and extending seaward to cover the coastwise sea lanes and focal points of shipping approaching the coast.

coast defense All measures taken by naval, artillery, and other military forces, to provide protection against any form of attack at or near a shoreline.

Coast Guard The United States Coast Guard founded in 1790 as "The Revenue Marine" and part of the Department of Transportation except when operating as part of the Navy in time of war. The Coast Guard, the primary maritime law enforcement agency of the United States, has a major role in the suppression of smuggling and illicit drug trafficking.

coast-in point The point when a coast is first reached heading inbound to a target or objective.

coaxial machine gun A MACHINE GUN mounted in the turret of a TANK in such a way that its line of fire is exactly parallel to that of the CANNON set on the same mounting.

code Any system of communication in which arbitrary groups of symbols represent units of PLAIN TEXT of varying length. Codes may be used for brevity or for security.

code, brevity See BREVITY CODE.

code, enciphered See ENCIPHERED CODE.

code book A book containing various cipher codes and their PLAIN TEXT equivalents.

code group A group of letters, numbers, or both, assigned (in a code system) to represent a PLAIN TEXT element.

code message A cryptogram that has been produced by means of a code.

code of conduct The rules governing how a soldier should conduct himself if captured by the enemy. These vary from nation to nation, but usually imply a duty to try to escape, a refusal to make disloyal statements, and a refusal to give information beyond personal identification.

code panel See AIR-GROUND LIAISON CODE.

code word 1. A word that has been assigned a classification and a classified meaning to safeguard intentions and information about a classified plan or OPERATION. 2. A cryptonym used to identify sensitive INTELLIGENCE data. 3. A word or phrase whose use in a political context alters its meaning. Code words are often used when it is not politic or respectable to address an issue directly. For example, in early 1986, when world oil prices began to fall dramatically, politicians from oil-producing states started talking about the need for "stable" oil prices; in this context "stable" became a code word for "higher" prices.

codress A type of message in which the entire address is contained only in the encrypted text. See also PLAINDRESS.

coexistence 1. An international relationship wherein states with differing social systems and conflicting ideologies refrain from war. Coexistence is less than peace, but preferable to war. It is often used to refer to strained relations between the Soviet Union and the West. See Hugh Gaitskill, *The Challenge of Co-Existence* (London: Methuen, 1957). 2. Any contentious relationship where genuine rivals (political, organizational, etc.) purposely refrain from a direct confrontation that might otherwise be logically expected of them.

coil up To assemble a march column, especially in ARMOR, during a halt in field or fields, to minimize the distance from the front to the rear.

coincidence adjustment A range adjustment in a COINCIDENCE RANGEFINDER.

coincidence rangefinder An optical instrument for determining distances. By this adjustment, separate images seen through the two eyepieces can be made to coincide. A reading of the adjustment gives the distance between two points.

cold 1. Not under enemy FIRE. 2. Something that is perfect, as in "he got him cold." 3. Premeditated.

cold launch The method by which a MISSILE that has BOOSTER rockets is launched without damaging its SILO. The missile is first ejected from the silo so that its boosters can be ignited at a location safely distant from the silo walls. A cold launch is distinguished from a hot launch only in that steps are taken to avoid severe damage to the silo. Also called a pop-up launch.

cold war 1. War by other than military means, which emphasizes ideological conflict, BRINKMANSHIP, and a consistent high level of international tension. 2. The hostile but nonlethal relations between the United States and the Soviet Union in the post-World War II period. The phrase was first used by Herbert Bayard Swope (1882–1958) in speeches he wrote for Bernard Baruch (1870–1965). After Baruch told the Senate War Investigating Committee on October 24, 1948: "Let us not be deceived—today we are in the midst of a cold war," the press picked up the phrase and it became part of everyday speech. Setting dates for, and measuring the intensity of, the cold war is largely open to debate. It is often suggested that the era of detente in the 1970s was a temporary truce in the cold war, which some think began again at the end of that decade with the Soviet invasion of Afghanistan and the breakdown of ARMS-CONTROL negotiations after the second round of Strategic Arms Limitations Talks (SALT II). However, such distinctions only highlight the strongly subjective nature of the concept itself of cold war. The obvious question to ask of those who adhere to the cold-war theory is just what their perception of real peace would be. For most of recorded European history, major powers have vied with each other and used every tool available to manuever for relative advantage between wars: yet these conditions, when no actual combat was taking place, have been seen as the normal conditions of peace. Compare to: CONTAINMENT. See Joseph R. Starobin, "Origins of the Cold War: The Communist Dimension," *Foreign Affairs* Vol. 47, No. 4 (July 1969); Michael Leigh, "Is There a Revisionist Thesis on the Origins of the Cold War?" *Political Science Quarterly* Vol. 89, No. 1 (March 1974); J. L. Black, *Origins, Evolution, and Nature of the Cold War: An Annotated Bibliography* (Santa Barbara, CA: ABC-Clio, 1985).

collateral damage Damage to structures (such as schools, hospitals, or other military targets) or personnel in the vicinity of a TARGET that are not part of the STRIKE objective. Such damage may be known to be unavoidable. Usually, the inevitable but undesired civilian damage caused by a nuclear attack on a military target. See Paul Bracken, "Collateral Damage and Theater Warfare," *Survival* (September–October 1980).

collecting point A place designated for the assembly of PERSONNEL, CASUALTIES, STRAGGLERS, disabled MATERIEL, salvage, and other items or persons for further movement to collecting stations or rear INSTALLATIONS.

collective call sign Any CALL SIGN that represents two or more military facilities, commands, authorities, or units. The collective call sign for any of these includes the commander thereof and all subordinate commanders therein.

colonel The field-grade OFFICER below a BRIGADIER general and above a LIEUTENANT COLONEL. Because their INSIGNIA is a silver eagle, full colonels have a variety of bird-related nicknames, such as bird colonel, chicken colonel, and full bird colonel. It was Napoleon who noted: "There are no bad regiments: there are only bad colonels."

colonel, lieutenant See LIEUTENANT COLONEL.

colonel, light See LIGHT COLONEL.

colonel-general A rank equivalent to a four star general; not used in the U. S. Army.

color The FLAG or flags of a DISMOUNTED unit. "To the color" is a bugle call sounded as a salute to the color, the president, the vice president, an ex-president, or a foreign chief of state. Service "with the colors" means active service.

color bearer One who carries the COLORS or standard at formal REVIEWS and ceremonies.

color-coding standard A uniform color-coding system for the various types of AMMUNITION used by the U.S. Department of Defense.

color guard A guard of honor (a ceremonial escort) that carries and escorts the colors or standard at formal reviews and ceremonies.

color patches Pieces of material of various shapes and colors that can be temporarily applied to the surface of an object in order to CAMOUFLAGE equipment so as to adjust its appearance to suit differing natural environments.

color salute A salute made by dipping a COLOR or standard. The national color is never dipped in salute.

column Any military force, whether consisting of soldiers, tanks, ships, or other entities, that is deployed one behind another in a single file.

column, fifth See FIFTH COLUMN.

column, flying See FLYING COLUMN.

column cover The cover given to a military COLUMN by aircraft in radio contact therewith, providing for its protection by RECONNAISSANCE or ATTACK on air or ground targets that threaten the column.

column formation Any formation in which the individual elements are placed one behind the other; a single file.

column half left (right) 1. A change of 45 degrees to the left (right) in the direction of a COLUMN. 2. A preparatory command to make such a change of direction.

column left (right) 1. A change by a full 90 degrees turn to the left (right) in the direction of a column. 2. A preparatory command to make such a change.

column of ducks Soldiers in a double FILE.

combat 1. Fighting; any effort to close with or kill an enemy. 2. Any military action in a designated COMBAT ZONE; thus, all the forces involved can be said to have been "in combat," even though only a relative few do the actual killing. Thomas Carlyle described combat in *Past and Present* (1843): "Under the sky there is no uglier spectacle than two men with clenched teeth and hellfire eyes, hacking one another's flesh; converting precious living bodies, and priceless living souls, into nameless masses of putrescence useful only for turnip-manure." See T. N. Dupuy, *Understanding War: History & Theory of Combat,* (New York, NY: Paragon House Publications, 1986); Wade B. Becnel, "The Five Functions of Land Combat," *Military Review* 66, 4 (April 1986).

combat close See CLOSE COMBAT.

combat, hand-to-hand See HAND-TO-HAND COMBAT.

combat air patrol A standing aircraft PATROL provided over an OBJECTIVE AREA, over the force protected, over the critical area of a COMBAT ZONE, or over an AIR DEFENSE AREA, for the purpose of intercepting and destroying hostile aircraft before they reach their target.

combatant 1. In INTERNATIONAL LAW, all of the individual members of belligerent forces subject to the laws, rights, and duties of war. 2. A soldier or unit assigned to duty as an active fighter or fighting unit, as distinguished from other service duty in any of the administrative, supply, or medical branches of the armed forces.

combat area A restricted area (air, land, or sea) that is established to prevent or minimize mutual interference between friendly forces engaged in combat operations. See also COMBAT ZONE.

combat arm A branch of an army whose officers are directly involved in the conduct of actual fighting; for example, the INFANTRY or ARTILLERY; as opposed to quartermaster.

combat crew The flying crew of a combat aircraft, or the operating crew of a combat vehicle.

combat day of supply The total amount of supplies required to support one day of combat, calculated by applying the INTENSITY FACTOR to a standard day of supply. See also ONE-DAY'S SUPPLY.

combat developer The agency responsible for doctrine, concepts, requirements (both materiel and nonmateriel), and organization of a military force.

combat drill A DRILL conducted for the purpose of training a small unit in FORMATIONS and movement designed for use in battle. Combat drill is usually conducted at extended intervals and distances. See also CLOSE-ORDER DRILL.

combat echelon Part of an organization that engages in combat, as distinguished from troops engaged in supply or administration.

combat efficiency The effectiveness of a force upon engaging the enemy; a unit that inflicted significant casualties on the enemy while sustaining none of its own would be considered highly efficient, while a unit that merely traded casualties with the enemy is less so. David M. Shoup, who would later become commandant of the U. S. Marine Corps, gave this situation report while fighting the Japanese on Betio Island during World War II: "Casualties many: percentage of dead not known; combat efficiency: we are winning."

combat element Those TROOPS that actually take part in fighting, as distinguished from troops engaged in supply or administration.

combat exercise A MANEUVER or DRILL in fighting technique.

combat firing practice A form of military training wherein UNITS solve a tactical problem involving a situation in which AMMUNITION must be fired at targets representing the enemy.

combat formations Extended FORMATIONS, intended specifically for rifle squads and PLATOONS but adaptable to any type of unit, that are designed to promote efficient control and tactical handling of small units in combat.

combat happy A slang phrase for mental problems caused by the strain of being in combat for excessive lengths of time.

combat information 1. Unevaluated data, gathered by or provided directly to a tactical commander which, due to its highly perishable nature or the criticality of the situation, cannot be processed into TACTICAL INTELLIGENCE. 2. Data that can be used for FIRE or MANEUVER decisions as received.

combat information center The agency in a ship or aircraft that is manned and equipped to collect, display, evaluate, and disseminate tactical information for the use of the embarked flag officer, commanding officer, and certain control agencies. Certain control, assistance, and coordination functions may be delegated by command to the combat information center. Also called "action information center."

combat intelligence Information about an enemy, the weather, and geographical features required by a commander in the planning and conduct of combat operations.

combat jump An act of leaving an aircraft in flight and returning to the ground by parachute in hostile territory.

combat liaison Any system of maintaining contact and communication between units during fighting, in order to secure their proper cooperation.

combat loading The arrangement of personnel and the stowage of equipment and supplies in a manner designed to conform to the anticipated tactical operation of an embarked organization. Each individual item is stowed so that it can be unloaded at the required time.

combat-maneuver forces Those forces that use fire and movement to engage the enemy, as distinguished from forces that engage the enemy with INDIRECT FIRE or otherwise provide COMBAT SUPPORT and COMBAT SERVICE SUPPORT. Such forces are primarily INFANTRY, ARMOR, CAVALRY (air and armored), and attack helicopter units.

combat orders Orders pertaining to operations in the field. They include OPERATION ORDERS, ADMINISTRATIVE ORDERS, and LETTERS OF INSTRUCTION.

combat pay Extra pay for troops in an officially designated combat area.

combat patrol 1. For ground forces, a TACTICAL UNIT sent out from the main body of forces to engage in independent fighting. It may be used to provide security or to harass, destroy, or capture enemy troops, equipment, or installations. 2. A DETACHMENT assigned to protect the front, flank, or rear of the main body. Also called a fighting patrol. See also COMBAT AIR PATROL; PATROL; RECONNAISSANCE PATROL.

combat phase That period during which a military force is actively engaged with the enemy.

combat power The total means of destructive or disruptive force that a military UNIT or FORMATION can apply against an opponent at a given time.

combat radius The distance between the POINT OF DEPARTURE of a fully loaded aircraft and the farthest point it can reach while still retaining enough fuel to complete a round trip.

combat ratio The proportioned relationship between defending and attacking forces. Thus a ratio of one to one means that the sides are evenly matched; two to one means that one side has twice as many troops as the other.

combat ready 1. As applied to organization or equipment, a term meaning available for combat operations. 2. As applied to personnel: qualified to carry out combat operations in the unit to which they are assigned.

combat reconnaissance Reconnaissance of the enemy while maintaining immediate contact with one's own forces, preliminary to or during their contact with the enemy.

combat service support The assistance provided to OPERATING FORCES primarily in the fields of administrative services, chaplain services, civil affairs, finance, legal service, health services, MILITARY POLICE, supply, maintenance, transportation, construction, troop construction, acquisition and disposal of real property, facilities engineering, topographical and geodetic engineering functions, food service, GRAVES REGISTRATION, laundry, dry cleaning, bath, property disposal, and other LOGISTIC services.

combat support FIRE SUPPORT and other types of operational assistance provided to combat elements. It may include support from artillery, air defense, aviation (less air cavalry and attack helicopter), engineering, military police, signal, and electronic-warfare units. See William T. McDaniel, Jr., "Combat Support Doctrine: Coming Down to Earth," *Air Force Journal of Logistics,* 11, 2 (Spring 1987).

combat support arm A branch of an army that provides operational assistance to the army's COMBAT ARMS; for example, a SIGNAL CORPS or an INTELLIGENCE CORPS.

combat survival Those measures to be taken by service personnel when involuntarily separated from friendly forces in combat, including procedures relating to individual survival, evasion, escape, and conduct after capture.

combat team see REGIMENTAL COMBAT TEAM.

combat tire A pneumatic tire of heavy construction that in an emergency is designed to operate without air pressure for a limited distance.

combat trains The portion of UNIT TRAINS that provides the COMBAT SERVICE SUPPORT required for immediate response to the needs of forward tactical elements.

combat troops Those units or organizations whose primary mission is the destruction of enemy forces or installations. See also TROOPS.

combat unit A unit trained and equipped for fighting as an independent tactical element.

combat vehicle A vehicle, with or without ARMOR, designed for a specific fighting function. Armor protection or armament mounted as supplemental equipment on noncombat vehicles will not change the classification of such vehicles to combat vehicles.

combat zone 1. That area required by combat forces for the conduct of operations. 2. The territory forward of the army REAR AREA boundary. See also COMBAT AREA; COMMUNICATIONS ZONE.

combined arms 1. More than one tactical branch of an army used together in operations. 2. Military activity, whether it be planning, training, or warfare, in which the separate functional sections of a military force operate in close integration under a single plan and commander. JOMINI wrote in his 1838 *Precis on the Art of War* that: "It is not so much the mode of formation as the proper combined use of the different arms which will insure victory." Modern armies assume that all operations must be of a combined-arms nature, but it was not until World War II that this integration of tactical branches became clearly necessary. Traditionally, divisions would consist purely of one type of fighting force. Infantry, for example, was entirely separate from an armored division, which would have only tanks. A modern army differentiates between divisions according to the balance of each type of troops, as each contains all necessary units. See Jon Erickson, "Trends in the Soviet Combined-Arms Concept," *Strategic Review,* 5, 1 (Winter 1977); Tommy L. Whitton, "The Changing Role of Air Power in Soviet Combined Arms Doctrine," *Air University Review* 34, 3 (March–April 1983).

combined arms team Two or more armed units, such as tank, infantry, attack helicopter, field artillery, or other units mutually supporting one another.

combined doctrine The fundamental principles that guide the use of the forces of two or more nations in coordinated action toward a common objective. It is ratified by the participating nations. See also JOINT DOCTRINE, MULTI-SERVICE DOCTRINE.

combined force A military force composed of elements of two or more allied nations. Both the NORTH ATLANTIC TREATY ORGANIZATION (NATO) and the WARSAW PACT are combined forces.

combined operation 1. An operation conducted by forces of two or more allied nations acting together for the accomplishment of a single mission. 2. An operation conducted by forces of two or more armed services of the same nation. A typical example is the invasion of a hostile country by seaborne landing of army units. Here the navy and army, and almost certainly the air force as well, have to integrate their planning completely and work on a minutely-detailed timetable. Combined operations of this type are notoriously difficult for a host of technical reasons. Different communications methods, probably different and possibly incompatible communications machinery, rival traditions, and ambiguity about relative rank structures and authority relations all interact to maximize the chance of failure. Apart from technical difficulties, serious tensions often arise between the commanders of the different services because of contradictory priorities. The naval commander knows that ships are most vulnerable while near land, and wishes to pull them away, while the army commander wants time to ensure that his troops are safely on the beachhead and that all equipment has been transferred. Most countries make some effort to prepare for these problems by having joint planning staffs to accustom officers from the different services to working together, but these do not usually amount to very much. Part of the difficulty is that interservice rivalry often dictates that the route to top positions is via experience with a single service, and time spent on combined-operations staff training is time taken away from the pursuit of career advancement.

combined staff A staff composed of personnel of two or more allied nations. See also INTEGRATED STAFF; JOINT STAFF.

combined training The training of any unit with the military branch or branches with which it would normally cooperate; for example, combined training of artillery, engineering units, infantry, armored units, and air units.

come-as-you-are war 1. A war in which there would be no time for the resupply (or augmentation) of PERSONNEL or MATERIEL. 2. A contingency of sufficient gravity to require RESERVE units to deploy in a peacetime configuration without the benefit of additional training or personnel and equipment fill.

comfort halt A rest period on a march or trip by vehicle so that troops may use the resting place as a restroom.

COMINT See COMMUNICATIONS INTELLIGENCE.

command 1. The authority vested in an individual of the ARMED FORCES for the direction, coordination, and control of military forces. While one individual may be said to have command, it is always a team effort. Even JOMINI wrote in his 1838 *Precis on the Art of War* that: "The best means of organizing the command of an army . . . is to: (1) Give the command to a man of tried bravery, bold in the fight and of unshaken firmness in danger. (2) Assign as his chief of staff a man of high ability, of open and faithful character, between whom and the commander there may be perfect harmony." 2. An order given by a commander; that is, the will of the commander expressed for the purpose of bringing about a particular action. 3. A UNIT or units, ORGANIZATION, or an AREA under the command of one individual. 4. To dominate by a field of weapon fire or by observation from a superior position. See also AREA COMMAND; BASE COMMAND; FULL COMMAND; NATIONAL COMMAND; OPERATIONAL COMMAND; UNIFIED COMMAND, SPECIFIED COMMAND. See Michael J. Deane, Ilana Kass and Andrew G. Porth, "The Soviet Command Structure in Transformation," *Strategic Review* 12, 2 (Spring 1984); Martin Van Creveld, *Command in War* (Cambridge, MA: Harvard University Press, 1985); John Keegan, *The Mask of Command* (New York: Viking, 1987).

command and control (C^2) The exercise of authority and direction by a properly designated commander over assigned forces in the accomplishment of a MISSION. Command and control functions are performed through an arrangement of personnel, equipment, communications, facilities, and procedures employed by a commander in planning, directing, coordinating, and controlling forces and operations in the accomplishment of the mission. See Daniel Ford, *The Button: The Pentagon's Strategic Command and Control System* (New York: Simon and Schuster, 1985); Timothy L. McMahon, "The Key to Success: Developing a C^2 Philosophy," *Military Review* 65, 11 (November 1985); Joe Halloran, "Command and Control Interoperability," *Military Review* 66, 10 (October 1986).

command, control, and communications countermeasures (C^3CM) The integrated use of OPERATIONS SECURITY, MILITARY DECEPTION, JAMMING, and physical destruction, supported by INTELLIGENCE, to deny information to, influence, degrade, or destroy adversary command, control, and communications (C^3) capabilities and to protect friendly C^3 against such actions. There are two divisions within C^3CM:

1. Counter-C^3—That division of C^3CM comprising measures taken to deny adversary commanders and other decision-makers the ability to command and control their forces effectively.
2. C^3-protection—That division of C^3CM comprising measures taken to maintain the effectiveness of friendly C^3 despite both adversary and friendly counter-C^3 actions. See Doyle E. Larson, "C^3CM: Progress and Outlook," *Defense Management* 18, 3 (1982); Doyle E. Larson, "C^3CM in the 21 Century," *Defense Science & Electronics* (July 1983); Gene E. Layman, "C^3CM-A Warfare Strategy," *Naval War College Review* 38, 2 (March–April 1985).

command, control, communications, and intelligence assets (C^3I) Usually referred to by the acronym C^3I, pronounced "C cubed I," this represents the combined capacity to deliver orders to military units, to continually monitor and control their presence, movements, and status, to be well-informed of enemy movements and intentions, and to be able to relay and receive messages reliably, quickly, and secretly. The C^3I function is more crucial to the successful conduct of warfare today than ever before. This is particularly so because the mass and detail of information involved is far more than any general, admiral, or political leader can hope to take in and act on rapidly, and now has to be handled by computer. The speed of movement of military units, the ranges over which weapons can be fired, and the size of the areas they affect have all increased massively since the last major war. At the same time, the need for detailed and infallible control of interlocking operations has increased. The doctrine of controlled escalation, for example, makes it imperative that orders for the use of different forms of weapon be obeyed with absolute precision. See Jerry W. Betts, "Logistics and C^{3}I," *Army Logistician* 12, 4 (July–August 1980); Gary D. Brewer and Paul Bracken, "Some Missing Pieces of the C Cubed I Puzzle," *Conflict Resolution* 28, 3 (September 1984); Stephen J. Cimbala, "U.S. Strategic C^{3}I: A Conceptual Framework," *Air University Review* 36, 1 (November–December 1984).

command, full See FULL COMMAND.

command, operational See OPERATIONAL COMMAND.

command, preparatory See PREPARATORY COMMAND.

command, tactical See TACTICAL COMMAND.

commandant 1. The functional title, though not a rank, of any commanding officer. The head of the United States Marine Corps is called a commandant. 2. The head of a military school. 3. At West Point and the Air Force Academy, the head of training and discipline as opposed to the dean of faculty who is in charge of academics. Both report to the superintendent.

command decision 1. A decisive order by a commander that is especially consequential because of the dangers it implies or its indication of a new policy direction. 2. A rather pompous way of referring to any decision by an appropriate military authority.

commander, crossing area See CROSSING AREA COMMANDER.

commander, executing See EXECUTING COMMANDER.

commander, lieutenant See LIEUTENANT COMMANDER.

commander, national See NATIONAL COMMANDER.

commander, releasing See RELEASING COMMANDER.

commander, relief See RELIEF COMMANDER.

command group The commander and a few staff assistants who normally move forward of the COMMAND POST, with appropriate COMMUNICATIONS, in order to see and generally supervise combat action at a critical point.

command guidance The steering and control of a MISSILE by transmitting commands to it.

command liaison Close contact maintained between commanders for sharing information and working together effectively.

command net A communications network that connects an ECHELON of command with some or all of its subordinate echelons for the purpose of command control.

command of execution The second part of a command, at which the order is carried out. In "forward, march," "forward" is the PREPARATORY COMMAND, and "march" is the command of execution.

command of the sea Admiral Alfred T. Mahan's (1840–1914) concept of the object of naval strategy: the free use of the sea and the denial of its free use to the enemy.

commando operations A British term for raids conducted in strength and generally for a strategic purpose by specially trained military forces against objectives located in enemy territory. The American equivalent of the British commandos is the RANGERS. See David Thomas, "The Importance of Commando Operations in Modern Warfare, 1939–1982," *Journal of Contemporary History* 18, 4 (October 1983); Bruce Hoffman, "Commando Warfare and Small Raiding Parties as Part of a Counterterrorist Military Policy," *Conflict* 7, 1 (1987).

command post (CP) A unit's or subunit's HEADQUARTERS where the commander and his staff operate. In combat, three CPs are usually established: main, rear, and tactical. The main command post (main CP) is the major headquarters of a unit. It is concerned with planning, preparing, and sustaining elements of the command, control, and communications function and with collating and integrating information and intelligence. All those staff elements necessary to develop current and future plans and to acquire and coordinate COMBAT SUPPORT and COMBAT SERVICE SUPPORT for the force are normally represented. It is usually located in an area beyond the range of enemy medium artillery. The rear command post (rear CP) is concerned with sustaining a tactical force. It usually consists of personnel, logistic, and special staff sections, and contains any other staff elements not required at the main or tactical command posts. The tactical command post (TAC CP) is the location from which commanders and a small supporting staff normally direct a battle. Its size and electronic SIGNATURE should be no larger than that of its next subordinate element.

command post, airborne See AIRBORNE COMMAND POST.

command post, alternate See ALTERNATE COMMAND POST.

command post exercise (CPX) A training simulation involving commanders and their staffs but not troops. It is often designed to test communications channels between a headquarters and its command posts.

command post, forward See FORWARD COMMAND POST.

command procedure, direct See DIRECT COMMAND PROCEDURE.

commander-in-chief 1. The military or naval officer in charge of all allied forces in a THEATRE OF WAR. VEGETIUS wrote in his 378 *De Re Militari* that: "A commander-in-chief, whose power and dignity are so great and to whose fidelity and bravery the fortunes of his country men, the defense of their cities, the lives of soldiers, and the glory of the state, are entrusted, should not only consult the good of the army in general, but extend his care to every private soldier in it." 2. The authority granted under Article III, Section 2, of the U.S. Constitution that "the president shall be commander-in-chief of the army and the navy of the United States and of the militia of the several states when called into the actual service of the United States." The last president to exercise his authority as commander-in-chief in order to command troops in the field was James Madison during the War of 1812. At Bladenburg, Maryland, the Americans under their president met the British and were soundly defeated. The British then marched on Washington to burn the White House and all other public buildings. No subsequent president, while in office, has sought to lead men in battle. 3. The officer in charge of a branch of a SERVICE, or of all services in a given area. For example, CINCPAC is the commander in chief, Pacific, commanding all United States military and naval units in the Pacific Region; CINCSAC is the Officer Commanding the Strategic Air Command. See Eberhard P. Deutsch, "The President as Commander in Chief," *American Bar Association Journal* Vol. 57, (January 1971); R. Gordon Hoxie, "The Office of the Commander in Chief:

An Historical and Projective View," *Presidential Studies Quarterly* Vol. 6 (Fall 1976); Edwin Timbers, "The Supreme Court and the President as Commander-in Chief," *Presidential Studies Quarterly* Vol. 16 (Spring 1986).

commander of the guard The senior officer or NONCOMMISSIONED OFFICER who is responsible for the instruction, discipline, and performance of duty of the GUARD at a military base. He is the member of the guard next junior to the OFFICER OF THE DAY.

commander's estimate of the situation A logical process of reasoning by which a commander considers all of the circumstances affecting the military situation and arrives at a decision as to a course of action to be taken to accomplish the mission. While this is a logical process, it is also to some extent an art. General George S. Patton, Jr., wrote in *War as I Know It* (1947): "It is sad to remember that, when anyone has fairly mastered the art of command, the necessity for that art usually expires—either through the termination of the war or through the advanced age of the commander." A commander's estimate that considers a military situation so far in the future as to require major assumptions is called a commander's long-range estimate of the situation. It was T. E. Lawrence (of Arabia) who said: "With two thousand years of examples behind us, we have no excuse when fighting, for not fighting well." To paraphrase, with two thousand years of examples of estimating situations behind us, there is no excuse for not estimating well. Also called appreciation of the situation. See also LOGISTIC ESTIMATE OF THE SITUATION.

commercial vehicle A vehicle selected from civilian production lines for military use.

commissar In the Soviet Union, a member of the Communist Party assigned to a military unit to teach party principles and ensure the commanders' loyalty to the policies of the government. The commissar often functions as a deputy commander. See Timothy J. Colton, *Commissars, Commanders, and Civilian Authority: Structure of Soviet Military Politics* (Cambridge: Harvard University Press, 1979).

commission 1. To put in or make ready for service or use, as to commission an aircraft or a ship. 2. A written order giving a person RANK and authority as an officer in the armed forces. 3. The rank and the authority given by such an order. See also ACTIVATE; CONSTITUTE.

commission, military See MILITARY COMMISSION.

commissioned officer An officer in any of the armed services who holds grade and office under a commission issued by the president. (In the United Kingdom the commission comes from the monarch.) In the U.S. Army, Air Force, and Marine Corps, a person who has been appointed to the grade of second LIEUTENANT or higher is a commissioned officer. A four-star general is currently the highest commissioned grade. In the navy, ensign is the lowest commissioned rank; a four star admiral is the highest commissioned rank. See also OFFICER.

committed force A force in contact with an enemy or deployed on a specific mission or course of action that precludes its employment elsewhere.

commodity loading A method of loading in which various types of cargoes are loaded together, such as ammunition, rations, or boxed vehicles, in order that each commodity can be discharged without disturbing the others.

commodore 1. In the United States Navy the lowest grade flag officer, above captain and below rear admiral (in effect, rear admiral, junior grade); at the same level as a brigadier general in the army. 2. The unofficial title for a U.S. or British naval officer in temporary command of a squadron. 3. The naval officer in command of a convoy of merchant ships.

commodore, air See AIR COMMODORE.

common infrastructure The INFRASTRUCTURE essential to the training of North Atlantic Treaty Organization (NATO) forces or to the implementation of NATO operational plans which, owing to its degree of common use or interest and its compliance with criteria laid down from time to time by the North Atlantic Council, is commonly financed by NATO members.

common-mode failure A type of system failure in which diverse components are disabled by the same single cause.

commonality A quality that applies to MATERIEL or SYSTEMS: 1. having like and interchangeable characteristics enabling each to be utilized, or operated and maintained, by personnel trained on the others without additional specialized training. 2. having interchangeable repair parts, components, or both. 3. applying to consumable items interchangeably equivalent without adjustment.

communication, basic See BASIC COMMUNICATION.

communication, direct See DIRECT COMMUNICATION.

communications 1. Routes and means of transportation for moving TROOPS and SUPPLIES, especially in a theater of operations. See also LINES OF COMMUNICATION. 2. The means of giving and receiving ORDERS and INTELLIGENCE. The classic comment on what to do if there is a communications foul-up, as is often the case in war, belongs to the British Admiral Horatio Nelson, who said before his great victory at the 1805 Battle of Trafalgar against the French: "But, in case signals can neither be seen or perfectly understood, no captain can do very wrong if he places his ship alongside that of an enemy."

communications cover The technique of concealing or altering the characteristics of communications patterns for the purpose of denying the enemy information that would be of value to him.

communications intelligence (COMINT) Technical and intelligence information derived from the monitoring of foreign communications such as the monitoring of shortwave radio, telephone, microwave, satellite, voice, teletyping, facsimile, binary data or multiplexed signals. The first significant use of COMINT was the British Admiralty's radio-monitoring service, which gave advance notice that the German Grand Fleet was putting to sea just before the World War I Battle of Jutland in 1916. It was the failure of the U. S. Navy to take seriously the signals intelligence reports of Japanese naval movements in December 1941 that led to the U. S. Pacific fleet being completely surprised by the attack on Pearl Harbor.

communications jamming ELECTRONIC COUNTER MEASURES taken to deny the enemy use of his communications. See JAMMING.

communications net An organization of STATIONS capable of intercommunications but not necessarily on the same CHANNEL.

communications security (ComSEC) The protection resulting from all measures designed to deny unauthorized persons information of value that might be derived from the possession and study of telecommunications, or to mislead unauthorized persons in their interpretation of the results of such possession and study. Communications security includes:

1. **Cryptosecurity**—The component of communications security that results

from the provision of technically sound CRYPTOSYSTEMS and their proper use.
2. **Transmission security**—The component of communications security that results from all measures designed to protect transmissions from interception and exploitation by means other than CRYPTANALYSIS.
3. **Emission security**—The component of communications security that results from all measures taken to deny unauthorized persons information of value that might be derived from the interception and analysis of compromising emanations from CRYPTO-EQUIPMENT and telecommunications systems.
4. **Physical security**—The component of communications security that results from all physical measures necessary to safeguard CLASSIFIED information against access or observation by unauthorized persons.

communications zone The rear part of theater of operations (behind but contiguous to the COMBAT ZONE), which contains the lines of communications, establishments for supply and evacuation, and other agencies required for the immediate support and maintenance of field forces. See also REAR AREA.

community relations program That command function which evaluates public attitudes, identifies the mission of a military organization with the public interest, and executes a program of action to earn public understanding and acceptance of the military organization. Community relations programs are conducted at all levels of command, both in the United States and overseas, by military organizations having a community-relations area of responsibility. Community relations programs include, but are not limited to, such activities as liaison and cooperation with associations and organizations and their local affiliates at all levels; armed-forces participation in international, national, regional, state, and local public events; installation open houses and tours; embarkations in naval ships; orientation tours for distinguished civilians; people-to-people and humanitarian acts; cooperation with government officials and community leaders; and encouragement of armed-forces personnel and their dependents to participate in activities of local schools, churches, fraternal, social, and civic organizations, sports and recreation programs, and other aspects of community life to the extent that they are feasible and appropriate.

company The basic administrative and tactical unit in most arms and services of the U.S. Army. A company is on a command level below a BATTALION and above a PLATOON, and is equivalent to a BATTERY of ARTILLERY. In other countries the equivalent of a company outside the infantry will often be a SQUADRON (armor) or battery (artillery). A typical battalion will have three or four companies plus a HEADQUARTERS COMPANY. Each company will typically contain four platoons.

company clerk The assistant to a company's FIRST SERGEANT. The company clerk mainly deals with routine paperwork, and assists the company's officers in preparing reports.

company grade Classification of those officers normally serving in a company. It is applied to LIEUTENANTS and CAPTAINS in pay grades 01 to 03. The British equivalent is subaltern.

company punishment Minor punishment decided by a company commander for minor infractions; punishment without a court martial. Compare to NONJUDICIAL PUNISHMENT.

company team A team formed by the attachment of one or more non-ORGANIC tank, mechanized, or infantry PLATOONS to another tank, mechanized, or infantry platoon, either in exchange for, or in addition to, organic platoons.

company transport Vehicles that form an ORGANIC part of company equipment and are directly available to the company commander for tactical use.

compartment An area of terrain bounded on two opposite sides by features that limit observation and OBSERVED FIRE into the area from points outside the area.

compartmentation 1. The establishment and management of an INTELLIGENCE organization so that information about the personnel, organization, or activities of one component is made available to any other component only to the extent required for the performance of assigned duties. 2. Effects of relief and drainage on avenues of approach so as to produce areas bounded on at least two sides by terrain features such as woods, ridges, or ravines that limit observation or observed fire into the area from points outside the area. 3. In UNCONVENTIONAL WARFARE, the division of an organization or activity into functional segments or cells to restrict communication between them and prevent them from knowing the identity or activities of other segments except on a NEED-TO-KNOW basis. 4. Restricting the use of specific cryptovariables to specific users for the purpose of limiting access to the information protected by these cryptovariables and limiting the adverse impact of a compromise of these variables. Also called compartmentalization.

compartment of terrain A land area bounded on at least two opposite sides by terrain features such as woods, ridges, or villages, which limit observation and observed fire into the area from points outside the area.

compass, aperiodic See APERIODIC COMPASS.

compass, lensatic See LENSATIC COMPASS.

compass azimuth The angle measured clockwise from a northerly point determined by means of a compass reading. Because of variations of the compass, it may not agree with an azimuth measured from true magnetic north.

compass bearing A direction or bearing as given by a compass reading.

compass compensation A method of adjusting a compass to compensate for the magnetic forces exerted by nearby metals, parts of an aircraft or ship's structure, or other objects.

compass course The course of an aircraft, tank, person, or other entity as indicated by the horizontal angle between the north-south line, as shown by a compass, and the direction of motion of the entity.

compass declination The angle from the true north in the pointing of a compass.

compass error 1. A false compass-needle reading induced by DEFLECTION because of nearby metallic objects, static electricity, or improper adjustment of the compass. 2. The total difference between the reading of north on the compass and true north.

competitive strategies A method of strategic thinking that evaluates United States national defense strategy in terms of long-term relations with the Soviet Union; its objective is to enhance deterence. This technique employs a chess match (see GAME THEORY) approach that aligns enduring United States strengths against enduring Soviet weaknesses in a move-response-counterresponse sequence. It calls for avoiding escalation to central nuclear war if the Warsaw Pact attacks NATO by countering Soviet numerical strengths with the West's technological edge, and perhaps resorting to short-range nuclear weapons. Competitive strategies is an attractive concept for NATO because it recognizes that the West can never match the Soviet conventional edge and seeks to cope with the problem by the creative use of EMERGING TECHNOLOGY.

complement 1. The full, authorized strength of a military UNIT or POST, including OFFICERS, ENLISTED PERSONNEL, and MATERIEL. 2. Extra units of various types attached to a given unit for adding to the services or operations of that unit.

complementary angle of site The correction needed to compensate for the error made in assuming rigidity of the trajectory of a PROJECTILE.

complete penetration 1. In the army, the penetration obtained when a projectile enters entirely into its target or light can be seen through the target from the rear of the target. 2. In the navy, the penetration obtained when a projectile passes through the target intact, or a major portion of the projectile passes through the target.

complete round A term applied to an assemblage of explosive and nonexplosive components designed to perform a specific function at the time and under the conditions desired. Examples of complete rounds of ammunition are:

1. SEPARATE-LOADING ammunition, consisting of a PRIMER, PROPELLING CHARGE, and, except for blank ammunition, a PROJECTILE and a FUSE.
2. FIXED or SEMI-FIXED AMMUNITION consisting of a primer, propelling charge, CARTRIDGE CASE, a projectile, and, except when solid projectiles are used, a fuse.
3. BOMBS, consisting of all of the component parts required to drop the bomb and have it function.
4. MISSILES, consisting of a complete WARHEAD section and a missile body with its associated components and propellants.
5. ROCKETS, consisting of all of the components necessary to proper function.

component force Each SERVICE element of a JOINT FORCE is called a component force and is titled naval, land or air component as appropriate.

composite armor A protective covering consisting of two or more materials, as distinguished from a single plate or piece or laminated structure with all of the laminae consisting of the same material. A composite-armor structure may consist of laminae of different materials or a matrix of one material in which are inbedded pieces of particles of one or more different materials.

composite defense In AIR DEFENSE artillery, a defense that employs two or more types of FIRE UNITS that are integrated into a single defense, such as antiaircraft fire and surface-to-air missiles.

composition C4 A PLASTIC EXPLOSIVE consisting of RDX (a water-insoluble high explosive) and plasticizing materials; usually used in DEMOLITION BLOCKS.

compound helicopter A helicopter with an auxiliary propulsion system that provides thrust in excess of that which the rotor(s) alone could produce, thereby permitting increased forward speeds; wings may or may not be provided to reduce the lift required from the rotor system.

compromise The known or suspected exposure of clandestine personnel, installations, or other assets, or of CLASSIFIED INFORMATION or material, to an unauthorized person.

compromised A term applied to classified matter the knowledge of which has, in whole or in part, passed to an unauthorized person or persons, or which has been subject to the risk of such passage.

computing gunsight A gunsight that compensates for some variable in WEAPON aiming.

computing sight A type of gun sight that includes an electrical or mechanical means for computing the proper angle between the line of sight to the target and the LINE OF DEPARTURE of a PROJECTILE. It usually includes means for automatically establishing the angle.

COMSEC See COMMUNICATIONS SECURITY.

concealment The fraction of a TARGET hidden (or concealed) from a given observer position by vegetation or urban development. The fraction concealed cannot be detected visually but may, under certain circumstances, be hit by direct fire.

concentrated fire 1. A volume of INDIRECT FIRE placed on an area within a limited time. 2. An area designated and numbered for future reference as a possible target. 3. The massing of maximum fire or COMBAT POWER at a certain point and time. 4. The fire of the BATTERIES of two or more ships directed against a single target. 5. Fire from a number of weapons directed at a single point or small area. See also MASSED FIRE.

concentration area 1. An area, usually in a THEATER OF war, where troops are assembled before beginning active operations. 2. A limited area on which a volume of gunfire is placed within a limited time.

concentration of force The gathering together of all troops, ships, or other units used to strike a decisive blow at one point against an enemy whose forces are distributed over a large area. In this way it can be hoped to have local superiority in numbers, despite being outnumbered overall. The opposite tactic, to divide an army or navy into a set of small units guarding particular points, or carrying out independent operations, runs the risk of having them defeated by a concentrated enemy. Concentration of force generally puts an attacker at an advantage over a genuinely peacefully inclined enemy. The defender has to cover a whole border, or at least all of its vulnerable points, while the attacker can choose the target area and force a way through the defender's lines.

Concentration has long been considered the single most important principle of war. Machiavelli wrote in 1521, "One should never risk one's whole fortune unless supported by one's entire forces." Napoleon wrote that "the essence of strategy is, with a weaker army, always to have more force at the crucial point than the enemy." This was restated by Confederate General Nathan Bedford Forrest, who said: "I always make it a rule to get there first with the most men." Indeed, according to B.H. LIDDELL HART, writing in 1944: "The principles of war could, for brevity, be condensed into a single word—'concentration'." Compare to: PRINCIPLES OF WAR. See Theodore C. Taylor, "Tactical Concentration and Surprise-in Theory," *Naval War College Review* 38, 4 (July–August 1985).

concept of operations A verbal or graphic statement, in broad outline, of a commander's assumptions or intent in regard to an operation or series of operations. The concept of operations is often embodied in CAMPAIGN plans and OPERATION plans; the latter is particularly used when the plans cover a series of connected operations to be carried out simultaneously or in succession. The concept is designed to give an overall picture of the operation. It is included primarily for additional clarity of purpose. Frequently referred to as commander's concept, it is described in sufficient detail for the commander's staff and subordinate commanders to understand what they are to do and how to fight a battle in the absence of further instructions.

concertina wire Coiled wire with razor-sharp edges (or barbs) used as a barrier instead of or in addition to fences.

concussion detonator kit A group of items, including a BLASTING CAP and a mechanical FIRING DEVICE, designed to be actuated by the concussion wave of a nearby blast.

concussion wave The force of air pressure following an explosion. A concussion grenade is designed to cause damage solely by the force of the concussion wave caused by its detonation rather than the destruction of fragments (as in a fragmentation grenade).

condensation cloud A mist or fog of minute water droplets that temporarily surrounds the FIREBALL following a NUCLEAR (or atomic) DETONATION in a comparatively humid atmosphere. The expansion of the air in the negative phase of the BLAST WAVE from the explosion results in a lowering of the temperature, so that water vapor present in the air condenses and a cloud forms. The cloud is soon dispelled when the pressure returns to normal and the air warms up again.

condensation trail A visible cloud streak, usually brilliantly white, that trails behind an aircraft or missile in flight under certain conditions. Also known as a contrail.

conduct of fire The technique by which effective fire is placed on a selected target.

cone of dispersion 1. The cone-shaped pattern formed by the paths of a group of shots fired from a gun with the same sight setting. The shots follow different paths as a result of gun vibration, variation in ammunition, and other factors such as changes in wind. Also called CONE OF FIRE and sheaf of fire.

cone of fire 1. The cone-shaped volume of coverage from a flexible gun emplacement. 2. The fire coverage produced by the fire of two or more FIXED GUNS converging on a single point.

confederation 1. A military alliance. 2. A league of sovereign states that delegates powers on selected issues to a central government. In a confederation, the central government is deliberately limited, and thus may be inherently weak because it has few independent powers. The United States was a confederation from 1781 to 1789; so were the Confederate States of America from 1861 to 1865.

Conference on Security and Confidence-Building Measures and Disarmament in Europe (CDE) An international conference convened in Stockholm in 1984 to seek agreements limiting the size of military exercises that can take place in Europe without the participating countries informing the other members of the CDE. The purpose is to remove the fear that traditionally attends major exercises, which is that they are really a cover for troop mobilization prior to war. To this end the CDE talks have had to face up to the problem of verification. In 1986 a plan was accepted to allow inspection by teams of airborne observers, as long as their aircraft were flown by pilots from the countries carrying out the exercises. Although this is only a modest achievement, it is a valuable first step, especially since it is the first time that the Soviet Union has agreed to any such "intrusive" inspections. Compare to: CONFIDENCE-BUILDING MEASURES.

Confidence-Building Measures (CBMs) A form of arms control designed to reduce the fear and risk of surprise attack as well as to make it hard to mass for such an attack. CBMs entail the required notification of opposing forces of any plans for large-scale exercises. Thus international crises are less likely to develop into war because no side would feel the need to make a pre-emptive strike. Armies have always feared that major military exercises could be used to mask troop concentrations intended to launch an attack. By letting a potential enemy know that an exercise is planned, and inviting them to send observers to check that it really is an exercise, this fear is alleviated. Under the auspices of the Conference on Security and Co-operation in Europe, real progress has been made in developing CBMs since 1982.

But the real problem remains that it is very hard to see how the CBM approach can be extended to cover nuclear arms. The form of inspection which would be required—checking, for example, that nuclear weapons had not been moved into a forbidden area, or put on board ships—would be far too intrusive to be accepted presently.

confidence course An OBSTACLE COURSE used in training to instill in troops the "confidence" to accomplish demanding tasks.

confinement, close See CLOSE CONFINEMENT.

confinement officer The correctional officer, appointed by the commanding officer of a military installation, who is charged with the custody, administration, and treatment of prisoners.

confirmation of information The reporting of an INTELLIGENCE information item for the second time, preferably by another independent source whose reliability is considered when confirming the information item.

confusion agent An individual who is dispatched by a sponsor for the primary purpose of confounding the INTELLIGENCE or COUNTERINTELLIGENCE apparatus of another nation rather than for the purpose of collecting and transmitting information.

confusion reflector A reflector of ELECTROMAGNETIC RADIATION used to create echoes for confusion purposes. RADAR confusion reflectors include such devices as CHAFF, ROPE and corner reflectors.

congressional medal of honor See MEDAL OF HONOR.

connection survey A survey that ties together a TARGET AREA SURVEY and POSITION AREA SURVEY.

conquer To defeat militarily an army and its nation so that the entire society is in the control of the conquerers. As Genghis Khan (1167–1227) was pleased to say: "Happiness lies in conquering one's enemies, in driving them in front of oneself, in taking their property, in savoring their despair, in outraging their wives and daughters." Julius Caesar would have agreed. He wrote in *The Gallic War* (51 B.C.): "War gives the right to the conquerors to impose any condition they please upon the vanquished." This is, in effect, still true today—except that it is not considered polite or politic to say so.

conscription The system by which young men, and sometimes women, are required to serve a period in their nation's armed forces. This is still the basic method of military recruitment for most of the world's armies. In the North Atlantic Treaty Organization (NATO), for example, the only major nations not relying on conscription are the United Kingdom, which abolished national service in the early 1960s, and the United States, which replaced "the draft" with an all-volunteer force after the Vietnam War, and Canada. These nations were already unusual in not having a long historical tradition of conscription in peacetime.

consolidation of position The organization and strengthening of a newly captured POSITION so that it can be used against the enemy; the occupying force also prepares for succeeding operations.

consolidation psychological operation A psychological operation conducted toward populations in friendly areas of operations or in territory occupied by friendly military forces with the objective of facilitating operations and promoting maximum cooperation among the civil population.

constellation size The number of SATELLITES required to realistically monitor enemy MISSILE launch sites, or provide worldwide communications or navigation data.

constitute To provide the legal authority for the existence of a new unit of the armed services. The new unit is designated and listed, but it has no specific existence until it is activated. See also ACTIVATE; COMMISSION.

constructive engagement A diplomatic phrase for maintaining political and economic ties with regimes with whom a nation has many disagreements, in the hope that the continuation of ties will gradually lead to changes in that nation's objectionable policies and practices. It has often been used to describe the relationship of the United States toward South Africa. See Sanford J. Ungar and Peter Vale, "South Africa: Why Constructive Engagement Failed," *Foreign Affairs* Vol. 64, No. 2 (Winter 1985–86); Michael Clough, "Beyond Constructive Engagement," *Foreign Policy* No. 61 (Winter 1985).

consumer A person or agency that uses INFORMATION or INTELLIGENCE produced by either its own staff or other agencies.

consumer logistics That part of logistics concerning the reception, storage, inspection, distribution, transport, maintenance (including repair and serviceability), and disposal of MATERIEL, and the provision of SUPPORT and SERVICES. In consequence, consumer logistics includes the determination of materiel requirements, follow-on support, stock control, provision or construction of facilities (excluding any materiel element and those facilities needed to support production logistics activities), movement control, codification, reliability and defect reporting, storage, transport, and handling safety standards, and related training.

consumption issues Issues of an item that is expended or consumed in use, such as expendable supplies.

consumption rate The average quantity of an item consumed or expended during a given time, expressed in prescribed units of measurement.

contact 1. To encounter the enemy. 2. To fire on or to receive fire from the enemy. It was ALFRED THAYER MAHAN who wrote in his *The Influence of Sea Power Upon History* (1890) that contact was the "word which perhaps better than any other indicates the dividing line between tactics and strategy."

contact fire, delayed See DELAYED-CONTACT FIRE.

contact mine A MINE detonated by physical contact.

contact patrol A PATROL detailed to maintain contact with adjoining units.

contact point 1. In land warfare, a point on the terrain, easily identifiable, where two or more units are required to make physical contact with one another. 2. In air operations, the position at which a mission leader makes radio contact with an air-control agency. See also CHECKPOINT; CONTROL POINT; COORDINATING POINT.

contact reconnaissance The location of isolated units, out of contact with a main force.

contact report A report indicating detection of the enemy.

contact team An element of a command organization or unit designated to visit another organization for the purpose of providing SERVICE or INTELLIGENCE; for example, a DETACHMENT from a maintenance company sent forward to deliver SUPPLIES or make repairs on ORDNANCE materiel of units needing assistance.

contain To stop, hold, or surround the forces of the enemy, or to cause the enemy to center activity on a given FRONT and to prevent his withdrawing any part of his forces for use elsewhere.

containment The official United States foreign policy towards the Soviet Union since 1947, developed by the Truman administration as a response to the apparent expansionist intentions of the Soviet Union under Josef Stalin. It was first espoused by George F. Kennan in the June 1947 *Foreign Affairs* article entitled "The Sources of Soviet Conduct," in which he asserted that "Soviet pressure against the free institutions of the Western World is something that can be contained by the adroit and vigilant application of counterforce." (The official author of this article was "X," because Kennan wrote it while serving as a an officer in the U.S. Foreign Service, but it was never a secret who the actual author was.) In many ways, containment represented a moderate approach, because at the time there was considerable support in the United States for an attempt to push Soviet influence out of Eastern Europe. The immediate consequences of the containment doctrine were the military commitment of the United States to Western Europe, out of which the North Atlantic Treaty Organization (NATO) developed, and the financial and military support to countries such as Greece and Turkey, which were under Soviet threat. The Korean War, although technically a United Nations action, was perhaps the first test of American resolve to hold the Soviet Union to its postwar boundaries. See Charles Gati, "What Containment Meant," *Foreign Policy* No. 7 (Summer 1972); William Welch, "Containment: American and Soviet Versions," *Studies in Comparative Communism* Vol. VI, No. 3 (Autumn, 1973); Robert E. Osgood, "The Revitalization of Containment," *Foreign Affairs* Vol. 60, No. 3 (1982); Howard J. Wiarda, "Updating U.S. Strategic Policy: Containment in the Caribbean Basin," *Air University Review* 37, 5 (July–August 1986).

contamination 1. The deposit or absorption of radioactive material, or of BIOLOGICAL or CHEMICAL AGENTS on and by structures, areas, personnel, or objects. See also INDUCED RADIATION; RESIDUAL RADIATION. 2. The making of food, water, or both unfit for consumption by humans or animals because of the presence of environmental chemicals, radioactive elements, bacteria or other organisms.

contamination, residual See RESIDUAL CONTAMINATION.

contingency force A force designed for rapid deployment into a theater of operations, from bases normally located outside of it. See also RAPID DEPLOYMENT FORCE.

contingency plan 1. Actions to be taken if a prior plan fails or if the situation changes. 2. An emergency plan.

contingent effect 1. Something that happens because something else happened first. 2. The effects associated with a NUCLEAR DETONATION, both desirable and undesirable, other than the primary effects.

contingent zone of fire An area within which a designated ground unit or FIRE-SUPPORT ship may be called upon to deliver fire. See also ZONE OF FIRE.

continuity of command The maintenance of vested authority by an individual of the armed forces for the direction, coordination, and control of military forces. Thus if the general in command is killed in action, his second in command (perhaps a lower ranked general or colonel) assumes command and continuity is maintained.

continuity of operations The degree or state of being continuous in the conduct of functions, tasks, or duties necessary to accomplish a military action or mission in carrying out a national MILITARY STRATEGY. It includes the functions and duties of the commander, as well as the supporting functions and duties performed by the staff and others acting under the authority and direction of the commander.

continuous fire 1. Fire conducted at a normal rate without interruption for the application of adjustment corrections or for other causes. 2. In FIELD ARTILLERY and naval gunfire support, loading and firing at a specified rate or as rapidly as possible consistent with accuracy within the prescribed rate of fire for the weapon. The firing continues until terminated by the command "END OF MISSION," or is temporarily suspended by the commands "CEASE LOADING" or "CHECK FIRING."

continuously pointed fire A system of FIRE CONTROL that supplies FIRING DATA to a gun crew continuously instead of at fixed intervals.

continuous operations Military operations that allow for no relief, sleep, or rest, and require constant fighting. This is an inherently temporary tactic because after only a few days of it, the necessary human efficiency becomes seriously impaired.

contour, approximate See APPROXIMATE CONTOUR.

contour flying Flight at low altitudes conforming generally, and in close proximity, to the contours of the earth. This type flight takes advantage of available COVER and CONCEALMENT in order to avoid observation or detection by RADAR of the aircraft and its points of departure and landing. It is characterized by a constant AIRSPEED and a varying altitude, as vegetation and obstacles dictate.

contour map A map showing altitude above sea level by lines that connect points on a land surface that have the same altitude.

contour matching, magnetic See MAGNETIC CONTOUR MATCHING.

contraband 1. Commodities that are formally declared to be illegal to import or export. 2. Smuggling; illegal trade.

contraband of war Materials that a neutral state cannot supply to a belligerent state except at the risk of seizure and confiscation by an opposing belligerent state. Contraband can range from food to munitions; it is anything a belligerent defines as useful to an enemy. Under the doctrine of continuous voyage (or continuous transportation) goods carried on neutral ships bound for neutral ports have been seized if the ultimate destination was an enemy port.

Contras The United States-backed "democratic resistance movement" in Nicaragua. The Contras oppose the Communist Sandinista government. They are called Contras by their government after the Spanish word for counter revolutionaries: "contrarrevolucíonaríos." President Ronald Reagan preferred to call them "freedom fighters," and had consistent problems obtaining funding for their guerrilla operations from the U.S. Congress. See Christopher Dickey, *With the Contras: A Reporter in the Wilds of Nicaragua* (New York: Simon & Schuster, 1985).

contravallation DEFENSIVE POSITIONS surrounding another fortified position created to isolate the BESIEGED forces and protect the attackers.

control 1. Authority that may be less than full COMMAND, exercised by a commander over part of the activities of subordinate or other organizations. See also ADMINISTRATIVE CONTROL and OPERATIONAL COMMAND. 2. In mapping, charting, and photogrammetry, a collective term for a system of marks or objects on the earth or on a map or a photograph, whose POSITIONS or ELEVATIONS, or both, have been or will be determined. 3. Physical or psychological pressure exerted with the intent to assure that an agent or group will respond as directed. 4. An indicator governing the distribution and use of documents, information, or material. Such indicators are the subject of INTELLIGENCE COMMUNITY agreement, and are specifically defined in appropriate regulations. 5. The degrees of authority exercised by a military commander over a civil population, government, or economy.

control-and-assessment team A provisional TASK ORGANIZATION used within combat organizations, such as BATTALIONS, GROUPS, BATTLE GROUPS, or DIVISIONS, when a subordinate unit has been subjected to a nuclear, chemical, biological, or radiological attack or natural disaster. The team determines the operational effectiveness of the unit to which it has been sent, assumes control of the unit if required, executes AREA DAMAGE CONTROL operations, and takes action to resume the primary mission of the unit. The team is employed in case of damage to either tactical or administrative support elements of a division, or to the tactical elements of other combat organizations, when the assigned commander of the affected unit is unable to act.

control-and-reporting center A subordinate air control element of the tactical air control center from which radar control and warning and ground control intercept operations are conducted within a given area of responsibility.

control-and-reporting system An organization set up for the early warning, tracking, and recognition of aircraft and tracking of surface craft; and for the control of all active air defenses. It consists primarily of a chain of radar reporting stations, control centers, and human observers together with the necessary communications network.

control measures Assignments of BOUNDARIES, OBJECTIVES, COORDINATING POINTS, CONTACT POINTS, LINES OF DEPARTURE, ASSEMBLY AREAS, and other strategic locations made graphically or orally by a commander to subordinate commands in order to delegate responsibilities, coordinate FIRE and MANEUVER, and control combat operations.

control officer An officer, usually the executive or second in command, who moves at the head of a COLUMN or ELEMENT thereof to regulate the rate of march and maintain the proper direction.

control officer, central See CENTRAL CONTROL OFFICER.

control point 1. A position along a route of march at which men are stationed to give information and instructions for the regulation of supply or traffic. 2. A position marked by a buoy, boat, aircraft, electronic device, conspicuous terrain feature, or other identifiable object which is given a name or number and used as an aid to navigation or control of ships, boats, or aircraft. 3. A point located by ground survey with which a corresponding point on a photograph is matched as a check, in making mosaics (photographs matched together to make one larger photograph).

controlled-effects nuclear weapons NUCLEAR WEAPONS designed to achieve variation in the intensity of specific effects other than normal BLAST EFFECT.

controlled exercise An EXERCISE characterized by the imposition of constraints on some or all of the participating units by planning authorities, with the principal intention of provoking types of interaction. See also FREE PLAY EXERCISE.

controlled forces Military or paramilitary forces (such as local police) under effective and sustained political and military direction.

controlled fragmentation The emission of fragments of a predetermined size, shape, density, velocity, and pattern upon detonation through the planned design and fabrication of a PROJECTILE, MINE, GRENADE, or BOMB.

controlled information Information conveyed directly or indirectly to an adversary in a DECEPTION OPERATION to evoke desired APPRECIATIONS.

controlled interception An AIR INTERCEPTION action wherein friendly aircraft are controlled from a ground, ship, or airborne station.

controlled mine A mine which after laying can be controlled by the user, to the extent of either making the mine safe or live, or of firing the mine.

controlled net A group of stations on a common channel of communication with one station designated as the controlling station and all of the other stations transmitting only when granted permission to do so.

controlled passing A traffic movement procedure whereby two lines of traffic traveling in opposite directions are enabled to alternately traverse a point or section of a route that can take only a single line of traffic at a time.

controlled pattern The dropping of parachuted supplies, weapons, or other materiel from aircraft in flight, with the prevention of their dispersal by connecting them into one group with webbing, rope, or by other means.

controlled port A harbor or anchorage at which military authorities control the entry and departure of vessels, assignment of berths, and traffic within the harbor or anchorage.

controlled response A military response, selected from a wide variety of feasible options, that will provide the greatest advantage under a given set of circumstances.

controlled supply rate The amount of AMMUNITION, expressed in rounds per weapon per day which may be requisitioned by a unit. A controlled supply rate is established by the commander of a military force for subordinate units, in order to control consumption, and normally lists only those types of ammunition on which restrictions have been placed.

control zone A controlled airspace extending upwards from the surface of the earth to a specified upper limit.

control ship In an amphibious operation, this is the navy ship from which the ship-to-shore movement of troops, equipment, and supplies is directed.

convention 1. An international agreement on matters less significant than those regulated by treaty. The best known conventions are probably the GENEVA CONVENTIONS of 1864, 1906 and 1949, which concern the treatment of PRISONERS OF WAR. 2. A political meeting of the members of one party. 3. A constitutional convention.

conventional mine 1. An industrially manufactured MINE that has a predictable, designed effect, a standard arming/disarming procedure, and a standard size and shape. 2. A HIGH-EXPLOSIVE mine as opposed to a nuclear mine.

conventional war 1. A conflict fought by regular military forces using any weapons short of NUCLEAR WEAPONS. 2. Any major conflict, whether fought by irregular or regular forces, that does not use nuclear weapons. See Myra Struck McKitrick, "A Conventional Deterrent for NATO: An Alternative to the Nuclear Balance of Terror," *Parameters: Journal of the US Army War College* Vol. 13, No. 1 (March 1983); Richard K. Betts, "Conventional Deterrence; Predictive Uncertainty and Policy Confidence," *World Politics* Vol. 37, No. 2 (January 1985).

conventional weapon Any instrument of battle that has no nuclear, biological, or chemical component. This term encompasses all levels of sophistication; from the slingshot to the machine gun to the main battle tank. Since 1973 NATO and the Warsaw Pact have been discussing conventional force reductions at the MUTUAL AND BALANCED FORCE REDUCTION talks. See David D. Finley, "Conventional Arms in Soviet Foreign Policy," *World Politics* 33, 1 (October 1980); James M. Garrett, "Conventional Force Deterrence in the Presence of Theater Nuclear Weapons," *Armed Forces and Society* 11, 1 (Fall 1984).

conventional wisdom What is generally believed to be true. However, writers almost always use the phrase to mean "that which most people believe to be true, but really isn't." The construction first gained currency after John Kenneth Galbraith used it in his *The Affluent Society* (Boston: Houghton Mifflin, 1958) when he observed: "Only posterity is unkind to the man of conventional wisdom, and all posterity does is bury him in a blanket of neglect." The conventional wisdom is often the point of departure in stragetic and tactical analyses. See R. James Woolsey, "Planning a Navy: The Risks of Conventional Wisdom," *International Security* Vol. 3, No. 1 (Summer 1978); Aaron D. Rosenbaum, "Discard Conventional Wisdom," *Foreign Policy* No. 49 (Winter 1982–83).

converged sheaf Planes of fire that intersect at a given point, as the result of a lateral distribution of fire from two or more pieces. See also OPEN SHEAF; PARALLEL SHEAF; SHEAF; SPECIAL SHEAF.

convergent lines of operation Troop dispositions of differing origination that later converge enough about a common place that the converging forces are close enough to support each other.

convoy 1. A number of merchant ships or naval auxiliaries (supply ships), or both, usually escorted by warships, aircraft, or both assembled and organized for the purpose of passage together. 2. A group of vehicles organized for the purpose of CONTROL and orderly movement, with or without ESCORT protection.

convoy, military See MILITARY CONVOY.

convoy dispersal point The position at sea which a CONVOY breaks up, each ship proceeding independently thereafter.

convoy escort 1. A naval ship(s) or aircraft in company with a convoy and responsible for its protection. 2. An ESCORT to protect a convoy of vehicles from being scattered, destroyed, or captured.

convoy loading The loading of TROOP units with their equipment and supplies in vessels of the same movement group, but not necessarily in the same vessel.

cook off The firing of a chambered ROUND OF AMMUNITION, initiated by the heat of the weapon in which the round is loaded.

cooperative logistics The LOGISTIC SUPPORT provided a foreign government or agency through its participation in the U.S. Department of Defense logistic system, with reimbursement to the United States for the support provided.

coordinated attack See ATTACK, DELIBERATE.

coordinated defense An air defense of two or more vulnerable areas which are too far apart to form an INTEGRATED DEFENSE, but are organized to effect economy of MATERIEL and provide greater effectiveness with mutual support.

coordinated fire line A line beyond which all surface-to-surface FIRE-SUPPORT means (MORTAR, FIELD ARTILLERY, and naval gunfire) may fire at any time within a specified zone without additional coordination. The line separates acceptable from unacceptable targets.

coordinates 1. Linear or angular quantities that designate the position that a point occupies in a given reference frame or system. 2. A general term to designate the particular kind of reference frame or system, such as plane rectangular coordinates or spherical coordinates.

coordinates, grid See GRID COORDINATES.

coordinating altitude An altitude determined by the area AIRSPACE MANAGEMENT authority below which air-force activity must be coordinated with army facilities and above which army activity must be coordinated with air-force facilities.

coordinating authority A commander or individual assigned responsibility for coordinating specific functions or activities involving forces of two or more SERVICES or two or more forces of the same service. The commander or individual has the authority to require consultation between the agencies involved, but does not have the authority to compel agreement. In the event that essential agreement cannot be obtained, the matter shall be referred to the appointing authority.

coordinating line, final See FINAL COORDINATING LINE.

coordinating point A designated point at which, in all types of combat, adjacent UNITS or FORMATIONS must make contact for purposes of control and coordination.

coordination area, airspace See AIRSPACE COORDINATION AREA.

coordination with In consultation with. Agencies coordinated with each other must concur and participate actively in a project. If concurrence is not obtained, the disputed matter is referred to the next higher authority. See also COORDINATING AUTHORITY.

coppering Metal fouling left in the BORE of a WEAPON by the ROTATING BAND or JACKET of a PROJECTILE.

cordon 1. A military force enclosing an area to deny the enemy passage. 2. To form a cordon around.

cordon sanitaire The French term for a sanitary line or a buffer state between two rivals. In this century, Poland has often been an example. See Stanley F. Gilchrist, "The Cordon Sanitaire—Is It Useful? Is It Practical?" *Naval War College Review* 35, 3 (May–June 1982).

corporal The most junior noncommissioned officer below a SERGEANT and above a PRIVATE.

corporal, lance See LANCE CORPORAL.

corporal punishment Physical punishment inflicted upon the body of the offender; whipping, for example.

corporal's guard A small group assigned to a task, under the command of a CORPORAL. As the corporal is the lowest ranking NONCOMMISSIONED OFFICER, his would be the smallest of units.

corps 1. A functional branch of an army, e.g., the SIGNAL CORPS. 2. An operational unit usually consisting of two or three DIVISIONS, normally commanded by a lieutenant-general.

corps, army See ARMY CORPS.

corps de chasse An exploitation unit to be used after an initial attack to pursue and destroy the enemy.

corpsman An enlisted person the U.S. Navy or Marine Corps who is trained in combat first aid. The comparable army term is MEDIC.

corps troops Troops assigned or attached to a CORPS, but not a part of one of the DIVISIONS that make up the corps.

corrected azimuth The AZIMUTH of the axis of the BORE of a gun firing on a moving target, after allowances have been made for atmospheric, MATERIEL, and other variable conditions.

corrected deflection The horizontal angle between the LINE OF SIGHT from a weapon to the target and the axis of the BORE of the weapon after allowances have been made for atmospheric, materiel, and other variable conditions.

corrected elevation A firing table elevation corresponding to the corrected range.

corrected range The range of a target, with allowances made for weather conditions, variation in ammunition, wear in the gun, or any other variations from standard conditions, so that the projectile fired from the gun will carry to the target.

correction 1. In FIRE CONTROL, any change in FIRING DATA to bring the MEAN POINT OF IMPACT or BURST closer to the target. 2. A communication proword to indicate that an error in data has been announced and that corrected data will follow.

corrector In artillery, an arbitrary figure used to indicate a change in the FUZE settings, ordered to guns to compensate for changes in weather, MUZZLE VELOCITY, and POSITION.

correlation In AIR DEFENSE, the determination that an aircraft appearing on a radarscope, plotting board, or visually, is the same as that on which information is being received from another source.

correlation of forces The Soviet Union's concept of comparative capacity to compete with the West in every important way—the combined economic, military, diplomatic, and ideological strength of one Warsaw Pact country against its opponents. Military force is seen much more as just one Factor in an essentially political conflict between Soviet Communism and the capitalist world, whereas Western attitudes tend more to separate a nation's various forms of power and influence. It is often noted by Western analysis that all Soviet theorists since Lenin himself have recorded their agreement with Clausewitz's dictum that war is "nothing but the continuation of politics by other means." To the Soviet military thinker this is axiomatic, and therefore any particular piece of military hardware is also considered as, in part, an ideological and diplomatic weapon. See James H. Hansen, *Correlation of Forces: Four Decades of Soviet Military Development* (New York: Praeger, 1987).

corridor 1. A COMPARTMENT of terrain, the longer axis of which is parallel to, or extends in, the direction of movement of a FORCE. 2. An airlane assigned to certain aircraft formations to prevent their being attacked by friendly air, ground, or naval forces.

council of war An assemblage of the most senior officers of a military force, initiated by the commander, to seek advice or consensus on how to deal with an emergency or combat problem. Councils of war have a long history. VEGETIUS wrote in his 378 *De Re Militari:* "It is the duty and interest of the general frequently to assemble the most prudent and experienced officers of the different corps of the army and consult with them on the state both of his own and the enemy's forces." But in recent centuries, councils of war have been disdained as the refuge of a poor commander. Napoleon said: "Councils of war will be what the effect of these things has been in every age: they will end in the adoption of the most pusillanimous or (if the expression be preferred) the most prudent measures, which in war are almost uniformly the worst that can be adopted." General Henry Halleck advised General George Meade after the 1863 Battle of Gettysburg: "Call no council of war. It is proverbial that councils of war never fight."

countdown The step-by-step process leading to the initiation of MISSILE testing, launching, and firing. It is performed in accordance with a pre-designated time schedule.

counterair An air operation of a tactical air command conducted to attain and maintain a desired degree of air superiority by the destruction or neutralization of enemy forces. Both air-offensive and air-defensive actions are involved. The former range throughout enemy territory and are generally conducted at the initiative of friendly forces. The latter are conducted near or over friendly territory and generally react to the initiative of the enemy air forces. See also ANTIAIR WARFARE. See Robert D. Rasmussen, "The Central Europe Battlefield: Doctrinal Implications for Counterair-Interdiction," *Air University Review* 29, 5 (July–August 1978).

counterair operation, offensive See OFFENSIVE COUNTERAIR OPERATION.

counterattack An attack by part or all of a defending force against an enemy attacking force, for such specific purposes as regaining ground lost or cutting off or destroying enemy advance units, and with the general objective of denying to the enemy the attainment of his purpose in attacking. In sustained defensive operations, a counterattack is undertaken to restore the BATTLE POSITION, and is directed at limited objectives.

counterbarrier operations Those actions taken to counteract an enemy OBSTACLE or BARRIER SYSTEM.

counterbattery Artillery positioned to counter the fire of an enemy battery by directly firing onto the enemy's artillery.

countercity strike See COUNTERVALUE.

counter-countermeasures Defensive measures taken to defeat offensive COUNTERMEASURES.

counterdeception Efforts to negate, neutralize, diminish the effects of, or gain advantage from a foreign DECEPTION OPERATION. Counterdeception does not include the INTELLIGENCE function of identifying foreign deception operations. See also DECEPTION.

counterespionage That aspect of counterintelligence designed to detect, destroy, neutralize, exploit, or prevent ESPIONAGE activities through the identification, penetration, manipulation, deception, and repression of individuals, groups, or organizations conducting or suspected of conducting espionage activities.

counterfire Fire intended to destroy or neutralize enemy weapons, especially INDIRECT FIRE systems. Most commonly, ARTILLERY force aimed at destroying the enemy's artillery units.

counterforce The employment of STRATEGIC air and missile forces in an effort to destroy, or render impotent, selected military capabilities of an enemy force under any of the circumstances by which hostilities may be initiated. A counterforce target is, in its broadest meaning, any military target, or, in a narrower sense, a target that is part of the enemy's nuclear arms system. In its more narrow meaning, counterforce refers to enemy missiles, C^3I facilities, and other heavily protected or vital operating bases, such as airfields and nuclear missile-carrying submarine ports. Counterforce targets can also include OTHER MILITARY TARGETS such as troop concentrations, arms factories, and logistic support bases. Considerable effort is taken to reduce the COLLATERAL DAMAGE to the civilian economic and social structure in planning counterforce strikes. The counterforce/COUNTERVALUE distinction is crucial in strategic theory because the two concepts are, to many, morally different and have contrasting escalation risks and values. Counterforce targets are at the center of strategies involving LIMITED NUCLEAR OPTIONS.

counterguerrilla warfare Operations and activities conducted by armed forces, paramilitary forces, or nonmilitary agencies against guerrillas.

counterinsurgency Those military, paramilitary, political, economic, psychological, and civic actions taken by a government to defeat INSURGENCY. The classic work on this is Robert Thompson's *Defeating Communist Insurgency* (London: Chatto & Windus, 1966). Also see: Douglas S. Blaufarb, *The Counterinsurgency Era: U.S. Doctrine and Performance, 1950 to the Present* (New York: Free Press, 1977).

counterintelligence Those activities concerned with identifying and counteracting the threat to security posed by hostile intelligence services or organizations or by individuals engaged in espionage, sabotage, or subversion. See Arnold Beichman, "Can Counterintelligence Come in from the Cold?" *Policy Review* 15 (Winter 1981); Melvin Beck, *Secret Contenders: The Myth of Cold War Counterintelligence* (New York: Sheridan Square, 1985).

counterintelligence, tactical See TACTICAL COUNTERINTELLIGENCE.

countermeasures 1. The employment of devices, techniques, or both to impair the operational effectiveness of enemy activity. 2. In the context of the STRATEGIC DEFENSE INITIATIVE (SDI), measures taken by the offense to overcome aspects of a BALLISTIC MISSILE DEFENSE (BMD) system. See Louis Rene Beres, *Mimicking Sisyphus: America's Countervailing Nuclear Strategy* (Lexington, Mass.: Lexington Books, 1983).

countermeasures, electronic See electronic countermeasures.

countermining 1. Tactics and techniques used to detect, avoid, breach, or neutralize enemy land MINES and the use of available resources to deny the enemy the opportunity to employ mines. 2. The detonation of naval mines by nearby explosions, either accidental or deliberate.

countermobility operations The construction of OBSTACLES and the reinforcement of terrain to delay, disrupt, and destroy the enemy. The primary objective of such operations is to slow or divert the enemy, to increase time for TARGET ACQUISITION, and to increase weapon effectiveness.

countermove An operation undertaken in reaction to or in anticipation of a move by the enemy.

counteroffensive A large-scale OFFENSIVE undertaken by a defending force to seize the initiative from the attacking force. See also COUNTERATTACK.

counterpreparation fire Intensive PREARRANGED FIRE delivered when the imminence of an enemy attack is discovered. It is designed to break up enemy formations; disorganize the enemy's systems of command, communications, and observation; decrease the effectiveness of artillery preparation; and impair the enemy's offensive spirit.

counterrecoil The forward movement of a gun returning to firing position after recoil.

counterreconnaissance All measures taken to prevent hostile observation of a force, area, or place.

countersabotage That aspect of COUNTERINTELLIGENCE designed to detect, destroy, neutralize, or prevent SABOTAGE activities through the identification, penetration, manipulation, deception, and repression of individuals, groups, or organizations conducting or suspected of conducting sabotage activities.

counterscarp The outer boundary of a ditch or TRENCH that faces the actual or anticipated direction of an enemy; the inner boundary is called the ESCARP.

countersign A secret challenge and its reply. See also CHALLENGE; PASSWORD.

countersubversion Action designed to detect and counteract SUBVERSION.

countersurveillance All measures, active or passive, taken to counteract hostile SURVEILLANCE.

counterterrorism Efforts to prevent and if necessary to punish for terrorist acts, in the expectation that retaliation will discourage further terrorism. See William V. O'Brien, "Counterterrorism: Lessons from Israel," *Strategic Review,* 13, 4 (Fall 1985).

countervailing strategy The strategic posture initiated in 1974 by then U. S. Secretary of Defense James Schlesinger. It indicated that the United States would maintain its capacity to match any Soviet threat at any particular level, so that it could not hope for any advantage. The posture was an attempt to produce a more subtle alternative to MUTUAL ASSURED DESTRUCTION in a world where strategic superiority is elusive. To avoid any argument about whether a nuclear war could be "won," the document used "countervailing" and at least implied that the United States would not "lose". This doctrine relied on what came to be known as LIMITED NUCLEAR OPTIONS, ensuring that the United States would never be in the position of having to choose between doing nothing or responding too massively were the Soviet Union, for example, to launch a restricted strike. This doctrine later became part of President Carter's PRESIDENTIAL DIRECTIVE-59.

countervalue A nuclear attack aimed at non-military targets. A countervalue target is one that does not have a specifically military nature. In most targeting theories, countervalue targets are economic, political, and industrial structures—factories, power plants, mines, transport systems, agricultural storage depots, government offices, police headquarters, and so on. Countervalue STRIKES figure heavily in DETERRENCE strategies, and require notably less sophisticated MISSILES, in terms of power and accuracy, than COUNTERFORCE strikes. However, the latter may not cause such low levels of COLLATERAL DAMAGE to sufficiently differentiate them from the former in practice despite the assertions of strategic planners.

count off 1. The call out of one's numerical position in a line in successive order from a given starting point. 2. The command to count off.

country, host See HOST COUNTRY.

country desk The unit within the U.S. Department of State that has the daily responsibility of monitoring and analyzing the activities of a given foreign country.

country team In a foreign country this is the coordinating and supervisory body headed by the chief of the United States diplomatic mission, usually an ambassador, and composed of the senior member of each represented United States department or agency, such as AID, MAAG and CIA.

coup de grace From the French: a "stroke of mercy." 1. A final and decisive end to a battle. 2. The pistol shot to the head of a firing squad's target, delivered, if the target is not yet dead, after the squad has fired.

coup de main From the French: a "stroke of the hand"; a sudden and vigorous ATTACK that successfully obtains an OBJECTIVE. But Napoleon warned that "The success of a coup-de-main depends absolutely upon luck rather than judgment."

coup d'etat From the French: a "stroke or sharp blow to the state"; a change in the leadership of a government brought about by force by those who already hold some form of power (either military or political). This technically differs from a REVOLUTION in that revolutions are usually brought about by those who are not presently in power. For theoretical analyses of coups d'etat, see: P. A. J. Waddington, "The Coup d'Etat—An Application of a Systems Framework," *Political Studies* Vol. XXII, No. 3 (September 1974); Rosemary H. T. O'Kane, "A Probabilistic Approach to the Causes of Coups d'Etat," *British Journal of Political Science* Vol. 11, No. 3 (July 1981); Rosemary H. T. O'Kane, "Towards an Examination of the General Causes of Coups d'Etat," *European Journal of Political Research* Vol. 11, No. 1 (March 1983).

coup d'oeuil From the French: a "stroke of the eye." The art of rapidly distinguishing the weak points of an enemy's POSITION and of discerning the advantages and disadvantages of terrain; the ability to quickly make an accurate estimate of a combat problem. Jean-Charles de Folard (1669–1752) wrote in his 1724 *Nouvelles Decouvertes Sur La Guerre* that: "The coup d'oeuil is a gift of God and cannot be acquired: but if professional knowledge does not perfect it, one only sees things imperfectly and in a fog, which is not enough in these matters where it is so important to have a clear eye . . . To look over a battlefield, to take in at the first instance the advantages and disadvantages is the great quality of a general." NAPOLEON agreed; he felt it was "inborn in great generals." 2. According to the 1747 *Instructions of Frederick the Great for His Generals:* the coup d'oeuil is also the "judgment that is exercised about the capacity of the enemy at the commencement of a battle."

course of action 1. Any sequence of activities that an individual member of an armed force or a military unit may follow. 2. A possible plan open to an individual or commander which would accomplish, or is related to the accomplishment of his MISSION. 3. The scheme adopted to accomplish a job or mission. 4. A line of conduct in an ENGAGEMENT.

court-martial 1. A military court established to try those who have committed offenses against MILITARY LAW. 2. To force someone to be tried by such a military court. Field Marshall Sir William Slim (1891–1970) wrote in a 1962 memoir, *Unofficial History,* that: "The popular conception of a court martial is half a dozen bloodthirsty old Colonel Blimps, who take it for granted that anyone brought before them is guilty . . . and who at intervals chant in unison, 'Maximum penalty—death!' In reality courts martial are almost invariably composed of nervous officers, feverishly consulting their manuals; so anxious to avoid a miscarriage of justice that they are, at times, ready to allow the accused any loophole of escape. Even if they do steel themselves to passing a sentence, they are quite prepared to find it quashed because they have forgotten to make something "A" and attach it to the proceedings."

court-martial, general See GENERAL COURT-MARTIAL.

court-martial, special See SPECIAL COURT-MARTIAL.

court-martial, summary See SUMMARY COURT-MARTIAL.

court-martial order An order that is published to promulgate the result of trial by a GENERAL or SPECIAL COURT-MARTIAL.

court-martial orders, general See GENERAL COURT-MARTIAL ORDERS.

court of inquiry A board of three or more officers (and a recorder) appointed by the president of the United States or another competent authority to examine the nature of any transaction of, or accusation or imputation against, military personnel.

court of military appeals A court composed of three civilian judges appointed by the president of the United States and confirmed by the U.S. Senate, which exercises the appellate functions over the Armed Forces as to records of trial by COURTS-MARTIAL required by the

UNIFORM CODE OF MILITARY JUSTICE. See Captain John T. Willis, "The United States Court of Military Appeals: Its Origin, Operation and Future," *Military Law Review* 55 (Winter 1972).

cover 1. Protection from possible enemy action rendered by land, air, or sea forces of friendly forces by OFFENSE, DEFENSE, or threat of either or both. 2. Covert measures necessary to protect a person, plan, OPERATION, FORMATION, or INSTALLATION from enemy INTELLIGENCE efforts. 3. The act of maintaining a continuous radio receiver watch with a transmitter that is calibrated and available, but not necessarily available for immediate use. 4. Shelter or protection, either natural or artificial. 5. Photographs or other recorded images that show a particular area of ground. 6. A MISSION assigned to a FORCE operating apart from the MAIN BODY to intercept, deceive, engage, delay, or disorganize the enemy, and to force the enemy to deploy and reveal the direction of his main effort before he attacks the covered main body; or, in OFFENSIVE operations, to ascertain the enemy's DISPOSITIONS so that the main body may attack under the best conditions. 7. An assumed identity.

cover code That fraction of a target hidden (or covered) from a given observer position by local relief (i.e., land mass). The fraction covered can neither be detected visually nor hit by DIRECT FIRE.

covered approach 1. Any route that offers protection against enemy observation or fire. 2. An approach made under the protection furnished by other forces or by natural COVER.

covered movement A movement of TROOPS when adequate security is provided by friendly forces.

covered way A protected line of COMMUNICATIONS.

covering barrier A BARRIER located beyond the forward edge of a battle area which is selected to assist in the delaying actions of COVERING and security forces.

covering fire 1. Fire from artillery or other rearward units used to protect troops when they are within range of enemy SMALL ARMS. This allows attacking troops to move forward to engage the enemy by temporarily preventing the enemy from returning fire. 2. In amphibious usage, fire delivered prior to a landing to cover preparatory operations such as underwater demolition or MINESWEEPING. 3. Protective fire by part of an infantry unit to enable the rest of the unit to advance.

covering force 1. A force operating apart from a MAIN FORCE for the purpose of intercepting, engaging, delaying, disorganizing, and deceiving the enemy before he can attack the force covered. 2. Any body or detachment of troops that provides security for a larger force by observation, RECONNAISSANCE, ATTACK, or DEFENSE, or by any combination of these methods.

covering-force area The area forward of the FORWARD EDGE OF THE BATTLE AREA out to the assessed positions of enemy forces. It is here that covering forces execute their assigned tasks.

cover off 1. To take position directly behind the person in front in a CLOSE ORDER DRILL. 2. A command to straighten the FILES in a FORMATION that is out of line.

covert operations MILITARY, POLICE, or INTELLIGENCE activities that are planned and executed so as to conceal the identity of, or permit plausible denial by, the sponsor. They differ from CLANDESTINE OPERATIONS in that emphasis is placed on concealing the identity of the sponsor rather than concealing the operation itself. Compare to: OVERT OPERATIONS. See Stephen R. Weissman, "CIA Covert Action in Zaire and Angola: Patterns and Consequences," *Political Science Quar-*

terly Vol. 94, No. 2 (Summer 1979); Stephen D. Wrage, "A Moral Framework for Covert Action," *Fletcher Forum* Vol. 4, No. 2 (Summer 1980).

CPX See COMMAND POST EXERCISE.

cradle guides That part of a gun carriage on which the gun slides when recoiling just after being fired.

crash boat A high-speed motorboat kept ready for rescue work in the event of crashes or forced landings of aircraft in water.

crater analysis A process by which the direction of hostile shelling is determined from an analysis of the BURST pattern of a shell crater. At the same time, shell fragments are collected and studied to determine their CALIBER.

crawl trench A shallow connecting TRENCH.

credibility 1. In diplomacy, the believability of a party's threats or promises. One side in a negotiation can be said to have credibility if the other side believes that it is not bluffing, even if it is bluffing. 2. A military posture that would allow one side to inflict unacceptable damage on an aggressor, even after having absorbed a FIRST STRIKE. Thus credibility functions as a deterrent.

credit system of supply A system whereby military units, organizations, and installations are allocated definite quantities of SUPPLIES for a prescribed period with the supplies furnished on call and credited.

creep The forward movement of FUZE parts caused by the deceleration of a PROJECTILE during flight.

creeping barrage A BARRAGE in which the fire of all units participating remains in the same relative position throughout, and which advances in steps at the same time. This is usually to allow troops to advance toward the enemy without suffering undue casualties from unsuppressed opposing fire.

creeping mine In naval mine warfare, a buoyant mine held below the surface by a weight, usually in the form of a chain, that is free to creep along the seabed under the influence of a stream or current.

crest clearance Elevation of a GUN at such an angle that its PROJECTILE will not strike an obstacle between the MUZZLE and the TARGET.

crimping The process by which a CARTRIDGE CASE is secured to a BULLET or PROJECTILE, and a BLASTING CAP to a FUZE.

crisis 1. An unstable situation ripe for decisive change. 2. A foreign policy problem involving a threat to state security and dealt with by the highest level of government forced to make crucial decisions within a short period of time.

crisis management Modulating foreign policy reactions during an unstable situation so as to maximize one's own interest while not forcing an opponent into an ACTION-REACTION cycle. In many ways crisis management is simply good diplomacy, in contrast to BRINKMANSHIP—taking a tough line and calling the opponent's bluff. A good crisis manager will avoid raising the temperature in the situation as much as possible. While tough or threatening behavior may be necessary to give a clear signal of resolve, it should be used carefully. Stress must be placed on absolute clarity of international expression to avoid causing the opponent to panic. The essence of this attitude is summed up by President John F. Kennedy's remark, reported by his brother Robert in his *Thirteen Days: A Memoir of the Cuban Missile Crisis* (New York: Norton, 1969): "If anybody is around to write after this, they are going to understand that we made every effort to find peace and every effort to give our adversary room to move." See Leo Hazlewood, John J. Hayes, and James R. Brownell, Jr., "Planning for Problems in Crisis Management: An Analysis of Post-1945 Behavior in the US Department of Defense," *International Studies Quarterly* 20, 1 (March 1977); Alexander L. George, "Crisis Management: The Interaction of Political and Military Considerations," *Survival* 26, (September–October 1984).

crisis stability A condition in which parity, or an overall balance of military power, exists, so that there is no reason for one power which could potentially come into conflict with another to fear that the opponent will think it advantageous to launch a major strike without warning. If there is no such fear, then the first nation should not be tempted into a preemptive strike to avoid such an attack.

critical intelligence Intelligence that is crucial and requires the immediate attention of the commander of a military force. It is required to enable the commander to make decisions that will provide a timely and appropriate response to actions by a potential or real enemy.

critical mass The minimum amount of fissionable material capable of supporting a nuclear chain reaction under precisely specified conditions.

critical node An ELEMENT, POSITION, or COMMUNICATIONS entity whose disruption or destruction immediately degrades the ability of a force to command, control, or effectively conduct combat operations.

critical point 1. A key geographic point or position important to the success of an OPERATION. 2. A crisis or a turning point in an operation. 3. A selected point along a LINE OF MARCH used for reference in giving instructions. 4. A point where there is a change of direction or change in slope in a ridge or stream. 5. Any point along a line of march where interference with a troop movement may occur. 6. That point between two bases from which it will take the same time to fly to either base.

critical zone The area over which a bombing aircraft engaged in HORIZONTAL or GLIDE BOMBING must maintain straight flight so that the bomb sight can be operated properly and bombs dropped accurately.

cross attachment The exchange of subordinate units between units for a temporary period. A tank company detached from a tank battalion and subsequently attached to a mechanized infantry battalion, in exchange for a mechanized company detached from the mechanized infantry battalion and then attached to the tank battalion, is one example of this.

cross-compartment A COMPARTMENT OF TERRAIN, the long axis of which is generally perpendicular to the direction of movement of a FORCE.

crossing area commander The officer responsible for all actions within a river-crossing area from the completion of the ASSAULT PHASE of an attack until he is relieved by the appointing commander.

crossing site The location along a water obstacle at which a crossing can be made using AMPHIBIOUS VEHICLES, assault boats, rafts, bridges, or fording vehicles.

cross level 1. To level a weapon or instrument, such as a mortar or surveyor's transit, at right angles to the LINE OF SIGHT. 2. The transfer of excess equipment from one unit to another to balance amounts on hand or in inventory.

cross loading A system of loading troops so that they may be disembarked or dropped at two or more LANDING OR DROP ZONES, thereby achieving UNIT INTEGRITY upon delivery. See also LOADING.

crossover point A pretermined range in an air-warfare area at which a target ceases to be an air intercept target and becomes a SURFACE-TO-AIR MISSILE target.

cross tell The transfer of information between facilities at the same OPERATIONAL LEVEL. See also TRACK TELLING.

cruise missile A pilotless jet aircraft that flies relatively slowly, by a combination of radar guidance and pre-set computer control, over ranges that can vary from less than one hundred miles to several thousand miles. It can be armed with either nuclear or conventional WARHEADS, and used in either a strategic or tactical role. It is capable of very great accuracy, but currently is relatively vulnerable to ordinary antiaircraft defenses. A second generation of supersonic cruise missiles has been developed by the United States and is soon to be deployed. There are three basic types of cruise missile: the GROUND-LAUNCHED CRUISE MISSILE (GLCM), fired from mobile launchers and intended to be used, most probably with a nuclear warhead, on theatre targets, or conceivably on strategic targets in the western Soviet Union; the air-launched cruise missile (ALCM); and the SEA-LAUNCHED CRUISE MISSILE (SLCM). These last two are far more numerous and have a diversity of roles. See Kenneth P. Werrell, "The Cruise Missile: Precursors and Problems," *Air University Review* 32, 2 (January–February 1981); A. White, *Symbols of War: Pershing II and Cruise Missiles in Europe* (London: Merlin Press, 1983); Joel Wit, "Soviet Cruise Missiles," *Survival* (November–December 1983).

cruiser A class of traditional warship second in size only to battleships; so called because they were built for high speed and long "cruising" distances. Today's cruisers are typically used as escorts for aircraft carriers. Compare to: DESTROYER, FRIGATE.

cryptanalysis The study of ENCRYPTED TEXTS. The steps or processes involved in converting encrypted text into PLAIN TEXT without initial knowledge of the KEY employed in the ENCRYPTION.

cryptoancillary equipment 1. Equipment designed specifically to facilitate the efficient or reliable operation of CRYPTOEQUIPMENT, but which does not perform any of the functions of cryptoequipment. 2. Equipment designed specifically to convert information to a form suitable for processing by cryptoequipment.

cryptocommunication Any communication whose intelligibility has been disguised by ENCRYPTION.

cryptocompromise A compromise of CRYPTOINFORMATION, or recovery of the PLAIN TEXT of encrypted messages by unauthorized persons through cryptoanalytic methods.

cryptocorrespondence Letters, messages, memoranda, and other communications that contain CRYPTOINFORMATION, but which, unlike CRYPTOMATERIAL, do not contain information essential to the ENCRYPTION, DECRYPTION, or AUTHENTICATION process.

cryptocustodian The individual designated by proper authority to be responsible for the custody, handling, safeguarding, and destruction of CRYPTOMATERIAL.

cryptodate The date that determines the specific KEY to be employed for encrypted material.

cryptodevice A device that contains no CRYPTOPRINCIPLE, but which may be used with appropriate keying material to simplify ENCRYPTION and DECRYPTION.

cryptoequipment Any equipment employing a CRYPTOPRINCIPLE.

cryptofacility A facility used for the operation, maintenance, research, development, testing, evaluation, and storage of registered CRYPTOMATERIAL.

cryptographic Of, pertaining to, or concerned with CRYPTOGRAPHY; normally abbreviated as "crypto" and used as a prefix.

cryptography 1. The art or science that treats the principles, means, and methods for rendering plain information unintelligible by ENCRYPTION, and for reconverting encrypted information into intelligible form. 2. The design and use of CRYPTOSYSTEMS.

cryptoguard 1. A procedure by which a test is decrypted, then encrypted into another CRYPTOSYSTEM for relaying telecommunications for other activities that do not hold compatible cryptosystems. 2. To provide secure telecommunications services for other activities.

cryptoinformation Information that would make a significant contribution to the cryptanalytic solution of ENCRYPTED text or a CRYPTOSYSTEM.

cryptology The science that treats hidden, disguised, or encrypted communications. It embraces COMMUNICATIONS SECURITY and COMMUNICATIONS INTELLIGENCE.

cryptomaterial All material, including documents, devices, or equipment, that contains CRYPTOINFORMATION and is essential to the ENCRYPTION, DECRYPTION or AUTHENTICATION of telecommunications.

cryptonet Two or more activities that hold a CRYPTOSYSTEM in common and possess a means of intercommunication.

cryptoperiod A specific time period during which a particular set of cryptovariables may be used.

cryptoprinciple A deterministic logic by which information may be converted into an unintelligible form and reconverted into an intelligible form.

cryptoproduction equipment Equipment and the components thereof that is specifically designed for, and used in, the manufacture and associated testing of cryptovariables.

cryptosystem The associated items of CRYPTOMATERIAL that are used as a unit and provide a single means of ENCRYPTION and DECRYPTION. See also CIPHER; CODE; DECRYPTION; ENCIPHER; ENCRYPTION.

cryptosystem, high grade See HIGH GRADE CRYPTOSYSTEM.

cryptovariables The letters and or numbers or sets of letters and/or numbers used in transmitting coded messages.

Cuban missile crisis The 1962 confrontation between the United States and the Soviet Union over the Soviet placement of nuclear missiles in Cuba. President John F. Kennedy demanded the removal of the missiles, imposed a naval blockage on Cuba, and waited for the Soviet response. The Soviets removed their missiles in exchange for a promise by the United States not to invade Cuba and an understanding that the United States would also remove its nuclear missiles in Turkey. For some time it was publicly believed that nuclear war had been a real alternative for the USSR, and Kennedy's stance, called BRINKMANSHIP, has been judged, favorably or otherwise, in that light. However, it has subsequently become known that the available ICBM force at the command of the Soviet leadership was still extremely small at that period, a fact almost certainly known to the U. S. administration. So nuclear warfare at that time was not a real option for the Soviet Union. Many analysts think that this was the nearest the world has come to a global war since World War II. The crisis demonstrated, among other things, the way in which an American president can ignore the other elected branches of government and commit United States forces. See Graham T. Allison, Jr., *Essence of Decision; Explaining the Cuban Missile Crisis* (Boston: Little, Brown and Co., 1971); Richard Ned Lebow, "The Cuban Missile Crisis: Reading the Lessons Correctly," *Political Science Quarterly* Vol. 98, No. 3 (Fall 1983); Thomas G. Paterson and William J. Brophy, "October Missiles and November Elections: The Cuban Missile Crisis and American Politics, 1962," *Journal of American History* Vol. 73, No. 1 (June 1986).

cultivation The deliberate and calculated development of a relationship with a person for the purpose of intelligence agent recruitment, obtaining information, or gaining control for these or other purposes.

cult of personality A concentration of political power and authority in one individual rather than in the office that is occupied. The phrase was first used in the 1956 meeting of the Russian Communist Party, when Joseph Stalin (1879–1953), the Soviet dictator from 1924 until his death, was denounced for his excesses in office.

cult of the offensive The military doctrine which holds that attack rather than defense is almost always the superior way of fighting. The classic example of the cult (or the ideology) of the offensive in influencing strategy is found in the French army after its defeat in the Franco-Prussian war. It lead to a massive death toll in the first World War when French generals, totally imbued with the idea that French soldiers were only useful when attacking, continually launched massive offensives even though trenches, wire obstacles, and machine-guns made the war pre-eminently a defensive one. Currently, the army of the Soviet Union is the primary devotee in the cult of the offensive. Most Warsaw Pact training and planning is made on the assumption that if NATO attacks, the most effective way to respond is to go immediately on the offensive and seize as much West European ground as possible. Ideas of the superiority of the offensive can stem either from theories about how soldiers can best be motivated and controlled, or from strategic theories about the most efficient use of forces in specific contexts. See Jack Snyder, "Civil-Military Relations and the Cult of the Offensive, 1914 and 1984," *International Security,* 9, 1 (Summer 1984); Stephen Van Evera, "The Cult of the Offensive and the Origins of the First World War," *International Security,* 9, 1 (Summer 1984).

culture Man-made features in an area of terrain. Included are such items as roads, buildings, and canals; boundary lines, and, in a broad sense, all names and legends on a map.

curb weight The weight of a ground vehicle, including its fuel, lubricants, coolant, and on-vehicle materiel, but excluding cargo and operating personnel.

currency The up-to-dateness of a map or chart as determined by comparison with the best available information at a given time.

current force The force that exists at present. The current force represents the actual force structure, manning, or both available to meet potential contingencies. It is the basis for OPERATIONS, CONTINGENCY PLANS and ORDERS.

current intelligence 1. INTELLIGENCE of all types of immediate interest, which is usually disseminated without the delays necessary to its complete evaluation or interpretation. 2. Intelligence that reflects the current situation at either the strategic or tactical level.

custer ring A relatively small (in diameter) PERIMETER DEFENSE used under desperate conditions; so called because it is commonly thought that General George Armstrong Custer used this arrangement during his "last stand" at the battle of the Little Big Horn.

customs of the service Military customs, practices, and procedures not prescribed by law or regulation, but which by tradition and practice have become of binding force; in effect, the common law of a military service.

cut-off The deliberate shutting off of a REACTION ENGINE.

cut-off velocity The velocity attained by a MISSILE at the point of CUT-OFF.

cut orders To issue formal orders; so called because orders were once produced on mimeograph machines from cut stencils.

cutout An intermediary or device used to obviate direct contact between members of a clandestine organization.

CV 1. An aircraft carrier. See ATTACK AIRCRAFT CARRIER. 2. A vehicle in Earth orbit that carries the sensors, and space-based interceptors of a BALLISTIC MISSILE DEFENSE.

CVN A nuclear-powered aircraft carrier.

CVS An anti-submarine aircraft carrier.

D

daily intelligence summary (DISUM) A report prepared in message form at a joint force component command headquarters that provides higher, lateral, and subordinate headquarters with a summary of all significant intelligence produced during the previous 24-hour period.

damage assessment 1. The determination of the effect of attacks on targets. 2. A determination of the effect of a compromise of CLASSIFIED INFORMATION on national security. See Andrew Hampton, "Battlefield Damage Assessment and Repair," *Army Logistician* 17, 6 (November-December 1985).

damage, collateral See COLLATERAL DAMAGE.

damage, effective See EFFECTIVE DAMAGE.

damage, fractional See FRACTIONAL DAMAGE.

damage, materiel See MATERIEL DAMAGE.

damage assessment, military See MILITARY DAMAGE ASSESSMENT.

damage assessment, nuclear See NUCLEAR DAMAGE ASSESSMENT.

damage assessment, tactical See TACTICAL DAMAGE ASSESSMENT.

damage assessment, technical See TECHNICAL DAMAGE ASSESSMENT.

damage control In naval usage, the measures necessary aboard a ship after it has sustained damage from enemy action or by accident, to preserve and reestablish watertight integrity, stability, maneuverability, and offensive power; to control list and trim; to effect rapid repairs of materiel; to limit the spread of and provide adequate protection from fire; and other steps.

damage control, area See AREA DAMAGE CONTROL.

damage estimation The analysis of data to estimate the damage that a specific WEAPON will cause to a TARGET.

damage estimation, poststrike See POSTSTRIKE DAMAGE ESTIMATION.

damage limitation A strategic concept concerning measures taken to reduce the amount of damage one's own country would suffer in nuclear attack. Although it could refer to damage of any type, the term usually suggests the limitation of COUNTERVALUE damage, i.e., to prevent strikes against industrial and population centers. There are a variety of possible TARGETING strategies that might accomplish this, but they have one common theme: not to attack the enemy's cities, but to hold in reserve a secure SECOND-STRIKE CAPACITY that could do this. In this way every incentive is given to the enemy to avoid attracting the sort of destruction one is trying to limit on one's own territory.

damage radius In naval mine warfare, the average distance from a ship within which a mine containing a given weight and type of explosive must detonate if it is to inflict a specified amount of damage.

damage threat The probability that a ship passing once through a MINEFIELD will explode one or more mines and sustain a specified amount of damage.

dan buoy A temporary marker buoy used during MINESWEEPING operations to indicate boundaries of swept paths, swept areas, known hazards, and other locations or reference points.

DANC solution A solution consisting of a chemical (RH-195) and a solvent (acetylene tetrachloride) used to decontaminate objects or areas contaminated with a BLISTER AGENT.

danger close In ARTILLERY AND NAVAL GUNFIRE SUPPORT, information in a CALL FOR FIRE to indicate that friendly forces are within 600 meters of the target.

danger space That space between a weapon and its target at which the trajectory of a projectile does not rise 1.8 meters above the ground (the average height of a standing human). This includes the area encompassed by the BEATEN ZONE.

dart A training target towed by a jet aircraft and fired at by fighter aircraft.

data, basic See BASIC DATA.

data, firing See FIRING DATA.

date, effective See EFFECTIVE DATE.

date break The date on which a change in CRYPTOGRAPHIC procedure, KEYS, CODE, and other items takes place.

date of rank The date on which an OFFICER was promoted to present rank (see COMMISSION); this is often important in determining seniority within grades.

day fighter A fighter aircraft designed for AIR INTERCEPTION puposes, primarily in visual meteorological conditions. It may or may not carry electronic devices to assist in interception and in aiming its weapons.

day of supply, combat See COMBAT DAY OF SUPPLY.

day room A recreation room (or even an entire building) designated for enlisted personnel.

dazzling In the context of the STRATEGIC DEFENSE INITIATIVE (SDI), the temporary blinding of a sensor by overloading it with an intense signal of electromagnetic radiation; for example, from a laser or a nuclear explosion.

DB 1. DISCIPLINARY BARRACKS. 2. Daily bulletin; a newsletter containing information on upcoming events, the name of the officer of the day, etc.

D-day 1. The unnamed day on which a particular OPERATION commences. Such a designation could also mark a commencement of hostilities. If more than one such event is mentioned in a single PLAN, the secondary events will be keyed to the primary event by adding or subtracting days as necessary. 2. The time reference system used in military plans, involving a letter that shows the unit of time employed and figures, with a minus or plus sign, to indicate the amount of time before or after the referenced event; e.g., "D" is for a particular day, "H" for an hour. Similarly, D+7 means 7 days after D-day, H+2 means 2 hours after H-hour. If the figure becomes unduly large, for example, D-day plus 90, the designation of D+3 months can be used. 3. June 6, 1944, the day the Allies landed on the beaches of Normandy, France, and began the campaign that ended World War II in Western Europe. See Cornelius Ryan, *The Longest Day* (New York: 1959); Max Hastings, *Overlord: D-Day and the Battle for Normandy, 1944* (New York: Simon & Schuster, 1984).

deactivate The act of rendering an explosive device harmless or inert. Compare to: INACTIVATE.

deadman A log, rail, or similar object that is buried in the ground and used for anchoring a line.

dead reckoning Finding one's position by means of a compass and calculations based on speed, time elapsed, and direction from a known position. Dead reckoning is used for desert travel, coastwise shipping, and air navigation.

dead space 1. An area within the MAXIMUM RANGE of a weapon, radar, or observer that cannot be covered by fire or observation from a particular position because of intervening obstacles, the nature of the ground, or the limitations of the pointing capabilities of a weapon. 2. An AREA or ZONE that is within range of a radio transmitter, but in which a SIGNAL is not received. 3. The volume of space above and around a GUN or GUIDED MISSILE system into which it cannot fire because of mechanical or electronic limitations.

dead time The interval required from the time of observation of a target to the instant at which a gun may be fired.

death gratuity A sum paid to beneficiaries of military personnel who died while in the service.

debarkation The unloading of troops, equipment, or supplies from a ship or aircraft.

debarkation net A specially prepared type of cargo net employed for the DEBARKATION of troops over the side of a ship.

debarkation schedule A schedule that provides for the timely and orderly DEBARKATION of troops, equipment, and emergency supplies for a waterborne ship-to-shore movement.

debriefing 1. The process through which any military member or civilian employee of an armed force, who has possession of data of a sensitive nature, is instructed in the safeguarding of such information prior to passing from the control of the agency in whose service the information was acquired. 2. The procedure of extracting from a given serviceman or employee facts, comments, or recommendations concerning his or her previous assignments or experience.

decay curves Graph lines representing the decrease of RADIOACTIVITY over the passage of time.

decay rate The rate of disintegration of radioactive material within a given period of time.

deception Those measures designed to mislead an enemy by the manipulation, distortion, or falsification of evidence to induce him to react in a manner detrimental to his interests. Deception is inherent in warfare. SUN TZU wrote that all warfare was based on it. NICCOLO MACHIAVELLI found that "Though fraud in other activities be detestable, in the management of war it is laudable and glorious, and he who overcomes an enemy by fraud is as much to be praised as he who does so by force." See also

COUNTERDECEPTION; MILITARY DECEPTION; STRATEGIC DECEPTION. See Charles Greig Cruickshank, *Deception in World War II* (New York: Oxford University Press, 1979); William H. Baugh, "Deceptive Basing Modes for Strategic Missiles: An Exercise in the Politics of an Ambiguous Nuclear Balance," *Western Political Quarterly* 33, 2 (June 1980).

deception, military See MILITARY DECEPTION.

deception operation A military operation conducted to mislead the enemy. A unit conducting a deception operation may or may not make contact with the enemy. Includes DEMONSTRATIONS, DISPLAYS, FEINTS, and RUSES.

deceptive basing Concealing the location of critical equipment, such as radar or missiles, by constantly moving them from one shelter to another to make it more difficult for the enemy to pinpoint and destroy them.

decimation 1. Literally, the loss of one tenth of a military force. In ancient times it was not uncommon for every tenth man of a mutinous or captive force to be killed as punishment. 2. Any heavy CASUALTIES.

decisive engagement 1. Any all-out battle that effects the course of a war in a major way. During World War II in the Pacific, the 1942 Battle of Midway was decisive because thereafter the Japanese were continuously on the defensive. In Europe, the 1944 Battle of the Bulge was decisive because thereafter the Germans were unable to mount an effective offensive in Western Europe. 2. A COMBAT engagement in which a unit is considered to be fully committed and is not free to MANEUVER or extricate itself. In the absence of outside assistance, the action must be fought to a conclusion with the forces at hand.

declaration 1. A document stating a COURSE OF ACTION and usually the reasons for its adoption, to which the signatories (either individuals or states) bind themselves. Perhaps the most famous of all political declarations is the American Declaration of Independence of 1776. 2. A DECLARATION OF WAR (or neutrality) which expresses to the world a nation's military intentions. 3. Statements of intentions by a political leader. The leader can either be in power and speak for the government or out of power and speak for an opposition party.

declaration of war 1. The legal obligation (under INTERNATIONAL LAW) on the part of a state to formally notify another sovereign state that a "state of war" exists between them if the first state intends to commence hostilities. The last time the United States declared war was during World War II when President Franklin Roosevelt called December 7, 1941 "a date which will live in infamy" and asked "that the Congress declare that since the unprovoked and dastardly attack on Sunday, December 7th, 1941, a state of war has existed between the United States and the Japanese empire." The United States was then at war with both Japan and Germany because Germany, as an ally of Japan, promptly declared war on the United States. Formal declarations of war seem to be rapidly becoming quaint relics of diplomatic history. The phrase "undeclared war" was used by American policymakers in the 1930s to define the illegality of the Japanese action against China. Yet President Truman's actions in KOREA proved that when a war was not a war, it could also be a "police action." President Eisenhower's request for a congressional resolution empowering action in defense of Formosa set the stage for President Lyndon Johnson's GULF OF TONKIN RESOLUTION, that is, the prior approval by Congress for a war action. There has been a real transformation of the concept of war power in U.S. law, culminating in the WAR POWERS RESOLUTION. 2. The constitutionally authorized power of the U.S. Congress (under Article I, Section 8) "to declare war."

declaratory policy 1. The announced foreign policy positions of a nation. 2. The publicly-announced doctrine about when and how a country will use its nuclear forces. Declaratory policy does not give a reliable account of what might be expected to happen in the event of nuclear war. Rather, it highlights the political aims that nuclear weapons will be used to protect or achieve, and presents a justification for their use.

declassification The determination that in the interests of national security, CLASSIFIED INFORMATION no longer requires any degree of protection against unauthorized disclosure, and the subsequent removal or cancellation of the "classified" designation.

declination 1. The angle formed between a magnetic compass needle and the geographical meridian, or true north. 2. The angular distance to a body on the CELESTIAL SPHERE measured north or south through 90 degrees from the celestial equator along the great circle of the body.

declination constant A constant correction applied to the readings of a compass instrument which represents the clockwise angle between true or grid north and magnetic north as indicated by that instrument.

declination protractor A map-orienting device, consisting of a horizontal degree scale and a pivot point, by which a magnetic-north line can be drawn across the face of a map.

declinator A magnetic instrument used to orient or check the orientation of another instrument or machine. A declinator measures declination or pointing error.

DeConcini reservation A treaty reservation sponsored by U. S. Senator Dennis DeConcini of Arizona, attached to the Panama Canal Neutrality Treaty that was ratified by the U. S. Senate on March 16, 1978. The DeConcini reservation specifies that the United States as well as the Republic of Panama has the independent right to take military steps if necessary to prevent the canal from being closed or otherwise interfered with. This means that if a future government of Panama or any other power tries to interfere with the normal operations of the Panama Canal, the United States has the right, according to the treaty, to militarily intervene.

decontamination The process of making any person, object, or area safe by absorbing, destroying, neutralizing, making harmless, or removing, CHEMICAL or BIOLOGICAL AGENTS, or by removing radioactive material clinging to or around it.

decoppering agent A metallic material (tin) placed in a PROPELLING CHARGE to remove the copper coat left by the ROTATION BANDS of a PROJECTILE on the surface of a weapon BORE.

decoration A formal recognition of valor or distinguished service that takes the form of a distinctively designed mark of honor, usually a ribbon or medal. When he created the Legion of Honor in 1802, NAPOLEON said: "Show me a republic, ancient or modern, in which there have been no decorations. Some people call them baubles. Well, it is by such baubles that one leads men." The highest level United States decoration is the MEDAL OF HONOR. The parallel British award is the Victoria Cross. The French have the Legion of Honor as their highest award.

decoy 1. An imitation in any sense of a person, object, or phenomenon that is intended to deceive enemy surveillance devices or otherwise mislead the enemy. See also CHAFF. Decoys are often used as FORCE MULTIPLIERS in the field. For example, NATO now deploys fake canvas tanks that include generators to create false heat SIGNATURES to mislead enemy missiles. Every $3,000 decoy that is hit by an enemy missile theoretically saves a $3 million tank. 2. A PENETRATION AID carried as part of the PAYLOAD of a BALLISTIC MISSILE, constructed to look and behave like a nuclear-weapons-carrying WARHEAD, but which is less costly, much less massive, and can be deployed in large numbers to complicate defenses. The simplest type of decoy is a metallic-coated balloon packed inside the front end of a ballistic missile, along with the RE-ENTRY VEHICLES (RVs) or warheads. Decoys are ejected from the missile at the same time as the RVs. Inflated by compressed air, they would give the same radar signature as a real warhead, and travel at the same speed because of zero gravity, presenting a defensive system with the problem of distinguishing between real and false targets. 3. Any object, person, or force used to lure an enemy force into a trap or ambush.

decoy, reaction See REACTION DECOY.

decoy ship A ship camouflaged as a noncombatant ship with its armament and other fighting equipment hidden and with special provisions for unmasking such weapons quickly. Also called a Q-ship.

decryption The conversion of encrypted (coded) text into its plain English equivalent.

dedicated battery A DIRECT SUPPORT ARTILLERY mission to provide FIRE SUPPORT for a specific MANEUVER unit—normally a COMPANY or COMPANY TEAM. Dedication is normally used only during a MOVEMENT-TO-CONTACT condition when the enemy situation is vague.

deep attack Any attack significantly behind enemy lines, as opposed to one near the FRONT dividing two hostile forces.

deep fording capability The ability of a self-propelled gun or ground vehicle equipped with built-in waterproofing or a special waterproofing kit, to negotiate a water obstacle with its wheels or tracks in contact with the ground. See also FLOTATION; SHALLOW FORDING CAPABILITY.

deep minefield An antisubmarine MINEFIELD that is safe for surface ships to cross.

deep six To bury or discard something at sea.

deep space Altitudes greater than 3,000 nautical miles (about 5,600 kilometers) above the Earth's surface.

deep strike The United States doctrine for fighting a war on the North Atlantic Treaty Organization's (NATO's) central front, best thought of as the air-force equivalent of the U.S. Army's doctrine of AIRLAND BATTLE CONCEPT. As with all current theories on countering the WARSAW PACT's conventional numerical superiority, it prescribes the INTERDICTION of Warsaw Pact armies as far from the battlefield as possible. In particular, the deep strike concept calls for rapid action to make Warsaw Pact airfields unusable, and for establishing air superiority in order to make the land battle winnable. See also FOLLOW ON FORCES ATTACK. See Bernard W. Rogers, " 'Strike Deep': A New Concept for NATO," *Military Technology* VIII, 5 (1983); Trevor N. Dupuy, "Why Strike Deep Won't Work," *Armed Forces Journal* (January 1983); Daniel Goure' and Jeffrey R. Cooper, Conventional Deep Strike: A Critical Look," *Comparative Strategy* 4, 3, (1984).

deep supporting fire FIRE directed on objectives not in the immediate vicinity of friendly forces, for neutralizing and destroying enemy reserves and weapons, and interfering with enemy command, supply, communications and observations. See also CLOSE SUPPORTING FIRE; DIRECT SUPPORTING FIRE; SUPPORTING FIRE.

default unit A unit that cannot meet the assigned REQUIRED DELIVERY DATE in support of a specific operation plan.

DEFCON See DEFENSE READINESS CONDITIONS.

defeat 1. The failure of a military unit to achieve its objective. 2. The failure of a nation to successfully obtain its war aims. 3. The capitulation of a military force or a nation.

defeat in detail To subdue the remaining portions of the MAIN BODY of a force after it has been divided and before it can be reinforced.

defector One who deserts a cause, a party, or a nation in order to join a rival; a person who repudiates his or her country, when beyond its jurisdiction or control.

defend The conducting of a DEFENSE; a MISSION that requires a unit to destroy or contain an attacking force, or force it to withdraw. The mission may also require a unit to retain a specific LINE, locale, SECTOR, or terrain feature (such as a hill).

defend in sector A mission, normally nonrestrictive in nature, that requires a defending unit to destroy, contain, or force the withdrawal of an attacker anywhere forward of the defending unit's rear boundary. The mission may become restrictive if directed to retain a specific line, locale, zone, or terrain feature.

defense 1. The employment of all means and methods available to prevent resist, or destroy an enemy ATTACK. The purpose of defensive operations may be to cause an enemy attack to fail, to protect forces, facilities, and installations, to control terrain or gain time, or to concentrate forces elsewhere. 2. Offense; a traditional military maxim holds that "the best defense is a good offense." CLAUSEWITZ wrote in *On War* (1832): "A swift and vigorous transition to attack—the flashing sword of vengeance is the most brilliant point of the defensive." 3. The United States Department of Defense. 4. The British Ministry of Defence. 5. A nation's total military, industrial, and political capacity to resist regression.

defense, active See ACTIVE DEFENSE.

defense, air See AIR DEFENSE.

defense, angle of See ANGLE OF DEFENSE.

defense, area See AREA DEFENSE.

defense, base See BASE DEFENSE.

defense, chemical See CHEMICAL DEFENSE.

defense, civil See CIVIL DEFENSE.

defense, coalition See COALITION DEFENSE.

defense, coast See COAST DEFENSE.

defense, composite See COMPOSITE DEFENSE.

defense, coordinated See COORDINATED DEFENSE.

defense, deliberate See DELIBERATE DEFENSE.

Defense, Department of (DOD) The federal agency created by the NATIONAL SECURITY ACT amendments of 1949, which is responsible for providing the military forces needed to deter war and protect American security. The major elements of these forces are the U.S. Army, Navy, Marine Corps, and Air Force, consisting of over two million men and women on active duty. Of these, almost half, including about 50,000 on ships at sea, serve outside the United States. They are backed, in case of emergency, by the 2.5-million members of the RESERVE components. In addition, there are over 1 million civilian employees in the Defense Department.

defense, dust See DUST DEFENSE.

defense, exoatmospheric See EXOATMOSPHERIC DEFENSE.

defense, extended See EXTENDED DEFENSE.

defense area For any particular COMMAND, the area extending from the FORWARD EDGE OF THE BATTLE AREA to its rear boundary. It is here that the decisive defensive battle is fought.

defense area, forward See FORWARD DEFENSE AREA.

defense area, main See MAIN DEFENSE AREA.

defense-assistance operation, internal See INTERNAL DEFENSE ASSISTANCE OPERATION.

defense, hasty See HASTY DEFENSE.

defense, industrial See INDUSTRIAL DEFENSE.

defense, integrated See INTEGRATED DEFENSE.

defense, passive See PASSIVE DEFENSE.

defense, perimeter See PERIMETER DEFENSE.

defense, point See POINT DEFENSE.

defense, position See POSITION DEFENSE.

defense, preferential See PREFERENTIAL DEFENSE.

defense, satellite See SATELLITE DEFENSE.

defense, social See SOCIAL DEFENSE.

defense, strategic See STRATEGIC DEFENSE.

defense fire FIRE delivered by supporting units to assist and protect a unit engaged in a defensive action. It may include fire delivered before the enemy forms for the attack (which may include HARASSING and INTERDICTION FIRE), fire delivered after he forms for the attack (COUNTERPREPARATION FIRE), and fire planned during the attack (PROTECTIVE FIRE).

defense, forward See FORWARD DEFENSE.

defense gun, first See FIRST DEFENSE GUN.

defense in depth The siting of mutually supporting defensive positions designed to absorb and progressively weaken an attack, to prevent initial observation by the enemy of the full defensive situation, and to allow the commander to maneuver his reserve. It is usually contrasted with a LINEAR DEFENSE, which stresses the complete prevention of enemy incursion. Defense in depth treats the FRONT LINE that separates one's own forces from the enemy as merely the forward border of an entire area to be defended. Several lines of prepared positions are set up, with tactical strongholds positioned throughout, and with one's forces distributed over the whole of the area. Although the enemy may find it easier to penetrate through the first line of such a defense, he will not then be free to move at will behind that line. Any attempt to exploit a BREAKTHROUGH will continually encounter attacks from well-prepared positions on the enemy force's flanks.

defense in place A system of defense based on firm resistance without RETREAT, as opposed to delaying actions in successive positions. During World War II, when a German invasion of England seemed a real possibility, Winston Churchill explained to the British people in a July 1940 radio broadcast his policy of defense in place: "We shall defend every village, every town and every city. The vast mass of London itself, fought street by street, could easily devour an entire hostile army; and we would rather see London laid in ruins and ashes than that it should be tamely and abjectly enslaved." Had the Germans actually invaded, Churchill was prepared to offer his people the motto: "You can always take one with you."

Defense Intelligence Agency (DIA) A unit of the United States Department of Defense created in 1961 to provide a single military voice on intelligence matters. The DIA is responsible for producing and disseminating foreign military intelligence for the Secretary of Defense and the Joint Chiefs of Staff, for coordinating the intelligence activities of the military services, and for managing the military attaches assigned to United States embassies around the world.

defense, internal See INTERNAL DEFENSE.

defense, layered See LAYERED DEFENSE.

defense, linear See LINEAR DEFENSE.

defense, military See MILITARY DEFENSE.

defense, mine See MINE DEFENSE.

defense, mobile See MOBILE DEFENSE.

defense plan A coordinated plan for preventing or defeating an enemy attack. A defense plan includes plans for elements of the defense, such as tactical organization, fire, AIR DEFENSE, SECURITY, AIR SUPPORT, GROUND ORGANIZATION, COUNTERATTACK, COMMUNICATIONS, and SUPPLIES.

defense position, forward See FORWARD DEFENSE POSITION.

defense readiness condition (DEFCON) A number or code word indicating the READINESS posture of a unit for actual operations or exercises. Also called state of readiness. The JOINT CHIEFS OF STAFF uses a system of progressive alert postures to match situations of varying military severity: DEFCON 5 is normal readiness; DEFCON 1 is maximum readiness (war may be imminent).

Defense Security Assistance Agency The Pentagon unit that administers arms sales to foreign governments.

defense system, low-altitude See LOW-ALTITUDE DEFENSE SYSTEM.

defensive, strategic See STRATEGIC DEFENSIVE.

defensive fire, close See CLOSE DEFENSIVE FIRE.

defensive minefield 1. In naval mine warfare, a MINEFIELD laid in international waters or international straits with the declared intention of controlling shipping in defense of sea communications. 2. In LAND MINE WARFARE a minefield laid in accordance with an established plan to prevent enemy penetration between positions and to strengthen the defense of the positions themselves.

defensive position An area occupied by troops organized in a system of mutually supporting DEFENSE AREAS or fortified TACTICAL LOCALITIES. Also called battle position.

defensive satellite weapon (DSAT) Any device intended to defend SATELLITES in space by destroying attacking ANTI-SATELLITE WEAPONS (ASAT's).

defensive zone A belt of terrain, generally parallel to the FRONT, that includes two or more organized, or partially organized, BATTLE POSITIONS.

deferred maintenance Maintenance intended to eliminate an existing fault in a device, but which can be safely postponed because the fault does not prevent continued successful operation.

deferred unit A unit whose REQUIRED DELIVERY DATE in support of a specific operation plan has been postponed.

defilade 1. Protection from hostile observation and fire provided by an OBSTACLE such as a hill, ridge, or bank. 2. The height of such an obstacle or vertical distance that conceals a position from enemy observation. 3. To shield from enemy fire or observation by using natural or artificial obstacles. Also see SMOKE-AND-FLASH DEFILADE.

defilade, dismounted See DISMOUNTED DEFILADE.

defilade, flash See FLASH DEFILADE.

defilade, hull See HULL DEFILADE.

defile A narrow passage that tends to constrict the movement of troops.

deflection 1. The distance by which a projectile is off target due to wind, variations in ammunition, etc. 2. The setting on a sight that compensates for deflecting factors.

deflection, corrected See CORRECTED DEFLECTION.

deflection, zero See ZERO DEFLECTION.

deflection change A change in AZIMUTH setting applying to all guns in a battery when the target moves, or when a shift is made from one target to another. Deflection change does not include the DEFLECTION DIFFERENCE that allows for the difference in positions of the various guns firing at the same target.

deflection correction A correction that must be applied to the AZIMUTH or shift measured on a FIRING CHART so that the line of fire will pass through the target.

deflection difference The amount (the angle) that a PIECE in a BATTERY is TRAVERSED toward or away from a given piece (usually the base piece) to vary the width of the SHEAF.

deflection error The distance to the right or left of a target of the burst of a PROJECTILE or the MEAN POINT OF IMPACT of a SALVO.

deflection scale A scale on a sight, marked in mils, for applying correction in deflection or for laying the PIECE in direction.

deflection shift A change in the DEFLECTION setting.

defoliant operation The employment of DEFOLIATING AGENTS on vegetated areas in support of military operations.

defoliating agent A chemical that causes trees, shrubs, and other plants to shed their leaves prematurely.

degaussing The process whereby a ship's magnetic field is reduced by the use of electromagnetic coils, permanent magnets, or other means in order to protect the ship from MAGNETIC MINES.

degrade 1. To reduce someone in grade, rank, or status. Thus a sergeant can be degraded to private. 2. To impair in a physical regard. A military unit can be degraded by transferring half its troops to another unit; and an enemy can be degraded (weakened) by a bombardment before an attack.

degree of risk As specified by a military commander, the risk to which friendly forces may be subjected from the effects of the detonation of a NUCLEAR WEAPON used in attacking a close-in enemy target; acceptable degrees of risk under differing tactical conditions are *emergency, moderate,* and *negligible.*

delay A mission that requires a military force to trade space for time without losing freedom to MANEUVER or risking PENETRATION or being BYPASSED. The delaying force may ATTACK, DEFEND, AMBUSH, RAID, or use any other tactic necessary to accomplish the mission.

delay action 1. The predetermined, delayed explosion of an ammunition item after activation of the FUZE. 2. A DELAYING OPERATION.

delayed-action bomb A BOMB equipped with FUZE that is adjusted to explode the CHARGE at a set time after contact with a target.

delayed-action mine An explosive charge designed to go off some time after planting, and often left behind by a retreating enemy to harass or destroy pursuing forces.

delayed-contact fire A firing system arranged to explode a MINE at a set time after it has been touched or disturbed.

delayed opening 1. Accidental temporary failure of a parachute to function. 2. The deliberate delay of a parachute functioning until personnel or equipment, are close to the ground.

delaying action A DELAY OPERATION.

delay fuze A fuze that has a delay element incorporated in it permitting a missile to penetrate its target by a distance corresponding to the delay. Such fuzes are used to permit PENETRATION of the target before detonation.

delaying arming mechanism A device designed to prevent a mine from ARMING until a preset time interval after planting.

delaying operation See DELAY OPERATION.

delay on alternate positions The conduct of a DELAYING OPERATION by deploying a FORCE, one ELEMENT behind the other, and, after holding for as long as possible or for a specified time, passing the forward element through the rearward element to another position.

delay on successive positions The conduct of a DELAYING OPERATION by fighting rearward from one position to another, holding each as long as possible or for a specified time.

delay operation An operation in which space is traded for time and maximum punishment is inflicted on the enemy without accepting DECISIVE ENGAGEMENT, PENETRATION, or being BYPASSED. Delay operations are normally conducted to deny the enemy access to a specified area for a specified time.

delay position A position taken to slow the advance of the enemy without becoming decisively engaged.

delegation of authority The action by which a commander assigns part of his authority to a subordinate commander. However, ultimate responsibility cannot be relinquished.

deliberate attack An attack planned and coordinated with all concerned elements on the basis of reconnaissance, evaluation of all available intelligence, and relative combat strength, analysis of various COURSES OF ACTION, and other factors. It is generally conducted against a well-organized defense when a HASTY ATTACK is not possible or has been conducted and failed. Also called a COORDINATED ATTACK.

deliberate breaching The creation of a LANE through a MINEFIELD or a clear route through a BARRIER or FORTIFICATION, which is systematically planned and carried out.

deliberate crossing A crossing of a river or stream that requires extensive planning and detailed preparations. See also HASTY CROSSING.

deliberate defense A defense normally organized when out of contact with the enemy or when contact with the enemy is not imminent and the time exists for organization. It normally includes an extensive fortified zone incorporating PILLBOXES, FORTS, and COMMUNICATIONS systems. See also HASTY DEFENSE.

deliberate field fortification A TRENCH, GUN, EMPLACEMENT, or OBSTACLE constructed before contact with the enemy. A deliberate field fortification is generally more elaborate than a hasty field fortification, which is constructed under fire or threat of attack.

deliberate fire Fire delivered at a rate that is intentionally less than normal to permit adjustment corrections, meet specific tactical requirements or conserve ammunition.

delivery error The inaccuracy associated with a given WEAPON SYSTEM, resulting in a dispersion of shots about the AIMING POINT. See also CIRCULAR ERROR PROBABLE; DISPERSION; DISPERSION ERROR; HORIZONTAL ERROR.

delivery vehicles A catch-all term to signify any system capable of delivering nuclear WARHEADS on an enemy. They can be INTERCONTINENTAL BALLISTIC MISSILES (ICBMs), theater or INTERMEDIATE-RANGE, BOMBERS, CRUISE MISSILES or whatever else is capable of doing the job. This concept is necessary in many ARMS-CONTROL negotiations, where an agreement may not only determine the quantity of warheads, but the various means of delivering them on a target.

delta 1. The U.S. Army's special force unit trained in counterterrorism methods. See Charlie A. Beckwith and Donald Knox, *Delta Force: The Army's Elite Counterterrorist Unit* (New York: Dell, 1985). 2. The NATO designation for a class of Soviet ballistic missile submarines. 3. The Mekong River Delta in Vietnam.

demilitarization 1. The act of destroying the military applications inherent in certain types of equipment and material. The term comprehends mutilation, dumping at sea, scrapping, burning, or alteration designed to prevent the further use of such equipment and material for its originally intended military or lethal purpose. 2. To eliminate the military character of something. For example, Japan was demilitarized after World War II when it was occupied by the United States and forced to accept a new constitution that forbade the maintenance of traditional military forces. 3. To supplant military rule with civilian rule.

demilitarized zone (DMZ) A defined area in which the stationing or concentration of military forces, or the retention or establishment of any military installation, is prohibited. For example, there is a demilitarized zone between North and South Korea.

demolition, preliminary See PRELIMINARY DEMOLITION.

demolition belt A selected land area sown with explosive charges, mines and other available obstacles to deny use of the land to enemy operations, and as a protection to friendly troops.

demolition block An EXPLOSIVE CHARGE, usually in a nonmetallic container used for demolition purposes.

demolition guard A local force positioned to ensure that a target is not captured by an enemy before its demolition has been successfully accomplished.

demolition munition, atomic See ATOMIC DEMOLITION MUNITION.

demolition target, reserved See RESERVED DEMOLITION TARGET.

demolition team, underwater See UNDERWATER DEMOLITION TEAM.

demonstration A show of force on a front where no engagement is sought, made with the aim of deceiving the enemy. It is similar to a FEINT with the exception that no contact is attempted. See also DIVERSION.

deniability The prearranged insulation of a political executive from a decision that he or she actually made, but is later able to plausibly deny, because there is no paper or other trail that would lead to the top. Arrangements for deniability are important parts of covert actions and diplomacy.

denial measure An action to hinder or deny the enemy the use of space, personnel, or facilities. It may include destruction of equipment, removal of supplies, CONTAMINATION, or the erection of OBSTACLES.

denial objects/areas Areas, facilities, or installations that military units are ordered to protect from enemy seizure or control.

denial operation Any operation designed to prevent or hinder enemy occupation of, or benefit from, areas or objects having tactical or STRATEGIC value.

denial target A facility, area, or installation to be destroyed or otherwise denied to the enemy. Examples of denial targets include power plants, railroad facilities, and storage tanks.

departure The east-west component, usually measured in nautical miles, of the RHUMB LINE distance between two points on the earth.

deployed nuclear weapons 1. When used in connection with the transfer of weapons between the U. S. Department of Energy and the Department of Defense, this term describes those weapons transferred to and in the custody of the Department of Defense. 2. Those nuclear weapons specifically authorized by the JOINT CHIEFS OF STAFF to be transferred to the custody of storage facilities or carrying, or delivery units of the armed forces.

deployment 1. The positioning of troops and equipment in preparation for a possible battle. 2. The act of extending battalions and smaller units in width, in depth, or in both width and depth to increase readiness for contemplated action. 3. In naval usage, the change from a cruising approach or contact DISPOSITION to a disposition for battle. 4. In a strategic sense, the relocation of forces to desired AREAS OF OPERATION. 5. The designated location of troops and troop units as indicated in a troop schedule. 6. The series of functions that transpire from the time a packed parachute is placed in operation until it is fully opened and is supporting its load. See Stuart L. Perkins, *Global Demands, Limited Forces: U.S. Army Deployment* (Washington, D.C.: National Defense University Press, 1984); Bruce A. Rothwell, "Supporting Deployed Forces with the Combat Supply System," *Air Force Journal of Logistics* 11, 3 (Summer 1987).

deployment capability The ability to deliver troops to a distant part of the world in fighting condition.

deployment phase, bus See BUS-DEPLOYMENT PHASE.

depot A place, whether a formal building or an open field, where military supplies of all kinds are stored. See James S. Welch and Tim Simmons, "Measuring Depot Performance," *Army Logistician* 11, 2 (March–April 1979).

depth bomb A steel container filled with a HIGH-EXPLOSIVE charge that is used against underwater targets. Depth bombs are set off by HYDROSTATIC FUZES, which operate by water pressure at the depth for which they are preset. These bombs are dropped from naval ships, ship-based helicopters, or aircraft. They can be CONVENTIONAL or NUCLEAR weapons. Also called depth charge.

descending branch That portion of the TRAJECTORY traced while a PROJECTILE is falling.

desertion Unlawful departure from a military unit without any intention of returning. This is always more serious when done during combat operations. An absence is usually not considered desertion until after 30 days. Compare to AWOL. See Edward A. Shils, "A Profile of the Military Deserter," *Armed Forces and Society* 3, 3 (Spring 1977); Gordon W. Tingle, "Desertion: Not Always the End of the Line," *Army* 31, 9 (September 1981).

desired ground zero The ground point on, above, or below which it is desired that a NUCLEAR WEAPON be detonated. The AIMING POINT for the weapon, or the spot where the theoretical center would be if the weapon was perfectly accurate. In practice, the uncertainty resulting from both the CIRCULAR ERROR PROBABLE and BIAS of strategic missiles means that the true ground zero for RE-ENTRY VEHICLES is likely to be some hundreds of meters from the designated ground zero for at least 50 percent of WARHEADS. Such discrepancies between designated and actual ground zero would be of little practical importance unless the intention was to destroy a HARD TARGET.

destroy To render an enemy force or its equipment incapable of performing its assigned mission or function.

destroyed The condition of a target so damaged that it cannot function as intended or be restored to a usable condition. In the case of a building, all vertical supports and spanning members are damaged to such an extent that nothing is salvageable. In the case of bridges, all spans will have dropped and all piers require replacement.

destroy enemy in zone An order to eliminate organized resistance from an assigned zone by isolating enemy forces, preventing their escape, and subsequently killing or capturing most of them. It also includes destroying, damaging, or capturing enough equipment to make the enemy force ineffective as a TACTICAL UNIT.

destroyer A warship smaller than a CRUISER but bigger than a FRIGATE, able to operate independently or as an ESCORT for larger naval ships and convoys.

destroyer-escort A smaller version of the traditional DESTROYER that is designed mainly for CONVOY escort.

destruct 1. The intentional destruction of a missile or similar vehicle for safety or other reasons. 2. To so destroy a missile or similar vehicle.

destruct system A system which, when operated by external command or preset internal means, destroys a missile or similar vehicle.

destruction fire FIRE delivered for the sole purpose of destroying MATERIEL objects (usually POINT TARGETS).

destructor An explosive or other device for intentionally destroying a missile, aircraft, component thereof, or other equipment, either for safety reasons or to prevent COMPROMISE.

detachment 1. A part of a unit separated from its main organization for duty elsewhere. 2. A temporary military or naval unit formed from other units or parts of units.

detachment, advance See ADVANCE DETACHMENT.

detachment, casual See CASUAL DETACHMENT.

detachment, headquarters See HEADQUARTERS DETACHMENT.

detail 1. An assigned DUTY. 2. A small group selected for a specific task, usually temporary.

detained Any CASUALTY other than one that is captured or interned, and who is known to have been taken into enemy custody against his will while apparently alive, and for whom there is no conclusive evidence of death after being taken into custody.

detection 1. The discovery by any means of the presence of a person, object, or phenomenon of potential military significance. 2. In tactical operations, the perception of an object of possible military interest but unconfirmed by recognition. 3. In SURVEILLANCE, the determination that an event has occurred and transmission of this information by a surveillance system. 4. In ARMS CONTROL, the first step in the process of ascertaining the occurrence of a violation of an arms-control agreement.

detector crayon A chalklike crayon that detects the presence of liquid or high vapor concentrations of BLISTER AGENTS (except NITROGEN MUSTARDS) by a color change from pink to blue.

detector paper Paper impregnated with a chemical compound that turns dark green, yellow, or red when in contact with V SERIES agents, G SERIES agents, or MUSTARD, in liquid form. It does not detect vapor.

detente 1. From the French, meaning "the easing of strained relations." In diplomatic usage this refers to the lessening of military and diplomatic tensions between two countries. Compare to: ENTENTE. 2. The ongoing process of Soviet-American relations in the 1970s, which include political SUMMIT conferences, economic agreements leading to increased trade, and agreements coming in STRATEGIC ARMS LIMITATIONS TALKS (SALT). In many ways, proclamations of both the birth and supposed death of detente are exaggerated, or at least highly subjective and partly media-created. When the extreme tension of the COLD WAR period, from 1948 to the early 1960s is used as a benchmark, anything measured against it could be called a detente. In reality, there was no significant reduction in superpower conflict during the 1970s, nor has there been any dangerous clash between the superpowers in the 1980s. The concept of detente is especially valued by those, mainly in Western Europe, who do not share the ideology of either superpower and fear both more or less equally as potential originators of war. The countries for whom detente has probably meant the most in practical terms would be the two Germanies, where the improved communications between the superpowers encouraged East and West German leaders to establish much stronger economic and diplomatic relations with each other than had previously been possible. Compare to: COLD WAR, CONTAINMENT. See Raymond L. Garthoff, *Detente and Confrontation: American-Soviet Relations from Nixon to Reagan* (Washington, D.C.: The Brookings Institution, 1985); Peter Wallensteen, "American-Soviet Detente: What Went Wrong?" *Journal of Peace Research* Vol 22, No. 1 (1985).

deterrence The prevention of an action by creating a fear of its consequences. Deterrence is a state of mind brought about by the existence of a credible threat of unacceptable counteraction. The foundation of American defense policy, the basic argument of deterrence is that so long as any potential enemy believes that the United States is capable of responding to an attack with a devastating COUNTERATTACK, there will be no war. Therefore, a massive defense establishment is essential to maintain the peace. In some fashion, deterrence has been the role of all military forces except those specifically intended for wars of conquest. In its broadest use, deterrence means any strategy, force position, or policy intended to persuade a potential enemy not to attack. See also MUTUAL ASSURED DESTRUCTION. See Steven E. Miller, *Strategy and Nuclear Deterrence,* (Princeton: Princeton University Press, 1984); John J. Mearsheimer, "Prospects for Conventional Deterrence in Europe,"

Bulletin of the Atomic Scientists 41, 7 (August 1985); Robert Powell, "The Theoretical Foundations of Strategic Nuclear Deterrence," *Political Science Quarterly* Vol. 100, No. 1 (Spring 1985).

deterrence, chemical See CHEMICAL DETERRENCE.

deterrence, extended See EXTENDED DETERRENCE.

deterrence, intrawar See INTRAWAR DETERRENCE.

deterrence, minimum See MINIMUM DETERRENCE.

deterrence, proportional See PROPORTIONAL DETERRENCE.

deterrence by denial The traditional role of military defense. Its argument is that one can prevent war by having such strong defenses that a potential enemy knows that an aggressive war cannot be won. Thus the enemy is being "denied" victory, or the territory that might be gained, and is in some general sense being "deterred" from trying.

deterrent triad See TRIAD.

detonating charge Charge used to set off a high explosive charge.

detonating net Network of detonating cord that is interlaced in a mesh design. Detonating nets are used for clearing paths through mine fields by exploding the mines over which the nets are placed and detonated. Also called a primacord net.

detonation, nuclear See NUCLEAR DETONATION.

detonation, sympathetic See SYMPATHETIC DETONATION.

detonator A device containing a sensitive, easily ignited explosive used to make another more powerful substance explode.

detonator kit, concussion See CONSUSSION DETONATOR KIT.

deuce-and-a-half A 2½-ton capacity military truck.

developer, combat See COMBAT DEVELOPER.

DEW See DIRECTED ENERGY WEAPONS.

DEW line See BALLISTIC MISSILE EARLY WARNING SYSTEM.

DI Drill instructor; a noncommissioned officer in charge of a unit's basic training.

diagram, beach See BEACH DIAGRAM.

diamond formation 1. A diamond shaped arrangement made up of four or five parts of a UNIT to permit maneuvering. When five components are used the fifth part of a diamond formation is in the center. The diamond formation is used especially by MECHANIZED units. 2. A SQUAD formation often used when the situation requires readiness for action in any direction.

dictator The ancient Roman Republic's term for the leader to whom extraordinary powers were given in times of crisis. The Roman office was inherently temporary until Julius Caesar and the Caesars who followed him gave the term dictatorship its modern definition as a government in which one person or party controls all political power. In this century the "classic" dictators, Adolph Hitler (1889–1945) of Germany and Benito Mussolini (1883–1945) of Italy, came to power as the leaders of mass movements. Others, such as Joseph Stalin (1879–1953) of the Soviet Union, rose to power by taking over a party that was already in control of a government. One should also differentiate between the dictators mentioned above, whose power was based on their personalities and control of force, and many modern dictators, such as the current leaders of China and the Soviet Union, who tend to be just the "first among equals" within a ruling elite. President Dwight D. Eisenhower once wrote: "In a half century of national service, I have yet to meet the American military officer who viewed himself as a budding Napoleon, or even a Rasputin, and I suggest it is worthy of note than in recent world history the three major dictators, Hitler, Mussolini and Stalin, came from civil life." See William P. Bundy, "Dictatorships and American Foreign Policy," *Foreign Affairs* Vol. 54, No. 1 (October 1975).

dictatorship of the proletariat The Marxist concept used to justify the dominant role of a communist party after the overthrow of capitalism. The party must exercise political and economic control in a "dictatorship of the proletariat" until the masses grasp the true meaning of communist ideology. Once this true socialist understanding is reached, the party's supreme power will be unnecessary; indeed the whole state will "wither away" leaving a peaceful cooperative society. Until then democracy could only hold the people back; in fact selfishness and conflict would be rife unless kept down by forceful central control on the part of those who, having been admitted to the communist party, are known to have a correct understanding of scientific socialism. There are many theoretical problems in accepting this idea.

difference chart A table by which the RANGE and AZIMUTH of a target from a GUN or STATION are obtained when the range and azimuth from some other gun or station are known.

differential effects The effects upon the elements of the TRAGECTORY of a projectile or weapon due to variations from standard conditions.

differential leveling A method by which the difference in ELEVATION between two points is determined, without regard to distance, by direct readings on graduated rods viewed through a leveled surveying instrument.

diktat The German word for a diplomatic or military settlement that is imposed by force. The Germans often used this word to refer to the Treaty of Versailles. The rest of the world then used the word to refer to the essence of German diplomacy just before and during the first part of World War II.

diplomacy 1. A state's foreign policy. While this is the most popular usage of the term, a policy in itself is not diplomacy. Foreign policies, made by governments or heads of state, represent the ends or goals of a state's diplomacy. 2. The formal relationships that independent states maintain with each other. In effect, all of the normal and idiosyncratic intentional communications that states have with each other short of war. Indeed, it is often said that diplomacy has "failed" when war begins. On the other hand, many states throughout history have taken CLAUSEWITZ's attitude that war is only the continuation of diplomacy "by other means." 3. The art of maintaining and conducting international relations and negotiations. 4. According to Ambrose Bierce's *The Devil's Dictionary* (1911) "The patriotic art of lying for one's country." 5. Skillful negotiations in any area. See Harold Nicolson, *Diplomacy*, 3rd edition (New York: Oxford University Press, 1939, 1964); Howard Jones, *The Course of American Diplomacy: From the Revolution to the Present* (New York: Franklin Watts, 1985).

diplomacy, dollar See DOLLAR DIPLOMACY.

diplomacy, gunboat See GUNBOAT.

direct action 1. Politically inspired violence. 2. Obstructing the political process to within the limits of the law; civil disobedience. 3. A French terrorist organization.

direct action fuze See IMPACT ACTION FUZE.

direct action mission In SPECIAL OPERATIONS, a specified act involving OVERT, COVERT, CLANDESTINE, or LOW-VISIBILITY OPERATIONS conducted primarily by a sponsoring power's SPECIAL OPERATIONS forces in hostile or denied areas.

direct aerial fire support Fire delivered from army air platforms as part of army forces conducting combat operations. Such fire is integrated with the fire of the ground commander's weapons, is delivered from within the three dimensional envelope of the battlefield, and is controlled by the ground commander in the same manner as the fire of his ground weapons.

direct air support center 1. A subordinate component of a tactical air control system designed for the control and direction of close air support and other tactical air support operations; normally located with fire support coordination elements. 2. An airborne aircraft equipped with the necessary staff personnel, communications, and operations facilities to function as a direct air support center.

direct command procedure A procedure for the conduct of INDIRECT FIRE according to which the observer sends corrections directly to the weapon.

direct communication 1. A communication authorized to pass directly from the sender to the receiver without going through the hands of any intermediate officers. 2. Radio contact established without aid of an intermediate relay point. Direct communication differs from indirect communication, which goes through a relay point.

directed energy Energy in the form of particle or laser beams than can be sent long distances at nearly the speed of light.

directed energy weapon (DEW) A weapon that kills or blinds its target by delivering energy to it at or near the speed of light; such weapons include LASERS and PARTICLE BEAM WEAPONS. See Donald L. Lamberson, "DOD's Directed Energy Program," *Defense* (June 1983).

directed net A NET in which no STATION other than the net control station can communicate with any other station, except for the transmission of urgent messages, without first obtaining the permission of the net control station.

direct exchange A supply method of issuing serviceable MATERIEL in exchange for unserviceable materiel on an item-for-item basis.

direct fire Gunfire delivered on a target, with the target itself as a visible point of aim for either the gun or the director. See also FIRE.

direct fire sights 1. Sights that permit laying fire directly on a visible target, as distinguished from those used to lay fire on an AIMING POINT. 2. Sights used with AIR DEFENSE guns when the DIRECTOR SIGHT is not available.

direct hit A strike by a PROJECTILE exactly on its target.

directing point A point of known location, normally at the geometric center of the middle gun of an AIR-DEFENSE artillery battery, for which the FIRING DATA are computed. If the directing point is a gun, it is called the base piece or directing gun.

directional gyroscope A gyroscope instrument for indicating direction, containing a free gyroscope that holds its position in AZIMUTH and thus indicates angular deviation from a present heading.

direction board A circular board marked off in 100-mil units with a pointer pivoted at the center. A direction board is used to assist an observer in identifying sounds by their direction.

direction finding See RADIO FIX.

direction of attack A specific direction or route to be taken by a MAIN ATTACK or the main body of a unit that will follow. It is a restricted control measure. Units are required to attack as indicated and are not normally allowed to bypass the enemy. The direction of attack is used primarily in COUNTERATTACKS and night attacks; or to insure that SUPPORTING ATTACKS make a maximal contribution to the main attack.

directive 1. A military communication in which policy is established or a specific action is ordered. 2. A plan issued in order to be in effect when so directed, or in the event that a stated contingency arises. 3. Broadly speaking, any communication that initiates or governs action, conduct, or procedure.

director A computer that takes data from radar as well as other sources to help naval and anti-aircraft guns fire accurately on a target.

directorate That component of a major command or installation headquarters that has primary responsibility for staff coordination. Its responsibilities normally include policy development, the establishment of controls, and review of the effectiveness of operations such as personnel and intelligence.

direct order 1. An order to do something given in person directly from a superior to a subordinate. 2. Any unambiguous order.

direct plotting A method of determining FIRING DATA for a moving target when calculation of its distance and AZIMUTH from a single gun position will meet the needs of a BATTERY of guns. OFFSET PLOTTING is used when data must be given for more than one gun.

direct pointing The pointing of an artillery PIECE according to both range and direction by means of a sight directed at the target.

direct pressure All action other than an encircling movement taken against a retreating enemy in order to achieve his ultimate destruction by denying him opportunity to reorganize his forces or reconstitute his defenses.

direct pressure force A force employed in a PURSUIT that orients on the enemy MAIN BODY to prevent enemy disengagement or defensive reconstitution prior to envelopment by the ENCIRCLING FORCE. It normally conducts a series of HASTY ATTACKS to slow the enemy's retirement by forcing him to stand and fight.

directrix The CENTER LINE of the FIELD OF FIRE of a gun.

direct support The support provided by a unit or formation not attached to or under the command of a supported unit.

direct support ammunition service A procedure wherein AMMUNITION is issued directly to the using organization.

direct support artillery Artillery whose primary task is to provide fire requested by a supported unit.

direct supporting fire FIRE delivered in support of part of a force, as opposed to general supporting fire, which is delivered in support of the force as a whole. See also CLOSE SUPPORTING FIRE; DEEP SUPPORTING FIRE; SUPPORTING FIRE.

dirty nuclear weapon A NUCLER WEAPON that causes damage (in addition to explosion and radiation emission) via radioactive byproducts. It is intended to do more harm to flesh than to inanimate objects. Also called a SALTED WEAPON. Initially, most nuclear weapons were dirty because the technology to make "clean" weapons was lacking; a situation that has been reversed today.

dirty tricks Some of the COVERT OPERATIONS of an INTELLIGENCE agency.

disarm 1. To remove the detonating device or FUZE of a BOMB, MINE, MISSILE, or other piece of explosive ORDNANCE; or to render an explosive device incapable of exploding in its usual manner. 2. To remove weapons from captured troops. 3. To abolish the military forces of a defeated power.

disarmament The recurring efforts of the major powers to put limits on their war-making capabilities. General and complete disarmament has never been seriously pursued, although the United Nations does have it as an official policy goal. Clearly, the idea of complete disarmament runs into tremendous problems of definition as well as being simply utopian in its impracticality. No disarmament movement has ever been known to succeed, giving rise to the common doctrine that nothing can be "uninvented." If scientists discover how to make a weapon, someone, somewhere, will make it. The classic statement on disarmament comes from the *Bible:* "They shall beat their swords into plowshares, and their spears into pruning-hooks; nation shall not lift up sword against nation, neither shall they learn war any more." (Isaiah, 11,4). WINSTON CHURCHILL is usually credited with having told the following fable on disarmament: "When the animals had gathered, the lion looked at the eagle and said gravely, 'We must abolish talons.' The tiger looked at the elephant and said, 'We must abolish tusks.' The elephant looked back at the tiger and said, 'We must abolish claws and jaws.' Thus each animal in turn proposed the abolition of the weapons he did not have, until at last the bear rose up and said in tones of sweet reasonableness: 'Comrades, let us abolish everything—everything but the great universal embrace.' Compare to: ARMS CONTROL. See James Clotfelter, "Disarmament Movements in the United States." *Journal of Peace Research* 23 (June 1986).

discharge 1. To fire a weapon. 2. To be formally released from military service. Most armed services have a variety of discharges that range from honorable to dishonorable, indicating the quality of the dischargee's service.

discharge, bad conduct See BAD CONDUCT DISCHARGE.

discharge, dishonorable See DISHONORABLE DISCHARGE.

discharge, general See GENERAL DISCHARGE.

discharge, honorable See HONORABLE DISCHARGE.

disciplinary barracks A military correctional treatment facility for the confinement, retraining, and restoration of prisoners to honorable duty status or return to civil life.

disciplinary exercise An exercise in a military DRILL that is intended not for physical development but for training in alertness, promptness in carrying out orders, and morale. Disciplinary exercise includes marching in formation, and other drill procedures.

discipline 1. The unified obedience of a military unit, responsiveness to orders and adherence to military regulations. Without discipline, an army is no more than a mob. This has been understood since ancient times. Thucydides wrote in his *History of the Peloponnesian Wars* (404 B.C.) that: "The strength of an army lies in strict discipline and undeviating obedience to its officers." 2. Punishment for violating regulations. Edward Gibbon wrote in his 1776 *The Decline and Fall of the Roman Empire* that: "It was an inflexible maxim of Roman discipline that a good soldier should dread his own officers far more than the enemy."

discipline, light See LIGHT DISCIPLINE.

disclaimer, boundary See BOUNDARY DISCLAIMER.

disengagement 1. The breaking of contact with the enemy by a military unit and its movement to a point at which the enemy can neither observe nor engage it by DIRECT FIRE. 2. In ARMS CONTROL, a general term for proposals that would result in the geographic separation of opposing nonindigenous (non-local) forces without directly affecting indigenous (local) military forces.

dishonorable discharge A formal release from military service without honor and without veterans' benefits. It can only be given a soldier upon his conviction and sentence by a GENERAL COURT-MARTIAL.

disinformation 1. A term used in INTELLIGENCE work to refer to the purposeful lies that a government overtly or covertly releases to the international mass media in order to mislead adversary nations. For example, it was a Soviet disinformation campaign that spread the rumor that the deadly AIDS virus was created in a U. S. military laboratory and is now being spread around the world by U. S. servicemen. The Americans use disinformation, too. After the 1986 U. S. air raid on Libya the Reagan administration purposely planted news stories that the Libyan leader, Colonel Muammer Qaddafi, had top aides that were plotting a coup and that the United States was planning another raid; all in an effort to destabilize the Libyan government. 2. The secret transmitting of false information to rival intelligence agencies. 3. The lies a government tells its own people in order to hide actions that would be considered unacceptable, and possibly would be "checked" by another branch of government if known. Compare to: PROPAGANDA. See Hans Graf Huyn, "Webs of Soviet Disinformation," *Strategic Review* 12, 4 (Fall 1984); Dennis Kux, "Soviet Active Measures and Disinformation: Overview and Assessment," *Parameters: Journal of the U.S. Army War College* Vol. 15, No. 4 (Winter 1985); James E. Oberg, "The Sky's No Limit to Disinformation," *Air Force*, 69, 3 (March 1986).

dismissal The release of an OFFICER or CADET from the service without honor upon sentence of dismissal by a COURT-MARTIAL or MILITARY COMMISSION.

dismissed Order given to a unit to break ranks after a DRILL, ceremony, or FORMATION of any kind.

dismounted Originally a soldier not on horseback; now any soldier not in a vehicle, whether that vehicle be a tank, an armored fighting vehicle, a helicopter, or some other. See William P. Baxter, "Dismounted Soviet Soldier a Vulnerable Adversary," *Army* 31, 8 (August 1981).

dismounted defilade A concealment sufficient to hide a DISMOUNTED man from observation.

dispersal 1. In the context of nuclear war and CIVIL DEFENSE, the moving of people from high-density population areas to low-density population areas. By distributing the population over a larger geographical area, the measure is designed to increase the percentage to survive nuclear attack. This concept can be compared to the diffusion of living cells in a microscopic area, and is one example of the similarity of both humans and germs at war. 2. Deliberately moving troops or vehicles into dispersed positions to present a less compact, and therefore less rewarding, target.

dispersal airfield An airfield, military or civil, to which aircraft might move just before H-HOUR on either temporary duty or a permanent change of station basis, and be able to conduct operations.

dispersal point, convey See CONVOY DISPERSAL POINT.

dispersion 1. A scattered pattern of hits around the MEAN POINT OF IMPACT of bombs and projectiles dropped or fired under identical conditions. 2. In antiaircraft gunnery, the scattering of shots in RANGE and DEFLECTION about the mean point of explosion. 3. The spreading out or separating of TROOPS, MATERIEL, ESTABLISHMENT, or ACTIVITIES that are vulnerably concentrated in limited areas. 4. In chemical or biological operations, the dissemination of agents in liquid or aerosol form. 5. In airdrop operations, the scatter of personnel or cargo on the DROP ZONE. 6. In naval control of shipping, the reberthing of a ship on the periphery of the port area or in the vicinity of the port to minimize its risk of damage from attack.

dispersion error The distance from the POINT OF IMPACT or BURST of a round to the MEAN POINT OF IMPACT or burst.

dispersion pattern The distribution of the points of bursts or impact around a point called the MEAN POINT OF IMPACT after a series of ROUNDS has been fired from one WEAPON or group of weapons under conditions as nearly identical as possible.

dispersion rectangle A table that shows the probable distribution of a succession of shots made with the same FIRING DATA. A dispersion rectangle consists of a diagram made up of 64 zones, in each one of which is shown the percentage of shots that may be expected to fall within it.

displace To leave one position and take another. Military forces may be laterally displaced to concentrate COMBAT POWER in threatened areas.

displaced person A civilian who has been expelled from, or forced to flee his country. See also REFUGEES.

displacement 1. In air intercept, separation between enemy and friendly interceptor flight paths maintained to position an interceptor in such a manner as to provide sufficient maneuvering and ACQUISITION space. 2. The amount of water (by weight or volume) a ship displaces when afloat. This is normally expressed in terms of displacement tons, each of which equals 2240 pounds of, or 35 cubic feet of, seawater. This is the standard way in which the size of ships are gauged. 3. Moving artillery from one position to another.

displacement, echeloned See ECHELONED DISPLACEMENT.

display Any DECEPTION technique designed to mislead an enemy's visual senses, including his observation by radar, camera, infrared device, or the human eye. Compare to DECOY.

display board A blackboard or other board on which FIRING DATA are marked up for the information of a gun crew. ARTILLERY units use an AZIMUTH display board.

disposition 1. The distribution of the elements of a COMMAND within an area, usually the exact location of each unit HEADQUARTERS and the deployment of the forces subordinate to it. 2. A prescribed arrangement of the STATIONS to be occupied by the several formations and single ships of a FLEET, or major subdivisions of a fleet, for any purpose, such as cruising, approach, maintaining contact, or battle. 3. A prescribed arrangement of all the TACTICAL UNITS composing a flight or group of aircraft. See also DEPLOYMENT; DISPERSION. 4. The removal of a patient from a medical treatment facility by reason of return to duty, transfer to another treatment facility, death, or other termination of the patient's medical case.

disruptive pattern An arrangement of suitable, colored, irregular shapes which, when applied to the surface of an object, is intended to enhance its CAMOUFLAGE. For example the technical name for a

British army camouflaged jacket is D.P.M., for "disruptive pattern materiel."

distant early warning line See BALLISTIC MISSILE EARLY WARNING SYSTEM.

distilled mustard A virtually odorless, pale yellow liquid that injures the eyes and lungs and blisters the skin. See also MUSTARD.

distributed fire FIRE so dispersed as to most effectively engage an AREA TARGET.

distribution 1. The arrangement of troops for any purpose, such as a battle, march, or maneuver. 2. A planned pattern of projectiles around a point. 3. A planned spread of fire to cover a desired FRONTAGE or depth. 4. An official delivery of anything, such as orders or supplies. 5. That functional phase of military logistics that embraces the act of dispensing MATERIEL, FACILITIES, and SERVICES. 6. The process of assigning military personnel to ACTIVITIES, UNITS, or BILLETS.

distribution point A point at which supplies, ammunition or both, obtained from supporting supply points by a division or other unit, are broken down for distribution to subordinate units. Distribution points usually carry no stocks; the items drawn from them are issued completely as soon as possible.

DISUM See DAILY INTELLIGENCE SUMMARY.

ditching The controlled landing on water of a damaged aircraft.

dive bombing A type of bombing in which the bomb is released when the bombing aircraft is in a dive at an angle of usually more than 60 degrees from the horizontal.

diversion 1. Drawing the attention and forces of an enemy from the point of a principal operation; an ATTACK or FEINT that diverts attention. See also DEMONSTRATION. 2. A change made in a prescribed route for operational or tactical reasons. A diversion order does not constitute a change of destination. 3. A rerouting of cargo or passengers to a new transshipment point or destination, or onto a different mode of transportation prior to arrival at an ultimate destination. 4. In naval mine warfare, a route or channel bypassing a dangerous area. A diversion may connect one channel to another, or may branch from a channel and rejoin it on the other side of the danger. 5. In air-traffic control, the act of proceeding to an airfield other than one at which a landing was intended. See also ALTERNATE AIRFIELD.

diversionary landing An operation in which troops are landed for the purpose of diverting enemy reaction away from a main landing.

divert 1. An order to an aircraft to "proceed to an alternate field or carrier as previously specified." 2. To change the target, mission, or destination of an airborne flight.

divide and conquer 1. To gain victory by dividing the forces of an enemy so that they can be destroyed in turn. 2. To gain victory because of internal enemy quarrels that divide its forces. This has often happened with allies who have a falling out and are thus more easily conquered by a common enemy.

division 1. A major ADMINISTRATIVE and TACTICAL UNIT in which is combined the necessary arms and services required for sustained combat; it is larger than a REGIMENT or BRIGADE and smaller than a CORPS. A typical division will have three regiments plus supporting units and be commanded by a major (two star) general. Divisions are usually defined as the smallest units in an army that can operate independently. Although a division may consist predominantly of one ARM of the service, such as infantry, most divisions are mixed. Thus the difference between an infantry division and an armored division is not that one has only foot soldiers and the other only tanks, but rather that each has proportionally more of one or the other. Divisions are the "building blocks" of armies, and assessments of one's own forces and the enemy's are often presented in terms of the number of divisions in the ORDER OF BATTLE. Ever since the founding of the NORTH ATLANTIC TREATY ORGANIZATION (NATO) there has been a tendency to assume that the WARSAW PACT's conventional forces are much stronger than NATO's because of their greater number of divisions. But an army's number of divisions reflects organizational doctrine, rather than either personnel levels or firepower. For example, as many as half of all Soviet army divisions are cadre (skeleton) divisions, maintained at only about 30 percent of their full strength. Such divisions could only be used after an extensive call up of reserves, and would be far less well trained and integrated than a combat-ready division. Despite such obvious arithmetic, the number of divisions is still frequently used as a comparative measure of East and West bloc armed might, often with seriously misleading results. 2. A number of naval vessels of similar type grouped together for operational and ADMINISTRATIVE CONTROL. 3. A TACTICAL UNIT of a naval aircraft SQUADRON, consisting of two or more SECTIONS. 4. An air division is an air combat organization normally consisting of two or more WINGS, with appropriate SERVICE UNITS. The combat wings of an air division will normally contain similar type units. The Air Force equivalent to an Army division in terms of operations and organization is a wing. 5. An organizational part of a HEADQUARTERS that handles military matters of a particular nature, such as personnel, intelligence, plans, and training, or supply and evacuation. 6. A number of personnel of a ship's COMPLEMENT grouped together for operational and administrative command.

division artillery Artillery that is permanently an integral part of a DIVISION. For tactical purposes, all artillery placed under the command of a division commander is considered division artillery.

division, light See LIGHT DIVISION.

division, mechanized See MECHANIZED DIVISION.

division columns Groups of military vehicles moving on roads, which comprises a division on the march. The actual groupings of vehicles are usually battalion-sized, about 40 to 60 vehicles per column, with spacing between columns using the same road.

division support command An ORGANIC divisional unit responsible for providing division level supply, transportation, maintanence, medical, and miscellaneous services for all assigned and attached elements of the division.

divisional unit A unit of the type organically assigned to a division, as opposed to the constituent COMBAT UNITS such as BATTALIONS or REGIMENTS.

DMZ See DEMILITARIZED ZONE.

dock landing ship See LANDING SHIP, DOCK.

doctrinal templating An analytical INTELLIGENCE technique used to assist a commander in predicting an enemy COURSE OF ACTION based on a comparison of known enemy tactical doctrine with enemy force composition, FRONTAGE, depth, echelon spacing, and disposition of supporting units.

doctrine 1. A legal principle or rule. 2. A foreign policy such as the MONROE DOCTRINE or the TRUMAN DOCTRINE. See also BREZHNEV DOCTRINE, CARTER DOCTRINE, EISENHOWER DOCTRINE, NIXON DOCTRINE, REAGAN DOCTRINE. 3. The principles by which military forces guide their actions in support of objectives. Morris Janowitz wrote in his *The Professional Soldier* (1960): "Generals and admirals stress the central importance of 'doctrine.' Military doctrine is the 'logic' of their professional behavior. As such, it is a synthesis of scientific knowledge and expertise on the one hand, and of traditions and political assumptions on the other." To equate doctrine with "theory," as some do, is perhaps misleading, because a theory, whether in politics or physics, is a body of principles from which one can derive solutions to new, previously unconsidered problems. A military doctrine, in contrast, tends to be somewhat non-theoretical and to consist of "correct" answers or ready-constructed solutions to anticipated problems. An army's doctrine is the general set of rules intended at least to guide, and possibly to control, combat decisions at all unit levels from the platoon to the entire army corps. Although often set out in lengthy documents, and subjected to endless debate in professional military journals, doctrines tend to be based on a few simple ideas. The doctrine of the French army in 1914, assiduously taught in staff colleges for the preceding 30 years, was that "attack" was the only suitable mode of fighting, because it matched the "elan vital" held to characterize the French military tradition. The current emphasis in U.S. Army conventional war doctrine is on rapidity of action, mobility, and taking the initiative. The doctrine it replaced was usually characterized as stressing ATTRITION—digging in strongly, massing troops, equipment, and munitions, and not committing to combat until overwhelming superiority had been built up. See also COMBINED DOCTRINE; JOINT DOCTRINE; MULTI-SERVICE DOCTRINE. See Jonathan S. Lockwood, *The Soviet View of U.S. Strategic Doctrine: Implications for Decision Making* (New Brunswick, NJ: Transaction Books, 1983); Robert S. Wang, "China's Evolving Strategic Doctrine," *Asian Survey* 24, 10 (October 1984); Jonathan R. Adelman, "The Evolution of Soviet Military Doctrine, 1945–84," *Air University Review* 36, 3 (March–April 1985).

Doctrine, Brezhnev See BREZHNEV DOCTRINE.

Doctrine, Carter See CARTER DOCTRINE.

doctrine, combined See COMBINED DOCTRINE.

Doctrine, Drago See DRAGO DOCTRINE.

Doctrine, Eisenhower See EISENHOWER DOCTRINE.

doctrine, joint See JOINT DOCTRINE.

doctrine, logistic See LOGISTIC DOCTRINE.

doctrine, multi-service See MULTI-SERVICE DOCTRINE.

Doctrine, Nixon See NIXON DOCTRINE.

doctrine, no cities See NO CITIES DOCTRINE.

Doctrine, Reagan See REAGAN DOCTRINE.

Doctrine, Schlesinger See SCHLESINGER DOCTRINE.

Doctrine, Soviet See SOVIET DOCTRINE.

doctrine, tactical air See TACTICAL AIR DOCTRINE.

Doctrine, Truman See TRUMAN DOCTRINE.

DOD DEFENSE, DEPARTMENT OF.

dogface World War II slang for infantryman.

dogfight 1. Any battle between aircraft. 2. A duel between two fighter aircraft.

dogleg 1. An intentional temporary divergence from the particular direction of a planned route. 2. One section of a course toward an object, which deliberately follows a zig-zag direction rather than a straight path.

dog robber A scrounger who obtains needed supplies or luxuries through informal, not necessarily legal means.

dogs of war Shakespeare's phrase for unrestrained destruction. It comes from *Julius Caesar* (Act III, Scene 1), when Mark Antony is planning military action to revenge Caesar's death:

And Caesar's spirit raging for revenge,
With Ate by his side come hot from hell,
Shall in these confines with a monarch's voice
Cry 'Havoc,' and let slip the dogs of war;
That this foul deed shall smell above the earth,
With carrion men, groaning for burial.

dog tags Metal identification tags worn by military personnel around their necks. In case of death, one tag stays with the corpse and the other is used on a gravemarker for reporting purposes. Dog tags contain name, rank, identification number, religious preference, and blood type.

dollar diplomacy 1. The expansion of American business overseas. 2. A pejorative term for those diplomatic and military efforts that seek to help American business penetrate into foreign markets. 3. The foreign policy of the Taft Administration (1909–1913), which actively sought to expand American trade in Latin America.

domestic intelligence Intelligence relating to activities or conditions within a state that threaten INTERNAL SECURITY and might require the employment of troops.

dominant terrain Terrain which, because of its elevation, proportions, or location, commands a view of and may offer FIELDS OF FIRE over surrounding terrain.

domino theory The notion that if a critically situated country becomes controlled by Communists, its neighbor states will soon follow it into Communism. The theory held that Communism spread from one country to another almost as if it were a disease epidemic. Thus, if one Third World nation fell to a Communist revolution, the nations bordering it would fall in due course, and then the nations bordering them, and so on. Explicitly or otherwise, this fear of radicalism by contagion lay behind the U.S. intervention in Vietnam. Successive American presidential administrations, from President Truman until the present, have felt it necessary to shore up anti-Communist regimes to prevent others from ultimately falling under the rule of Communist parties. Most recently, the Reagan Administration has analyzed the dangers presented by the left-wing government in Nicaragua in the same way: if Nicaraguan Communism remains unopposed, the rest of Central America will follow, and eventually Mexico as well, leaving the United States with an enemy at its Southern border.

doolie A first-year cadet at the U.S. Air Force Academy.

doomsday machine HERMAN KAHN's concept of an ultimate nuclear deterrent from his *On Thermonuclear War* (Princeton, NJ: Princeton University Press, 1961). He proposed that one nation should construct an enormously powerful nuclear device capable of annihilating the world's population. This machine would be automatically programmed to explode if the nation that built it were the subject of a nuclear attack. Finally, he proposed that the automatic program be impossible to turn off, even by the builders. Thus no aggressor could risk a nuclear strike under such a threat because any nuclear attack would be, in effect, suicide.

door bundle A bundle of supplies for manual EJECTION in flight that normally has an attached parachute.

Doppler effect The perceived change in the frequency of a sound or radio wave, caused by a time rate of change in the length of the path of travel of the wave between its source and the point of observation. In effect, the frequency increases as the source approaches, and decreases as it moves away.

dosimetry The measurement of RADIATION DOSES. It applies to both the devices (dosimeters) and techniques used.

double 1. A second shot fired automatically by a SEMIAUTOMATIC WEAPON, when only the first shot is intended. 2. "On the double" or "at the double" means in double time.

double action A method of fire in revolvers and old style RIFLES, PISTOLS, and SHOTGUNS in which a single pull of the trigger both cocks and fires the weapon; in contrast to a single action mechanism, in which the hammer must be cocked by hand before firing.

double agent An AGENT in contact with two opposing INTELLIGENCE services of different states, only one of which is aware of the double contact.

double-apron fence An obstacle consisting of a fence with a network of barbed wire entanglements extending out on each side.

double envelopment A form of enveloping maneuver executed by forces that move around both FLANKS of an enemy position to attack the flanks or OBJECTIVES in the rear of the enemy. The enemy is normally fixed in position by a supporting frontal attack, or HOLDING ATTACK, or by indirect and/or aerial fire. Compare to: CANNAE, ENVELOPMENT, SINGLE ENVELOPMENT.

double flash The two bursts of light in sequence that follow, and are unique to the detonation of a NUCLEAR WEAPON.

double-staggered column A two-lane column of vehicles moving in the same direction, so spaced that the vehicles in one lane are opposite the space between vehicles in the other lane.

double tent A shelter tent set up in the field by buttoning together the square ends of two single shelter tents. Also called double shelter tent.

double time 1. Marching (really running) at the rate of 180 steps, each 36 inches in length, per minute. 2. A PREPARATORY COMMAND to march at this rate.

Douhet, Giulio (1869–1930) The Italian founder of much modern strategic military theory. He developed the early DOCTRINE that a war could be won by direct strategic bombardment of the enemy's civilian population from the air. This doctrine of STRATEGIC BOMBARDMENT influenced the British Royal Air Force from its inception, as well as the U.S. Army Air Forces in both the European and Asian theaters during World War II. Douhet drew less on experience (there had been very little that even approached strategic bombardment during World War I) than on logic. It seemed evident to him and to many leading airmen of his time that terror brought about by continuous aerial bombardment could destroy the civilian morale of a combatant nation, seriously weakening that nation's government and bringing the end of its ability to wage war. This was simply an adaptation to modern circumstances of a tried and tested technique of shortening a war. General William Sherman's famous march through Georgia, during the American Civil War, for example, was an intentional infliction of terror on civilians to help bring about the overthrow of the Confederate government. Douhet proposed a way to deliver this terror by air without having to defeat the enemy's armies first. As with all such theories, however, the establishment of AIR SUPREMACY is essential to guarantee the effectiveness of the bombing. Therefore, the first task of an AIR FORCE in a conventional war is, in fact, to fight a battle against the enemy's air force, rather than immediately to deliver the supposed war-shortening strike against the civilian population. Besides, as experiences in Great Britain and Germany during World War

II and in North Vietnam during the Vietnam War show, the bombing of civilians may only fortify civilian support for a war effort. See Louis A. Sigaud, *Air Power and Unification: Douhet's Principles of Warfare and Their Applications to the United States* (Harrisburg, PA: Military Service, 1949); John F. Shiner, "Reflections of Douhet," *Air University Review* 37, 2 (January–February 1986); Michael J. Eula, "Giulio Douhet and Strategic Air Force Operations," *Air University Review* 37, 6 (September–October 1986).

down 1. A term used in a CALL FOR FIRE to indicate that the target is at a lower ALTITUDE than the REFERENCE POINT used to identify the target. 2. A correction used by an observer/spotter to indicate that a decrease in HEIGHT OF BURST is desired.

downgrade To reduce the SECURITY CLASSIFICATION of a classified document or an item of classified matter or material.

down range 1. In the direction of the targets on a FIRING RANGE. 2. In the LINE OF FIRE.

down time 1. The interval between the arrival of an empty ammunition train at an AMMUNITION SUPPLY POINT and its departure with a load. 2. The interval between the receipt of a request for supplies at a supply depot and their delivery to the troops. 3. The time during which any military equipment is not available for use because of maintenance requirements. 4. The time during which a computer system is inoperative.

downwash The downward flow of gases under pressure, resulting from the application of power to the lifting or propelling of vertiplanes, helicopters, rockets, guided missiles, and other airborne vehicles.

Drago Doctrine The policy of states not to use their armed forces to force the payment of debts owed by the citizens of foreign countries. This doctrine, which has been widely endorsed, was first formulated in 1902 by Luis M. Drago (1859–1921), the foreign minister of Argentina.

dragon's teeth Reinforced concrete tank OBSTACLES which when used in large numbers can slow the advance of ARMOR.

dress 1. To form a straight line in a DRILL formation. 2. A command to take this line. 3. The arrangement of flags and pennants on the masts of a ship for formal occasions.

dressing station A medical aid station close to the locus of combat.

dress left (right) Preparatory command at which soldiers turn their heads and eyes to the left (right) and straighten a line, each individual in relation to the person at his or her left (right).

dress right See DRESS LEFT (RIGHT).

dress uniform A uniform authorized for wear at social, ceremonial, and official military occasions.

drift In BALLISTICS, a shift in projectile direction due to gyroscopic action, which results from gravitational and atmospherically induced torques on the spinning projectile.

drift signal A floating signal dropped from an aircraft over water to provide a reference point in determining oceanic drift for navigation purposes, or for marking an area or object for the aid of surface ships. Fire, smoke, or metallic powder is released when the signal strikes the water surface.

drill 1. Practice and training to learn or retain combat-related skills. VEGETIUS wrote in *De Re Militari:* "What is necessary to be performed in the heat of action should constantly be practiced in the leisure of peace." 2. A CLOSE ORDER DRILL in which soldiers march in formation. Even though this is a tactic no longer used in modern war, all armies still drill their soldiers to impart two critical lessons: that orders must be obeyed instantly, and that individuality must be subordinated to the group effort.

drill, casualty See CASUALTY DRILL.

drill, combat See COMBAT DRILL.

drill, gun See GUN DRILL.

drill ammunition Ammunition without an EXPLOSIVE CHARGE, used in training and practice.

drill call A bugle call to turn out for DRILL.

drill sergeant A NONCOMMISSIONED OFFICER qualified to instruct and supervise basic combat and advanced individual trainees.

drone 1. A land, sea, or air vehicle, unmanned, that is remotely or automatically controlled. 2. A pilotless aircraft used for RECONNAISSANCE over enemy territory that is essentially a flying television camera. See Gene Bigham, "The Future of Drones: A Force of Manned and Unmanned Systems," *Air University Review* 29, 1 (November–December 1977).

drop 1. A parachute jump, individually, or en masse, or a supply delivery by parachute from an aircraft in flight, or the act of making such a jump or delivery. 2. In artillery and naval gunfire support, a correction used by an observer/spotter to indicate that a decrease in range along a SPOTTING LINE is desired.

dropmaster 1. An individual qualified to prepare, perform the acceptance inspection of, load, lash, and eject material for an AIRDROP. 2. An aircrew member who, during parachute operations, will relay any required information between the pilot and JUMPMASTER.

drop message A message dropped from an aircraft to a ground or surface unit.

dropping angle The angle between the line from an aircraft to a target and the vertical line from the aircraft to the ground at the instant a bomb is released. Also called the RANGE ANGLE.

drop zone A specific area upon which airborne troops, equipment, or supplies are AIRDROPPED.

drum 1. An adjustment device and scale for making fine settings on certain types of gunsights. A COARSE SETTING is made with a device called a plateau. 2. A cylinder MAGAZINE, from which CARTRIDGES are fed to certain MACHINE GUNS and RECOILLESS RIFLES. 3. A metal container for liquids. In this meaning, usually preceded by the quantity, e.g., a 5-gallon drum or a 55-gallon drum.

drum out 1. To expel from a military service. 2. A formal ceremony, involving drums, during which an unworthy soldier is disgraced and expelled from a military camp.

dry run Any practice test or session.

DSAT See DEFENSIVE SATELLITE WEAPON.

D-to-P concept A LOGISTIC planning concept by which the gross MATERIEL readiness requirement in support of approved forces at planned wartime rates for conflicts of indefinite duration will be satisfied by a balanced mix of assets on hand at D-DAY and assets to be gained from production through P-DAY when the planned rate of production deliveries to the users equals the planned wartime rate of expenditure (consumption) of the produced supplies.

dual agent An individual who is simultaneously and independently employed by two or more INTELLIGENCE agencies.

dual-capable unit A nuclear-certified delivery unit (such as a missile, aircraft, or artillery) capable of executing both conventional and nuclear missions.

dual key 1. A nuclear weapons safety measure that requires two different keys, which are each held by two different officers, for the arming or launch of nuclear missiles or other devices. This procedure is designed to ensure that no lone person can ever loose a nuclear weapon. 2. An arrangement by which the United States shares control of a NUCLEAR WEAPONS system with the nation in which it is based, or the national military service primarily responsible for operating it. An early example was the deployment of Thor (an INTERMEDIATE-RANGE BALLISTIC MISSILE) in the United Kingdom during the late 1950s and early 1960s. These missiles were provided by the United States but maintained and operated by the Royal Air Force (RAF). However, they could not be launched without the consent of both governments. Officers from both services were required to carry out the launch sequence, in this case literally by turning keys. Many of NATOs battlefield nuclear weapons are held under a version of this system.

dual-warning phenomenology Deriving warning information from two systems observing separate physical phenomena (e.g., RADAR/infrared light or visible light/X ray) associated with the same events to attain high CREDIBILITY while being less susceptible to false reports.

dual- (multi)-capable weapons 1. Weapons, weapon systems, or vehicles capable of selective equipage with different types or mixes of ARMAMENT or FIREPOWER. 2. Weapons capable of handling either nuclear or non-nuclear munitions. An example is the U.S. F-111 fighter-bomber, which can carry nuclear or conventional bombs. Another example is the standard 155-mm artillery cannon used by most NORTH ATLANTIC TREATY ORGANIZATION (NATO) armies, which is also capable of firing nuclear shells. Such dual-capable weapons pose particular problems in ARMS-CONTROL negotiations because it is virtually impossible to find a way of verifying that such a system is not, in fact, being readied for a nuclear role.

dual- (multi)-purpose weapons Weapons that have the capability for effective application in two or more basically different military functions or levels of conflict; particularly weapons designed for effective fire against air or surface targets.

dud An explosive MUNITION that has not been armed as intended, or which has failed to explode after being armed.

dud, absolute See ABSOLUTE DUD.

dud, dwarf See DWARF DUD.

dud, flare See FLARE DUD.

dud, nuclear See NUCLEAR DUD.

duffle A soldier's personal possessions such as would fit into a DUFFLE BAG.

duffle bag A long, cylindrical cloth bag for a soldier's clothing and personal possessions (his DUFFLE).

dugout An underground shelter for PERSONNEL or MATERIEL.

dumb bomb A bomb that is simply dropped on, as opposed to being guided toward, a target; the antonym of SMART BOMB.

dummy message A message sent for some purpose other than its content, which may consist of groups of random letters and numbers or may have a meaningless text.

dummy minefield In naval mine warfare, a MINEFIELD containing no live mines and presenting only a psychological threat.

dummy run Any simulated firing practice, particularly a DIVE BOMBING approach made without the release of a bomb. Identical to a DRY RUN.

dump A temporary storage area, usually in the open, for bombs, ammunition, equipment, or supplies.

dust defense A technique for protecting land-based missiles by planting nuclear explosives underground all around a missile base. Several minutes before enemy missiles arrived at the site, the underground explosives would generate a huge cloud of dust from the dirt surrounding them. The dust particles would then scrape off the heat-protective tips of the incoming missiles, causing them to burn up from the heat generated by their own speed before reaching the base. It is alternatively called an environmental defense.

dustoff A helicopter expedited means for the aero MEDICAL EVAUCATION of wounded personnel. "Dustoff" was the radio call-sign of the first aeromedical helicopter to be shot down during the Vietnam War (in 1964). This slang term then informally evolved as a tribute to the bravery of that fallen crew in coming into a HOT landing zone.

duty 1. A prescribed task to which military personnel are assigned. General George S. Patton wrote during World War II, "Any commander who fails to attain his objective, and who is not dead or severely wounded, has not done his full duty." 2. A legal obligation to do something because of an office one holds, a profession one practices, or the like. 3. A tax imposed on imported products. 4. Extra assignments beyond normal military responsibilities; tasks that must be done on "off-duty" hours. See Lewis Sorley, "Beyond Duty, Honor, Country," *Military Review* 67, 4 (April 1987).

duty, active See ACTIVE DUTY.

duty assignment A group of closely related tasks and responsibilities that are normally assumed by one individual assigned to a military unit. A MILITARY OCCUPATIONAL SPECIALTY usually qualifies an individual for a variety of duty assignments. Also known as duty position.

duty branch The branch (infantry, artillery, etc.) in which an officer is serving as a result of appointment, assignment, or DETAIL.

duty officer An officer detailed to be constantly available for call in emergencies during a specific period; he or she is usually in command when most of the unit is off duty, and during holidays and vacations. Often equivalent to "OFFICER OF THE DAY," or, in naval usage, "officer of the deck."

duty position See DUTY ASSIGNMENT.

duty roster A list of the personnel of a unit, showing the duties each man has performed, such as GUARD, FATIGUE, or KITCHEN POLICE. It is kept to determine the date of an individual's next TOUR OF DUTY and to insure fair distribution of duties among the personnel of the unit.

duty station A military establishment or post to which an officer or enlisted man has been assigned for duty.

dwarf dud A NUCLEAR WEAPON that, when launched at or emplaced on a target, fails to provide a YIELD within a reasonable range of that which could have been anticipated from normal operation. This constitutes a DUD only in a relative sense.

dwell at/on In ARTILLERY AND NAVAL GUNFIRE SUPPORT, a term used when fire is to continue for an indefinite period at a specified time or on a particular target or targets.

dynamic reconfiguration The means whereby a battle-management system for the STRATEGIC DEFENSE INITIA-

TIVE (SDI) can adapt to changing circumstances, such as the destruction of some defensive components.

dynamite, military See MILITARY DYNAMITE.

DZ A DROP ZONE for airlifted personnel or supplies.

E

E 1. A designation for an aircraft that is equipped with special electronic devices. For example, the AWACS aircraft is the E-3. 2. The designation for "enlisted" in reference to the United States Department of Defense pay grades. For example, a private will start at pay grade E-1; a master sergeant will be at grade E-8.

E-date See EFFECTIVE DATE.

eagle colonel See COLONEL.

eagle flight A type of tactical operation employing helicopters to reconnoiter likely enemy positions, and deposit troops once enemy units are located.

early resupply The shipping of supplies during the period between D-DAY and the beginning of "planned resupply."

early warning Early notification of the launch or approach of unknown weapons or weapon carriers; particularly early detection of an enemy BALLISTIC MISSILE launch, usually by means of surveillance SATELLITES and long-range RADAR. See also BALLISTIC-MISSILE EARLY-WARNING SYSTEM; TACTICAL WARNING.

early warning system, ballistic missile See BALLISTIC MISSILE EARLY WARNING SYSTEM.

earthworks Field FORTIFICATIONS made essentially of earth, such as TRENCHES or earthen BARRIERS.

east-west The Soviet Union and its Eastern European satellites are the East; the United States and its Western European allies (plus allied Asian countries) are the West.

ECCMS Electronic counter-countermeasures. See ELECTRONIC WARFARE.

echelon 1. A subdivision of a HEADQUARTERS, i.e., FORWARD ECHELON, REAR ECHELON. 2. A separate level of command. As compared to a REGIMENT, a DIVISION is a higher ECHELON, a BATTALION is a lower echelon. 3. A fraction of a command to which a principal combat mission is assigned; i.e., attack echelon, support echelon, reserve echelon. 4. A FORMATION of troops or vehicles whose subdivisions are placed one behind another, with a lateral and even spacing to the same side. 5. A WAVE.

echelon, advance by See ADVANCE BY ECHELON.

echelon, assault See ASSAULT ECHELON.

echelon, combat See COMBAT ECHELON.

echelon, follow-on See FOLLOW-ON ECHELON.

echelon, follow-up See FOLLOW-UP ECHELON.

echelon, forward See FORWARD ECHELON.

echelon, rear See REAR ECHELON.

echelon, sea See SEA ECHELON.

echelon, service See SERVICE ECHELON.

echelon, support See SUPPORT ECHELON.

echeloned attack An attack in several WAVES or ECHELONS. This strategy allows a commander to avoid having too many troops at one place on the FRONT LINE, which would present an easy target for BATTLEFIELD NUCLEAR WEAPONS. Since the mid-1960s, Soviet doctrine has called for the first echelon of the Warsaw Pact to attack along the entire length of the NORTH ATLANTIC TREATY ORGANIZATION'S (NATO) front line. Even if the attack was repulsed, the weakened NATO forces would then face renewed attack from the second echelon, arriving fresh and at full strength perhaps 48 to 72 hours after the first echelon. By then, gaps or weak spots would be expected to have opened up in the NATO defenses, through which forces held in reserve could infiltrate to attack deep into NATO's rear area in BLITZKRIEG fashion. The entire doctrine seeks to take advantage of the WARSAW PACT'S superiority in CONVENTIONAL TROOPS and WEAPONS, without having to prompt NATO escalation to the use of BATTLEFIELD or THEATER NUCLEAR FORCES. NATO's own doctrinal response to this stragegy is the FOLLOW ON FORCES ATTACK. See Christopher Donnelly, "The Development of the Soviet Concept of Echeloning," *NATO Review* 32, 6 (December 1984).

echeloned displacement A movement of a military unit from one POSITION to another without discontinuing performance of its primary function. Normally, the unit divides into two functional ELEMENTS (base and advance). While the base element continues to operate, the advance element displaces to a new site where, after it becomes operational, it is joined by the base element.

echelonment The arrangement of personnel and equipment into ASSAULT, COMBAT FOLLOW-UP, and rear components or groups.

echo The signal received by a RADAR set as a result of the reflection of a transmitted pulse from objects in the field of scan. See also PIP.

ECMs Electronic countermeasures. See ELECTRONIC WARFARE.

economic action The planned use of economic (as opposed to political or military) measures designed to influence the policies or actions of a real or potential enemy state by: 1. hurting its economy (such as with a trade embargo); or 2. helping the economy of a friendly power (such as with loans or favorable terms of trade).

economic mobilization Organizing a domestic economy in response to a wartime emergency.

economic potential The total capacity of a nation to produce goods and services.

economic potential for war That share of the total economic capacity of a nation that can be used for the purposes of war.

economic warfare 1. An aggressive use of economic means to achieve national objectives. There are many levels of intensity to economic warfare. They range from freezing a foe's assets and confiscating its property during a formally declared war to using clandestine methods to destablize an opponent's economy during a cold war. 2. The normal peaceful competition between nations for markets and trade advantages. See Thomas A. Wolf, *U.S. East-West Trade Policy: Economic Warfare versus Economic Welfare* (Lexington, MA: Lexington Books, 1973); H. Clayton Cook, Jr. "Soviet Economic Warfare and U.S. Maritime Policy: A Critique," *Conflict* 6, 3 (1985).

economic zone, exclusive See EXCLUSIVE ECONOMIC ZONE.

economy of force The military principle (see PRINCIPLES OF WAR) dictating that only minimum forces be used for secondary efforts, so that maximum forces are conserved for concentration on primary efforts. French Field Marshal Foch defined it in 1919 as the "art of pouring out *all* one's resources at a given moment on one spot."

economy of force measure The use of the minimum force necessary to accomplish a mission. To free troops and permit their concentration elsewhere.

EEZ See EXCLUSIVE ECONOMIC ZONE.

effective 1. A soldier fit for service. 2. FIRE hitting its target.

effective beaten zone The section of a target area in which a very high percentage of shots fall.

effective damage That damage necessary to render a target inoperative, unserviceable, nonproductive, or uninhabitable.

effective date The specific date of a formal change in unit status. Also known as E-date.

effective parity A measure of the overall balance of destructive capability. One country might, for example, have superiority in LAUNCHERS and MEGATONS, another in WARHEADS and accuracy. If both were thus able to carry out their respective war plans, this situation could represent effective parity or essential equivalence.

effective range That RANGE at which an average soldier has a 50 percent probability of hitting a target with a SMALL ARMS weapons.

eighty-two percent zone See EFFECTIVE BEATEN ZONE.

Eisenhower Doctrine The 1957 statement of United States policy on the Middle East, which asserted that the United States would use force "to safeguard the independence of any country or group of countries in the Middle East requesting aid against aggression" from a Communist country. This differs from other presidential doctrines in that it was adopted as a Joint Resolution of the U.S. Congress. See Harry N. Howard, "The Regional Pacts and the Eisenhower Doctrine," *Annals of the American Academy of Political and Social Science* Vol. 401 (May 1972).

ejection 1. Escape from an aircraft by means of an independently propelled seat or capsule. 2. In air armament, the process of forcefully separating an aircraft store (an externally carried bomb, rocket, etc.) from an aircraft.

élan The high ESPRIT DE CORPS and self-confidence of a fighting force. Having it does not guarantee victory (as the French have demonstrated), but it can't hurt.

electric primer A metallic device containing a small amount of a sensitive explosive that is actuated by energizing an electric circuit. It is used in certain weapon systems to initiate the PROPELLING CHARGE.

electric squib An item designed for the electric firing of burning type ammunition. Also used to position, through piston action, the internal components of an electric fusing system. It consists essentially of a tube containing a flammable material, and a small charge of powder compressed around a fine resistance wire that is connected to electric leads or terminals. Compare to SQUIB.

electromagnetic camouflage The use of electromagnetic SHIELDING, absorption, or enhancement techniques to minimize the possibility of detection and identification of troops, materiel, equipment, or installations by hostile SENSORS employing radiated electromagnetic energy.

electromagnetic pulse (EMP) A phenomenon of nuclear explosions caused by the secondary reactions that occur when gamma radiation is absorbed into the air or ground. Its effects are very much like those of an intensely-magnified radio wave, but the magnification is so great that it destroys or temporarily disables the same aerials and circuits that receive radio transmissions. There are two main differences between an electromagnetic pulse and a radio wave. First, a radio wave produces an electrical field of perhaps only one-thousandth of a volt in a receiving antenna, but an EMP can cause a field running to several thousands of volts. Second, an EMP is much faster than other electromagnetic waves, such as a lightning flash, because the entire energy in a pulse disappears in a fraction of a second. Thus, equipment meant to protect antennae and internal circuits from lightning cannot cope with an EMP overload, and the delicate electronics of radars, radios and short-wave telephone communications can be literally burned out by an EMP. See Richard E. Fitts, ed., *The Strategy of Electromagnetic Conflict* (Los Altos, CA: Peninsula Publishing, 1980).

electromagnetic radiation A form of propagated energy, arising from electrical charges in motion, that produces a simultaneous wavelike variation in electrical and magnetic fields in space. The highest frequencies for such radiation are possessed by GAMMA RAYS, which originate from processes within atomic nuclei. As one goes to lower frequencies, the electromagnetic spectrum includes X-rays, ultraviolet light, visible light, infrared light, microwaves, and radio waves. See Frank Henry Read, *Electromagnetic Radiation* (New York: John Wiley, 1980).

electronic-beam weapon A theorized future type of gun that would shoot streams of electrons at in-flight missiles and disable their electronic guidance systems, causing them to miss their targets. See Jeff Hecht, *Beam Weapons: The Next Arms Race* (New York: Plenum, 1985).

electronic counter-countermeasures (ECCMs) See ELECTRONIC WARFARE.

electronic countermeasures (ECMs) Methods of blinding radars and of blanking out radio frequencies to prevent target acquisition and interrupt communications to cause them to be ineffective. See ELECTRONIC WARFARE JAMMING.

electronic deception The simulation or manipulation of friendly ELECTROMAGNETIC RADIATIONS, and the imitation of enemy electromagnetic radiations for the purpose of deceiving the enemy.

electronic deception, imitative See IMITATIVE ELECTRONIC DECEPTION.

electronic intelligence (ELINT) All forms of INFORMATION-gathering based on radio and radar, but principally the monitoring of an opponent's radio and radar emissions. Ever since World War I, the process of listening in to the enemy's radio traffic has been crucial as a way of discovering his position, intentions, and movements. The ELINT process can be usefully broken down into two activities: 1. the interception and recording of communications between enemy units and, indeed, communications inside the opponent's governing and administrative structures, in peacetime as much as in wartime. See also SIGNALS INTELLIGENCE; 2. the gathering of information on an opponent's radio and radar procedures, wavelengths, and electronic characteristics to identify the particular emissions and RADAR SIGNATURES of specific types of ships, aircraft, or other transmitters that are operating at certain ranges, for certain purposes, and so on. All major powers carry out such intelligence gathering operations, often involving taking ships or submarines close to an opponent's shores, even sometimes deliberately triggering an ALERT in order to record the resulting signal characteristics and operating procedures. In recent years ELINT has been used to refer solely to the interception of non-communications signals. Traditional ELINT of the World Wars variety is called COMMUNICATIONS INTELLIGENCE.

electronic order of battle Intelligence pertaining to the deployment of enemy electronic emitters in a given area.

electronic warfare (EW) Military action using electromagnetic energy to determine, exploit, reduce or prevent the hostile use of the electromagnetic spectrum, or action that retains friendly use of the electromagnetic spectrum. There are three divisions within electronic warfare:

1. Electronic countermeasures (ECMs) That division of electronic warfare involving actions taken to prevent or reduce an enemy's effective use of the electromagnetic spectrum. This includes: (1) electronic JAMMING—the deliberate radiation, reradiation, or reflection of electromagnetic energy for the purpose of disrupting enemy use of electronic devices, equipment, or systems; and (2) ELECTRONIC DECEPTION—the deliberate radiation, reradiation, alteration, suppression, absorption, denial, enhancement, or reflection of electromagnetic energy in a manner intended to convey misleading information and to deny valid information to an enemy or to enemy electronics-dependent weapons.
2. Electronic counter-countermeasures involves electronic warfare employing actions taken to insure friendly effective use of the electromagnetic spectrum despite the enemy's use of electronic countermeasures.
3. Electronic warfare support measures involves electronic warfare employing actions taken under the direct control of an operational commander to search for, intercept, identify, and locate sources of radiated electromagnetic energy for the purpose of recognizing an immediate threat. Such support measures provide a source of information required for immediate decisions involving electronic countermeasures, electronic counter-countermeasures, and other tactical employment of forces. See Alfred Price, *Instruments of Darkness: The History of Electronic Warfare* (New York: Charles Scribner's Sons, 1980); Don E. Gordon, *Electronic Warfare: Element of Strategy and Multiplier of Combat Power* (New York: Pergamon Press, 1981); Terry E. Bibbens, "Simulating the Modern Electronic Battlefield," *Defense Management* 18, 3 (Third Quarter 1982).

element 1. Staff or operational organizations (OFFICES, DIRECTORATES, DIVISIONS, branches, etc.) that form the principal structure of, and are immediately subordinate to the next larger organization. 2. A portion of an airborne or air-landed unit described by its method of entry into the combat area, such as parachute element, airplane element, seaborne element, or overland element.

elements of national power All of the means (military, economic, political, etc.) available for use in the pursuit of NATIONAL OBJECTIVES such as geography, natural resources, population, and quality of diplomacy.

elephant steel shelter A shelter for personnel or materiel made from steel arch sections of large size. This shelter is the largest type of steel arch shelter, and serves as a SPLINTER-PROOF SHELTER.

elevate 1. To raise up the MUZZLE of a gun. 2. Often used to mean the aiming of a gun, whether or not the muzzle goes up or down. General Andrew Jackson, at the Battle of New Orleans, is supposed to have told his artillerymen to: "elevate those guns a little lower."

elevating arc An upright, geared arc, attached to a gun or gun carriage, along which the gun is raised or lowered.

elevation 1. The vertical angle between horizontal and the axis of the BORE of a weapon required for a PROJECTILE to reach a prescribed RANGE. 2. In air-defense artillery, the angular height of a weapon. Dials on some equipment, which indicate angular height, are marked "elevation."

elevation, adjusted See ADJUSTED ELEVATION.

elevation circle A circular scale showing the QUADRANT ELEVATION of the gun barrel.

elevation indicator An electrical instrument on some guns that shows the quadrant elevation to be used; it is part of a remote control system.

elevation of security The minimum ELEVATION permissible for firing above friendly troops without endangering their safety. This term can only be applied to certain equipment having a flat TRAJECTORY.

elevation post A point with a known horizontal and vertical position with respect to some defined reference system.

elevation quadrant See QUADRANT.

elevation scale A scale on a GUN CARRIAGE that shows the ELEVATION of the gun.

elevation stop 1. A structural unit in a gun or other equipment that prevents it from being elevated or depressed beyond certain fixed limits. 2. A FIRING TABLE giving a list of RANGES, with the corresponding QUADRANT ELEVATION setting to be applied to a gun.

ELF See EXTREMELY LOW FREQUENCY.

elicitation The acquisition of information from an interview or conversation with a person or group in a manner that does not disclose the intent of the interviewer. This is a technique of HUMAN-INTELLIGENCE collection, generally overt, unless the identity of the collector is other than what he purports it to be.

ELINT See ELECTRONIC INTELLIGENCE.

elite troops 1. Military units that have been given special training or weapons to make them more effective than regular units under certain conditions. 2. Military units that think of themselves as superior to others of the same armed force because of a distinguished history, fancy emblems, or conceited attitudes.

elliptical orbit A non-circular egg-shaped ORBIT.

embargo A government prohibition of the import or export of commodities to or from, or of vessels from specific nations. An embargo is a mildly hostile act more related to foreign policy than trade policy. For example, shortly after the Communists came to power in Cuba, the United States embargoed sugar from Cuba in an effort to disrupt the Cuban economy. And after the Soviet Union invaded Afghanistan, the United States embargoed grain shipments to the Soviets in an effort to impress upon them the American displeasure over their aggressive actions. See Alan Abouchar, "The Case for the U.S. Grain Embargo," *World Today* Vol. 37, No. 7–8 (July–August 1981); Clas Bergstrom, Glenn C. Loury and Mats Persson, "Embargo Threats and the Management of Emergency Reserves," *Journal of Political Economy* Vol. 93, No. 1 (February 1985).

embarkation The loading of troops with their supplies and equipment into ships or aircraft at ports of embarkation in coastal waters or on airfields.

embarkation and tonnage table A consolidated table showing personnel and cargo, by troop or naval units, loaded aboard a combat-loaded ship.

embarkation area A shore area comprising a group of EMBARKATION points, in which final preparations for embarkation are completed and through which assigned personnel and loads for craft and ships are called forward to embark. See also MOUNTING AREA.

embassy 1. The highest class of diplomatic mission, headed by an ambassador. In this context, "the embassy" refers to all of the diplomatic staff as well as to all of their support personnel. 2. The physical building(s) that is used to house the office (the chancery) and personal quarters of an embassy staff and the ambassador. Also called the mission or compound. 3. The mission undertaken by an ambassador. Thus an ambassador's embassy might be to negotiate a treaty.

embrasure An opening cut into a wall to enable artillery to fire with minimum exposure to the gunners.

emergency anchorage An anchorage for naval vessels, mobile support units, auxiliaries, or merchant ships that may have a limited defense organization.

emergency medical treatment The application of medical procedures for a condition in which the life of the patient is in immediate danger or permanent crippling may result if treatment is delayed.

emergency powers The enlarged authority that a president of the United States is deemed to assume either by statute, the Constitution, or because of the nature of the problem at hand to deal with an emergency.

emergency priority A category of IMMEDIATE MISSION REQUEST that takes precedence over all other priorities, e.g., an enemy breakthrough.

emergency risk The measure involving the highest degree of risk to friendly troops, used in computing the MINIMUM SAFE DISTANCE from a NUCLEAR DETONATION. Five percent of the personnel exposed to emergency risk will become combat ineffective; a larger number will suffer less damaging effects. Emergency risk is accepted only when absolutely necessary. It is expressed in terms of risk to unwarned, exposed personnel; to warned, exposed personnel; and to warned, protected personnel.

emerging technology (ET) WEAPONS SYSTEMS and C^3I links that are currently being developed to solve the NORTH ATLANTIC TREATY ORGANIZATION's (NATO) inferiority in conventional forces. The hopes for ET are based on a combination of two technological developments, the first of which is PRECISION-GUIDED MUNITIONS (PGMs). It is believed possible to construct SMART BOMBS so that a missile could carry dozens of warheads which, when dropped over a large target such as a squadron of tanks, would steer themselves individually to each tank with a near-perfect probability of hitting them. This will supposedly be capable of accomplishment at launch ranges of several hundred kilometers.

Even if the weapons can be developed, they are obviously useless without the capacity to detect these very distant and mobile targets; thus a second development in real time information on remote TARGET ACQUISITION is essential. Airborne radar and infrared search equipment, and possibly pilotless DRONES, must be designed to be capable of detecting such targets and fixing their positions very precisely at great ranges. Even so, the demand on communications and control would be far too intense without a "real time" intelligence processing aspect. It is not enough to know that at a certain hour there *was* a suitable target at a specific location; such information must be instantly available to a commander before he can decide whether to launch a strike with these inevitably expensive weapons, which are certain to be in short supply. See Fen Osler Hampson, "NATO's Conventional Doctrine: The Limits of Technological Improvement," *International Journal* 41, 1 (Winter 1985–86); Robert L. Maginnis, "Selecting Emerging Technologies," *Military Review* 66, 12 (December 1986); Frank Barnaby, "How the Next War Will Be

Fought: Sophisticated New Missiles Are Making Tanks, Combat Aircraft, and Large Warships Obsolete," *Technology Review*, 89, 12 (October 1986).

EMP See ELECTROMAGNETIC PULSE.

emplacement 1. A prepared position for one or more weapons or pieces of equipment, for protection against hostile fire or bombardment, and from which they can execute their tasks. 2. The act of fixing a gun in a prepared position from which it may be fired.

EMT See EQUIVALENT MEGATONNAGE.

enceinte The surrounding walls of a FORTRESS.

encipher To convert text into unintelligible form by means of a CIPHER SYSTEM.

encirclement 1. The enemy control of all ground routes of EVACUATION and reinforcement available to a force, resulting in their loss of freedom to MANEUVER. 2. The post-World War II United States policy of CONTAINMENT of the Soviet Union and its satellites by encircling them with American allies and military forces to contain their expansion. See Uri Ra'anan, "The USSR and the 'Encirclement' Fear: Soviet Logic or Western Legend?" *Strategic Review*, 8, 1 (Winter 1980).

encircling force A force employed to envelop an enemy force that has lost its capability to defend or delay in an organized fashion. It seeks to cut off escape routes and, in conjunction with DIRECT PRESSURE FORCES, to attack and destroy the enemy force. See Michael H. Vernon, "Encirclement Operations," *Military Review* 66, 9 (September 1986).

encode 1. To convert plain text into unintelligible form by means of a code system. 2. That section of a CODE BOOK in which the PLAIN TEXT equivalents of the CODE GROUPS are in alphabetical, numerical or other systematic order.

encounter A spontaneous, unplanned fight between two opposing forces.

encryption The conversion of plain text into an unintelligible (coded) form by means of a cipher system.

endoatmospheric Taking place within the earth's atmosphere, generally considered to be below altitudes of 100 kilometers. An endoatmospheric interceptor reaches its target within the atmosphere. Compare to EXOATMOSPHERIC.

end of mission In ARTILLERY and NAVAL GUNFIRE SUPPORT, an order given to terminate firing on a specific target.

end strength The actual or authorized strength of an army, or subdivision thereof, at the close of a specific period (fiscal year, calendar year, month, or operation). See also INITIAL STRENGTH.

endurance The time during which an aircraft can continue flying, or a ground vehicle or ship can continue operating, under specified conditions, e.g., without refueling.

endurance distance The total distance over which a ground vehicle or ship can be self-propelled at any specified endurance speed.

enemy alien A person of enemy nationality who is outside the boundaries of his country and is in territory of, or territory occupied by, a belligerent power.

enemy capabilities Those COURSES OF ACTION of which an enemy is physically capable. Enemy capabilities include not only the general courses of action open to an enemy, such as ATTACK, DEFENSE, or withdrawal, but also all the particular means possible to accomplish each general course of action. Enemy capabilities are considered in the light of all known factors affecting military operations, including time, space, weather, terrain, and the strength and disposition of enemy forces. In strategic thinking, capabilities reflect national objectives in peace or war.

enemy in zone, destroy See DESTROY ENEMY IN ZONE.

enemy personnel, retained See RETAINED ENEMY PERSONNEL.

enemy state (nation) A state recognized or unrecognized, which is at war with, or engaged in armed conflict with, the United States.

energy, directed See DIRECTED ENERGY.

energy weapon, directed See DIRECTED ENERGY WEAPON.

enfilade 1. A flank position that permits unrestricted fire along the enemy's MAIN LINE OF RESISTANCE. 2. To rake with gunfire in a lengthwise direction; also called raking fire.

engageable target An identified target that can be engaged by a weapon because its visibility pattern does not preclude successful employment of the weapon.

engagement 1. Any hostile encounter between opposing forces. 2. In air defense, an attack with guns or air-to-air missiles by an interceptor aircraft, or the launch of an AIR DEFENSE MISSILE by air defense artillery and the missile's subsequent travel to intercept its target. 3. The fire of a weapon upon a target.

engagement, cease See CEASE ENGAGEMENT.

engagement, constructive See CONSTRUCTIVE ENGAGEMENT.

engagement, decisive See DECISIVE ENGAGEMENT.

engineer 1. A member of a corp of engineers. A soldier who performs engineering duties, including construction, demolition, surveying, road and bridge building, and camouflage. 2. The senior officer of the engineer troops in a large unit. He is a staff officer, and advises the commander on engineering matters.

engineering, general See GENERAL ENGINEERING.

engineer work line A line used to compartmentalize an operational area to indicate where specific ENGINEER units have primary responsibility for an engineering effort.

engraving Grooves cut into a ROTATING BAND of a projectile by the RIFLING of a gun tube.

enhanced-radiation weapon The neutron bomb, which emits an unusually large proportion of its released energy as radioactive particles, mainly neutrons. It effects little physical damage, but is intended to cause high human casualty rates. The usual objective for a nuclear weapon is to destroy physical structures—buildings, concrete silos, bridges and so on. Thus most NUCLEAR WEAPONS engineering has striven to maximize immediate BLAST EFFECT, and to cut down on energy release through RADIATION, both to increase the physical destruction through SHOCK WAVES and heat, and to reduce FALLOUT and longer-term radiation damage. Enhanced-radiation weapons, however, operate on precisely the opposite design philosophy. They limit blast and heat to the area immediately surrounding the point of detonation, and boost the emission of high-energy radiation, particularly particles with short HALF-LIVES.

Enhanced-radiation weapons are not intended for strategic operations against fixed enemy physical plants, but for tactical strikes against its armies. In this latter case it is precisely the destruction of the troops themselves that is sought. The United States has stocks of neutron bombs but has not deployed them in Europe because of the bad publicity associated with "the bomb that destroys people but not buildings." However, the French have started developing their own neutron weapons. See S. T. Cohen, "Enhanced Radiation Warheads: Setting the Record Straight," *Strategic Review,* 6, 1 (Winter 1978).

enlisted aide A servant to a high-ranking officer.

enlisted person A term used to include both male and female members of the armed forces below the grade of an OFFICER. Also called enlisted personnel.

enlistee An individual who voluntarily enrolls as a member of a military service for a period of ENLISTMENT.

enlistment The voluntary enrollment in a military service, as contrasted with induction whereby someone is drafted into service.

enriched nuclear fuel Uranium that has gone through the ENRICHMENT process and is suitable to fuel NUCLEAR WEAPONS.

enrichment The long, complex process by which nuclear fuel becomes ENRICHED NUCLEAR FUEL by increasing the ratio of U-235 to U-238. This ratio must be 9:1 or greater before the uranium is suitable for nuclear weapons.

ensign 1. The lowest grade commissioned officer in the U.S. Navy and Coast Guard (and the Royal Navy); parallel to a second lieutenant in an army. 2. A national flag.

entente From the French: "an understanding." In diplomatic usage it is more than a DETENTE but less than a formal TREATY. However, it may imply a tacit ALLIANCE. Great Britain, France, and Russia formed the pre-World War I "Triple Entente." Once the war started, they created a formal military alliance, but were still known as the "Entente Powers."

entrenching tool The small shovel carried by soldiers for digging TRENCHES and FOXHOLES. NAPOLEON said that there were five things a soldier should never be without: his musket, his ammunition, his rations, his knapsack, and his entrenching tool.

entrenchments TRENCHES to provide cover. JOMINI wrote in his 1838 *Precis on the Art of War* that: "To bury an army in entrenchments, where it may be outflanked and surrounded or forced in front even if secure from a flank attack, is manifest folly." Yet because of the machine gun, trenches played a dominant role during World War I.

entrucking table A form giving information and instructions about truck schedules, entrucking and detrucking points, and troops to be moved. It is usually annexed to a FIELD order.

envelop To attack on one or both FLANKS of the enemy, usually attacking his FRONT at the same time. See ENVELOPMENT.

envelopment An offensive maneuver in which the main attacking force passes around or over the enemy's principal defensive positions to secure objectives to the enemy's rear. It usually involves a HOLDING ATTACK directed against the enemy's front. VERTICAL ENVELOPMENTS may be conducted by AIRBORNE and AIRMOBILE assaults. General Douglas MacArthur said before the 1950 amphibious assault at Inchon, Korea: "The deep envelopment based on surprise, which severs the enemy's supply lines, is and always has been the most decisive maneuver of war. A short envelopment which fails to envelop and leaves the enemy's supply system intact, merely divides your own forces and can lead to heavy loss and even jeopardy." See also CANNAE, TURNING MOVEMENT. See Milan Vego, "The Soviet Envelopment Option on the

Northern Flank," *Naval War College Review* 39, 4 (Autumn 1986).

envelopment, double See DOUBLE ENVELOPMENT.

envelopment, single See SINGLE ENVELOPMENT.

envelopment, vertical See VERTICAL ENVELOPMENT.

environmental defense See DUST DEFENSE.

epaulement Any defensive wall built to protect a FLANK position.

ephemeris A collection of data about the predicted positions (or apparent positions) of celestial objects, including artificial SATELLITES, at various times in the future. A satellite ephemeris might contain the orbital elements of satellites and their predicted changes.

equal section charge A PROPELLING CHARGE made up of a number of charges equal in size. The number of sections used determines the MUZZLE VELOCITY and RANGE of the PROJECTILE being fired.

equatorial orbit An orbit above the earth's equator.

equilibrator A device that balances the weight of a weapon TUBE or LAUNCHER beam so that it can be elevated without difficulty.

equipment, cryptoancillary See CRYPTOANCILLARY EQUIPMENT.

equipment, cryptoproduction See CRYPTOPRODUCTON EQUIPMENT.

equivalent megatonnage (EMT) A unit of measurement for the combined explosions of two or more NUCLEAR WEAPONS on the same target. The theory behind the EMT system is that multiple small weapons will always cause more damage to a single target than one large weapon of the same total weight. For example, three 300-KILOTON WEAPONS dropped on the same city would do more damage than one 900 kiloton weapon could.

error 1. A unintentional deviation from the truth or accuracy. 2. A deviation from a true measurement resulting from instrumental or mechanical defects (or limitations), as opposed to resulting from mistakes. 3. In gunnery, the divergence of a point of impact from the CENTER OF IMPACT.

error, absolute See ABSOLUTE ERROR.

error, deflection See DEFLECTION ERROR.

error, delivery See DELIVERY ERROR.

error, dispersion See DISPERSION ERROR.

error, gross See GROSS ERROR.

error, meters See METERS ERROR.

error, mils See MILS ERROR.

error, parallax See PARALLAX ERROR.

error probable, circular See CIRCULAR ERROR PROBABLE.

error range, probable See PROBABLE ERROR RANGE.

error, range See RANGE ERROR.

error height of burst, probable See PROBABLE ERROR HEIGHT OF BURST.

error in height, probable See PROBABLE ERROR IN HEIGHT.

escalade An attack against a fortified wall using ladders or other ascending devices.

escalation An increase in the scope or violence of a conflict, whether deliberate or unpremeditated. In modern strategy, escalation has a vital role both as an analytical category and an actual policy. Using the analogy of a ladder, strategists think of a potential conflict as arrayed in a series of steps from minimum to maximum violence. Even before an actual war between the NORTH ATLANTIC TREATY ORGANIZATION (NATO) and the WARSAW PACT were to break out, one would perhaps expect an "escalating" succession of increasingly hostile acts. Escalation is used as a strategy when a power is able to dominate the enemy at each step in the ladder, forcing him to decide whether to move up to the next stage of violence and destruction, where he knows that he will again be at a disadvantage. This has led to the belief that NATO must at least match Warsaw Pact capacity at each level, so that the latter would not be tempted to try to seek victory by moving on to the use of theatre nuclear weapons, for example. With matching capacities the Warsaw Pact, if it could not prevail at the conventional level, would be deterred from increasing the risks, and escalation would have been controlled. The theory would be somewhat more suitable were it not NATO, rather than the Warsaw Pact, that is liable to have to "go nuclear." Escalation has its attraction to Western strategic thinking because it acts as an extension of DETERRENCE into war itself. Western thought has tended to concentrate almost entirely on war-deterrence doctrine, at the cost of having very little to say about what to do if deterrence fails. Another reason for escalation seeming an appropriate model for a future war is probably that the major NATO partner, the United States, is geographically separate from the likely MAIN BATTLE AREA and could avoid sustaining any direct damage from a CENTRAL FRONT war, as long as it does not escalate too far. Finally, the official doctrine of NATO, FLEXIBLE RESPONSE, itself assumes a ranking of possible military reactions rather than a predetermined strategic plan. See also FIRE-BREAK, HORIZONTAL ESCALATION, INTRA-WAR BARGAINING, INTER-WAR DETERRENCE. See Bernard Brodie, *Escalation and the Nuclear Option,* (Princeton, N.J.: Princeton University Press, 1966); Dennis M. Gormley and Douglas M. Hart, "Soviet Views on Escalation," *Washington Quarterly,* 7, 4 (Fall 1984).

escalation, horizontal See HORIZONTAL ESCALATION.

escalation control 1. The key concept in NORTH ATLANTIC TREATY ORGANIZATION (NATO) strategy for a possible future war with the WARSAW PACT. All combatants would prefer, all things being equal, to fight at a lower rather than a higher level of ESCALATION. For the sake of simplicity it is possible to think in terms of four successive levels of violence in a NATO–Warsaw Pact conflict. The first is a purely CONVENTIONAL WAR, the second involves the use of BATTLEFIELD NUCLEAR WEAPONS, and the third and fourth respectively see the use of longer-range theater missiles (see INTERMEDIATE NUCLEAR FORCES) and finally CENTRAL STRATEGIC WARFARE. The purpose of NATO's strategy of FLEXIBLE RESPONSE is to match the Warsaw Pact at each level of escalation. This would deter the latter from moving up a level of violence if it were not winning at the existing one. Thus a power theoretically "controls" an escalation ladder when it can increase the violence of its actions without the enemy being able to counter at that level. Such "control" deters escalation by posing an unacceptable risk to the other side. It is generally thought that NATO does not have the parity of force at either the conventional or short-range nuclear levels to be able to claim to exert escalation control at those stages on the ladder. This could present NATO with the serious choice of either escalating to a very advanced level of nuclear confrontation or admitting defeat. Despite its paper strategy of flexible response NATO has no real ability to "control" escalation. See ESCALATION DOMINANCE. 2. The limitations or restrictions placed on NUCLEAR EXCHANGES that would allow a nuclear war to be fought on such a limited scope that it would not destroy the whole human race, only most of it. Compare to ESCALATION MATCHING. See Richard Smoke, *War: Controlling Escalation* (Cambridge, MA: Harvard University Press, 1977).

escalation dominance When one side in a war has a superior force capacity not only at the current level of ESCALATION, but at successive levels. Thus escalation dominance is a "control" in the sense that the enemy cannot escape from a weak position in, for example, the conventional phase by moving on to the use of BATTLEFIELD NUCLEAR WEAPONS, because it would still be at a disadvantage. The dominant side, however, does have that choice: if it is suffering more casualties than it thinks acceptable at one level, it can try to terminate conflict by changing the intensity level to, perhaps, the use of THEATRE NUCLEAR FORCES, where its relative advantage may be even greater. The key to a strategy based on escalation domination is superiority at *every* level.

escalation ladder Herman Kahn's conception from his *On Escalation* (New York: Praeger, 1965) of the gradations of military escalation as a ladder of 44 rungs. Because Kahn conceived of first using nuclear weapons at rung 15, this implied that there were 30 rungs on the ladder involving nuclear weapons. This 30 different varieties approach to nuclear war gave Kahn's escalation ladder great notoriety.

escalation matching A military strategy or train of thought that intimidates an enemy by assuring him that his every move can be matched and overpowered. When used correctly, this technique can cause a smaller, more aggressive nation to "back down," thereby limiting the ESCALATION of a war. Compare to ESCALATION CONTROL.

escape and evasion See EVASION AND ESCAPE.

escape chit A means of identification carried by a soldier or airman operating in a foreign country, normally printed on cloth in the language of the area, and carrying a message promising a reward for assisting the bearer to safety. Replicas of the United States and Allied flags are sometimes superimposed on the escape chit as well. Also known as a BLOOD CHIT.

escape line A planned route to allow personnel engaged in CLANDESTINE OPERATIONS to depart from a site or area when possibility of COMPROMISE or apprehension exists.

escarp That part of a ditch or moat closest to the wall of a FORTIFICATION; the part that faces the enemy is the COUNTERSCARP.

escort 1. A combatant unit(s) or armed guard assigned to accompany and protect a FORCE, CONVOY, train, prisoners, etc. 2. Aircraft assigned to protect other aircraft during a MISSION. 3. To so accompany and protect. 4. An armed guard accompanying persons as a mark of honor. 5. A member of the armed forces assigned to accompany, assist, or guide an individual or group, e.g., an escort officer. 6. A warship with antisubmarine capacity that travels with other warships or cargo ships.

escort guard Those soldiers who supervise and guard PRISONERS OF WAR in camps, enclosures, on work details, during interrogation, and during evacuation or transfer.

escort of the color A ceremony of escorting the COLOR with a COLOR GUARD. Also called the escort of the standard.

espionage Actions designed to obtain government secrets, usually diplomatic or military, through CLANDESTINE OPERATIONS. Compare to: INTELLIGENCE. Henry S. A. Becket, *The Dictionary of Espionage: Spookspeak Into English* (New York: Stein and Day, 1986).

esprit de corps 1. The spirit or morale of a military group; the pride that soldiers take in their military units. Frederick the Great said in 1768 that it existed when a soldier had "a higher opinion of his own regiment than all the other troops in the country." 2. The soldier's belief that his army, and his countrymen, are superior to those of another nation. Lord Chesterfield wrote in a letter of February 7, 1749 that the "silly, sanguine notion, which is firmly entertained here, that one Englishman can beat three Frenchmen, encourages, and has sometimes enabled, one Englishman, in reality, to beat two." Compare to ÉLAN.

essential elements of information The critical items of information regarding the enemy and his environment that a commander needs (by a particular time) to compare with other available INFORMATION and INTELLIGENCE in order to assist him in reaching a logical decision.

essential equivalence 1. The condition that exists when two or more nations have roughly the same amount of nuclear damage potential available to them; although the numbers and sizes of their WARHEADS may differ greatly. Great numbers of missiles that are technologically outdated may be *essentially equivalent* to one completely modern missile. 2. A possible definition of NUCLEAR PARITY.

establishing authority For an amphibious operation, the JOINT FORCE commander who established an AMPHIBIOUS TASK FORCE as a joint force, and who issues the initiating directive for the operation.

establishment 1. An INSTALLATION, together with its personnel and equipment, organized as an operating entity. 2. The table setting out the authorized numbers of men and major equipment in a UNIT or series of FORMATIONS: sometimes called a table of organization or TABLE OF ORGANIZATION AND EQUIPMENT. 3. The collective holders of power in all segments of a society: political, military, social, academic, religious, literary, and so forth. It is always "the establishment" that revolutionaries, whether political, organizational, or otherwise, wish to overthrow so that they and their ideas can become the new establishment. See Adam Yarmolinsky, *The Military Establishment* (New York: Harper & Row, 1971); Leonard Silk and Mark Silk, *The American Establishment* (New York: Basic Books, 1980).

estimate 1. An analysis of a foreign situation, development, or trend that identifies its major elements, interprets the significance of, and appraises the future possibilities and prospective results of the various actions that might be taken. 2. An appraisal of the capabilities, vulnerabilities, and potential COURSES OF ACTION of a foreign nation or combination of nations, which would be consequent to a contemplated national plan, policy, decision, or course of action. 3. An analysis of an actual or contemplated CLANDESTINE OPERATION in relation to the situation in which it is or would be conducted in order to identify and appraise such factors as available and needed assets and potential obstacles, accomplishments, and consequences. See also INTELLIGENCE ESTIMATE. 4. In air interception, a code meaning, "Provide a quick estimate of the height/depth/range/size of designated contact," or "I estimate height/depth/range/size of designated contact as . . ."

estimated expenditure of ammunition The expected number of PROJECTILES fired in a given time period, in determining the FIREPOWER potential of an AREA FIRE weapon.

estimate of the situation, commander's See COMMANDER'S ESTIMATE OF THE SITUATION.

estimate of the situation, logistic See LOGISTIC ESTIMATE OF THE SITUATION.

ET See EMERGING TECHNOLOGY.

evacuation 1. The process of moving any person who is wounded, injured, or ill to or between medical treatment facilities. 2. The clearance of inhabitants, PERSONNEL, animals, or MATERIEL from a given locality. 3. The controlled process of collecting, classifying, and shipping unserviceable or abandoned materiel, to appropriate reclamation, MAINTENANCE, TECHNICAL INTELLIGENCE, or disposal facilities.

evacuation, aeromedical See AEROMEDICAL EVACUATION.

evacuation hospital A mobile medical treatment facility near a front line that provides major medical and surgical care for wounded who are then further evacuated to more permanent facilities.

evacuation policy 1. A COMMAND DECISION, indicating the length in days of the maximum period that inactive military patients may be held within a command for medical treatment. Patients who, in the opinion of responsible medical officers, cannot be returned to DUTY STATUS within the period prescribed are evacuated by the first available means, provided the travel involved will not aggravate their disabilities. 2. A command decision concerning the movement of civilians from the proximity of military operations for security and safety reasons, and including the arrangements for their movement, reception, care, and control. 3. A command policy concerning the evacuation of unserviceable or abandoned MATERIEL, and including the designation of CHANNELS and destinations for evacuated materiel, the establishment of controls and procedures, and the dissemination of condition standards and DISPOSITION instructions.

evaluation 1. In intelligence usage, the appraisal of an item of information in terms of CREDIBILITY, pertinency, and accuracy. Appraisal is accomplished at several stages within the INTELLIGENCE CYCLE, with progressively different contexts. Initial evaluations are focused on the reliability of the SOURCE and the accuracy of the information as judged by data available at or close to their OPERATIONAL LEVELS. Later evaluations, by intelligence analysts, are primarily concerned with verifying the accuracy of information and may, in effect, convert information into intelligence. Appraisal or evaluation of items of information or intelligence is indicated by a standard letter-number system. The evaluation of the reliability of sources is designated by a letter from A through F, and the accuracy of the information is designated by a numeral from 1 through 6. These are two entirely independent appraisals, and these separate appraisals are made in accordance with the system indicated below. Thus, information adjudged to be "probably true," received from a "usually reliable source," is designated "B-2" or "B2," while information of which the "truth cannot be judged," received from a "usually reliable source," is designated "B-6" or "B6."

Reliability of Source	Accuracy of Information
A—Completely reliable	1—Confirmed by other sources
B—Usually reliable	2—Probably true
C—Fairly reliable	3—Possibly true
D—Not usually reliable	4—Doubtful
E—Unreliable	5—Improbable
F—Reliability cannot be judged	6—Truth cannot be judged

2. A subjective determination, accomplished jointly by the several major subordinate commands of the utility or military value of a hardware item or system, real or conceptual, to the potential user. 3. A personal performance appraisal for military personnel.

evasion and escape The procedures and operations whereby military personnel and others emerge from an enemy-held or hostile area to areas under friendly control.

evasion and escape net The organization within enemy-held or hostile areas that operates to receive, move, and exfiltrate military personnel or selected individuals to friendly control.

event templating An analytical INTELLIGENCE technique used to help a commander predict enemy intentions by comparing enemy activities to time and space considerations. See also DOCTRINAL TEMPLATING.

EW See ELECTRONIC WARFARE.

exclusion area A restricted area containing a SECURITY interest or other matter of a vital nature.

exclusive economic zone (EEZ) A concept in INTERNATIONAL LAW that has evolved since World War II to distinguish that area of adjacent waters over which a nation has unilaterally extended its sovereignty in order to control its economic resources (mainly for fishing and mining). The 1982 United Nations (UN) Convention on the Law of the Sea formally codified the EEZ concept to allow coastal states exclusive jurisdiction to the seabed and waters up to a maximum for 200 miles from their shores. However, not all states, most notably the United States, accept the UN Convention and recognize the exclusive economic zones of other states. See David Attard, *The Exclusive Economic Zone in International Law* (Oxford: Oxford University Press, 1986).

exec An EXECUTIVE OFFICER.

executing commander A commander to whom NUCLEAR WEAPONS are released for delivery against specific targets or in accordance with approved plans. See also RELEASING COMMANDER.

executive 1. Any of the highest managers in an organization. 2. That branch of government concerned with the implementation of the policies and laws created by a legislature.

executive action 1. The last stage of the policymaking process. 2. An INTELLIGENCE community term for assassination.

executive agreement A term that covers a wide variety of international agreements and understandings that are reached by the governments concerned in the course of administering their relationships with one another. The executive-agreement device permits a president to enter into open or secret arrangements with a foreign government without the advice and consent of the U.S. Senate. Gary J. Schmitt, "Executive Agreements and Separation of Powers: A Reconsideration," *American Journal of Jurisprudence* Vol. 28 (1983).

executive officer The second in command of a naval ship or military organization. Often called the XO or exec, he usually manages the day to day routine of the ship or unit. In the Air Force or the Army the XO is usually more of an administrator; another officer is designated to be the deputy commander.

executive order An order issued by the president of the United States by virtue of the authority vested in him by the U.S. Constitution or by an act of the U.S. Congress. It has the force of law.

exercise A military maneuver or simulated wartime operation involving planning, preparation, and execution. It is carried out for the purpose of training and evaluation. It may be a combined, joint, or single service exercise, depending on participating organizations. The annual military exercises in NATO have been called Reforger (for Return of Forces to Germany). See also MANEUVER.

exercise, command post See COMMAND POST EXERCISE.

exercise, controlled See CONTROLLED EXERCISE.

exercise, disciplinary See DISCIPLINARY EXERCISE.

exercise, field See FIELD EXERCISE.

exercise, free play See FREE PLAY EXERCISE.

exfiltration The removal of personnel or units from areas under enemy control by stealth, deception, surprise, or clandestine means.

existence load The items other than those in the FIGHTING LOAD that are required to sustain or protect the combat soldier. They may be necessary for increased personal and environmental protection, and are not normally carried by the individual.

exit road A road leading out of a BEACHHEAD or LANDING AREA into the area of subsequent operations.

exoatmospheric Taking place outside the earth's atmosphere, generally considered as occurring at altitudes above 100 kilometers. An exoatmospheric interceptor reaches its target in space. Compare to ENDOATMOSPHERIC.

exoatmospheric defense A generic term for any system that intercepts and destroys incoming missiles from above the earth's atmosphere. Of all the theoretical systems for accomplishing this, the most highly publicized is the STRATEGIC DEFENSE INITIATIVE, also known as "Star Wars." The Soviet ABM, code named "Galosh" by the NORTH ATLANTIC TREATY ORGANIZATION (NATO), is an exoatmospheric defensive weapon.

Exocet A French-built anti-ship missile used extensively in the 1980s in both the Falklands War and the Persian Gulf.

expeditionary force An armed force organized to accomplish a specific OBJECTIVE in a distant place or foreign country. The United States forces in World War I were called the AEF for American Expeditionary Force.

expeditionary troops In amphibious operations, all troops of all services assigned to a joint expeditionary force for all operations ashore.

expert 1. A specialist in some skill or area of knowledge. 2. The United States Army's highest rating for ability with a rifle; above MARKSMAN or SHARPSHOOTER.

exploder A device designed to generate an electric current in a FIRING CIRCUIT after deliberate action by the user in order to initiate an EXPLOSIVE CHARGE or charges.

exploitation 1. Taking full advantage of success in battle and following up on initial gains. 2. Taking full advantage of any information that has come to hand for TACTICAL or STRATEGIC purposes. 3. An OFFENSIVE operation that usually follows a successful ATTACK and is designed to disorganize the enemy in depth. All too often a battle has been won and the war lost or prolonged because a commander failed to adequately exploit the situation. The ancient Roman historian Livy (59 B.C.–17 A.D.) wrote that Hannibal, after the Battle of CANNAE was urged by Maharbal, his cavalry commander, to exploit the victory and march on Rome at once. But Hannibal hesitated and never managed to enter Rome. Livy reports Maharbal saying: "In very truth the gods bestow not on the same man all their gifts; you show how to gain a victory, Hannibal: you know not how to use one." Hannibal is just one example of many commanders who failed to exploit a victory and thus lost the war. The same could be said of the Japanese after Pearl Harbor and the German Army at Dunkirk during World War II.

exploiting force A previously uncommitted unit ordered to capitalize on a successful attack by striking to seize deep objectives and command-and-control facilities, destroy enemy reserves, and deny the enemy an opportunity to escape or reorganize. Also see FOLLOW AND SUPPORT.

explosion Almost instantaneous combustion accompanied by a violent bursting of extreme gas pressure.

explosive A chemical composition that burns rapidly enough to cause an explosion.

explosive, binary See BINARY EXPLOSIVE.

explosive, high See HIGH EXPLOSIVE.

explosive, low See LOW EXPLOSIVE.

explosive, plastic See PLASTIC EXPLOSIVE.

explosive, sheet See SHEET EXPLOSIVE.

explosive charge Explosive used in firing a gun, whether a PROPELLING CHARGE which throws the projectile, or a bursting charge, which breaks the casing of a PROJECTILE to produce demolition, fragmentation, or chemical action.

explosive devices, improvised See IMPROVISED EXPLOSIVE DEVICES.

explosive ordnance All MUNITIONS containing EXPLOSIVES, NUCLEAR FISSION or FUSION materials, and BIOLOGICAL and CHEMICAL AGENTS. This includes BOMBS and WARHEADS, GUIDED and BALLISTIC MISSILES, ARTILLERY, MORTAR, ROCKET, and SMALL ARMS ammunition, MINES, TORPEDOES, DEPTH BOMBS, and other devices.

explosive projectile, high See HIGH-EXPLOSIVE PROJECTILE.

explosive train That portion of a fuze or a fuze system consisting of explosive components, such as PRIMER DETONATOR, BOOSTER, etc., necessary to cause functioning of a WARHEAD or DESTRUCTOR. Compare to: IGNITER TRAIN.

expropriation The confiscation of private property by a government, which may or may not pay a portion of its value in return.

extended defense A form of POSITION DEFENSE employed on a wide front. It is characterized by limited mutual support, great depth of position, and withholding of a strong reserve.

extended deterrence A strategy where one nuclear power seeks to deter another not only from attacking it, but also from attacking its allies. This is the basis of the NORTH ATLANTIC TREATY ORGANIZATION's (NATO) FLEXIBLE RESPONSE doctrine, which ultimately depends on the American NUCLEAR UMBRELLA. But the United States is not the only country capable of offering extended deterrence to its allies. The Soviet Union's nuclear capacity implicitly deters NATO from attacks on other members of the WARSAW PACT. A justification that is sometimes made for French and British independent nuclear forces is that if the United States for some reason were to remove the nuclear umbrella from other non-nuclear European members, NATO would benefit from the extension of Anglo-French deterrence.

If the United States were to withdraw its defense of Europe, it is highly questionable whether the much smaller nuclear forces of France or the United Kingdom could offer credibility as an overall replacement. In fact one of the principal justifications used by France for building its FORCE DE FRAPPE has been that no country can credibly extend deterrence. France maintains that it does not make sense to believe that a nuclear-armed nation, which might otherwise be safe from nuclear attack, would risk nuclear devastation on behalf of another nation. Instead of extended deterrence, this doctrine, which is by no means unique to French analysts, stresses that a country own nuclear weapons solely for the purpose of self-SANCTUARIZATION. See Paul Joseph, "Making Threats: Minimal Deterrence, Extended Deterrence, and Nuclear Warfighting," *Sociological Quarterly* Vol. 26, No. 3 (Fall 1985); Paul K. Huth, *Extended Deterrence and the Prevention of War* (New Haven, CT: Yale University Press, 1988).

exterior ballistics A subdivision of BALLISTICS which deals with the phenomena associated with the aerodynamic performance of missiles or projectiles.

exterior ballistic table A table containing data on the trajectories of projectiles under various conditions. See also BALLISTIC TABLE.

exterior lines See LINES OF OPERATION.

external reinforcing force A strengthening force that is stationed outside its intended AREA OF OPERATIONS, principally in peacetime.

extradition The surrender by one state to another of an individual accused or convicted of an offense in the second state.

extraterritoriality A state's exercise of its authority and laws outside its physical limits, such as on its ships at sea, over its own soldiers who commit crimes in foreign countries, or in the residences of its diplomats stationed abroad. See David Leyton-Brown, "Extraterritoriality in Canadian-American Relations," *International Journal* Vol. 36, No. 1 (Winter 1980–81).

extremely low frequency (ELF) The only frequency able to transmit messages to submarines that do not have radio-receiving equipment (such as antennas) protruding from their hulls. It is also preferred for underwater use because it is inaudible to most hostile TRACKING systems. However, the transmitting aerials must be so large, and the speed of transmission so slow, that such systems are extremely expensive and cumbersome.

exudation The emission of any substance (usually oil, tar, or a gas) from an EXPLOSIVE item, generally the results of chemical reaction or pressure due to thermal changes.

eyes left (right) A command given to troops in march formation to turn head and eyes to the left (right). The movement constitutes a SALUTE to a reviewing party.

eyes right See EYES LEFT (RIGHT).

F

F A designation for a fighter aircraft, such as the F-4 Phantom or F-15 Eagle.

Fabian strategy Avoiding a decisive battle with a superior force but constantly seeking to delay and harness it. This technique is named after the Roman general Fabius, who used it to defeat Hannibal during the Second Punic War (218–201 B.C.).

FAC See FORWARD AIR CONTROLLER.

facepiece The airtight part of a PROTECTIVE MASK (a gas mask), which fits over the face of the wearer. It consists of a faceblank, eyepieces, an outlet valve, a head harness, and other components, depending upon the type of mask.

facing distance The space between men standing in ranks, a distance of 14 inches, calculated as the smallest space in which a soldier can carry out DRILL facings (right, left, about face).

factor of safety The extra strength built into a structure or mechanism to provide a margin of safety for loads or stress above those normally expected. See also SAFETY FACTOR.

fail-safe 1. Used to describe NUCLEAR WEAPONS security procedures that attempt to prevent the unintentional launching of an attack. The possibility of an accidental nuclear war being triggered either by mechanical malfunction, unauthorized officers or a deranged person has always provoked horror. To protect against such a disaster there are numerous mechanical and human checks built into the launch-command system. These range from very complex CODE-WORDS and AUTHENTICATION procedures to mechanical requirements for two or more officers independently and simultaneously to turn certain keys before a missile can be armed. All such systems are set up to make launch impossible and render weapons harmless in the event of any failure in the correct authentication procedure—hence the term. 2. The term also applies to the fact that if the central government of a nation should *fail* (lose radio contact, be destroyed, etc.), the individual bases containing missiles would be *safe* because they would not launch in the absence of specific consent. 3. A term descriptive of FUZE design features whereby a component failure prevents the fuze from functioning. 4. In aircraft and some missile-launching equipment, an item designed to continue operating normally when its controlling device fails, so as not to increase the emergency. See Sidney Hook, *The Fail-Safe Fallacy* (New York: Stein & Day, 1963).

fail-deadly The opposite of FAIL-SAFE; a fail-deadly system allows nuclear weapons to be fired without authoritative consent. If the central government of a nation should *fail* (lose radio contact, be destroyed, etc.), the individual bases containing missiles would be allowed to be *deadly* by launching in spite of the absence of specific consent. This policy is not known to be in effect by any nuclear power.

faker A friendly aircraft that simulates a hostile aircraft in an AIR DEFENSE exercise.

fall in A command to form ranks, to create a formation.

fall out A command to break ranks, to leave a formation.

fallout 1. The precipitation to earth of radioactive particle matter from a NUCLEAR CLOUD. 2. The particulate matter itself in a nuclear cloud. Fallout is the secondary consequence of a nuclear explosion, consisting of the dust particles and water droplets made radioactive by the initial explosion and thrown up into the atmosphere. The amount of fallout, and the area that it covers, depends on a variety of factors, of which the principal one is whether the explosion was an AIRBURST or a GROUND BURST. An airburst produces considerably less fallout because its nuclear FIREBALL does not actually touch the ground, resulting in much less irradiation of dust and other material particles. A ground burst, on the other hand, causes a huge crater, vaporizing the soil and scattering the irradiated material, to create a very large amount of fallout. Some factors that mitigate against this simple picture of airbursts as "clean" and ground bursts as "dirty" are atmospheric conditions and wind patterns. In dry conditions with strong winds, that fallout an airburst does produce can be spread throughout the upper atmosphere over very large areas, whereas the fallout from a ground burst may be contained over a fairly small area around GROUND ZERO.

fallout pattern The geographic shape of the area in which FALLOUT will occur after a nuclear explosion. This pattern is influenced a great deal by landscape and weather. It is usually a cigar-shaped area, with the wider end over the target and the long tip downwind.

fallout safe height of burst The height of burst at or above which no militarily significant fallout will be produced as a result of a nuclear weapon detonation.

false parallax The apparent vertical displacement of an object from its true position when viewed stereoscopically, due to movement of the object itself as well as to change in the point of observation.

famished In AIR INTERCEPTION, a code meaning, "Have you any instructions for me?"

fascines Bundles of sticks used as defensive OBSTACLES.

fascism 1. A political philosophy or movement that advocates governance by a dictator at the head of a hierarchically organized, strongly ideological party, and that maintains social and economic regimentation through violence, intimidation, and the arbitrary use of power. 2. A mass-based reactionary political movement in an industrialized nation which, through the means of a charismatic leader, espouses NATIONALISM in the extreme. The main difference between totalitarian fascism and totalitarian communism is that fascism professes sympathy toward many aspects of capitalism and seeks to resolve conflicts between capital and labor by using the state to enforce their relationship in the interest of full employment and high productivity. The "classic" fascist regimes of the twentieth century were the World War II-era dictatorships of Hitler in Germany, Mussolini in Italy, and Franco in Spain. See also TOTALITARIANISM. See Walter Laqueur, editor, *Fascism: A Reader's Guide* (Berkeley: University of California Press, 1976); Stanley G. Payne, *Fascism: Comparison and Definition* (Madison: University of Wisconsin Press, 1980).

fast-burn booster A BALLISTIC MISSILE that can burn out much more quickly than current versions, possibly before exiting the atmosphere entirely. Such a rapid burnout would complicate BOOST-PHASE defenses against such missiles.

fast track 1. A railroad track for express trains. 2. By analogy a career path taken by those members of an organization who are moving rapidly to the top. For example, a military officer on such an informal fast track would start as a second lieutenant. After two years as a platoon leader he moves up to first lieutenant. Three years later he makes captain. After six years as a captain he makes major. Another five years and he is a lieutenant colonel; four years later a full colonel. If he is promoted to brigadier general three years later, he will have risen to general in 23 years after his commission.

fatigue Military duties that are not for the purpose of either training or combat and are usually associated with cleaning, moving supplies, and other functions.

fatigues The work uniform worn while on FATIGUE duty. Compare to BATTLE DRESS UNIFORM.

FDC See FIRE DIRECTION CENTER.

FEBA See FORWARD EDGE OF THE BATTLE AREA.

feed belt see AMMUNITION BELT.

feet dry In AIR INTERCEPTION, CLOSE AIR SUPPORT, and AIR INTERDICTION, a code meaning, "I am, or contact designated is, over land."

feet wet In AIR INTERCEPTION, CLOSE AIR SUPPORT, and AIR INTERDICTION, a code meaning, "I am, or contact designated is, over water."

feint An OFFENSIVE operation intended to draw the enemy's attention away from the area of a MAIN ATTACK, inducing the enemy to move his RESERVES or shift his FIRE SUPPORT in reaction to the feint. Feints must necessarily appear real, therefore, some contact with the enemy is required. B.H. LIDDELL HART wrote: "In war the power to use two fists is an inestimable asset. To feint with one fist and strike with the other yields an advantage, but a still greater advantage lies in being able to interchange them—to convert the feint into the real blow if the opponent uncovers himself." Compare to DEMONSTRATION.

fellow traveler 1. A person who is sympathetic to communist thinking but is not formally a Communist party member. 2. A person who is passively in agreement with a cause or group. See Lewis S. Feuer, "The Fellow-Travellers," *Survey* 20, No. 2–3 (Spring-Summer 1974).

ferret An aircraft, ship, or vehicle especially equipped for the detection, location, recording, and analyzing of ELECTROMAGNETIC RADIATION.

field 1. Any place away from one's normal work station. 2. An outdoor location used for training troops. 3. The locus of combat.

field army Administrative and tactical organization composed of a headquarters, certain organic Army troops, service support troops, a variable number of CORPS, and a variable number of DIVISIONS.

field artillery 1. Equipment, supplies, ammunition, and personnel involved in the use of CANNON, ROCKET, or SURFACE-TO-SURFACE MISSILE launchers. Field-artillery cannons are classified according to caliber as:

Light	120mm and less
Medium	121–160mm
Heavy	161–210mm
Very heavy	greater than 210mm

2. A basic branch and arm of an army. The branch name identifies personnel and units that employ cannons, rockets and missile systems capable of TARGET ACQUISITION to assist in land combat operations. 3. Artillery weapons that are sufficiently mobile to accompany and support infantry, mechanized, armored, airborne, and airmobile units in the field. Their primary mission is to engage ground targets.

See also DIRECT-SUPPORT ARTILLERY; GENERAL SUPPORT ARTILLERY.

field artillery cannon calibration The comparison of the MUZZLE VELOCITY of a given artillery piece with some accepted standard performance. That standard may be selected arbitrarily from the performance of a group of weapons being calibrated together, as in comparative CALIBRATION, or it may be the standard defined in the FIRING TABLES, as in absolute calibration.

field artillery observer A person who watches the effects of artillery fire, adjusts the CENTER OF IMPACT of that fire onto a target, and reports the results to the firing agency. See also NAVAL GUNFIRE SPOTTING TEAM; SPOTTER.

field artillery survey A survey consisting of topographical operations necessary to construct a firing chart for UNOBSERVED FIRE. The main object of the survey is to determine, with sufficient exactness, the relative locations of PIECEs and targets both horizontally and vertically.

field artillery tactical operations center A facility within which are merged targeting, operations, and fire control for field artillery support operations.

field exercise An exercise conducted in the field under simulated war conditions in which troops and armament of the friendly side represent themselves while those representing the enemy may be imaginary or have their roles carried out by other friendly troops.

field fortifications An EMPLACEMENT or shelter of a temporary nature that can be constructed with reasonable facility by units requiring no more than minor engineering supervision and participation.

field fortifications, deliberate See DELIBERATE FIELD FORTIFICATIONS.

field grade A classification of officers ranking above a CAPTAIN and below a BRIGADIER GENERAL. Field grade includes COLONELS, LIEUTENANT COLONELS, and MAJORS. The term comes from the last century, when only these ranks had sufficient experience to command a REGIMENT in the field.

field hospital A nonfixed medical-treatment facility.

field kitchen A mobile or temporary kitchen in the field or a temporary installation utilizing field-kitchen equipment.

field manual A manual containing instructional, informational, and reference material relative to military training and operations. It is the primary means of promulgating military doctrine, tactics, and techniques.

field marshal See MARSHAL.

field of fire The area that a weapon or group of weapons may cover effectively with fire from a given position.

field of view The total solid angle available to a gunner who is looking through the gunsight.

field rations Articles of food (not money) authorized for issue to troops in the field.

field strip 1. To dissassemble a weapon so that its parts can be cleaned. 2. To shred a cigarette butt so as to leave no noticeable residue.

field train A UNIT TRAIN not required for the immediate support of COMBAT ELEMENTS. Field trains may include kitchen and baggage trains, administrative trains, heavy maintenance and water vehicles, and those ammunition, fuel, and lubricants trucks not required for the direct support of troops in an immediate ENGAGEMENT. They are located rearward to prevent interference with the tactical operation. See also COMBAT TRAINS.

field type A term used to describe equipment, troops, or units utilized primarily to carry out a combat mission.

fieldworks The temporary defenses built by an army in the field; non-permanent FORTIFICATIONS.

fifth column Those traitors within a country who engage in espionage or sabotage while waiting to join forces with invading enemy soldiers. The term dates from 1936 during the Spanish Civil War, when a rebel general advanced on Madrid with four columns of troops and boasted that a "fifth column" awaited him within the city. During World War II the term came to be used as a description for anyone who was secretly sympathetic with the Germans.

fighter A military aircraft designed to intercept and destroy enemy aircraft, to protect friendly bomber or naval forces, or to provide close fire support to ground forces. The basic mission of a fighter force is to maintain TACTICAL air superiority.

fighter, all-weather See ALL-WEATHER FIGHTER.

fighter, day See DAY FIGHTER.

fighter-bomber A dual mission aircraft designed both for air-to-air combat and bombing missions usually in support of ground forces.

fighter cover The maintenance of a number of fighter aircraft over a specified area or force for the purpose of repelling hostile air activities and to protect friendly bombers.

fighter liaison officer A member of a TACTICAL AIR CONTROL PARTY who meets all the prerequisites for being a FORWARD

AIR CONTROLLER and is qualified to control AIR STRIKES. He advises the air liaison officer and ground commander's staff on the capabilities, limitations, and employment of CLOSE AIR-SUPPORT resources.

fighter sweep An offensive mission by FIGHTER aircraft to seek out and destroy enemy aircraft or TARGETS OF OPPORTUNITY in an allotted area of operations.

fighting War, according to CLAUSEWITZ. In *On War;* he wrote that: "War in its literal meaning is fighting, for fighting alone is the efficient principle in the manifold activity which in a wide sense is called war." Confederate General Thomas Jonathan "Stonewall" Jackson agreed, "War means fighting . . . the business of the soldier is to fight . . . to find the enemy and strike him; to invade his country, and do him all possible damage in the shortest possible time."

fighting load The items of individual clothing, equipment, weapons, and ammunition that are carried by, and are essential to, the effectiveness of the combat soldier and the accomplishment of the immediate mission of the unit when the soldier is on foot. See also EXISTENCE LOAD.

fighting through A DISMOUNTED infantry technique of attacking and overrunning an enemy position without resorting to a STANDUP ASSAULT. The attacking unit takes advantage of cover, concealment, and short rushes, thus reducing its vulnerability while suppressive fire continues from overwatching positions. The attacking force seeks to penetrate enemy defenses in a narrow SECTOR, and then overcome remaining defenses from the flank or rear.

file A COLUMN of soldiers who are situated one behind another.

file-closer The last soldier in a FORMATION.

filler personnel Individuals of suitable grade and skill initially required to bring a unit or organization to its authorized strength.

filtering The process of interpreting reported information on movements of aircraft, ships, and submarines in order to determine their probable true TRACKS and, where applicable, heights or depths.

final bomb-release line An imaginary line around a defended area or objective over which a BOMBER should release its last bomb in order to hit the far edge of the area or objective.

final coordinating line A line close to an enemy position used to coordinate the lifting and shifting of SUPPORTING FIRE with the final deployment of MANEUVER elements. It should be recognizable on the ground.

final protective fire An immediately available, pre-arranged barrier of fire designed to impede enemy movement across defensive lines or areas.

final protective line A line selected where an enemy assault is to be checked by interlocking fire from all available weapons. A final protective line may be parallel with, or oblique to, the front of a POSITION.

fine sight Adjustment of the sight of a gun so that only the tip of the front sight can be seen through the notch of the rear sight. A less accurate adjustment is called COARSE SIGHT.

Finlandization A term used to refer to the limitation of a smaller nation's sovereignty by the establishment of a *modus vivendi* with a larger and stronger close neighbor. The term originated from a 1948 treaty between Finland and the Soviet Union, through which Finland retained its independence and promised in turn to maintain Finnish neutrality, limit foreign-policy initiatives, and inhibit domestic political behavior. The potential

"Finlandization" of other close neighbors of the Soviet Union is frequently the subject of international speculation. See William Pfaff, "Finlandization," *Atlantic Community Quarterly* 18, 4 (Winter 1980–81); Fred Singleton, "The Myth of 'Finlandisation'," *International Affairs (Great Britain)* 57, 2 (Spring 1981); Paul Malone, " 'Finlandization' as a Method of Living Next Door to Russia," *International Perspectives* (July–August 1981).

fire 1. To shoot at a target or enemy. It was Admiral David Farragut who in 1863 wrote that: "The best protection against the enemy's fire is a well directed fire from our own guns." See also COUNTERFIRE. 2. The command given to discharge a WEAPON(s). 3. To detonate the main EXPLOSIVE CHARGE of a projectile by means of a firing system.

fire, accuracy of See ACCURACY OF FIRE.

fire, adjustment of See ADJUSTMENT OF FIRE.

fire, aimed See AIMED FIRE.

fire, area See AREA FIRE.

fire, assault See ASSAULT FIRE.

fire, band of See BAND OF FIRE.

fire, barrage See BARRAGE FIRE.

fire, base of See BASE OF FIRE.

fire, call See CALL FIRE.

fire, cease See CEASE FIRE.

fire, close support See CLOSE SUPPORTING FIRE.

fire, concentrated See CONCENTRATED FIRE.

fire, continuous See CONTINUOUS FIRE.

fire, counterpreparation See COUNTERPREPARATION FIRE.

fire, covering See COVERING FIRE.

fire, deep supporting See DEEP SUPPORTING FIRE.

fire, defense See DEFENSE FIRE.

fire, delayed contact See DELAYED CONTACT FIRE.

fire, deliberate See DELIBERATE FIRE.

fire, destruction See DESTRUCTION FIRE.

fire, direct See DIRECT FIRE.

fire, direct supporting See DIRECT SUPPORTING FIRE.

fire, enfilade See ENFILADE.

fire, frontal See FRONTAL FIRE.

fire, grazing See GRAZING FIRE.

fire, ground See GROUND FIRE.

fire, hang See HANG FIRE.

fire, harassing See HARASSING FIRE.

fire, hold See HOLD FIRE.

fire, indirect See INDIRECT FIRE.

fire, interdiction See INTERDICTION FIRE.

fire, limit of See LIMIT OF FIRE.

fire, marching See MARCHING FIRE.

fire, marking See MARKING FIRE.

fire, massed See MASSED FIRE.

fire, neutralization See NEUTRALIZATION FIRE.

fire, obscuration See OBSCURATION FIRE.

fire, observed See OBSERVED FIRE.

fire, plunging See PLUNGING FIRE.

fire, prearranged See PREARRANGED FIRE.

fire, precision See PRECISION FIRE.

fire, predicted See PREDICTED FIRE.

fire, preparation See PREPARATION FIRE.

fire, protective See PROTECTIVE FIRE.

fire, quick See QUICK FIRE.

fire, radar See RADAR FIRE.

fire, registration See REGISTRATION FIRE.

fire, ricochet See RICOCHET FIRE.

fire, scheduled See SCHEDULED FIRE.

fire, screening See SCREENING FIRE.

fire, searching See SEARCHING FIRE.

fire, seen See SEEN FIRE.

fire, shifting See SHIFTING FIRE.

fire, slow See SLOW FIRE.

fire, spreading See SPREADING FIRE.

fire, supporting See SUPPORTING FIRE.

fire, suppressive See SUPPRESSION.

fire, sweeping See SWEEPING FIRE.

fire, time See TIME FIRE.

fire, unobserved See UNOBSERVED FIRE.

fire, zone See ZONE FIRE.

fire and maneuver A tactical technique, usually an extension of bounding overwatch, used once contact with the enemy is gained. In this technique one element of a force moves while another provides a BASE OF FIRE. See also: MOVEMENT TECHNIQUES, COVERING FIRE.

fire area, free See FREE-FIRE AREA.

fire area, restrictive See RESTRICTIVE FIRE AREA.

firearm Any hand-held device that shoots off projectiles using a propellant (gunpowder) and a triggering mechanism. Rifles and pistols are the most typical examples.

fireball The luminous sphere of hot gases that forms a few millionths of a second after the detonation of a NUCLEAR WEAPON and that immediately starts expanding and cooling.

fire base A FIRE SUPPORT base; an area in a forward COMBAT ZONE from which artillery and logistic support is provided. Compare to FIRE-SUPPORT BASE.

fire bomb A container, sometimes an auxiliary aircraft fuel tank, filled with a gelled gasoline mixture and equipped with igniter assemblies. See NAPALM.

firebreak In the language of nuclear strategy, a fire-break is a point in time, or a stage during an ESCALATION process, during which the whole nature of a war might be negotiated. A fire-break might occur, for example, after the initial use of BATTLEFIELD NUCLEAR WEAPONS as an opportunity for intra-war bargaining to avoid moving on to the use of THEATRE NUCLEAR FORCES or INTERCONTINENTAL BALLISTIC MISSILES (ICBMs). At a higher level of escalation, a fire-break might occur after the exchange of two or three ICBMs; in such circumstances both sides might assess the sustained damage, consider their strategy, and seek further rounds of intra-war bargaining.

fire chart, observed See OBSERVED FIRE CHART.

fire control The control of all operations associated with the application of FIRE on a target.

fire-control equipment Equipment required and used to directly aim guns or controlled missiles at a particular target. Fire-control equipment includes all instruments used in calculating and adjusting the proper ELEVATION and DEFLECTION of guns or missiles in flight. Included are such items as RADARS, telescopes, RANGE FINDERS, PREDICTORS, DIRECTORS other computers, power plants, and communication-control systems connecting these elements.

fire-control grid A system of lines that divide a military map into squares, the distance between any two parallel lines representing 1,000 yards or 1,000 meters, depending on the type of map. Maps using the fire control grid are of sufficiently large scale to be useful in FIRE CONTROL.

fire-control party, shore See SHORE FIRE CONTROL PARTY.

fire-control radar RADAR used to provide target information input for a weapon FIRE-CONTROL system.

fire-control system, integrated See INTEGRATED FIRE CONTROL SYSTEM.

fire direction The tactical employment of FIRE-POWER; the exercise of TACTICAL COMMAND of one or more units in the selection of targets, the concentration or distribution of fire, and the allocation of ammunition for each mission. The term also applies to the methods and techniques used in FIRE-DIRECTION CENTERS to convert target information into appropriate fire commands.

fire direction center (FDC) That element of a COMMAND POST, consisting of gunnery and communication personnel and equipment, by means of which the commander exercises FIRE DIRECTION, FIRE CONTROL, or both. The fire-direction center receives target intelligence and requests for fire, and translates them into appropriate fire direction.

fire direction net The communications system linking ground observers, liaison officers, air observers, and firing batteries with the FIRE-DIRECTION CENTER for the purpose of FIRE CONTROL.

fire discipline 1. The ability to withhold FIRE until the enemy is close enough for the fire to be maximally effective. This is best exemplified by the 1775 Battle of Bunker Hill admonition of the American commander William Prescott: "Don't fire 'til you see the whites of their eyes." 2. To get troops to fire at all. After World War II, S. L. A. Marshall published a major analysis of the U.S. Army infantry and found that only 15 percent of riflemen fired at the enemy. In his *Men Against Fire* (New York: Morrow, 1947), he wrote that "out of an average one hundred men along the line of fire during the period of an encounter, only fifteen men on average would take any part with the weapons. . . . In the most aggressive infantry companies, under the most intense local pressure, the figure rarely rose above 25% of total strength from the opening to the close of an action." In contrast, crew-manned weapons, such as artillery, would almost always fire. The deciding factor was often whether the soldier was alone or in a group. Marshall's analysis, long accepted as classic, has recently been discredited since it has been found that "the systematic collection of data that made Marshall's ratio of fire so authoritative appears to have been invented." Roger Spiller, "S. L. A. Marshall and the Ratio of Fire," *Journal of the Royal United Services Institute* (December 1988). Also see Fredric Smoler, "The Secret of the Soldiers Who Didn't Shoot," *American Heritage* (March 1989).

fire fight 1. An exchange of fire between opposing units. 2. That phase of an attack usually following the APPROACH MARCH and DEPLOYMENT, and coming before the ASSAULT.

fire for effect 1. A specific assignment given to a FIRE UNIT, often after the BRACKETING of a target when the observer thinks that the next ROUNDS will be deposited directly on the target. 2. An order used to alert the WEAPON/BATTERY area and indicate that the message following is a CALL FOR FIRE.

fire lane A path cleared for gunfire in wooded or overgrown areas.

fire line, coordinated See COORDINATED FIRE LINE.

fire plan 1. A tactical plan to coordinate the use of a unit's or formation's weapons. The term primarily refers to ARTILLERY PREPARATION for a major assault. 2. Standing directions to the members of a unit for preventing, reporting, or extinguishing fires.

firepower A measure of the amount of MUNITIONS (bullets, shells, mortar bombs, and other devices) that a military unit can deliver on a target. In the recent history of warfare, the concept firepower has achieved increasing dominance, at the expense of maneuver, tactics, and even courage and morale. Although the trend first appeared with technical developments such as the breech-loading rifle and rapid-firing field gun in the last third of the nineteenth century, it was only during World War I that military experts became convinced that sheer firepower decided battles. The impossibility of advancing in strength against the machine guns of 1914–18, and the suppressive power of artillery bombardment, were the conclusive factors. Since then, the firepower of ordinary infantry units has increased enormously. All major armies equip their ordinary infantry with personal automatic WEAPONS capable of firing hundreds of rounds a minute. Small squads carry portable antitank and antiaircraft weapons as powerful as the specialist weapons that would have been available only to artillery units during World War II. See Robert S. Fairweather, Jr., "A New Model for Land Warfare: The Firepower Dominance Concept," *Air University Review* 32 (November-December 1980).

firepower umbrella An area of specified dimensions defining the airspace boundaries over a naval force at sea, within which the fire of antiaircraft weapons can endanger enemy aircraft, and within which special procedures have been established for the identification and operation of friendly aircraft. See also AIR-DEFENSE OPERATIONS AREA.

fire sights, direct See DIRECT-FIRE SIGHTS.

fire storm A stationary, mass incendiary fire, generally initiated in a built-up urban area, that generates strong, inrushing winds from all sides. The winds keep the fires from spreading while adding fresh oxygen to increase their intensity.

fire support 1. The collective employment of MORTARS, FIELD ARTILLERY, CLOSE AIR SUPPORT, and naval gunfire in support of a battle plan. 2. Assistance to those elements of a ground force closing with the enemy (such as infantry and armor units), rendered by field artillery fire, naval fire, and close air support. NAPOLEON wrote that, "The better the infantry, the more it should be economized and supported by good batteries. Good infantry is without doubt the sinews of an army; but if it has to fight a long time against very superior artillery, it will become demoralized and will be destroyed."

fire-support area An appropriate MANEUVER area assigned to FIRE-SUPPORT ships from which to deliver gunfire support for an AMPHIBIOUS OPERATION.

fire-support base A forward area where artillery and mortars are placed to provide close fire support to ground forces. The United States Army often provided such forward bases in Vietnam.

fire-support coordination center A single location in which communications facilities and personnel incident to the coordination of all forms of fire support are centralized. See also SUPPORTING ARMS, COORDINATION CENTER.

fire-support coordination line A line established by the appropriate ground commander to insure coordination of FIRE not under his control but which may affect current tactical operations. Supporting elements may attack targets forward of the fire support coordination line without prior coordination with the ground-force commander, provided the attack will not produce adverse surface effects on, or to the rear of, the line. Such a line is used to coordinate the firing of air, ground, or sea weapons systems using any type of ammunition against surface targets, and should follow well-defined terrain features.

fire-support coordinator The senior FIELD ARTILLERY officer at each echelon above PLATOON level who serves as the principal advisor to the commander for the coordination of all FIRE SUPPORT within a unit's area of responsibility.

fire-support element That portion of a TACTICAL OPERATIONS CENTER (at every echelon above COMPANY) that is responsible for the targeting, coordination, and integration of fire delivered on surface targets by FIRE-SUPPORT means under the control of or in support of a force.

fire-support group A temporary grouping of ships under a single commander, charged with supporting troop operations ashore by naval gunfire. A fire-support group may be further subdivided into fire-support units and FIRE-SUPPORT ELEMENTS.

fire-support officer In FIRE-SUPPORT operations, the officer who is the full-time coordinator of all fire support and is the FIELD ARTILLERY commander's representative at the supported headquarters.

fire-support plan A plan containing the information necessary for understanding how FIRE SUPPORT will be used to assist an operation. There should be a portion of the fire-support plan for each fire-support element involved in the operation.

fire-support station An exact location at sea within a FIRE-SUPPORT AREA from which a fire-support ship delivers fire.

fire support team A team provided by field artillery to each maneuver unit. It is responsible for planning and coordinating all INDIRECT FIRE means available to the unit, including MORTARS, FIELD ARTILLERY, CLOSE AIR SUPPORT, and naval gunfire.

fire trench A TRENCH from which soldiers can fire their rifles or other small arms, and in which they are relatively well protected.

fire unit A unit whose fire in battle is under the immediate and effective control of one leader.

firing, check See CHECK FIRING.

firing, predicted See PREDICTED FIRING.

firing, preliminary See PRELIMINARY FIRING.

firing battery 1. That part of a BATTERY actually at the firing position when a battery is prepared for action. It includes the PIECES, PERSONNEL, and equipment necessary for its operation. 2. A battery organized and equipped to fire FIELD-ARTILLERY weapons, as differentiated from a HEADQUARTERS or SERVICE battery. 3. An element of a field-artillery CANNON battery.

firing bay One of a series of short, straight sections of a FIRE TRENCH, set forward and joined to the next section by short trenches, and which makes an indentation in any of various shapes. A fire trench is divided into firing bays so that a bomb or shell falling in one bay does not cause destruction in the trenches on either side. Also called a fire bay.

firing chart A map, photo map, or grid sheet showing the relative horizontal and vertical positions of batteries, base points, base point lines, check points, targets, and other details needed in preparing firing data.

firing circuit In land operations, an electrical circuit or PYROTECHNIC loop designed to detonate connected charges from a firing point.

firing data All data necessary for firing an ARTILLERY PIECE at a given OBJECTIVE. Such data may be determined by computation and then transmitted by verbal commands, or may be applied electromechanically by one of the several types of directing devices.

firing device A metal case, containing an INITIATOR and a spring-propelled metal pin, designed to set off the MAIN CHARGE of explosives contained in BOOBY TRAPS, ANTIPERSONNEL MINES, ANTITANK MINES, and demolition charges. There are four types of firing device: pressure, pull, and release devices, and combinations thereof. See also PRESSURE FIRING DEVICE.

firing jack An adjustable device that stabilizes and levels certain mobile ARTILLERY weapons while the weapons are in firing position.

firing line 1. The positions from which troops fire upon an enemy or targets. 2. The troops who fire from the firing line. 3. "Where the action is" either in a combat or an administrative sense.

firing party See FIRING SQUAD.

firing pin 1. A plunger in the firing mechanism of a firearm that strikes the PRIMER and thus ignites the PROPELLING CHARGE. 2. That part of a FUZE that strikes the sensitive explosive in the fuze and sets it off; the device that sets off the action of a detonator or primer.

firing point That point in the FIRING CIRCUIT where the device employed to initiate the detonation of the charges is located.

firing practice, combat See COMBAT FIRING PRACTICE.

firing, record See RECORD FIRING.

firing range 1. The maximum distance a projectile will effectively travel when fired. 2. A target's distance from a weapon. 3. An area set aside for practice or target shooting.

firing squad 1. The small group of soldiers assigned to shoot a condemned person at a formal execution. 2. The small group of soldiers that fires volleys at a military funeral. This is more properly called a firing party.

firing system In demolition, a system composed of elements designed to fire the MAIN CHARGE or charges.

firing table A table or chart giving the data needed for firing a weapon accurately on a target under standard conditions, and also the corrections that must be made for special conditions such as wind or variations of temperature.

firing table elevation The angle between the axis of the BORE and the horizontal when an artillery piece is laid to fire at a given range under conditions that are accepted as standard.

first call A warning signal given before men are summoned to a FORMATION,

DUTY, or other activity. The first call is usually a bugle call.

first captain The commander of a brigade of military cadets.

first-class gunner 1. A classification, given for skill in the use of CANNONS and MACHINE-GUNS, ranking above the grade of second class gunner and below that of expert. The grade of first class gunner corresponds to that of a SHARPSHOOTER. 2. A soldier having this classification.

first-defense gun A MACHINE-GUN placed where it can cover the enemy from the time the enemy starts to attack until the enemy breaks through the front lines of a battle position.

first-fire mixture A quick-igniting mixture used to set off the MAIN CHARGE of a munition.

first lieutenant The JUNIOR OFFICER who is senior to only one other officer rank, the SECOND LIEUTENANT, and just below a CAPTAIN. The INSIGNIA is a silver bar.

first line 1. The most carefully trained and equipped troops. 2. Those troops positioned closest to the enemy. 3. The best of its kind, as in first-line equipment.

first salvo at In naval gunfire support, a portion of a ship's message to an observer or SPOTTER to indicate that because of proximity to troops, the ship will not fire at the target, but will instead offset the first SALVO a specific distance from the target.

first sergeant An occupational title for the chief NONCOMMISSIONED OFFICER of a company, battery, or similar unit. The British equivalent is a company, battery, or squadron sergeant-major, while in France it would be an adjutant-chief.

first sergeant's call The periodic assembly of FIRST SERGEANTS at battalion or regimental headquarters for the purpose of receiving administrative instructions.

first strike The first offensive move of a war. (Generally associated with nuclear operations.)

first-strike capability The ability of one nuclear power to destroy so much of the enemy's nuclear retaliatory forces with a surprise nuclear attack that the other side would be unable to retaliate effectively. The strategic nuclear policy of the United States is to maintain a deterrent SECOND STRIKE CAPABILITY; that is, to maintain nuclear forces in such numbers and variety that no first strike could prevent an American second strike that would inflict unacceptable damage on the aggressor. The original clarity of the first-strike/second-strike dichotomy depended on the simplicity of ideas about nuclear war in the 1950s and 1960s. At the time, strategists assumed that a nuclear war would take the form of a single exchange, in which each side would fire its entire arsenal in one SALVO. They also usually assumed that each side would aim its weapons at a comprehensive collection of target types, trying simultaneously to destroy the military, political, and industrial power of the enemy, and possibly the civilian population as well. The distinction began to be somewhat more important when the MUTUAL ASSURED DESTRUCTION (MAD) doctrine was developed in the 1960s. In order to ensure retaliatory destruction of the enemy, it was necessary to calculate what could be destroyed in one's own country by an enemy first strike. If it was assumed that such an attack would have to be "ridden out" before the "receiving" nation would be in a position to retaliate, a secure second-strike capacity would have to be guaranteed. Thus, the MAD policy dictated that enough deliverable nuclear warheads should remain after an attack had been received in order to inflict an amount of damage which would be unacceptable to the enemy. See Jack H. Nunn, *The Soviet First Strike Threat: The U.S. Perspective* (New York, N.Y.: Praeger, 1982); William M. Arkin, "The Drift toward First Strike," *Bulletin of the Atomic Scientists* 41, 1 (January 1985); Daniel Charles, *Nuclear Planning in NATO: Pitfalls of First Use* (Cambridge, Mass.: Ballinger Publishing Co., 1987).

first world The rich, industrialized Western democracies: the United States, Canada, Western Europe, Australia, New Zealand, and Japan. These countries are also referred to as the North, or the developed countries. Compare to SECOND WORLD, THIRD WORLD, FOURTH WORLD.

fishbone mine system A series of independent underground passages that military engineers cut out in the direction of the enemy, with branches for purposes of attack, flank protection, and listening. A fishbone system differs from a LATERAL MINE SYSTEM, which is an underground passage cut parallel to the front line from which galleries (small tunnels) are carried toward the enemy.

fishnet Net made of knotted cord, used to hold CAMOUFLAGE materials in place.

fishtail wind A wind that constantly changes direction.

fission See NUCLEAR FISSION.

fix 1. A position determined from terrestrial, electronic, or astronomical data. 2. The determination of one's position. 3. To prevent the enemy from withdrawing any part of his force from one area for its use elsewhere; to make it impossible for an enemy to MANEUVER.

fixed ammunition Ammunition in which the CARTRIDGE CASE is permanently attached to the PROJECTILE.

fixed echo A RADAR echo that is caused by reflection from a fixed object, such as a terrain form or building, that is visible to the radar set.

fixed-forward defense See FORWARD DEFENSE.

fixed gun An aircraft-mounted MACHINE GUN that can be aimed only by pointing the aircraft toward the target.

fixed pivot 1. A fixed point on which a line of troops turn when changing formation or direction of march. 2. One who is at the pivot point.

fixed-post system The assignment of SENTINELS to guard duty at fixed posts, where they must stand until relieved.

fixed-station patrol A PATROL in which each SCOUT (in this context, a surface ship, submarine, or aircraft) maintains a station relative to its assigned point on a BARRIER LINE while searching the surrounding area. Scouts are not stationary, but remain mobile and patrol near the center of their assigned stations.

flag A cloth, with a distinguishing color or design, that has a special symbolic meaning or serves as a signal. The flag of the United States, the white flag of truce, and weather flags are examples. In military service, the COLOR is the flag of a DISMOUNTED unit; an ENSIGN is a national flag; a pennant is a small triangular flag usually flown for the identification of a unit; and a GUIDON is a flag carried by army units for identification, especially in DRILLS and ceremonies.

flag, chaplain's See CHAPLAIN'S FLAG.

flag, garrison See GARRISON FLAG.

flag, show the To make your presence known by sending troops or ships, flying your flag, to participate in an action.

flag, strike the To indicate a willingness to surrender.

flag of convenience The flag flown by a merchant ship registered in a foreign country rather than in the country where it is owned or does business, in order to avoid high fees, safety requirements, or other requirements. The ship must legally fly this "flag of convenience," of the nation in which it is registered. This is different from reflagging, which is a diplomatic undertaking designed to qualify foreign merchant ships for naval protection from friendly powers, whose flag they then fly. See Adam Bozcek, *Flags of Convenience* (Cambridge, MA: Harvard University Press, 1962); Rodney Carlisle, *Sovereignty for Sale: The Origins and Evolution of the Panamanian and Liberian Flags of Convenience* (Annapolis, MD: Naval Institute Press, 1981).

flag of truce A white flag carried by an envoy with an important message, usually signifying the surrender of one side to the other. Sometimes a flag of truce is used as a strategem to gain time.

flag officer A term applied to an officer holding the rank of GENERAL, LIEUTENANT GENERAL, MAJOR GENERAL, or BRIGADIER GENERAL in the U.S. Army, Air Force, or Marine Corps, or an ADMIRAL, vice admiral, rear admiral, or commodore in the U.S. Navy or Coast Guard. In most other western countries, the term is restricted only to naval ranks, other such officers being known as "general officers" or "air officers." In 1988 all of the flag officers in the Armed Services of the United States totaled less than eleven hundred.

flag semaphore A system of signaling by which messages are spelled out with flags, especially used by navies for ship-to-ship communications when it is necessary to maintain radio silence.

flail tank A specially constructed TANK equipped with a flailing device, consisting of chain flails attached to a roller powered by the tank engine, which is employed to detonate ANTITANK MINES.

flak 1. Antiaircraft fire; originally a German acronym for *Flugzeug Abwehr Kanonen* (aircraft defense cannon). 2. The shellbursts from such fire. 3. By analogy, a bureaucratic obstructionism.

flak jacket 1. A heavy-fabric body ARMOR worn by air crews as protection against enemy fire of all kinds. 2. A broadly used term for any bulletproof vest.

flame A container (usually for messages) that is thrown from an aircraft to communicate with friendly units on the ground. The name came from the flame-like streamers attached to the container to make it more visible and easier to locate.

flame thrower A weapon that uses compressed nitrogen to project, throw, or squirt thickened incendiary fuel that ignites as it approaches its target. The flame thrower that a soldier wears on his back has an effective distance of only a few yards; in contrast, a flame thrower tank can send ignited fuel about 75 yards.

flank 1. The right or left ELEMENTS of a body of troops. 2. The terrain to the right or left of a body of troops. 3. To attack an enemy in a vulnerable position, presumably a lightly defended side. 4. To outsmart or deceive someone, whether it be an enemy or a superior officer. See also OUTFLANK.

flank, assailable See ASSAILABLE FLANK.

flank, northern See NORTHERN FLANK.

flank guard A security element operating to the FLANK of a moving or stationary force to protect it from enemy ground observation, DIRECT FIRE, and surprise attack.

flanking attack An offensive MANEUVER directed at the FLANK of an enemy. NAPOLEON wrote: "A well-established maxim of war is not to do anything which your enemy wishes—and for the single reason that he does so wish. You should, therefore, avoid a field of battle which he has reconnoitered and studied. You should be still more careful to avoid one which he has fortified and where he has entrenched himself. A corollary of this principle is, never to attack in front a position which admits being turned." German Field Marshall Alfred Von Schlieffen (1833–1913), agreed: "The flank attack is the essence of the whole history of war." See also FRONTAL ATTACK; INDIRECT APPROACH.

flanking obstacle An obstacle located to protect the flanks of a unit and to prevent or slow enemy PENETRATION and ENVELOPMENT.

flap A major disturbance or excitement about something; great confusion; loud warranted or unwarranted criticism, or some other ado.

flare A PYROTECHNIC item designed to produce a single source of intense light for purposes such as target illumination or identification of friendly units.

flare dud A NUCLEAR WEAPON launched at a target that detonates with its anticipated YIELD but at an altitude appreciably greater than intended. This is not a DUD insofar as yield is concerned, but it is a dud with respect to the effects on the target and the normal operation of the weapon.

flare ship A cargo aircraft modified to drop large numbers of aerial FLARES to light up a battlefield.

flash defilade 1. Condition in which the flash of firing from a gun position is concealed from enemy observation by an intervening obstacle, such as a hill or the side of a ravine. 2. The vertical distance by which the flash of a gun is concealed from enemy observation.

flash message A category of precedence reserved for initial enemy-contact messages or operational combat messages of extreme urgency.

flash ranging Finding the position of the BURST of a PROJECTILE, or of an enemy gun, by observing the flash of its explosion or muzzle fire.

flash suppressor A device attached to the MUZZLE of a weapon that reduces the amount of visible light or flash created by burning propellant gases.

flash-to-bang time The time interval between visual observation of the flash of a weapon being fired and the auditory perception of the sound of the discharge proceeding from the same weapon. Because of the speed of the projectiles fired by military weapons, soldiers' tradition has it that "you'll never hear the one that gets you!"

flash tube A tubular device used to transmit an igniting spark or flame from a fuse or detonating device to an EXPLOSIVE CHARGE.

flechette A small, fin-stabilized missile used in antipersonnel ammunition.

fleet A naval organization of ships, aircraft, marine forces, and shore-based fleet activities all under the command of a commander or commander-in-chief, who may exercise operational as well as administrative control. A fleet is the largest grouping of naval ships for tactical and other purposes. A major fleet is a permanent subdivision of the operating forces of a navy. The United States has two major fleets: the Pacific Fleet and the Atlantic Fleet. A numbered fleet (such as the Seventh Fleet) is a tactical unit immediately subordinate to a major fleet.

fleet ballistic missile submarine A nuclear-powered submarine designed to deliver BALLISTIC MISSILES against assigned targets from either a submerged or surfaced condition. Designated as SSBN.

fleet in being A FLEET that avoids decisive action, but because of its strength and location, causes or necessitates enemy counter-concentrations and so reduces the number of opposing units available for operations elsewhere.

fleeting target A moving target that remains within observing or firing distance for such a short period that it affords little time for deliberate adjustment and fire against it. Fleeting targets may be aircraft, vehicles, marching troops, etc. Also called a TRANSIENT TARGET.

flexibility 1. The desirable military posture of having a variety of modes of attack to choose from in a battle situation. 2. An attitude in ARMS-CONTROL talks that allows for COMPROMISE.

flexible response 1. The capability of military forces to make an effective reaction to any enemy threat or attack with a commensurate level of counter threat. Such a capability implies the option to use a variety of conventional forces or nuclear weapons as the circumstances warrant. 2. The official strategy of the NORTH ATLANTIC TREATY ORGANIZATION (NATO), as enshrined in the 1967 planning document MC 14/3. The NATO doctrine of flexible response, which rests heavily on ESCALATION theories, was originally developed to circumvent the strategic contradictions in maintaining a credibile, massive nuclear retaliation doctrine as well as building up an adequate conventional defense against the WARSAW PACT (as was planned at the time). The idea of flexibility was never really spelled out by NATO, which has made such uncertainty into a virtue by arguing that the Warsaw Pact would be unable to plan any attack properly if it could never know how the West will choose to respond. Yet MC 14/3 remains a paper strategy, because the nuclear option is its most credible component. No pretence has ever been made that a prolonged conventional war would be fought by NATO. The alliance does not, in fact, have sufficient war stocks to fight conventionally for more than a few weeks. Although this aspect could be remedied were Western European governments sufficiently motivated, Europeans continue to rely on the deterrent threat of nuclear escalation as a defense strategy because they do not view the prospect of a conventional war with modern weapons as at all acceptable. Thus, their original objection to flexible response was not just due to economic meanness; it was strongly felt that a reduction in the nuclear threat seriously increased the risk of war on European soil. Hence, flexible response has never been intended to remove the threat of escalation: it is largely a compromise between an American unwillingness to risk a nuclear counterattack on its cities to prevent a relatively minor invasion of Western Europe, and a European refusal to fight an updated version of World War II. See Robert C. Powers, "Flexible Response and External Force: A Contrast of U.S. and Soviet Strategies," *Strategic Review* 9, 1 (Winter 1981); David Dessler, " 'Just in Case'—the Danger of Flexible Response," *Bulletin of the Atomic Scientists* 38, 9 (November 1982); Bernard W. Rogers, "Greater Flexibility for NATO's Flexible Response," *Strategic Review* 11, 2 (Spring 1983).

flight 1. A specified group of aircraft usually engaged in a common mission. 2. The basic TACTICAL UNIT in the U.S. Air Force, consisting of four or more aircraft in two or more ELEMENTS. See SQUADRON. 3. A single aircraft airborne on a nonoperational mission. 4. An air-force equivalent of a PLATOON.

flight deck 1. In certain airplanes, an elevated compartment occupied by the crew for operating the airplane in flight. 2. The upper deck of an aircraft carrier that serves as a runway.

flight leader 1. The pilot in command of a specified group of aircraft. 2. An air force NONCOMMISSIONED OFFICER.

flight path The line connecting the successive positions occupied, or to be occupied, by an aircraft, missile, or space vehicle as it moves through air or space.

flight plan Specified information provided to air-traffic services units, relative to an intended flight or a portion of the flight of an aircraft.

flight profile The FLIGHT PATH of an aircraft expressed in terms of altitude, speed, range, and maneuver.

flight surgeon A physician specially trained in aviation medical practice, whose primary duty is the medical examination and medical care of aircrews.

flights, eagle See EAGLE FLIGHTS.

floating reserve In an amphibious operation, reserve troops that remain embarked until needed. See also GENERAL RESERVE.

FLOT See FORWARD LINE OF OWN TROOPS.

flotilla A naval administrative or tactical organization consisting of two or more squadrons of DESTROYERS or smaller types of vessels, together with such additional ships as may be assigned as flagships and tenders.

flying, contour See CONTOUR FLYING.

flying column A highly mobile ground force operating separately from the MAIN BODY of troops, often behind enemy lines. "Flying" is a term left over from the nineteenth century when anything rapid was called "flying."

flying officer The Royal Air Force (of Great Britain), and the Royal Canadian Air Force rank that corresponds to a FIRST LIEUTENANT in the United States Air Force.

flythrough The act of the target passing through the CONE OF FIRE.

FO See FORWARD OBSERVER.

FOFA See FOLLOW ON FORCES ATTACK.

fog of war The confusion inherent in combat and large-scale military operations; a breakdown in communications. The earliest reference to this now classic description of military blindness comes from Jean-Charles de Folard, who wrote in his 1724 *Nouvelles Decouvertes sur la Guerre* that "The COUP D'OEUIL is a gift of God and cannot be acquired; but if professional knowledge does not perfect it, one only sees things imperfectly and in a fog." See Bruce P. Schoch, "The 'Fog of War'," *Army Logistician* 16, 5 (September–October 1984); Derrik Mercer, *The Fog of War* (London: Heineman, 1987).

fog oil A special petroleum oil used in mechanical smoke generators.

follow and support The mission of a force that may be committed behind an EXPLOITING FORCE to secure lines of communication or relieve elements of the exploiting force left behind to contain bypassed enemy forces, and subsequently to eliminate such bypassed forces. The follow-and-support force may be required to provide limited COMBAT SERVICE SUPPORT to the exploiting force, to reinforce the leading force, or to assume the primary mission of the leading force. It is a fully committed unit and not part of a commander's reserve.

follow-on echelon In amphibious operations, that echelon of the assault troops, vehicles, aircraft equipment, and supplies which, though not needed to initiate the assault, is required to support and sustain the assault. See also FOLLOW-UP. See Kevin N. Lewis and Peter A. Wilson, "A Follow-On U.S. Echelon for NATO Defense," *Military Review* 66, 6, (June 1986).

follow-on forces The Warsaw Pact reserve units designed to be used as the second wave of an ECHELONED ATTACK.

follow-on forces attack (FOFA) The official NORTH ATLANTIC TREATY ORGANIZATION (NATO) doctrine for fighting a CENTRAL FRONT war against the WARSAW PACT. It is largely a development of the DEEP STRIKE tactic, and to a lesser extent the AIRLAND BATTLE CONCEPT devised by the United States. The FOFA concept is a response to the Soviet war plan of ECHELONED ATTACK, and like a deep strike, is partly dependent on the development of EMERGING TECHNOLOGY weapons. Because the Warsaw Pact has a considerable numerical superiority over NATO in troops, and even more so in tanks and other conventional weaponry, it might normally be expected to use its forces to hold the NATO line in most places and position one or two highly-concentrated and overpowering armies at selected points, in order to smash its way through, according to the traditional principle of CONCENTRATION OF FORCE. However, because of the danger of BATTLEFIELD NUCLEAR WEAPONS being used against such concentrations, the Soviet Union has developed an alternative approach, in which the full force of the Warsaw Pact will be arranged in two or three lines or echelons. These will attack NATO lines in sequential waves all along the front. The FOFA concept and the theories from which it derives is a plan to attack the second and subsequent echelons of Warsaw Pact forces by airpower when they are still days away from the main land battle; should this succeed, the NATO armies would only have to deal with the first-echelon troops. The doctrine was, somewhat unwillingly, accepted by the European NATO members in the mid-1980s. However, this acceptance may have little significance, since the emerging technology needed to carry out FOFA will, if it ever appears, be largely owned by the United States. See Bernard W. Rogers, "Follow-on Forces Attack (FOFA): Myths and Realities," *NATO Review* 32, 6 (December 1984); Thomas A. Cardwell III, "Follow-On Forces Attack: Joint Interdiction by Another Name," *Military Review* 66, 2 (February 1986).

follow-up In amphibious operations, the landing of reinforcements and stores after the ASSAULT ECHELON and FOLLOW-ON ECHELON have been landed.

follow-up echelon In air-transport operations, elements moved into the OBJECTIVE AREA after the ASSAULT ECHELON.

follow-up elements Elements following a march column whether for cleanup, prevention of straggling, maintenance and recovery of equipment, or other purposes.

follow-up supply That initial resupply that is delivered by air directly to forces at an AIRHEAD. It is prepackaged on a unit basis for automatic or ON-CALL delivery.

foot 1. A term referring to INFANTRY in general. 2. A foot soldier.

football The briefcase, carried by a military officer, that follows the President of the United States at all times. It contains the secret launch codes needed to initiate a nuclear attack.

forage 1. Food for animals, especially horses. 2. A military operation that takes food and other essentials from civilians in a battle area. 3. A search. 4. A small military RAID.

forage cap A 19th century military cap having a low, flat crown and a visor; worn by both sides in the American Civil War.

foray 1. A sudden ATTACK, RAID, or brief INVASION. 2. An attack on a SECONDARY TARGET that is not in the principal objective.

force 1. An aggregation of military personnel, weapon systems, vehicles, and necessary support forces, or a combination thereof. 2. A major subdivision of a FLEET. 3. The ruling principle in military matters; and often in foreign relations.

force, advance See ADVANCE FORCE.

force, breaching See BREACHING FORCE.

Force d'Action Rapide (FAR) A French army force of about 40,000 troops organized for rapid intervention in Europe or elsewhere. The Force d'Action Rapide, or rapid-reaction force, consists of specalist troops, marines, paratroops, and mountain troops who are not otherwise organized into the main French army corps structure. It also contains troops who already have an assignment in a TABLE OF ORGANIZATION, but who could be quickly mobilized and transferred to the command of the FAR. Although as yet inadequately equipped, particularly in the helicopters needed for rapid troop transport, the FAR is seen by France's allies as a demonstration of a much greater willingness of the French to come rapidly to the aid of NORTH ATLANTIC TREATY ORGANIZATION (NATO) forces in an emergency. The FAR also represents an increased French interest in having a FORCE PROJECTION capacity.

Force de Frappe The Force de Frappe was the original name for the French strategic nuclear force when it was first deployed, consisting of some 40 Mirage IV medium-range bombers, between 1964 and 1966. Since then the French have developed a full nuclear triad of bombers, intermediate-range ballistic missiles (IRBMs), and ballistic-missile submarines. The overall strategic force is now often described as the Force de Dissuasion, which can be translated very roughly as "Deterrent Force." See Pierre M. Gallois, "The Future of France's Force De Disuasion," *Strategic Review* 7, 3 (Summer 1979); David S. Yost, *France's Deterrent Posture and Security in Europe* (London: International Institute for Strategic Studies, 1984).

force list A total list of forces required by an OPERATION PLAN, including ASSIGNED FORCES, augmentation forces, and other forces to be employed in support of the plan.

force majeure 1. Superior military force; the phrase often indicates that a given political outcome is dependent on the exercise of irresistible force rather than on consent, agreement, or legal process. 2. Compelling circumstances. Force majeure is the means by which things have been often accomplished in international relations; it is an underlying factor in all negotiations. See Barry Nicholas, "Force Majeure and Frustration," *American Journal of Comparative Law* vol. 27, 2–3 (Spring–Summer 1979).

force multiplier Any piece of technology that allows a smaller body of soldiers, or a smaller set of tanks, artillery, fighter aircraft, or other equipment to defeat a larger force of similar type. Although the concept is new, the phenomenon is as old as warfare, particularly if innovations in tactical doctrine or military theory are also classed as force multipliers. The development of the phalanx formation by Greek spearmen, which allowed them to overcome much larger numbers of less well-coordinated enemies, is an early example. At the purely technological level, a familiar historical example would be the ease with which small units of British troops with bolt-action rifles and Maxim machine guns could defeat masses of primitively-armed natives in imperial battles during the nineteenth century. Force multipliers are currently much sought after by the NORTH ATLANTIC TREATY ORGANIZATION (NATO) as a way of offsetting the numerical superiority in troops, tanks, and artillery enjoyed by the Warsaw Pact. See also DECOY.

force planning The determination of the number, size, composition, weapons, and deployment of military units. See Benjamin S. Lambeth, "Pitfalls in Force Planning: Structuring America's Tactical Air Arm," *International Security* 10 (Fall 1985); Joshua M. Epstein, *Strategy and Force Planning: The Case of the Persian Gulf* (Washington, D.C.: Brookings Institution, 1987).

force projection The military capacity to transport army and air force units to a distant spot and, if necessary, land them under fire in order to engage the local enemy. Such capacity is the classic requirement of a country that wishes to be able to exert control on a world-wide scale. The concept of force projection is well summed up by the idea popular at the beginning of this century that the British army was a "bullet to be fired by the Royal Navy." As the rivalry between the United States and Soviet Union extends worldwide, and as both countries increasingly perceive danger to their vital interests away from the traditional European CENTRAL FRONT, there is growing concern with force-projection capacity on both sides of the Iron Curtain. The United States, which like the United Kingdom has traditionally been a sea power rather than a land power, has until recently had a decided advantage over the Soviet Union in this area, despite the WARSAW PACT's overall numerical superiority, because the Soviet Union has not had a force-projection capacity. This situation has been changing in the last two decades as the Soviet Union continues to develop its BLUE-WATER NAVY. Force-projection capacity is not merely built on the sheer size of naval or airlift facilities, as special types of ships are required. The American advantage, which the Reagan Administration planned to enhance, consists of having powerful aircraft carriers and a large, specially trained Marine Corps, equipped with amphibious-warfare ships and specifically-designed armor and artillery for easy transportation and landing. Naval air power may be the key, above all else, to force projection, because of the need to fight at great distances from home bases, against a local enemy whose entire military infrastructure is close by. See Richard Albert Wise, *A Mobile Missile Brigade for a Power Projection Role*, (Santa Monica, CA: Rand, 1980); Dennis M. Gormley, "The Direction and Pact of Soviet Force Projection Capabilities," *Survival* (November/December 1982).

force ratios The ratio of troops or equivalent pieces of equipment between two opposing nations. Force ratios are also the most commonly used measure for the relative strength of the WARSAW PACT and the NORTH ATLANTIC TREATY ORGANIZATION (NATO). Most of the evidence for the conventional superiority of the Warsaw Pact over NATO is expressed in terms of such rather simple-minded ratios. For example, the ratio of tanks between the two superpowers is about 3:1 in favor of the Warsaw Pact, and the ratio of troop forces is approximately 1.3:1. The problem is that such numbers are virtually meaningless taken outside of a specific context. The 3:1 tank ratio does most certainly mean that NATO would be defeated were it to try to invade Eastern Europe, but as NATO is expressly a defensive alliance, this factor is irrelevant. A more appropriate consideration is that traditional military wisdom calls for a 3:1 advantage as the minimum necessary for the success of an invading force, which suggests that the Warsaw Pact is not so superior to NATO in conventional forces that it would be assured of victory. This argument is not meant to be a substantive judgment, but to indicate the danger of attaching too much importance to force ratios unless backed by the complex scenario-dependent analyses of military operations research.

force readiness The readiness of an army as measured by its ability to man, equip, train its forces, and to mobilize, deploy, and sustain them as required to accomplish assigned missions.

force rendezvous A CHECKPOINT at which formations of aircraft or ships join and become part of a main force.

force sourcing The identification of units, their origins, ports of embarkation, and movement characteristics to satisfy the time-phased force requirements of the commander requesting this support.

forced crossing Going across a river in the face of enemy opposition.

forced issue A mandatory issue of something from a QUARTERMASTER, usually something not requested or something in place of what was requested.

forced march 1. A fast-paced MARCH over a long distance to reach a position in less time than the enemy or the superior of the marching unit might expect. 2. A movement toward battle that is significantly faster than what would be required under less than urgent conditions. See Robert E. Rogge, "Forced March: Armor," *Military Review* 66, 7 (July 1986).

forces, aggressor See AGGRESSOR FORCES.

forces, black See BLACK FORCES.

force, structure See MILITARY CAPABILITY.

forcing cone The tapered beginning of the bands at the origin of the RIFLING of a gun tube. The forcing cone allows the ROTATING BAND of the projectile to be gradually engaged by the rifling, thereby centering the projectile in the BORE of the gun.

ford 1. A shallow part of a body of water that can be crossed without bridging, boats, or rafts; a location in a water barrier where the physical characteristics of current, bottom, and approaches permit the passage of personnel or vehicles and other equipment without losing contact with the bottom. 2. To so cross a body of water.

fording, deep See DEEP FORDING.

foreign aid All official grants and loans in currency or in kind, that are broadly aimed at transferring resources from developed to less developed countries for the purposes of economic development, income distribution, or both. Foreign aid may be bilateral (from one country to another) or multilateral (distributed through international financial institutions such as the World Bank or the International Monetary Fund). Foreign aid, which is also referred to as "economic assistance," may be given as project aid (where the donor provides money for a specific project such as a dam or a school) or as program aid (where the donor does not know for which projects the money will be spent). Economic assistance consists of both "hard" loans (that is, at commercial-bank interest rates) and "soft" loans (concessional or low interest rates). The aid may be "tied" to multilateral aid agencies, or may be "tied" in bilateral arrangements (that is, the money must be spent on procurement in the donor country or must be transported on donor-country shipping). Countries give foreign aid for various reasons: for humanitarian purposes after wars or natural disasters; to strengthen allies militarily against external or internal threats; to promote the economic development of the recipient country; or to simply meet the basic human needs of the poor citizens of the recipient country. The first significant instance of foreign aid was that given by the United States to its allies during and right after World War I. But this was ad hoc. Only in the Truman Administration did foreign aid first become institutionalized and a continuous part of American foreign policy (see MARSHALL PLAN). It was and remains motivated both by humanitarian concerns and a desire to allow grantees to achieve the kind of economic and social growth that would allow their governments to withstand the efforts of Communists to take them over. See Peter Bauer and Basil Yamey, "Foreign Aid: What Is at Stake?" *Public Interest* No. 68 (Summer 1982); Manual F. Ayau, "The Impoverishing Effects of Foreign Aid," *Cato Journal* 4, 1 (Spring–Summer 1984).

foreign legion 1. A military force serving outside its home territory. 2. A military force composed of foreign nationals, not of the citizens of the nation it services. 3. The French Foreign Legion, which comprises both of these definitions. See Tony Geraghty, *March or Die: A New History of the French Foreign Legion* (New York: Facts on File, 1987).

foreign ministry (or foreign office) The cabinet agency in most parliamentary governments that corresponds to the U.S. Department of State. See Zara Steiner, "Foreign Ministries Old and New," *International Journal* 37, 3 (Summer 1982).

foreign national One who owes allegiance to a foreign nation, without regard to the formal status of citizen.

foreign service 1. Military service in foreign locations. 2. A corps of professional diplomats. The U.S. Foreign Service is the diplomatic corps responsible for administrating American foreign policies. Andrew L. Steigman, *The Foreign Service of the United States: First Line of Defense* (Boulder, CO: Westview Press, 1985).

fork The change in ELEVATION of an artillery piece required to shift the center of impact of its projectiles in range by four PROBABLE ERRORS. See HORIZONTAL ERROR.

formation 1. An ordered arrangement of troops, vehicles, or both for a specific purpose. 2. An ordered arrangement of two or more ships, units, or aircraft proceeding together under a commander. 3. A formal parade.

formation, diamond See DIAMOND FORMATION.

formations, combat See COMBAT FORMATIONS.

forming-up place The last position occupied by an ASSAULT ECHELON before crossing the LINE OF DEPARTURE. Also called ATTACK POSITION.

forms of maneuver The general orientation or direction of a force approaching an enemy. The two basic forms of maneuver are the PENETRATION and the ENVELOPMENT.

Forrestal, James V. (1892–1949) United States Secretary of the Navy during the latter part of World War II and the first Secretary of the new Department of Defense from 1947 to 1949. Although he personally opposed unification of the United States' armed services, he was a major participant in the development of the National Security Act of 1947 (which combined the services into one agency), and subsequently did much to strengthen the office of the Secretary of Defense at a time when it was considered weak vis à vis the military. See Robert Greenhaigh Albion, *Forrestal and the Navy* (New York: Columbia University Press, 1962); Arnold A. Rogow, *James Forrestal: A Study of Personality, Politics, and Policy* (New York: Macmillan, 1963).

fort 1. A permanent post as opposed to a CAMP, which is a temporary INSTALLATION. 2. The land area within which harbor defense units are located. 3. A strong, fortified building or place that can be defended against an enemy. 4. What a commander under fire should hold while waiting for reinforcements.

fortification 1. The art and science of making a position militarily defensible if atttacked. 2. Defensive works of any kind, permanent or temporary.

fortifications, field See FIELD FORTIFICATIONS.

fortress 1. A FORT. 2. A town or place of inhabitation surrounded by FORTIFICATIONS. 3. Any location so strongly fortified that it can resist an attack.

fortunes of war The fate of massed conflict on individuals: the luck of being killed or made a hero by dumb circumstances, or one's financial or physical condition at the close of a war. Johann Wolfgang von Goethe (1749–1832) wrote that: "The fortunes of war flow this way and that, and no prudent fighter holds his enemy in contempt."

forward air controller (FAC) An officer member of a tactical air control party who, from a forward ground or airborne position, controls aircraft in CLOSE AIR SUPPORT of ground troops.

forward area An area in proximity to combat.

forward arming and refueling point A temporary facility organized, equipped, and developed by an aviation unit commander. Located closer to the AREA OF OPERATIONS than the aviaton unit's combat service area it is intended to provide fuel and ammunition necesary for the employment of helicopter units in combat.

forward based Military units kept permanently in or near the area in which they are expected to fight should a confrontation occur. Thus, the British Army of the Rhine and the U.S. Seventh Army are forward based, as they are kept permanently in West Germany, rather than in their home countries. Forward basing is sometimes politically controversial, since it is often believed that it is more expensive to keep troops forward based than home based. From time to time, therefore, those wanting to cut defense budgets, yet unwilling to accept actual reduction in military preparedness, call for home basing rather than forward basing. This has, intermittently, been a particularly strong demand in the United States, and has also occurred in British politics. Compare to POMCUS.

forward command post The station of a unit's HEADQUARTERS, where the commander and staff work. In combat, a unit's headquarters is often divided into a FORWARD ECHELON and REAR ECHELON.

forward defense A specific defense plan advocated by the NORTH ATLANTIC TREATY ORGANIZATION (NATO) for a CENTRAL FRONT war. In a forward defense strategy, NATO armies would attempt to stop any WARSAW PACT invasion of Germany at the border, and try to win the war there, without surrendering any territory. This approach runs counter to many doctrines of war-fighting, holding fixed linear forward defenses—which commit troops to stay in place along the whole front—to be disastrous, yielding of far too much initiative to the enemy. The analogy most often drawn to this plan is the static trench warfare of World War I, although it must be said that the forward defense in that case was enforced on the opposing armies by stalemate; the generals on both sides tried desperately to create a war of maneuver. Compare to DEEP STRIKE, MAGINOT MENTALITY. See Benjamin F. Schemmer, "NATO's New Strategy: Defend Forward, But Strike Deep," *Armed Forces Journal* (November 1982); Phillip A. Karber, "The Defense of Forward Defense," *Armed Forces Journal* (May 1984); James A. Blackwell, Jr., "In the Laps of the Gods: The Origins of NATO Forward Defense," *Parameters: Journal of the US Army War College* 15, 4 (Winter 1985).

forward defense area The area in which the FORWARD DEFENSE POSITIONS are located in a MOBILE DEFENSE.

forward defense position 1. In a MOBILE DEFENSE, any combination of islands of resistance, STRONG POINTS, and OBSERVATION POSTS utilized by the defender to warn of impending attack, to canalize the attacking force into less favorable terrain, and to block or impede the attacking force. Forward defense positions are occupied by the minimum necessary forces, while the bulk of the defending force is employed in offensive action. 2. Any forwardly based island of resistance.

forward echelon That part of a HEADQUARTERS that is principally concerned with the tactical control of battle. Compare to REAR ECHELON.

forward edge of the battle area (FEBA) Wherever one's troops confront the enemy. The traditional concept of the "front line" in warfare, familiar from World War I, has become outdated. INTERDICTION tactics mean that aircraft and very long-range artillery, multiple rocket launchers, and even chemical and nuclear weapons will be targeted deep beyond any imaginary front line between ground forces. Indeed, because of the complex demands of C^3I and the danger from the enemy's own interdiction forces, targets such as command bunkers, supply depots, and airbases behind the lines will be possibly more important than the traditional targets of front-line troops. As a consequence, the image of a linear front between two armies, with the war zone limited to a narrow strip on each side, is anachronistic. Hence the concept of a "battle area," with the line where one's ground troops directly face the forces of the enemy being seen as the forward edge of a deep battlefield.

forward line of own troops (FLOT) A line that indicates the most forward positions of friendly forces in any kind of military operation at a specific time.

forward observer (FO) An artillery observer operating with front-line troops who is trained to adjust ground or naval gunfire and pass back battlefield information. In the absence of a FORWARD AIR CONTROLLER the observer may control CLOSE AIR SUPPORT strikes. See also SPOTTER.

forward obstacles OBSTACLES generally located along the initial and successive DEFENSE AREAS of forward units. They consist of natural and man-made obstacles employed in depth for CLOSE-IN SECURITY. See also INTERMEDIATE OBSTACLES; REAR OBSTACLES.

forward slope Any slope which descends towards the enemy.

forward tell The transfer of information to a higher level of command. See also BACK TELL and TRACK TELLING.

forward working limit An engineer work line to the rear of which all engineer effort is the responsiblity of a corps engineer.

fougasse A MINE constructed so that upon explosion of the CHARGE, pieces of metal, rock, gasoline, or other substances will be blown in a predetermined direction. The word, which is sometimes spelled "foo gas" or "fu gas" comes from the French word meaning a small land mine.

fouling Any deposit that remains in the BORE of a gun after it is fired.

found on post Any property found on a military installation, other than that which has been formally issued to a unit, which the unit puts to use. Such items may be abandoned or not yet abandoned; in which case "found on post" refers to thinly disguised stealing, but for a noble purpose.

four by four A term meaning a motor vehicle with four wheels, all of which are driving wheels, dual wheels being considered as one wheel. It is usually written 4×4.

four by two 1. A term meaning a motor vehicle with four wheels, of which two are driving wheels, dual wheels being considered as one wheel. It is usually written 4×2. 2. A piece of cloth 4×2 inches in size, used to clean a rifle barrel by being pulled through it at the end of a cord.

four-man team A group of four trainees formed into a team during their ADVANCED INDIVIDUAL TRAINING, who will complete such training and proceed together to a station or unit of ultimate assignment.

fourragere A shoulder cord, much like an AIGUILLETE, issued by some countries as a unit DECORATION; usually worn on the right shoulder.

Fourth World Those developing countries with very low per-capita incomes, little expectation of economic growth, and few natural resources; in effect, the poorest of the poor. Compare to FIRST WORLD, SECOND WORLD and THIRD WORLD.

fox away In AIR INTERCEPTION, a code meaning, "Missile has fired or been released from aircraft."

foxhole A small pit used for COVER, usually for one or two soldiers, and so constructed that the occupant can fire effectively from it. During World War II, an anonymous army chaplain took a survey and discovered there were no atheists in foxholes.

fractional damage Damage or casualty through nuclear attack to a fraction or percentage of the elements of a target.

fractionation The deployment of multiple WARHEADS on a single missile. Also called MIRV (MULTIPLE INDEPENDENTLY TARGETABLE REENTRY VEHICLE).

frag 1. A fragmentation GRENADE. 2. A fragment from a fragmentation grenade, bomb, or artillery shell. 3. An abbreviation for FRAGMENTARY ORDER. 4. To murder a fellow soldier with an explosive; commonly a fragmentation grenade. See Eugene Linden, "Fragging and Other Withdrawal Symptoms," *Saturday Review* (January 8, 1972).

fragmentary order An abbreviated form of an OPERATION ORDER, usually issued on a day-to-day basis, that eliminates the need for restating information contained in a basic operation order.

fragmentation bomb A bomb designed to break into many small metal pieces when exploded.

fragmentation, controlled See CONTROLLED FRAGMENTATION.

fragmentation grenade A grenade with a metal casing that when exploded shatters into many metal fragments traveling at lethal speed.

frangible Used to describe a brittle, plastic, or other nonmetallic BULLET or GRENADE used for practice and which, upon striking a target, breaks into powder or small fragments without penetrating.

fraternize 1. To mix with or associate with the native population of an occupied territory or conquered land. 2. Sexual relations with people of an occupied territory or conquered land. 3. Sexual or other interpersonal relations between officers and enlisted personnel of the opposite sex.

fratricide 1. Brother killing brother; a CIVIL WAR is often called a fratricide. 2. The concept that subsequent missiles may be less effective than the first missiles fired on the same target. Because debris is left in the air after initial explosions, this often damages the GUIDANCE SYSTEMS of immediately following missiles. Also, the ELECTRO-MAGNETIC PULSE (EMP) of a nuclear explosion from a first missile will probably burn out the arming circuits of closely subsequent WARHEADS.

free drop The dropping of equipment or supplies [or people] from an aircraft without the use of parachutes.

free fall A PARACHUTE maneuver in which the parachute is opened, either manually or automatically, at a predetermined minimally safe distance from the ground.

free-fire area A specific designated area into which any WEAPONS SYSTEM may fire without additonal coordination from the establishing headquarters. Also called FREE-FIRE ZONE.

free flight That portion of a missile's TRAJECTORY that is without thrust; may include GUIDED MISSILES as well as free-flight ROCKETS.

free lance 1. A MERCENARY 2. In AIR INTERCEPTION, a code meaning, "Self-control of aircraft is being employed."

free maneuver A practice MANEUVER in which each force acts as it chooses, and is limited only by the field orders received, by restrictions of area and time, and by the actions of the opposing force.

free net A NET in which any station may communicate with any other station in the same net without first obtaining permission from the net control station to do so.

free-play exercise An exercise to test the capabilities of forces under simulated contingency and/or wartime conditions, limited only by those artificialities or restrictions required by peace time safety regulations. See also CONTROLLED EXERCISE.

free rocket A ROCKET not subject to GUIDANCE or CONTROL in flight.

free tower A tower with mechanisms and equipment to permit the free descent of an inflated parachute. It is used for training in parachute jumping.

free-type parachute A PARACHUTE not attached to an aircraft, which is operated by the jumper at his discretion, or by an automatic device such as a timer, barometric device, or a combination of both.

free world 1. The Western European and North American democracies plus Japan and other isolated pockets of democracy such as Israel. 2. Those portions of the world not under Communist control.

freeze, nuclear See NUCLEAR FREEZE.

friction CLAUSEWITZ's concept as stated in *On War* that no matter how well planned an operation is, the reality of delays, misunderstandings, and other problems will inevitably make its execution less than ideal.

friendly 1. A military ELEMENT, whether consisting of personnel or materiel, that is of the same nation or a group of allied nations. 2. A contact positively identified as friendly.

friendly fire FIRE that inadvertently hits its own friendly troops.

frigate A warship usually smaller than a DESTROYER, capable of independent operations (or with others) in countering submarine, air, and surface threats.

frogman 1. A member of a clandestine UNDERWATER DEMOLITION TEAM. 2. Any combatant who performs duties underwater using self-contained breathing apparatus.

frock To assume legally the duties, rank, and insignia of a higher position before it is formally approved by appropriate authority. Frocked officers may exercise their new rank, but they are not eligible for their new level of pay until formally promoted. The word comes from the fact that MIDSHIPMEN were once allowed to assume the rank, duties, and frock coat of a naval lieutenant under certain conditions.

front 1. The lateral space occupied by a military UNIT, measured from the extremity of one FLANK to the extremity of the other flank. 2. The direction of the enemy. 3. The LINE OF CONTACT of two opposing forces. 4. The direction toward which the COMMAND is faced when a combat situation does not exist or is not assumed. 5. More generally, where the fighting is, as opposed to the relatively safe REAR AREAS.

front, action See ACTION FRONT.

front, central See CENTRAL FRONT.

front and center 1. A command to call certain men forward out of a FOR-

MATION to receive awards, promotions, special instructions, etc. 2. A command to "come here."

front end The section of a BALLIISTIIC MISSILE that remains after the propellant stages have been disengaged. It is this section that leaves the earth's atmosphere and continues the ballistic flight of the missile through the MIDCOURSE PHASE and the TERMINAL PHASE. The front end of a more advanced missile would consist of components such as WARHEADS, GUIDANCE computers and DECOYS. The unit that houses all of these constituents is known as the BUS.

frontage The width of the FRONT plus that distance beyond the FLANKS covered by OBSERVATION and FIRE by a unit in combat.

frontal attack 1. An OFFENSIVE maneuver in which the main action is directed against the front of the enemy forces. 2. In AIR INTERCEPTION, an attack by an interceptor aircraft that terminates with a heading crossing angle greater than 135 degrees.

front line 1. The foremost part of a military unit. 2. A LINE OF BATTLE. 3. A place where combat is ongoing. See also FORWARD EDGE OF BATTLE AREA.

front loading A disproportionate positioning of forces toward the front of an anticipated battle at the expense of normal reserve elements.

frontal fire FIRE delivered at right angles to the front of a target.

frustrated cargo Any shipment of supplies, equipment, or both that is en route to a destination and which is stopped prior to its receipt, and for which further DISPOSITION instructions must be obtained.

FST-1 The NATO designation for the "Follow-on Soviet Tank," the Soviet Union's newest and most powerful main battle TANK.

fuel, peptized See PEPTIZED FUEL.

fuel, unthickened See UNTHICKENED FUEL.

full The highest in a class; thus a COLONEL, as opposed to a LIEUTENANT COLONEL; a GENERAL as opposed to a LIEUTENANT, MAJOR, or BRIGADIER GENERAL; a professor as opposed to an associate or assistant professor.

full command The military authority and responsibility of a superior officer to issue orders to subordinates. It covers every aspect of military operations and administration, and exists only within national services. The term COMMAND, as used internationally, implies a lesser degree of authority than when it is used in a purely national sense. It follows that no North Atlantic Treaty Organization (NATO) commander has full command over the forces that are assigned to him. This is because nations, in assigning forces to NATO, assign only OPERATIONAL COMMAND or OPERATIONAL CONTROL.

Fuller, John Frederick Charles (1878–1966) The chief of staff of the British Tank Corps in World War I who in the interwar years published the theoretical analyses on tank warfare that would be adopted by the Germans in the BLITZKRIEG. Fuller posited a science of war guided by principles or laws, that could be studied as an art; the key to success lies in the application of these principles to changing conditions. Fuller also stressed that the aim of all strategy should be the dislocation and paralysis of the enemy's COMMAND AND CONTROL system. Fuller was far less influential than many of his contemporaries because of his well known extreme right-wing sympathies. See J. F. C. Fuller, *The Foundations of the Science of War* (London: Hutchinson & Co., 1925).

full step A thirty inch step taken in walking or marching.

full track A tracked vehicle such as a tank or tractor that moves on caterpillar tracks only.

fully mission capable A condition status of an item of equipment, military unit, or system signifying that it has all mission-essential subsystems installed and operating.

functional kill The destruction of a target by disabling vital components in a way not immediately detectable, but nevertheless able to prevent the target from functioning properly. An example is the destruction of electronics in a GUIDANCE SYSTEM by a neutral PARTICLE BEAM.

fusillade The simultaneous or rapid fire of multiple weapons.

fusion All NUCLEAR FUSION.

fusion, nuclear See NUCLEAR FUSION.

fuse An igniting or explosive device in the form of a cord, consisting of a flexible fabric tube and core of LOW or HIGH EXPLOSIVE. Used in blasting and demolition work and in certain MUNITIONS. Not to be confused with the term "fuze."

fuze A mechanical or electrical detonating device for setting off the bursting charge of a projectile, bomb, or torpedo.

fuze, antidisturbance See ANTIDISTURBANCE FUZE.

fuze, blasting See BLASTING FUZE.

fuze, boresafe See BORESAFE FUZE.

fuze, delay See DELAY FUZE.

fuze, hydrostatic See HYDROSTATIC FUZE.

fuze, igniting See IGNITING FUZE.

fuze, mechanical time See MECHANICAL TIME FUZE.

fuze, nonboresafe See NONBORESAFE FUZE.

fuze, nondelay See NONDELAY FUZE.

fuze, point detonating See POINT DETONATING FUZE.

fuze, proximity See PROXIMITY FUZE.

fuze, safety See SAFETY FUZE.

fuze, self-destroying See SELF-DESTROYING FUZE.

fuze, shuttered See SHUTTERED FUZE.

fuze, superquick See SUPERQUICK FUZE.

fuze, supersensitive See SUPERSENSITIVE FUZE.

fuze, time See TIME FUZE.

fuze cavity A well or cylindrical space in which a mechanical FUSE is installed.

fuze range The range at which a PROJECTILE will burst when the FUZE is set at a given time value.

fuze setter A device that sets the time for a time fuze to go off.

FYDP 1. Fiscal Year, Defense Department. 2. Five-Year Development Plan; a Defense Department budget planning document.

G

G 1. The symbol used to indicate acceleration due to gravitational force. 2.

The code letter for a functional section of a GENERAL STAFF.

G-agent A chemical agent in the G SERIES.

game theory The application of mathematical reasoning to problems of conflict and collaboration between parties who are assumed to be rational and self-interested. Strategic theorists use it as a mode of analyzing the interactions between powers in a potentially nuclear confrontation. The main point of game theory is that, given assumptions about parties' preferences, possession of information and psychological tendency to risk, it is possible to deduce how each will react to actions or possible actions of the other. Game theory is often demonstrated through a paradigm known as the "Prisoners' Dilemma." It supposes that two men have been arrested on suspicion of having jointly robbed a bank. They are held in separate cells by the police, who do not have enough evidence to prosecute unless one of the suspects confesses and implicates the other. Both men know this, but cannot communicate with each other. The dilemma facing the prisoners is that the best outcome, not being convicted, is only available if each can trust his partner. So if burglar X decides to trust Y, but Y fears X may not be trustworthy, Y may confess to get a lesser sentence and X gets the full sentence given to a criminal who does not cooperate with the police. The text book answer to the Prisoners' Dilemma is that both cooperate, to minimize the worst that can happen, rather than trying for the outcome that is maximum. This yields what is known as the Minimax strategy, taken by game theorists to be the most probable outcome in such a game interaction. The use of game theory in strategic analysis is not great. However, it does serve to focus attention on the way that the interaction of payoffs determine how decisions are made, and that any strategic decision must be taken with an eye to the opponent's most likely action. One reason that game theory is held to apply to nuclear strategy is that nuclear SCENARIOS often involve a quality present in the game theory example; the superpowers cannot wait to find out what the enemy actually has done before they make their own decision. See Louis A. Picard, "Use of Games Theory to Explain the SALT Process: The SALT I Negotiations: A Game Theory Paradigm," *Policy Studies Journal* 8, 1 (Autumn 1979); R. Harrison Wagner, "The Theory of Games and the Balance of Power," *World Politics* 38, 4 (July 1986).

gamma-ray High energy ELECTROMAGNETIC RADIATION emitted from atomic nuclei during a nuclear reaction.

gap 1. Any break or breach in the continuity of tactical dispositions or formations beyond effective SMALL ARMS coverage. 2. An area within a MINEFIELD or obstacle belt, free of live mines or obstacles, whose width and direction would allow a friendly force to pass through in tactical formation.

gap marker In LAND-MINE WARFARE, markers used to indicate a minefield GAP.

garland Strips of cotton cloth or burlap, fastened to a light wire framework, used to thicken overhead CAMOUFLAGE or natural COVER, or to hide the edge of a protecting net.

garnishing Any natural or artificial material applied to an object to achieve or assist CAMOUFLAGE.

garrison 1. A permanent military INSTALLATION. 2. The troops assigned to a permanent installation. 3. Descriptive of a soldier who is ignorant of the "real" army; one who has never been in combat. 4. Descriptive of peacetime equipment or items not taken into combat. 5. To assign troops to a place.

garrison flag The largest size United States flag. It is flown at posts only on holidays and other important occasions.

garrison force All units assigned to a base or area for defense, development, operation, and maintenance of facilities.

gas mask See PROTECTIVE MASK.

gasoline gels See THICKENED FUEL.

G-day The day on which the decision to deploy a specific force or forces is made by a competent authority.

general 1. All GENERAL OFFICERS. 2. A "full" or four-star general. The qualities needed in a good general have been discussed since antiquity. The ancient Greek historian Xenophon quotes Socrates: "The general must know how to get his men their rations and every other kind of stores needed in war. He must have imagination to originate plans, practical sense and energy to carry them through. He must be observant, untiring, shrewd, kindly and cruel, simple and crafty, a watchman and a robber, lavish and miserly, generous and stingy, rash and conservative. All these and many other qualities, natural and acquired, must he have. He should also, as a matter of course, know his tactics; for a disorderly mob is no more an army than a heap of building materials is a house."

general, inspector See INSPECTOR GENERAL.

general, lieutenant See LIEUTENANT GENERAL.

general, major See MAJOR GENERAL.

general air support Air operations in gaining and maintaining air superiority, as well as AIR INTERDICTION activities.

General Assembly 1. The largest unit of the United Nations, in which all member nations are represented and each has a single vote. While the General Assembly, which meets annually each fall, is a continuing international conference that has many United Nations housekeeping responsibilities and generates a goodly number of resolutions, it has no real power to affect the behavior of sovereign states. It functions mainly as a forum for international propaganda and debate. Compare to SECURITY COUNCIL. 2. A legislature in some state governments.

general court-martial The highest level of United States COURT-MARTIAL, consisting of not fewer than five members, not including the military judge, and having the power to try any offense punishable by the UNIFORM CODE OF MILITARY JUSTICE. It is generally used for the most serious offenses, and can impose the most severe penalties, including death. It is very much like a civil trial, in that both counsels must be lawyers and verdicts of guilty must be proved beyond a reasonable doubt and can be reviewed.

general court-martial orders Orders promulgating the results of trial by a GENERAL COURT-MARTIAL, and any subsequent action by the convening or higher authority of record for such trial.

general discharge A form of DISCHARGE, given to a soldier under honorable conditions, for service that is satisfactory but that does not qualify for an HONORABLE DISCHARGE.

general engineering Those engineering missions that do not directly contribute to the mobility, countermobility, or survivability of committed MANEUVER units, but which are essential for their FIREPOWER and LOGISTIC SUPPORT.

general officer An officer of the rank of BRIGADIER GENERAL or above in the U.S. Army or Air Force.

general orders 1. Permanent instructions, issued in order form, that apply to all members of a COMMAND, as compared with special orders, which affect only individuals or small groups. General orders are usually concerned with matters of policy or administration. 2. A series of permanent GUARD orders that govern the duties of a SENTRY on post.

general purpose bag An adjustable container that is used by an individual parachutist, and is designed for attachment to the parachutist's harness to carry individual or team equipment that must accompany him during an AIRBORNE OPERATION.

general purpose forces The vast bulk of the forces in a large nation's military; those units that are available for deployment or reployment at any time for any mission as opposed to those units that are dedicated to a particular mission and consequently cannot be readily redeployed. See William Schneider, Jr., "Soviet General-Purpose Forces," *Orbis* 21, 1 (Spring 1977); Alan Ned Sabrosky, "U.S. General-Purpose Forces: Four Essential Reforms," *Orbis* 24, 3 (Fall 1980).

general quarters A condition of readiness when naval action is imminent; all BATTLE STATIONS are fully manned and alert.

general reserve A reserve of troops under the control of an overall commander. See also FLOATING RESERVE.

general staff A group of officers in the headquarters of divisions or similar larger units that assist their commanders in planning, coordinating, and supervising operations. A general staff may consist of four or more principal functional sections: PERSONNEL (G-1), MILITARY INTELLIGENCE (G-2), operations and training (G-3), LOGISTICS (G-4), and CIVIL AFFAIRS/MILITARY GOVERNMENT (G-5). (A particular section may be added or eliminated by the commander, depending on the need that has been demonstrated.) In brigades and smaller units, staff sections are designated S-1, S-2, etc., with corresponding duties. The idea of a general staff originates principally from the Prussian military reforms that transformed an inefficient army into the foremost military machine in Europe by the middle of the nineteenth century. The Prussian, later German, general staff has been admired by military thinkers in most countries. It consisted of a small group of highly trained and intellectually able officers, drawn from the main officer corps relatively early in their careers, who then spent the rest of their professional lives in a central unit. This unit had the full-time job of planning for all the possible wars that Germany might face. Staff members were not combat commanders, and saw little active service; their professional loyalties were to the general staff, and therefore to the armed forces, or even the nation, as a whole. Because they worked and planned together on a permanent basis they were able, once war broke out, to communicate easily and with trust and understanding, thus avoiding the confusion, rivalries, and disagreements endemic to higher command in the armed forces of most nations. In World Wars I and II, the ability of the German military to co-ordinate its efforts and to put its plans into effect was in marked contrast to that of the Allied armies opposing them. But the German model has had few imitators. In most of North Atlantic Treaty Organization (NATO) nations, for example, staff posts are held for a few years at a time by officers who rotate through a variety of military occupations. Therefore real expertise and, even more crucially, the communal loyalty of a general staff never develops. See Waiter Goerlitz, *History of the German General Staff: 1657–1945* (New York: Praeger, 1953); J. D. Hittle, *The Military Staff: Its History and Development* (Harrisburgh: Stackpole, 1960).

general stopping power Used to describe a minefield: the percentage of a group of vehicles in battle formation likely to be stopped by mines when attempting to cross a MINEFIELD.

general support 1. That FIRE SUPPORT which is given to the supported force as a whole, and not to any particular subdivision thereof. 2. A mission that is frequently assigned to COMBAT SUPPORT and COMBAT SERVICE SUPPORT units. For example, a division field artillery battalion assigned to a general mission operates under control of the division artillery headquarters while supporting the entire division.

general support artillery Artillery that executes the FIRE directed by the commander of the unit to which it organically belongs or is attached. It fires in support of the operation as a whole rather than in support of a specific subordinate unit.

general support-reinforcing A tactical artillery mission; supporting a force as a whole and providing reinforcing fire for another artillery unit.

general unloading period In amphibious operations, that part of the ship-to-shore movement during which unloading is primarily logistic in character, and emphasizes speed and volume of unloading operations. It encompasses the unloading of units and cargo from the ships as rapidly as facilities on the beach permit. Compare to INITIAL UNLOADING PERIOD.

general war An armed conflict between major powers in which the total resources of the belligerents are employed, and the national survival of a major belligerent is in jeopardy.

Geneva Conventions Agreements made between a number of nations at Geneva, establishing rules for the treatment during war of the sick, wounded, prisoners of war, civilians, and other persons. A reference to the Geneva Convention now in effect, generally connotes the Geneva Convention of 1949.

geographic coordinates The degrees of latitude and longitude that define the position of a point on the surface of the earth.

geographic reference points A means of indicating a position, usually expressed either as double letters or CODE WORDS that are established in OPERATION ORDERS or by other means.

georef A worldwide position reference system that may be applied to any map or chart graduated in latitude and longitude regardless of projection. It is a method of expressing latitude and longitude in a form suitable for rapid reporting and plotting. (This term is derived from the words "The World Georgraphic Reference System.")

geostationary orbit An orbit at an altitude of 35,800 kilometers above the Earth's equator. A SATELLITE placed in such an orbit revolves around the Earth once per day, thus maintaining the same position relative to the surface of the Earth at all times. Stationary as such, it can be used as a COMMUNICATIONS relay or SURVEILLANCE post. Also called GEOSYNCHRONOUS orbit.

geosynchronous orbit See GEOSTATIONARY ORBIT.

GI 1. An ENLISTED PERSON in the U.S. Army. 2. A description of things that are uniquely military. 3. Supplies formally issued by the Army to all soldiers. Such *G*eneral *I*ssue items may be the origin of the acronym "GI." 4. Conforming to the strictly official or army way of doing things. 5. To clean up an area; a GI party is a cleaning detail. This is why many believe that the origin of GI comes from the "galvanized iron" trash cans and water buckets traditionally used by soldiers on a cleaning detail.

GIs Diarrhea from gastro-intestinal distress; a common problem for GIs in the field.

glacis 1. In a prepared DEFENSIVE POSITION, a clear slope extending downward from which FIRE could be directed. 2. The frontal armor on a TANK.

GLCM See GROUND-LAUNCHED CRUISE MISSILE.

glickum See GROUND-LAUNCHED CRUISE MISSILE.

glide bomb A bomb fitted with airfoils to provide lift, carried and released in the direction of a target by an airplane.

gliding metal A soft metal used to jacket a SMALL ARMS bullet; this metal can be readily engraved by the LANDS as the bullet moves down the bore.

globalism A description for United States foreign policy that implies active United States involvement, both politically and militarily, in all parts of the world; the opposite of ISOLATIONISM. American globalism has been more restrained since the VIETNAM WAR. See Stephen E. Ambrose, *Rise to Globalism: American Foreign Policy Since 1938* fourth edition (New York: Penguin Books, 1985).

going The classification of TERRAIN according to its ability to support the passage of vehicles.

go/no-go The condition or state of operability of a component or system. "Go" indicates proper functioning; "no-go" indicates improper functioning.

good offices The disinterested use of one's official position or office, in order to help others settle their differences; an offer to mediate a dispute. See B. G. Ramcharan, "The Good Offices of the United Nations Secretary-General in the Field of Human Rights," *American Journal of International Law* Vol. 76, No. 1 (January 1982).

gradeability The capability of a vehicle to negotiate a slope, either ascending or descending, measured in percent (not degrees).

grain A term that applied to a PROPELLANT means one piece (a small particle), which may be used separately, cemented to other grains, or collectively with other grains.

grand slam A term meaning "All enemy aircraft originally sighted are shot down."

grand strategy 1. The overall strategic policies of a nation or alliance. All military strategy logically follows from this overall national strategy. 2. Strategy at a higher level than that used for one THEATER OF WAR or CAMPAIGN. See John M. Collins, *Grand Strategy: Principles and Practices* (Annapolis, Maryland: Naval Institute Press, 1973); Edward N. Luttwak, Herbert Block and Seth W. Carus, *The Grand Strategy of the Soviet Union,* (New York: St. Martin's, 1983).

granulation A term referring to the size and shape of the grains of a PROPELLANT powder.

graphic intersection An INTERSECTION drawn by plotting direction lines that determines the position of an unknown location point.

graser See GAMMA-RAY LASER.

gratuity, death See DEATH GRATUITY.

graves registration Supervision and execution of matters pertaining to the identification, removal, and burial of the dead, and collection and processing of their effects.

gravity bomb 1. Any BOMB having just dropped from an airplane. 2. A nuclear bomb delivered by an airplane without further guidance.

gravity drop The vertical DEFLECTION of a BOMB OR PROJECTILE due to the action of gravity.

gravity drop angle The angle in the vertical plane between the gun line and a position in the TRAJECTORY due to gravity drop.

gray line, long See LONG GRAY LINE.

graze 1. A graze burst: the burst of a PROJECTILE on impact with the ground. 2. To pass close to the surface, as a shot that follows a path nearly parallel to the ground and low enough to strike a standing man.

graze sensitive The capability of a FUZE to be initiated by grazing; that is, when the projectile or missile containing the fuze strikes a surface obliquely and is deflected at an angle (e.g., 80–90 degrees from the normal).

grazing fire Fire approximately parallel to the ground, and the center of the CONE OF FIRE of which does not rise above the height of a man.

grease gun A SUBMACHINE GUN; some early models looked like automotive grease guns.

great circle A circle on the surface of the earth, the plane of which passes through the center of the earth.

great circle route The route that follows the shortest arc of a great circle between two points.

Great Patriotic War 1. The war of 1812, in which Russia defeated France. 2. The Russian phrase for World War II.

Green Berets See SPECIAL FORCES.

Greenwich Mean Time Mean solar time at the meridian of Greenwich, England, used as a basis for the standard time throughout the world. Normally expressed in four numerals, from 0001 through 2400. Greenwich Mean Time is always five hours ahead of Eastern Standard Time. Also called ZULU TIME.

Grenada, Invasion of The American military action of October 25, 1983, which took control of the Caribbean island nation of Grenada away from a Marxist military government that had seized power six days earlier. The Reagan administration acted in response to requests for military intervention from the Grenada Governor-General and the Organization of Eastern Caribbean States, and to guarantee the safety of the approximately one-thousand American citizens (mostly medical students) on the island. Within sixty days all American combat units were gone and the island was left in the control of a civilian council that would govern pending elections. See Ronald M. Riggs, "The Grenada Intervention: A Legal Analysis," *Military Law Review* Vol. 109 (Summer 1985); Michael Rubner, "The Reagan Administration, the 1973 War Powers Resolution, and the Invasion of Grenada," *Political Science Quarterly* Vol. 100, No. 4 (Winter 1985–86).

grenade A small EXPLOSIVE or chemical BOMB thrown by hand or fired from a RIFLE, CARBINE, or special grenade launcher.

grenade, chemical hand See CHEMICAL HAND GRENADE.

grenade, frangible See FRANGIBLE.

grenade, offensive See OFFENSIVE GRENADE.

grenade, rifle See RIFLE GRENADE.

grenade court A training ground laid out for instruction in throwing GRENADES.

grenade launcher 1. A special-purpose, shoulder-fired weapon or an attachment to a RIFLE or CARBINE that facilitates the firing of cartridge type grenades from a tube, or rifle and hand grenades from an extension attached to the MUZZLE. 2. A component of certain aircraft armament subsystems that permits automatic firing of cartridge-type grenades from aircraft.

grey-area systems Nuclear weapons that have proved difficult to classify during arms-control negotiations because they can be considered either strategic or tactical depending upon deployment. See

Karl Lautenschlager, "Theater Nuclear Forces and Grey Area Weapons," *Naval War College Review* 33, 5 (September–October 1980).

grey propaganda PROPAGANDA that does not specifically identify any originating source.

grid 1. Two sets of parallel lines intersecting at right angles and forming squares. A grid is superimposed on maps, charts, and other similar representations of the Earth's surface in an accurate and consistent manner to permit the identification of ground locations with respect to other locations, and the computation of direction and distance with respect to other points. 2. A term used in giving the location of a geographic point by grid coordinates.

grid, military See MILITARY GRID.

grid, military reference system See MILITARY GRID REFERENCE SYSTEM.

grid bearing The direction of an object from a point, expressed as a horizontal angle and measured clockwise with reference to grid north. It is proportional to the longitudinal difference between the point and the central meridian. See also MAGNETIC NORTH, TRUE NORTH.

grid convergence The horizontal angle at a place between true north and grid north. It is proportional to the longitudinal difference between a given place and the central meridian.

grid coordinates Coordinates of a grid coordinate system to which numbers and letters are assigned for use in designating a point on a gridded map, photograph, or chart. See also COORDINATES.

grommet 1. A device made of rope, plastic, rubber, or metal to protect the ROTATING BAND of a projectile. 2. A ring of stiffening material used in the inside circumference of the peak of a service cap (the military dress hat) to maintain its rigidity of shape.

groove One of the channels in the BORE of a rifled gun barrel.

gross error The detonation of a NUCLEAR WEAPON at such a distance from the DESIRED GROUND ZERO as to cause no nuclear damage to the target.

gross weight 1. The weight of a vehicle that is fully equipped and serviced for operation, including the weight of the fuel, lubricants, coolant, vehicle tools and spare parts, crew, personal equipment, and load. 2. The weight of a container or pallet, including freight and binding. See also NET WEIGHT.

ground alert That status of alert in which aircraft on the ground or deck of an aircraft carrier are fully serviced and armed, with combat crews in readiness to take off within a specified short period (usually 15 minutes) after receiving a mission order.

ground burst Any explosion of a PROJECTILE that has impacted on the ground, as opposed to the explosion of projectiles designed to explode at a preset distance above the ground in order to be maximally destructive. Ground bursts of nuclear weapons produce far more long-term radioactivity than AIR BURSTS.

ground clutter See RADAR CLUTTER.

ground controlled approach (GCA) A procedure whereby a pilot landing an aircraft is guided by instructions from a radar operator who is tracking the aircraft from a ground facility. GCA was developed by the military services and then adopted by commercial airlines.

ground control interceptor See AIR DEFENSE CONTROLLER.

ground-effect phenomenon The generation of an artificial air cushion to support a vehicle close to the ground, allowing it to ride free. Ground-effect machines (see ACV) utilize this phenomenon to eliminate ground friction.

ground fire SMALL ARMS ground-to-air fire directed against aircraft.

ground-launched cruise missile (GLCM) Politically the most controversial, but militarily the least important member of the family of CRUISE MISSILES. Although they can be equipped with conventional warheads, the GLCMs operated by NORTH ATLANTIC TREATY ORGANIZATION (NATO) forces were deployed entirely for a THEATRE NUCLEAR FORCE role. Ground-launched cruise missiles are intermediate-range weapons capable of travelling several thousand of miles, but are very slow compared to BALLISTIC MISSILES, and could be vulnerable to SURFACE-TO-AIR MISSILES and other defenses. An advantage of GLCMs is their high mobility. Fired from self-powered launch vehicles, they can be driven around the country and hidden relatively easily. The United States owned GLCMs deployed in Europe were at the center of the Euromissile debate and, in 1988, were designated for removal and destruction as a result of the INF TREATY.

ground-liaison officer An officer especially trained in air RECONNAISSANCE or OFFENSIVE AIR SUPPORT activities. Such officers are normally organized into teams, under the control of the appropriate ground-force commander, to provide liaison to air force and navy units engaged in training and combat operations. See also AIR LIAISON OFFICER.

ground-liaison party An army unit responsible for liaison with a TACTICAL AIR SUPPORT agency.

ground-liaison section A ground unit responsible for ground-air liaison under the control of a ground headquarters.

ground-observer team Small UNITS or DETACHMENTS deployed to provide information on enemy aircraft movements over a defended area, obtained either by aural or visual means.

ground organization The land based facilities and non-flying personnel necessary for the operation of military aircraft.

ground position The position on the Earth vertically below an aircraft.

ground readiness That status wherein aircraft can be armed and serviced and their personnel alerted to take off in them within a specified time after receiving orders.

ground shocks Tremors in the Earth that result from an underground nuclear explosion.

ground signals A visual signal displayed on an airfield to give local air traffic rules information to flight crews in the air.

ground speed The horizontal component of the speed of an aircraft relative to the Earth's surface.

ground-support equipment All equipment required to maintain aircraft and their associated equipment.

ground tactical plan An AIRBORNE or AIRMOBILE operational plan covering the conduct of operations in an objective area.

ground zero The point on the surface of the Earth at or vertically below or above the center of a planned or actual NUCLEAR DETONATION. Also called actual ground zero.

ground zero, desired See DESIRED GROUND ZERO.

group 1. A flexible ADMINISTRATIVE and TACTICAL UNIT composed of either two or more BATTALIONS or two or more SQUADRONS. The term also applies to combat support and service support units. 2. A number of ships or aircraft, normally a subdivision of a FORCE, assigned for a specific purpose. 3. A tactical unit larger than the usual size of its core element; for example, a battalion group or BRIGADE group.

grouping The size of the target area in which all ROUNDS have fallen during firing practice.

group of targets Two or more targets on which simultaneous FIRE may be desired.

groupthink The psychological drive for consensus at any cost, which tends to suppress both dissent and the appraisal of alternatives in small decision-making groups. Groupthink, because it refers to a deterioration of mental efficiency and moral judgment due to in-group pressures, has an perjorative connotation. For the basic work on the subject, see Irving L. Janis, *Victims of Groupthink: A Psychological Study of Foreign Policy Decisions and Fiascoes* (Boston: Houghton Mifflin, 1972); Steve Smith, "Groupthink and the Hostage Rescue Mission," *British Journal of Political Science* Vol. 15, No. 1 (January 1985).

grunt A slang term for American infantrymen, especially those who served in Vietnam. See Charles R. Anderson, *The Grunts* (Novato, CA: Presidio Press, 1987).

G series A group of chemical NERVE AGENTS that are highly toxic and practically odorless, and are used in chemical operations. These were the first nerve gases developed by the Germans. Large stocks of them were captured at the end of World War II. These new gases were 50 times more powerful than the World War I varieties. Compare to V SERIES.

guard 1. A SECURITY element whose primary task is to protect a main force by fighting to gain time, while also observing and reporting information. See also SCREEN. 2. A soldier who provides protective services of any kind. 3. The NATIONAL GUARD.

guard, advance See ADVANCE GUARD.

guard, flank See FLANK GUARD.

guard, interior See INTERIOR GUARD.

Guard, National See NATIONAL GUARD.

guard, rear See REAR GUARD.

guard, special See SPECIAL GUARD.

guardhouse A building occupied by men detailed for INTERIOR GUARD; whenever used, it includes a guard tent or any other designated location for the headquarters of the guard.

guard mount A ceremony at which a new guard is inspected and installed and the old guard relieved; used primarily by military police and interior guard units.

guard reserve, advance See ADVANCE-GUARD RESERVE.

guards Elite forces in most armies other than the U.S.

guerrilla warfare Military operations conducted by IRREGULAR FORCES in enemy-controlled territory. The term was first used to refer to the Spanish partisans who fought against NAPOLEON's troops in the early 1800s. Today, any armed uprising by the people of a nation against their oppressors is considered guerrilla warfare. Since guerrilla troops do not follow normal battle tactics or use standard weapons systems in open combat, they are much more difficult for official battle troops to control. Chairman Mao Tse-tung (1893–1976), leader of the People's Republic of China (1949–1976), summed up the essence of guerrilla strategy: "The enemy advances, we retreat; the enemy camps, we harass; the enemy tires, we attack; the enemy retreats, we pursue." By never forming into large units or allowing themselves to be trapped into PITCHED BATTLES, guerrillas are able to avoid the damage that the superior numbers of a regular army and its massed firepower could inflict. Very large armies can have their effective personnel substantially reduced by such tactics by being forced to GARRISON hundreds of villages and towns, to send guard detachments with every supply convoy, and generally to operate in an inconvenient environment. It is widely believed that a drawn-out guerrilla war will defeat an orthodox army, but history produces little evidence for this. The Napoleonic armies were only defeated by Wellington's British army, and the French Resistance was really of use only as a support to the 1944 Allied invasion. The U.S. Army was almost invariably successful in actual operations against the Vietcong, which was almost completely destroyed in the 1968 Tet Offensive; thereafter the Americans fought the regular army of North Vietnam, an army that had already shown its worth in the entirely orthodox campaign against the French in 1954. The reason for the apparent success of guerrilla warfare is its usual practice in countries where a foreign "army of occupation" has no support among the indigenous population. Although there is generally no significant military success in guerrilla tactics, the costs of continuing to garrison a foreign country and maintain a presence by force is often too much for the political will of the occupying power, which may decide to withdraw, or accommodate the guerrilla leaders, even though militarily undefeated. See Walter Laqueur, "The Origins of Guerrilla Doctrine," *Journal of Contemporary History* Vol. 10, No. 3 (July 1975); John J. Tierney, Jr., "America's Forgotten Wars: Guerrilla Campaigns in U. S. History," *Conflict: An International Journal for Conflict and Policy Studies* Vol. 2, No. 3 (1980); Gberard Chaliand, *Guerrilla Strategies: A Historical Anthology from the Long March to Afghanistan,* (Berkeley: University of California Press, 1982).

guidance 1. Military policy, direction, decision, or instruction having the effect of an ORDER when promulgated by a higher ECHELON. 2. The entire process by which TARGET INTELLIGENCE information perceived by a guided missile is used to effect proper flight control so that the missile makes timely direction changes for effective target interception.

guidance, homing See HOMING GUIDANCE.

guidance, inertial See INERTIAL GUIDANCE.

guidance, midcourse See MIDCOURSE GUIDANCE.

guidance-station equipment The ground-based portion of a missile guidance system necessary to provide guidance during missile flight.

guidance system (missile) A system that evaluates flight information relating to a missile, correlates it with target data, determines the desired flight path of the missile, and communicates the necessary commands to the missile flight control system. See also ILLUMINATOR RADAR.

guide 1. That member of a unit marching in formation who sets the pattern and pace of movement in response to orders; he is often at the extreme right or left front of the formation. 2. A civilian, who because he is knowledgeable of the local terrain, is attached to a military unit.

guided missile 1. An unmanned vehicle moving above the surface of the earth whose TRAJECTORY of flight is capable of being altered by an external or internal mechanism. Compare to AERODYNAMIC MISSILE; BALLISTIC MISSILE. 2. An air-launched guided missile for use against air targets. 3. A surface-launched guided missile for use against air targets. 4. A surface-launched guided missile for use against surface targets. The essential element of a guided missile is that with less technological weapons, no in-course correction could be made.

guided weapon, laser See LASER GUIDED WEAPON.

guide left (right) 1. A command to regulate march on the left (right) guide or the left (right) element of a FORMATION. 2. An order to the guide to move to the left (right) of a formation.

guide on me An order given by a BATTERY commander for a unit to follow his movements. This unit then becomes the guide for another unit.

guide right See GUIDE LEFT (RIGHT).

guides on line A command given by the commander of a marching unit at a review, parade, or drill, at which the guides of the unit leave their places and run up to the guide position on a line along which the unit will halt and form for the parade.

guides post A command given by the ADJUTANT of a unit at a review, parade, or drill, at which the guides of the subordinate units resume their proper places in line. The command is given after all the marching elements have reached, and halted on, the line of guides from which the review begins.

guidon 1. A FLAG with a swallow-tailed end, carried by units for identification. 2. The bearer soldier who carries the guidon.

Gulf of Tonkin Resolution The joint resolution of the U.S. Congress in 1965 that sanctioned the use of great numbers of American forces in Vietnam. It was in reaction to a presumed 1965 attack on American ships in the Gulf of Tonkin by North Vietnamese naval units. Subsequent investigations have conclusively shown that the information about the "attack" was fabricated by the Johnson Administration to encourage the passage of the resolution. See Joseph C. Goulden, *Truth is the First Casualty: The Gulf of Tonkin Affair—Illusion and Reality* (Chicago: Rand McNally, 1969).

gull In ELECTRONIC WARFARE, a floating radar reflector used to simulate a surface target at sea for deceptive purposes.

gun 1. Any FIREARM. 2. A CANNON with relatively long barrel, operating with a relatively low angle of fire, and having a high MUZZLE VELOCITY. 3. A cannon with a tube length of 30 calibers or more. See also HOWITZER and MORTAR.

gun, assault See ASSAULT GUN.

gun, fixed See FIXED GUN.

gun, grease See GREASE GUN.

gun, minute See MINUTE GUN.

gun, multiple See MULTIPLE GUN.

gun, primary See PRIMARY GUN.

gun, riot See RIOT GUN.

gun, saluting See SALUTING GUN.

gunboat A small warship.

gunboat diplomacy The term is used in international relations to imply that a great power is seeking to achieve its diplomatic goals through the threat or use of force. See James Cable, *Gunboat Diplomacy 1919–1979: Political Applications of Limited Naval Force,* revised edition (New York: St. Martin's Press, 1981); Robert Mandel, "The Effectiveness of Gunboat Diplomacy," *International Studies Quarterly* Vol. 30, No. 1 (March 1986).

gunbore line A reference line established by the linear extension of the BORE axis of a gun.

gun carriage A mobile or fixed support for a gun that sometimes includes elevating and traversing mechanisms.

gun density The number of guns that may be brought to bear on a given target.

gun direction The distribution and direction of the gunfire of a ship.

gun displacement 1. The distance from a target to the DIRECTING POINT or the BASE PIECE of a BATTERY. 2. The movement of a gun to a new firing position.

gun drill The operation and maintenance of a gun by personnel assigned to the GUN SECTION.

gunfire liaison team, naval See NAVAL GUNFIRE LIAISON TEAM.

gunfire operations center, naval See NAVAL GUNFIRE OPERATIONS CENTER.

gunfire spotting team, naval See NAVAL GUNFIRE SPOTTING TEAM.

gung ho Very enthusiastic. This is a contraction of a Chinese phrase meaning "industrial cooperation." It was popularized by the U.S. Marine Corps during World War II who sometimes used it as a battle cry.

gun jump The angle between the direction of the gun BORE at the instant a charge is fired and the LINE OF DEPARTURE of the projectile as it leaves the MUZZLE.

gunner 1. The artilleryman who actually aims a gun. 2. All artillerymen. 3. A shortened version of "machine gunner." 4. In the U.S. Navy, a warrant officer responsible for the maintenance of ordnance and the firing of ship's guns.

gunner's rule A method of determining the safe range for firing MACHINE GUNS and RECOILLESS RIFLES over the heads of friendly troops, when the range to the target is 900 meters or less for machine guns and 1,000 meters or less for recoilless rifles.

gunnery 1. The art and science of gun operations. 2. The firing of guns.

gunnery officer The officer aboard a warship responsible for the maintenance and firing of the ship's guns.

gunnery sergeant A Marine Corps enlisted grade between staff sergeant and first sergeant corresponding to chief petty officer in the Navy and sergeant first class in the Army.

guns, pom-pom See POM-POM GUNS.

gun section In FIELD ARTILLERY, a subdivision of a BATTERY consisting of a gun with proper personnel and equipment.

guns free In AIR INTERCEPTION, a weapons control status indicating that fire may be opened on all aircraft not recognized as friendly.

gunship 1. Fixed wing aircraft designed for close air support and interdiction of supply lines, filled with multibarreled machine guns and/or cannons. 2. Helicopter gunship. Either a conventional transport helicopter filled with machine guns and rocket pods or a helicopter especially designed for close support of ground operations. Both were major innovations during the Vietnam War.

gunsight Any device used to aim a gun toward a target; any stationary, mechanical or optical device that functions to aid the eye in aiming.

gunsight line The line of sight to the AIMING POINT through the fixed optical system of an aircraft's gunsight.

gunsight radius On a RING SIGHT, the radical distance between two concentric circles used to indicate both the target range and the deflection of fire.

gun slide 1. The portion of a gun that rests on the CRADLE GUIDES. 2. Part of a PLOTTING AND RELOCATING BOARD mechanism.

gun-target line An imaginary straight line from a gun(s) to its target. See also SPOTTING LINE.

guns tight In AIR INTERCEPTION, a weapons control status indicating not to open fire, or to cease firing on any aircraft (or BOGEY specified, or in section indicated) unless it is known to be hostile.

gun wave See MUZZLE WAVE.

Gustavus Adolphus (1594–1632) The King of Sweden who was the first major military commander to make effective use of massed and maneuverable FIREPOWER on the battlefield.

gyroscope, directional See DIRECTIONAL GYROSCOPE.

H

hachuring The representation of relief on a map by shading in short, disconnected lines drawn in the direction of the slopes of elevated and depressed areas.

hair trigger The state in which nuclear-potent nations are ready, willing, and able to start World War III at the slightest provocation.

half left (or right) A direction 45 degrees to the left (or right) of the original FRONT in which a soldier faces at the execution of a COMMAND. 2. Preparatory command to face in a direction 45 degrees to the left (right) of the original front.

half-life The time required for the RADIOACTIVITY of a radioactive isotope to decrease to exactly one half of its initial value due to radioactive decay. The half-life of any isotope is a trait that is unchangeable, and is independent of the amount or condition of the isotope. Depending on the elements involved, a half-life can be as short as a thousandth of a second or as long as one-hundred trillion years.

half-loaded An automatic arms belt or MAGAZINE inserted without its first cartridge actually in the chamber.

half-residence time The time required for the amount of radioactive weapon debris deposited in a particular part of the atmosphere to decrease to half of its initial value.

half right See HALF LEFT (OR RIGHT).

half thickness The thickness of absorbing material that will reduce the intensity of RADIATION that passes through it by one-half.

half-track vehicle A combination wheeled and tracked motor vehicle whose rear end is supported on and propelled by complete band tracks, while the front end is supported on and steered by wheels.

halving The division of the field of view, observed from a COINCIDENCE RANGEFINDER, into two equal portions, one the exact mirror image of the other.

hand grenade See GRENADE.

hand grenade, chemical See CHEMICAL HAND GRENADE.

handoff 1. The passing of responsibility for a battle from one commander to another. 2. The transfer of responsibility for a battle that occurs between units at all echelons of command during PASSAGE OF LINES and relief operations. 3. The passing of a TARGET from one

WEAPONS SYSTEM to another; for example, between two TANKS when one is repositioning, or between a SCOUT helicopter and an ATTACK helicopter, or between two FIGHTER aircraft during an engagement of enemy ARMOR. 4. The passing of operational control of units in transit.

handover The passing of the CONTROL authority for an aircraft from one control agency to another. Handover action is complete when the receiving controller acknowledges the assumption of control authority.

handover line A control feature (such as a river or road), preferably following easily defined terrain features, at which responsibility for the conduct of combat operations is passed from one force to another.

hand-to-hand combat The ultimate in closing with the enemy; fighting at such close quarters that soldiers can physically assault the enemy with BAYONETS, rifle butts, and bare hands. The demise of this tactic has been predicted for centuries. Nonetheless, Machievelli wrote in 1531: "As to the proposition advanced . . . that hereafter war will be made altogether with artillery, I consider that this observation is wholly erroneous . . . For whoever wishes to train a good army must . . . train his troops to attack the enemy sword-in-hand, and to seize hold of him bodily."

hang fire 1. An undesired delay in the functioning of a FIRING SYSTEM. 2. An order to HOLD FIRE pending further instructions.

harassing fire FIRE designed to disturb enemy troops, to curtail their movement and, by threat of losses, to lower their morale.

harassment 1. A military operation whose primary objective is to disrupt the activities of a UNIT, INSTALLATION, or ship, rather than to inflict serious casualties or damage in or on it. 2. An air attack that is not connected with INTERDICTION or CLOSE AIR SUPPORT on any target within the area of a land battle. Also called a harassing attack.

hard beach A portion of beach especially prepared with a hard surface extending into the water, employed for the purpose of loading or unloading troops and material directly into or from LANDING SHIPS or LANDING CRAFT.

harden To take measures preventing aircraft, ships, vehicles, or other military equipment from being easily damaged by any of the effects of a nuclear explosion.

hardened site A site constructed to provide protection against the effects of conventional and nuclear explosions. It may also be equipped to provide protection against a chemical or biological attack; it usually cannot be destroyed except by a direct hit from a nuclear weapon.

hard kill The total destruction of a target as opposed to the disabling of it; the destruction of a target in such a way as to produce unambiguous visible evidence of its neutralization. A target that is disabled but not destroyed is a "soft kill."

hard missile base A missile LAUNCHING SITE that is protected against an enemy nuclear explosion.

hard target 1. A target that is heavily protected against a nuclear blast. Typical hard targets are missile SILOS and command control BUNKERS. Such targets are buried deep underground, and have walls several feet thick built of steel-reinforced concrete. They can be destroyed only by enormous BLAST power, which would require a combination of extreme accuracy and very high YIELD from a nuclear WARHEAD in GROUND BURST detonations. As a rough guide, an ordinary brick or concrete building above ground would be seriously damaged by a blast with an "overpressure" of between 5 and 10 pounds per square inch above the normal atmospheric pressure. The command and control bunkers built by the NORTH ATLANTIC TREATY ORGANIZATION (NATO) in West Germany are reported to be safe up to blast overpressures of 2000 psi. The destruction of hard targets would be vital in a nuclear war, but would require many more committed warheads than would ordinary industrial or even most military targets. 2. In the context of conventional warfare, an ARMORED VEHICLE immune to an ordinary HIGH-EXPLOSIVE shell, in contrast to a soft or "thin-skin" target such as a truck or a human body.

hardware 1. The electronic components of computers themselves, as opposed to their software programs. 2. Major items of military equipment or their components.

Hart, Liddell See LIDDELL HART, B.H.

hasty attack An OFFENSIVE attack or operation usually conducted following an enemy MOVEMENT TO CONTACT, for which a unit trades preparation time for speed. It is conducted with the resources immediately available. See also DELIBERATE ATTACK.

hasty breaching The rapid creation of a route through a minefield, barrier or fortification by any expedient method.

hasty crossing A crossing of a river or stream using crossing means at hand or readily available, without pausing to make elaborate preparations. See also DELIBERATE CROSSING.

hasty defense A DEFENSE normally organized while in contact with the enemy or when contact is imminent and time available for the organization of a military force is limited. It is characterized by improvement of the natural defensive strength of the terrain by utilizing of FOXHOLES, EMPLACEMENTS, and OBSTACLES. See also DELIBERATE DEFENSE.

hatch list A list showing, for each HOLD section of a cargo ship, a description of the items stowed, their volume and weight, the consignee of each, and the total volume and weight of MATERIEL in the hold.

hawk A person inclined toward military action; a dove, a far more peaceful bird in metaphor and symbol since ancient times, is not. "Hawk" and "dove" emerged during the Vietnam period as ready-made terms to describe someone's attitude toward American participation in the war. Yet both terms were used much earlier on. Thomas Jefferson, in a letter to James Madison on April 26, 1798, used the term "war hawks" to describe those Federalists who wanted to instigate war with France. Later the term was applied to those who brought on the War of 1812 with England. "Dove" is of more recent vintage in American national politics; it was first used during the Kennedy Administration to describe those presidential advisors who advocated a policy of accommodation with the Soviet Union. The hardliners were, of course, hawks. See also CHICKEN HAWK.

H-bomb See HYDROGEN BOMB.

heading The direction in which the longitudinal axis of an aircraft or ship is pointed, usually expressed in degrees clockwise from north (true, magnetic, compass, or grid).

headquarters (HQ) The executive or administrative elements of a COMMAND unit. When Union General John Pope told the press in 1862 that "my headquarters will be in the saddle," his COMMANDER-IN-CHIEF, President Abraham Lincoln, responded by saying: "A better place for his hindquarters."

headquarters, alternate See ALTERNATE HEADQUARTERS.

headquarters, management See MANAGEMENT HEADQUARTERS.

headquarters, operational See OPERATIONAL HEADQUARTERS.

headquarters commandant The officer in charge of the housekeeping functions (food, office machines, and supplies, etc.) of a military headquarters.

headquarters company (battery) (troop) The administrative and tactical element of a BATTALION or larger unit, with personnel assigned to administration, intelligence, communications, and other necessary activities.

headquarters detachment Adinistrative and tactical element of a BATTALION or larger unit; it usually differs from a HEADQUARTERS COMPANY by having fewer personnel.

head space The distance between the face of the BOLT (fully closed) and the CARTRIDGE-seating shoulder of the chamber of a SMALL ARM.

HEAT See HIGH EXPLOSIVE ANTI-TANK PROJECTILES.

heavy 1. The largest and most powerful of a series of military devices. While all TANKS are heavy, the heavy tank is the most potent. 2. An enemy, deriving from the motion picture usage of the term "heavies" for villains. 3. A military unit with greater FIREPOWER capability than another of comparable size but less "heavy" equipment.

heavy drop The delivery of heavy supplies and equipment by PARACHUTE.

heavy infantry Those INFANTRY units that have heavier and more powerful equipment (such as ARMORED FIGHTING VEHICLES) than light infantry. Heavy infantry, as compared to light, is a more potent fighting force but less mobile.

heavy level of operations Combat involving more than 60 percent of all force maneuver ECHELONS and FIRE SUPPORT means engaged in all-out COMBAT. Such operations demand the application of a force's total strength over a period of time, and may include possible employment of resources at the next higher echelon to assure accomplishment of the force MISSION.

heavy machine gun A classification of MACHINE GUN, including the .30-caliber water-cooled machine gun and all larger caliber machine guns. In the British Army, the equivalent of the .30 caliber machine gun is regarded as a medium machine gun, and only guns of .50 caliber and above are "heavy."

heavy missile A BALLISTIC MISSILE considerably larger and heavier than those to which it is being compared. The exact weight that differentiates a heavy missile from a regular missile has always been a point of contention between the superpowers.

heavy shellproof shelter A bomb SHELTER that can protect against continuous bombardment by shells of 8-inch or larger caliber.

height of burst The vertical distance above the ground at which a NUCLEAR DETONATION occurs.

helicopter Any of a class of heavier-than-air motor-powered vehicles that lift and sustain themselves in flight by means of rotating blades turning horizontally. There are four general categories of military helicopters; (1) utility helicopters transport troops and supplies and evacuate wounded (when modified they can also function as GUNSHIPS); (2) observation helicopters are used for command and control, for adjustment of artillery fire, and for reconnaissance; (3) assault helicopters are equipped with a wide variety of weapons capable of dissipating enemy troop movements and destroying enemy armor; and (4) cargo helicopters are used to move the heaviest supplies and equipment.

helicopter, compound See COMPOUND HELICOPTER.

helicopter assault force A TASK ORGANIZATION combining helicopters, supporting units, and helicopter-borne troop units for use in helicopter-borne ASSAULT operations.

helicopter drop point A designated point within a HELICOPTER LANDING ZONE where helicopters are unable to land because of the terrain, but in which they can discharge cargo or troops while hovering.

helicopter flight An individual helicopter, or two or more helicopters grouped under a flight leader, and launched from a single helicopter transport ship or base at approximately the same time.

helicopter flight rendezvous An AIR-CONTROL POINT in the vicinity of a HELICOPTER TRANSPORT or base where helicopters are assembled into flights prior to proceeding to the WAVE rendezvous. It is designated by code name.

helicopter gunship See GUNSHIP.

helicopter landing site A designated subdivision of a HELICOPTER LANDING ZONE on which a single flight or wave of assault helicopters lands to embark or disembark troops, cargo, or both.

helicopter landing zone A specified ground area for landing assault helicopters to embark or disembark troops or cargo or both. A landing zone may contain one or more landing sites.

helicopter team The combat-equipped troops lifted in one helicopter at one time.

helicopter wave rendezvous An AIR-CONTROL POINT where helicopter flights are assembled into helicopter waves prior to executing a mission. It is designated by a code name.

helmet A metal hat; head ARMOR.

H-hour The specific hour on D-DAY at which a particular OPERATION commences. The hour may mark commencement of hostilities or the hour at which an OPERATION PLAN is executed or to be executed.

high airburst The FALLOUT-SAFE HEIGHT of burst for a NUCLEAR WEAPON that increases damage to or casualties on unprotected targets, or reduces induced radiation contamination at actual GROUND ZERO. See also types of burst.

high-altitude bombing HORIZONTAL BOMBING with the height of bomb release over 15,000 feet.

high-altitude burst The explosion of a NUCLEAR WEAPON at a height in excess of 100,000 feet (30,000 meters).

high-angle fire FIRE delivered at angles of elevation greater than the elevation that corresponds to the maximum RANGE of a gun and its ammunition; fire whose range decreases as the angle of elevation is increased.

high-burst ranging The adjustment of gunfire by observation of AIRBURSTS of its projectiles. Also called airburst ranging.

high explosive A term generally describing the bursting charges for BOMBS, PROJECTILES, GRENADES, MINES, and demolition charges. Defined by the U.S. Department of Transportation as materials susceptible to detonation by a BLASTING CAP.

high explosive anti-tank projectiles All forms of anti-tank projectiles except KINETIC ENERGY rounds (which can only be used in the very long cannons carried by other tanks). Ordinary high-explosive shells are useless against armored targets, because the chemical energy they release is too broadly distributed to damage steel armor. HEAT weapons work by concentrating the heat in the form of high-pressure gasses on a tiny portion of the armor, burning a small hole through, and filling the tank's interior with molten metal and chemical gasses that destroy the crew and sets fire to any combustible part of the tank's equipment. Because the energy is narrowly focused they are virtually useless against other targets, where a wide area needs to be covered by heat, blast, and metal fragments.

high-explosive projectile A PROJECTILE with a BURSTER of high explosive, used against personnel and materiel.

high frequency Those radio frequencies in the spectrum between 3 and 30 megahertz.

High Frontier A Washington, D.C., based organization that promotes space technology for commercial and military purposes. It is a strong advocate of the Strategic Defense Initiative.

high-grade cryptosystem A CRYPTOSYSTEM system designed to provide lasting security i.e., by inherently resisting solution for a comparatively long or indefinite period.

high-mobility multipurpose wheeled vehicle See HMMWV.

high-order detonation A complete and instantaneous explosion.

high port The position in which a rifle is carried while a soldier is charging or jumping. The rifle is carried diagonally across the body, with the left wrist in front of the left shoulder and the right wrist near the right hip.

high-value-asset control items Expensive items of supply identified for intensive inventory management control; special efforts are required to keep them from being stolen.

high velocity 1. A term describing the MUZZLE VELOCITY of an artillery projectile that travels from 3,000 feet per second to, but not including, 3,500 feet per second. See also HYPERVELOCITY. 2. Velocities of SMALL ARMS ammunition between 3,500 and 5,000 feet per second. 3. Velocities of tank CANNON projectiles between 1,550 and 3,350 feet per second.

higher law A value system to which individuals owe an even greater obligation than to the laws of a state. A "higher" law is often appealed to by those who wish to attack an existing law or practice which courts or legislators are unlikely or unwilling to change. Martyrs throughout the ages have asserted a higher law in defiance of the state. The classic representation of the struggle between an appeal to higher law and a state's judicial mandate can be found in Sophocles' (490–406 B.C.) play *Antigone* in which the heroine defies the king, asserts a higher law as her justification, and "forces" the king to have her killed. Because the courts of any state can only enforce the law of the land, appealing to a higher law is always chancy business. A recognition of a "higher law" is implicit in the notion of war crimes. Because of the concept of higher law, soldiers are expected not to commit WAR CRIMES, even though they may be ordered to do so.

highway transport lift The PAYLOAD tonnage or PERSONNEL that can be transported by a vehicle or by a truck in one trip.

hill shading A method of representing relief on a map by depicting the shadows that would be cast by high ground if light were shining from a certain direction.

HMMWV (high-mobility multipurpose wheeled vehicle) The HMMWV is a light, highly mobile, diesel-powered, four-wheel drive tactical vehicle that uses a common 1.25-ton payload chassis. It can be configured through the use of common components to become a cargo/troop carrier, armament carrier, two-or four-litter ambulance, or missile carrier. It is the replacement for the jeep. See George Taylor III, "On the Road to HMMWV," *Army Logistician* 15, 6 (November–December 1983).

hold 1. A cargo stowage compartment aboard a ship. 2. To maintain or retain possession of by force, as a POSITION or an AREA. 3. In an attack, to exert sufficient pressure to prevent movement or redisposition of enemy forces. 4. As applied to air traffic, to keep an aircraft within a specified space or location.

hold fire In AIR DEFENSE, an emergency order to stop firing. Guided missiles already in flight must not be permitted to continue to intercept their targets if it is possible to destroy or deflect them. See also CEASE ENGAGEMENT.

holding anchorage An anchorage where ships may lie if the ASSEMBLY, or port to which they have been assigned is full; when delayed by enemy threats or other factors from proceeding immediately on their next voyage; or when dispersed from a port to avoid the effects of a nuclear attack.

holding area The nearest covered and concealed position to the PICKUP ZONE or CROSSING SITE where troops are held until the time for them to move forward.

holding attack 1. An attack designed to hold the enemy in position, to deceive him as to where the main attack is being made, to prevent him from reinforcing the elements opposing the main attack, or to cause him to commit his reserves prematurely at an indecisive location. 2. An attack to hold the enemy in place and prevent maneuver while a flanking or enveloping attack is made by other forces. See also SUPPORTING ATTACK.

holding force A force assigned to hold a place or position; the force that carries out a HOLDING ATTACK.

holding line In RETROGRADE crossing operations, the outer limit of the area established between the enemy and the water obstacle to preclude direct and indirect FIRE into crossing areas.

holding station A medical treatment facility established by a medical unit at a railhead, AIRHEAD or PORT to provide temporary shelter and emergency medical treatment for patients who are awaiting further transportation.

holiday 1. In naval mine warfare, a gap left unintentionally during minesweeping or MINEHUNTING. 2. An unintentional omission in the photographic imagery coverage of an area. See IMAGERY SORTIE.

hollow charge An EXPLOSIVE CHARGE designed to pass through armor or concrete.

holy war A war that some participants believe to have been specifically ordained by God. The Christian crusades of the Middle Ages were considered holy wars, as are the Islamic Jihads of the present day.

homing device A device, mounted on a MISSILE, that uses SENSORS to detect the position or to help predict the future position of a target, and subsequently directing the missile to intercept the target.

homing guidance A system by which a missile steers itself towards a target by means of a self-contained mechanism that is activated by some distinguishing characteristics of the target.

homing guidance, active See ACTIVE HOMING GUIDANCE.

homing mine In naval mine warfare, a mine fitted with propulsion equipment that homes in on a target.

Honor, Medal of See MEDAL OF HONOR.

honor code As observed at WEST POINT and the U.S. Air Force Academy: "A cadet will not lie, cheat or steal or tolerate those who do." The code has come under increasing criticism in recent years because it requires a cadet not only to be personally honest in action and intent, but to report other cadets that he or she suspects might have violated the code. See Richard P. Hansen, "The Crisis of the West Point Honor Code," *Military Affairs* (April 1985).

honorable discharge The form of DISCHARGE given to a United States soldier whose service has been honest and faithful, and who has been given conduct ratings of at least good, has been given efficiency ratings of at least fair, has not been convicted by a GENERAL COURT-MARTIAL, and has not been convicted more than once by a SPECIAL COURT-MARTIAL.

honors 1. Battle or campaign STREAMERS attached to the standard or COLORS of a military unit. 2. Formal military ceremonies for visiting dignitaries.

honors of war Allowing a defeated force, in testimony of its valor and honorable actions during combat, a measure of dignity in surrender by allowing it to surrender while still armed or with colors flying.

horizontal action mine In LAND-MINE WARFARE, a mine designed to produce a destructive effect on a plane approximately parallel to the ground.

horizontal bombing Dropping bombs from an aircraft flying at a constant height over a target.

horizontal clock system A gun firing system used in describing the direction of the wind by reference to the figures on an imaginary clock dial. The FIRING POINT is considered the center of the clock and the target is at 12 o'clock. At 3 o'clock, wind comes directly from the right. At 9 o'clock, wind comes directly from the left. At 6 o'clock, wind comes directly from behind. Compare to CLOCK CODE POSITION and CLOCK METHOD.

horizontal error The error in RANGE, DEFLECTION, or radius that a weapon may be expected to exceed as often as not. The horizontal error of weapons whose projectiles make a nearly vertical approach to the target is described in terms of CIRCULAR ERROR PROBABLE. The horizontal error of weapons whose projectiles produce an elliptical DISPERSION PATTERN is expressed in terms of PROBABLE ERROR. See also DELIVERY ERROR; DEVIATION; DISPERSION ERROR.

horizontal escalation A strategic option particularly advocated by the U.S. Navy and other supporters of a global war strategy. It is a logical development from FLEXIBLE RESPONSE and other such doctrines that treat the means of responding to a conventional WARSAW PACT attack without having to "go nuclear." If, for example, the Warsaw Pact attacked in one area, usually assumed to be the CENTRAL FRONT in Europe, flexible response, the current NORTH ATLANTIC TREATY ORGANIZATION (NATO) doctrine would require a matching response, or possibly an escalated response, in that same area. But to so limit the range of retaliatory options would break an age-old military axiom: the enemy should not be allowed to take the initiative. Consequently, the advocates of flexible response argue that NATO, or at least the United States, should be prepared to attack the Soviet Union elsewhere in the world (i.e., escalate horizontally), as well as to match the original aggression. Thus, American troops might advance suddenly into North Korea, bombing raids might be launched against the Asian Republics of the Soviet Union, or Cuba might be invaded. The intent of this doctrine is to make it clear to the Soviet Union that they will not be allowed to choose a battleground safely away from their homeland or other vital interests. Such a strategy necessarily requires a FORCE-PROJECTION capacity, and also justifies the huge 600-ship navy planned by the Reagan administration. Rightly or wrongly, the implication of horizontal escalation is that any NATO—Warsaw Pact conflict would rapidly develop into global war.

horn In naval mine warfare, a projection from the mine shell of some CONTACT MINES which, when broken or bent by contact, causes the mine to fire.

horned scully An underwater OBSTACLE designed to tear holes in the bottoms of boats, consisting of a tapered block of concrete with steel rails, usually pointed, and projecting upward at an angle.

hospital, evacuation See EVACUATION HOSPITAL.

hospital, field See FIELD HOSPITAL.

hostage A person held as a pledge that certain terms or agreements will be kept. (The taking of hostages is forbidden under the GENEVA CONVENTIONS, 1949.)

hostage, nuclear See NUCLEAR HOSTAGE.

host country A nation in which representatives or organizations from another state are present by government invitation or international agreement.

hostile In AIR INTERCEPTION, a contact positively identified as an enemy. See also BOGEY and FRIENDLY.

hostile criteria A description of conditions under which an aircraft or vehicle may be identified as hostile for ENGAGEMENT purposes.

host nation A nation that receives the forces or supplies of allied nations or NORTH ATLANTIC TREATY ORGANIZATION (NATO) member nations to be located, to operate in, or to transit through their territory.

host-nation support Civil and military assistance rendered in peace and war by a host nation to allied forces and NORTH ATLANTIC TREATY ORGANIZATION (NATO) members that are located on or in transit through the host nation's territory. The basis of such assistance is commitments arising from the NATO alliance or from bilateral or multilateral agreements concluded between the host nation, NATO members, and (the) nation(s) having forces operating on the host nation's territory. See Herbert W. Mylks, "Host Nation Support," *Army Logistician* 11, 1 (January–February 1979).

hot 1. Furious or intense as of a battle. 2. Under fire from the enemy. For example, a LANDING ZONE for a helicopter is "hot" if the enemy is firing upon the position.

hot line 1. The Washington-to-Moscow telephone and teletype links between the White House and the Kremlin, established for instant communications should a crisis occur. 2. Any communications system that links chief executives of governments with each other. 3. The communications that link a chief executive with his or her military commanders. See William L. Ury and Richard Smoke, *Beyond the Hotline: Controlling a Nuclear Crisis* (Cambridge, MA: Nuclear Negotiation Project, Harvard Law School, 1984).

hot pursuit 1. The legal doctrine that allows a law-enforcement officer to arrest a fleeing suspect who has fled into another jurisdiction. 2. The doctrine of international maritime law that allows a state to seize a foreign vessel that has initiated an act of war on the territory of the invaded state, and is pursued into international waters. 3. Pursuit of an enemy while in sight of, or in contact with, it. 4. The pursuit of an enemy across international borders.

hot spot A region in a radioactively contaminated area in which the level of radioactive contamination is considerably greater than in neighboring regions in the area.

housekeeping supplies Items required for the shelter, health, welfare, and administration of PERSONNEL.

howitzer 1. A CANNON that combines certain characteristics of guns and mortars. The howitzer delivers PROJECTILES with medium velocities, either by low or high trajectories. It was originally designed to reach targets behind cover or in a trench; its high angle of fire allows it to place a shell in a steep arc so that it can fall almost vertically behind embankments. 2. Normally a cannon with a TUBE length of 20 to 30 calibers; however, the tube length can exceed 30 calibers. See also GUN and MORTAR.

howler An electrical device, similar to an automobile horn, placed at gun positions as a firing signal in a time-interval system of TRACKING a moving target.

HQ See HEADQUARTERS.

hull 1. The body of a ship, exclusive of its engines, superstructure, and masts. 2. The part of the body of a seaplane that supports the aircraft while it is resting on water. 3. The massive armored body of a TANK, exclusive of its tracks, motor, turret, and armament.

hull defilade An offensive system consisting of TANKS that are dug into defensive positions up to their HULLS.

hull-down The positioning of an ARMORED VEHICLE in such a manner that the MUZZLE of the vehicle's GUN or LAUNCHER is the lowest part of the vehicle exposed to the front. See also: TURRET-DOWN.

human-factors engineering The application of scientific principles concerning human physical and psychological characteristics to the design of military equipment so as to increase speed and precision of military operations, provide maximum maintenance efficiency, reduce fatigue, and simplify operations.

human intelligence (HUMINT) A category of INTELLIGENCE regarding information collected and provided by human sources. Also called human resources intelligence.

HUMINT See HUMAN INTELLIGENCE.

hung bomb A bomb that accidentally remains attached to an aircraft after its intended release from the bomb rack.

hung striker The defective STRIKER of a grenade FUZE, which fails to strike the PRIMER and explode the grenade.

hunter/killer submarine A submarine whose prime mission in wartime would be to hunt for and kill (destroy) enemy ships (including other submarines), in contrast to ballistic missile carrying submarines whose mission would be to attack enemy land targets.

hunting 1. A RADAR antenna motion by which the antenna oscillates about rather than stopping smoothly at the point determined by the setting of the control. 2. A rapid up-and-down movement of a tank gun TUBE caused by an over-sensitized gyrostabilizer control unit.

hydrogen bomb A THERMONUCLEAR bomb; the vastly more powerful, successor to the atomic bomb in the field of NUCLEAR WEAPONS. It relies on NUCLEAR FUSION, whereas the atomic bomb depends on the weaker, but technically much simpler, NUCLEAR FISSION process. There is a relatively low limit to the power that can be generated by a fission bomb because the technique involves the rapid bringing together of two pieces of radioactive material (usually uranium, but alternatively plutonium). Separately the two radioactive constituents are sub-critical (stable), but together they constitute a CRITICAL MASS. Clearly the upper limits of the power of an atomic bomb are set both by the size to which a piece of fissionable material can remain subcritical, and the distance apart that two such masses can feasibly be kept in a portable device such as a bomb. A hydrogen bomb suffers from no such inherent limits, and theoretically could be made to yield as much energy as required. Certainly hydrogen bombs of above 50 megatons have been tested, although almost certainly never produced for deployment by the Soviet Union. There was considerable debate in the American nuclear weapons fraternity about whether the thermonuclear weapons should be built at all. It was suggested by many that there was no military use for it, and that resources should go to the production of atomic weapons for tactical use. After the Soviet Union's detonation of its first atomic bomb in 1949 President Truman authorized the development of thermonuclear weapons. It was possible to move rapidly because much of the theoretical work had already been done during the MANHATTAN PROJECT. See Jonathan B. Stein, *From H-Bomb to Star Wars: The Politics of Strategic Decisionmaking* (Lexington, MA: Lexington Books, 1984).

hydrographic chart A nautical chart showing the depths of water, nature of bottom, contours of bottom and coastline, and tides and currents in a given sea or sea and land area.

hydrographic reconnaissance Reconnaissance of an area of water to determine depths, beach gradients, the nature of the bottom, and the location of coral reefs, rocks, shoals, and manmade obstacles.

hydrographic section A section of a BEACH PARTY whose duties are to clear the beach of damaged boats, conduct hydrographic reconnaissance, assist in removing underwater obstructions, act as stretcher bearers for causalties and furnish relief boat crews.

hydropneumatic Pertaining to, or operated by means of, a liquid and a gas; a term ordinarily used in connection with certain artillery RECOIL and EQUILIBRATOR MECHANISMS that provide variable absorption of energy or thrust. See also RECOIL CYLINDER.

hydrospring Pertaining to, or operated by means of, a liquid and springs; a term ordinarily used in connection with certain artillery RECOIL and EQUILIBRATOR MECHANISMS which provide variable absorption of energy or thrust. See also RECOIL CYLINDER.

hydrostatic fuze A FUZE employed with DEPTH BOMBS or depth charges to cause underwater detonation at a predetermined depth.

hypabaric chamber A chamber used to induce a decrease in ambient pressure, such as would occur when ascending in altitude. This type of chamber is primarily used for training and experimental purposes. Also called altitude chamber; decompression chamber.

hyperbaric chamber A chamber used to induce an increase in air pressure, such as would occur in descending below sea level. This type of chamber is used for training, experimentation, medical treatment, and the controlled decompression of divers. It is the only type of chamber suitable for use in the treatment of decompression sickness in flying or diving. Also called compression chamber; diving chamber; recompression chamber.

hypersonic Of or pertaining to speeds equal to, or in excess of, five times the speed of sound.

hypervelocity 1. An artillery projectile MUZZLE VELOCITY of 3,500 feet per second or more. 2. A SMALL ARMS projectile muzzle velocity of 5,000 feet per second or more. See also HIGH VELOCITY. 3. A muzzle velocities of a tank CANNON projectile in excess of 3,350 feet per second.

hypocenter The point on the ground directly below a nuclear explosion. See also GROUND ZERO.

hypsobaric chamber A chamber used to induce a decrease in air pressure, such as would occur in ascending to altitude. This type of chamber is primarily used for training and experimental purposes. Also called a decompression chamber and an altitude chamber.

hypsographic map A topographic map on which elevations are given in relation to sea level. Sometimes called a hypsometric map.

hypsometric diagram A small-scale diagram representing RELIEF by different patterns or degrees of shading for specified levels of elevation.

hypsometric tinting A method of showing RELIEF on maps and charts by coloring, in different shades, those parts that lie between selected levels. Sometimes referred to as elevation tint, altitude tint, or layer tint.

I

ICAF See INDUSTRIAL COLLEGE OF THE ARMED FORCES.

ICBM See INTERCONTINENTAL BALLISTIC MISSILE.

icemining The breaking of river or lake ice by the action of ANTITANK or ANTIPERSONNEL MINES to deny passage to the enemy. Actuation of the mines may be effected by control, passage of time, or enemy initiation.

ideal bomb An imaginary BOMB whose fall is not subject to air resistance.

identification 1. The process of determining the friendly or hostile character of an unknown detected contact. See CHALLENGE AND REPLY AUTHENTICATION. 2. In ARMS CONTROL, the process of determining which nation is responsible for the detected violations of any arms-control measure. 3. In ground combat operations, discrimination between recognizable objects as being friendly or enemy, or the name that belongs to the object as a member of a class (such as a tank as opposed to an armored fighting vehicle.) 4. In IMAGERY INTERPRETATION, the discrimination between objects within a particular type or class.

identification, friend or foe (IFF) A system using electromagnetic transmissions, to which equipment carried by friendly forces automatically responds, for example, by emitting pulses, thereby distinguishing themselves from enemy forces. An IFF system is mainly used to allow combat aircraft to be identified on RADAR as either friendly or not.

identification code, international See INTERNATIONAL IDENTIFICATION CODE.

identification smoke Smoke employed to identify targets, supply and evacuation points, friendly unit perimeters, and for prearranged battlefield communications.

igloo space An area in an earth-covered structure of concrete, steel, or both, designed for the storage of AMMUNITION and EXPLOSIVES. See also STORAGE.

igniter 1. A device designed to produce a flame or a flash that is used to initiate a FIRING CIRCUIT. 2. A device containing a readily burning composition, usually a form of black powder, used to amplify the initiation of a PRIMER in the functioning of a FUZE. An igniter may be used to assist in the initiation of a PROPELLING CHARGE and in some types of projectile bursting charges. 3. A device containing a spontaneously combustible material, such as white phosphorous, used to ignite the fillings of incendiary bombs at the time of rupture of the bomb casing. 4. A device used to initiate the burning of the fuel mixture in a ramjet or ROCKET combustion chamber.

igniter train A step-by-step arrangement of charges in PYROTECHNIC bombs, shells, and other devices by which the initial fire from the PRIMER is transmitted and intensified until it reaches and sets off the main charge. An explosive bomb, projectile, or other device uses a similar series, called an EXPLOSIVE TRAIN.

igniting fuze A grenade FUZE that ignites the filler with a small quantity of BLACK POWDER.

igniting primer A charge of BLACK POWDER that carries the fire from a primer to the PROPELLING CHARGE in certain types of ammunition.

ignition The action of a device used as the first element of an EXPLOSIVE TRAIN which, upon receipt of the proper impulse, causes the rapid burning of a PROPELLANT or PYROTECHNIC item.

ignition cartridge 1. An IGNITER in cartridge form that may be used alone or with additional propellant increments as a PROPELLING CHARGE for certain MORTAR ammunition. 2. An assembly consisting of a PRIMER and igniter used to provide sufficient impulse to start certain forms of diesel engines.

illuminator radar That part of a GUIDED MISSILE weapon system used to track and illuminate the target. The illuminating RADAR energy is reflected by the target, detected by the missile, and used by the missile in homing on the target. In active homing GUIDANCE SYSTEMS, the illuminator radar is on board the missile, whereas in semiactive homing systems this radar may be aboard a ship, an aircraft, or on land.

image interpreter An INTELLIGENCE specialist qualified to recognize, identify, locate, describe, and analyze objects, activities, and terrain represented on imagery, and to extract intelligence information from them.

imagery Collectively, the representations of objects reproduced electronically or by optical means on film, electronic display devices, or other media.

imagery intelligence (IMINT) INTELLIGENCE derived from visual photography, infrared SENSORS, LASERS, electro-optics, RADAR sensors, or other media.

imagery interpretation 1. The process of location, recognition, identification, and description of objects, activities, and terrain represented on imagery. 2. The extraction of information from photographs or other recorded images.

imagery sortie A flight by a single aircraft for the purpose of recording air imagery.

imaging radar A theoretical system that would be used defensively to distinguish real MISSILES from DECOYS.

I method The transmission of a message from one station to another in such a way that other stations for which it is intended may receive it without having to acknowledge its receipt.

IMINT See IMAGERY INTELLIGENCE.

imitative electronic deception Imitating enemy electromagnetic radiations (predominately communications) in order to deceive or to disrupt enemy operations.

immediate air support AIR SUPPORT to meet specific requests that arise during the course of a battle and which by their nature cannot be anticipated in advance.

immediate message A category of PRECEDENCE reserved for messages relating to situations that gravely affect the security of a nation's or its allies' forces or a populace, and which require immediate delivery to the addressee(s).

immediate mission request 1. A request for an AIR STRIKE on a target or a RECONNAISSANCE mission in an area which, by its nature, cannot be identified sufficiently in advance to permit detailed mission coordination and planning.

Compare to: PREPLANNED MISSION REQUEST.

immediate operational readiness The state in which an armed force is ready in all respects for instant combat.

immediate permanent incapacitation dose (IP) A RADIATION ABSORBED DOSE (determined to be 8,000 rads) which causes incapacitation within 5 minutes of exposure and causes death within 1 to 2 days.

immediate suppressive fire Rapid fire delivered in response to enemy fire or the disclosure of enemy positions.

immediate transient incapacitation dose (IT) A RADIATION ABSORBED DOSE (determined to be 3,000 rads) that causes incapacitation within 5 minutes of exposure. Exposed personnel will then partially recover but will be functionally impaired until death, which occurs in 4 to 6 days.

impact, normal See NORMAL IMPACT.

impact-action fuze A FUZE that is set in action by the striking of a projectile or bomb against an object, e.g., percussion fuze or a contact fuze. Synonymous with direct action fuze.

impact area An area having designated boundaries within the limits of which all ORDNANCE delivered by all means should detonate or impact.

impact point See POINT OF IMPACT.

impermeable protective clothing Clothing made of material that prevents the passage of toxic CHEMICAL AGENTS in any physical form, and which can be worn for only short periods of time because of its excessive retention of body heat.

implied tasks Those additional tasks not specifically stated in an operations plan or order that a commander would identify as necessary to insure the accomplishment of a mission.

implosion weapon A weapon in which a quantity of fissionable material, less than a CRITICAL MASS at ordinary pressure, is designed to be suddenly compressed in volume by chemical explosives, so that it become SUPERCRITICAL, thereby producing a nuclear explosion.

impregnated clothing See IMPERMEABLE PROTECTIVE CLOTHING.

impregnite Material used to impregnate clothing that will afford protection against all forms of CHEMICAL AGENTs such as vapors, aerosols, and small droplets.

improvised explosive devices Devices normally devised from non-military components and incorporating destructive, lethal, noxious, pyrotechnic or incendiary chemicals, designed to destroy, disfigure, distract or harass.

impulse A mechanical jolt delivered to an object.

impulse kill The destruction of a target by vibrations set up by an initial burst of energy. The actual shot to the target does no damage, but only serves to set up the vibrations that will cause it to collapse. The intensity of the energy directed at the target may be so great that the surface of the target violently and rapidly boils off, delivering a mechanical SHOCK WAVE to the rest of the target that causes its structural failure.

inactivate To remove a military unit from the active service list. Compare to DISBAND; DEACTIVATE.

inactive status Being officially connected with a military service, but not on active duty.

incapacitating agent A CHEMICAL AGENT that produces temporary disablement that (unlike the effects of RIOT CON-

TROL AGENTS) can be physical or mental, and persist for hours or days after exposure to the agent has ceased. Medical treatment, while not usually required, facilitates a more rapid recovery.

incapacitating dose, median See MEDIAN INCAPACITATING DOSE.

incident Brief clash or other military disturbance, generally of a transitory nature and not involving protracted hostilities.

incoming Enemy FIRE, usually artillery fire, upon one's position.

incompetence The demonstrated failure to meet minimal standards of military performance. Unless incompetence is gross, it is difficult to prove, especially when it involves judgments made under the stress of combat in the "FOG OF WAR." See Norman F. Dixon, *On the Psychology of Military Incompetence* (London: Jonathan Cape 1976).

increment 1. An amount of PROPELLANT added to, or taken away from, a propelling charge of semifixed or separate loading ammunition to allow for differences in range. 2. An additional amount of pay for longevity, promotion, etc.

incursion 1. A more politic word for a military invasion of another state's territory. 2. A sudden hostile but short-lived raid.

independent deterrent A phrase used to describe the British, and sometimes the French, strategic nuclear forces. All five publicly acknowledged members of the NUCLEAR CLUB actually have independent nuclear forces, but this point does not need stressing with the United States, the Soviet Union, or China, as their forces are self-designed and built entirely from their own resources. The French and British deterrents, to a greater or lesser extent, were created with American help.

independent mine A MINE that is not controlled by the user after laying.

indeterminate change of station An assignment to temporary duty away from a permanent station, on orders that provide that the individual or unit that has been moved will not return to the former permanent station, but will be ordered to a new station, to be determined later.

index contour line A contour line on a map that is accentuated by a heavier marking to distinguish it from intermediate contour lines. Index contour lines are usually shown as every fifth contour line along with their assigned values, to facilitate reading elevations.

indicated altitude The altitude of an aircraft as indicated or shown by an altimeter.

indicator In INTELLIGENCE usage, an item of information that reflects the intention or capability of a potential enemy to adopt or reject a COURSE OF ACTION.

indicator regulator An instrument that shows FIRING DATA, such as fuze settings, azimuth, and elevation, that are transmitted by the FIRE-CONTROL SYSTEM of a gun. The gun is pointed in accordance with the data provided by the instrument.

indigenous personnel The local inhabitants at a foreign location; usually used for persons employed at military bases.

indirect air support All forms of air support provided to land or naval forces and which do not immediately assist those forces in a tactical battle.

indirect approach A FLANKING attack that seeks to force the enemy to move from a strong position by threatening his lines of communication. Sometimes called a STRATEGIC ENVELOPMENT or end run. The classic analysis of the indirect approach is B. H. LIDDELL HART, *Strategy*, 2nd edition (New York: Praeger, 1954, 1967). Also see: William J.

Dalecky, "The Strategy of the Indirect Approach Applied to NATO," *Air University Review* 38, 1 (November–December 1986).

indirect fire FIRE delivered at a target that cannot be seen by the aimer.

indirect laying Aiming a gun either by sighting at a fixed object (the AIMING POINT) instead of at the target, or by using a means of pointing other than a sight, such as a gun DIRECTOR, when the target cannot be seen from the gun position.

indirect laying position A gun position masked by some feature of the ground surface that hides the enemy target from direct view. Compare to DEFILADE.

individual reserves The supplies carried by a soldier, animal, or vehicle for individual use in an emergency.

inductee A person who has been induced (drafted) into military service.

induction The formal process by which a volunteer or draftee is taken into a military service and made subject to military law.

induction field locator A, small battery-powered radio HOMING DEVICE for use in locating PARACHUTE bundles that have been dropped.

Industrial College of the Armed Forces (ICAF) The senior service college dedicated to the study of management of resources for national security. A division of the NATIONAL DEFENSE UNIVERSITY (NDU), the College is chartered to conduct senior-level courses of study and research in order to enhance the military preparation of selected military officers and senior federal civilian officials. The other division of the NDU is the NATIONAL WAR COLLEGE.

industrial defense All nonmilitary measures to assure the uninterrupted productive capability of vital facilities and the attendant resources essential to mobilization.

industrial mobilization The transformation of industry from its peacetime activity to the industrial program necessary to support national military objectives. It includes the mobilization of materials, labor, capital, production facilities, and volunteered items and services.

industrial preparedness The state of preparedness of industry to produce essential MATERIEL for the support of national military objectives. See Nancy M. Hoesly, "Automated Planning for Industrial Readiness," *Army Logistician* 18, 2 (March–April 1986).

industrial security That portion of INTERNAL SECURITY that is concerned with the protection of CLASSIFIED INFORMATION in the hands of private industry.

inert ammunition AMMUNITION, or a component thereof, that contains an inactive filler in lieu of a SERVICE AMMUNITION filler (explosive, pyrotechnic, or chemical).

inert filling In INERT AMMUNITION, a prepared, non-explosive filling for a given type of ammunition that has the same weight as the explosive filling.

inertial guidance A GUIDANCE SYSTEM designed to project a missile over a predetermined path, wherein the path of the missile is adjusted after launching by devices wholly within the missile and independent of outside information. These are essentially based on a set of gyroscopic devices that measure minute changes in speed and orientation. See also DIRECTIONAL GYROSCOPE.

inertial navigation system An aircraft's self-contained navigation system using inertial detectors, which automatically provides vehicle position, heading and velocity.

inert mine An inert replica of a standard MINE. It is used for instructional purposes.

INF See INTERMEDIATE NUCLEAR FORCES.

INF Treaty The 1987 treaty (ratified in 1988) between the United States and the Soviet Union on the elimination of their intermediate-range and shorter-range nuclear missiles. It covers U.S. and Soviet land-based nuclear missiles with ranges from about 300 to 3,400 miles (500 to 5,500 kilometers). It bans all production and flight testing of INF missiles, eliminates all U.S. and Soviet INF missile systems (within three years), and eliminates all facilities to deploy, store, repair, and produce INF missile systems once all U.S. and Soviet INF missile systems are eliminated. For the first time, an entire class of missiles on both sides will be eliminated. The treaty embodies the principle of unequal reductions; the Soviets will eliminate deployed systems capable of carrying about four times as many warheads as those eliminated by the United States. The treaty also establishes the most stringent and comprehensive verification system in the history of arms control, including several kinds of on-site inspections. Nevertheless, nuclear weapons will remain an essential part of NATO's deterrent. NATO will continue to have about 6,000 nuclear weapons in Europe—short-range missiles and nuclear artillery, as well as nuclear-capable aircraft that are able to reach deep into Soviet territory. In addition, numerous ballistic missile warheads on United States submarines are dedicated to NATO. See also PERSHING II, SS-20.

infantry The most basic branch of an army; those PERSONNEL and UNITS who close with the enemy by means of FIRE and MANEUVER in order to destroy or capture him, or to repel his assault by fire, CLOSE COMBAT, and COUNTERATTACK. Personnel and units so identified fight DISMOUNTED on foot or MOUNTED, according to the means of mobility provided. They are only provided with relatively light weapons and must depend on ARMOR and ARTILLERY support. Infantry is one of the three traditional branches of an army, together with CAVALRY and artillery. Increasingly, the distinction of infantry as foot soldiers is less clear; with the tendency to put troops into ARMORED FIGHTING VEHICLES—they no longer have to walk everywhere, and may not even dismount from their transports to fight. Most military thinkers would still believe, however, that infantry retains a vital and distinct role in military operations, which is often described as "holding the ground" once it has been taken, even if armor and airpower are the predominant forces that drive enemy forces away from the battlefield in the first place. Traditional infantry forces, with little or no mechanized transport and no armored protection, have largely vanished from modern armies, except in special roles such as AIRBORNE troops. British Field Marshall Sir Archibald Wavell (1883–1950) once wrote: "One well-known Brigadier always phrases his requirements of the ideal infantryman as 'athlete, stalker, marksman.' I always feel a little inclined to put it on a lower plane and say that the qualities of a successful poacher, cat burglar, and gunman would content me." Compare to AIR CAVALRY. See J. I. H. Owen, *NATO Infantry and Its Weapons* (Boulder, CO: Westview Press, 1976).

infantry, armored See ARMORED INFANTRY.

infantry division, mechanized See MECHANIZED INFANTRY DIVISION.

infantry, heavy See HEAVY INFANTRY.

infantry, straight leg See STRAIGHT LEG INFANTRY.

infill In cartography, the filling of an area or feature with color, e.g., roads, town shapes, lakes, etc.

infiltration 1. The movement through or into an area or territory occupied by either friendly or enemy troops or organizations. The movement is made either by small groups or by individuals, at extended or irregular intervals. When used in connection with the enemy, it infers that contact is avoided. 2. In INTELLIGENCE usage, the placing of an AGENT or other person in hostile territory. This usually involves crossing a frontier or other guarded line. Methods of infiltration are: black (clandestine); grey (through legal crossing point but under false documentation); and white (legal). 3. When used in conjunction with a TACTICAL vehicular march, the dispatch of vehicles individually or in small groups at irregular intervals to reduce traffic density and prevent undue massing of vehicles.

inflight reliability The percentage of MISSILES launched that deliver their WARHEADS on target.

influence mine A MINE actuated by the effect of a target on some physical condition in the vicinity of the mine (such as vibrations).

influence-release sinker That sinker which holds a moored or rising mine at the sea-bed and releases it when actuated by a suitable surface (ship-related) influence.

influence sweep A mine SWEEP designed to produce an influence similar to that produced by a ship, and thus actuate mines.

information In intelligence usage, unprocessed data of every description which may be used in the production of intelligence. See also INTELLIGENCE CYCLE.

information classified See CLASSIFIED INFORMATION.

information, combat See COMBAT INFORMATION.

information, controlled See CONTROLLED INFORMATION.

information, nondefense See NONDEFENSE INFORMATION.

information center, combat See COMBAT INFORMATION CENTER.

information report Report used to forward raw information collected to fulfill intelligence requirements.

information requirements Those items of INFORMATION about an enemy and his environment that need to be collected and processed in order to meet the intelligence requirements of a commander. See also PRIORITY INTELLIGENCE REQUIREMENTS.

infrared detector A thermal device for observing and measuring infrared radiation (heat) from a mechanical object, such as a missile *re-entry* VEHICLE.

infrared film Film carrying an emulsion especially sensitive to the "near-infrared" portion of the infrared spectrum. Used to photograph through haze, because of the penetrating power of infrared light, and in CAMOUFLAGE detection to distinguish between living vegetation and dead vegetation or artificial green pigment.

infrastructure 1. A general term for a locality's fixed assets such as bridges, highways, tunnels, and water treatment plants, etc. 2. A political party's or a government's administrative structure, as well as the people and processes that make it work. 3. The institutional framework of a society that supports the educational, religious, and social ideology that in turn supports the political order. 4. The permanent installations and facilities for the support, maintenance, and control of naval, land, or air forces. The NORTH ATLANTIC TREATY ORGANIZATION (NATO) with its large forward based strength deployed in West Germany and made up of the forces of other countries (and with the prospect of even larger reinforcement needs during mobilization), is necessarily deeply concerned with infrastructure provisions. Air bases must be provided with HARDENED shelters for aircraft, and C^3I bunkers must be prepared, regardless of the absence of any immediate likelihood of their being used. Supply depots and the special storage shelters needed for the United States's forward-positioned POMCUS equipment, all add to the outlay. The problem is that while infrastructure is vital, and could even mean the difference between defeat and victory in a war, it is an "unglamorous" defense expenditure. Neither the active-duty military nor the civilian voters of a nation are likely to show much enthusiasm for its acquisition. See Bruce B. Geibel, "The Atlantic Alliance—the Infrastructure Story," *Naval Civil Engineer* 24, 3 (Fall 1984).

infrastructure, bilateral See BILATERAL INFRASTRUCTURE.

infrastructure, common See COMMON INFRASTRUCTURE.

infrastructure, national See NATIONAL INFRASTRUCTURE.

inhabited-building distance The minimum distance that may be expected to protect buildings or structures from the substantial damage of BLAST EFFECT.

initial-entry training BASIC TRAINING for new recruits.

initial issues Newly authorized items of materiel not previously furnished to an individual or organization, or the materiel issued to new military inductees and newly activated organizations.

initial point (IP) 1. A well defined point, easily distinguishable visually, electronically, or both, used as a starting point for an attack on a target. 2. The first point at which a moving target is located on a plotting board. 3. In AIRBORNE OPERATIONS, a point close to the landing area where SERIALS (troop carrier air formations) make final alterations in course to pass over individual DROP or LANDING ZONES. 4. In helicopter operations an AIR CONTROL POINT in the vicinity of the landing zone, from which individual flights of helicopters are directed to their prescribed landing sites. 5. Any designated place at which a COLUMN or element thereof is formed by the successive arrival of its various subdivisions, and comes under the control of the commander ordering the move. See also TARGET APPROACH POINT.

initial radiation The radiation, essentially neutrons and gamma rays, resulting from a NUCLEAR BURST and emitted from the FIREBALL within one minute after the burst. See also INDUCED RADIATION; RESIDUAL RADIATION.

initial requirements All supplies needed to equip soldiers or organizations when they are put on ACTIVE DUTY.

initial reserves In an AMPHIBIOUS OPERATIONS, those supplies that are normally unloaded immediately following the assault waves. Such supplies are usually for the use of the beach organization, BATTALION LANDING TEAMS, and other elements of REGIMENTAL COMBAT TEAMS to initiate and sustain combat until higher supply installations are established. See also RESERVE SUPPLIES.

initial resupply The ships in transit that are already loaded with the cargoes that will serve OPERATIONS requirements after D-DAY.

initial source of supply The point to which requisitions are sent for supply or approval and necessary action. This point may be a depot, inventory control point, supply/stock point, head of a procuring agency, or procurement office, depending upon the circumstances.

initial strength The actual or AUTHORIZED STRENGTH of a military force, or subdivision thereof, at the beginning of a specific time period (fiscal year, calendar year, month, or operation). See also END STRENGTH.

initial unloading period In AMPHIBIOUS OPERATIONS, that part of the ship-to-shore movement in which unloading is primarily TACTICAL in character and must be instantly responsive to LANDING-FORCE requirements. See also GENERAL UNLOADING PERIOD.

initial vector The initial command heading to be assumed by an interceptor after it has been committed to intercept an airborne object.

initiator A small quantity of a very sensitive and powerful EXPLOSIVE used to start the detonation of another, less sensitive explosive.

in-place force A NATO assigned force of the NORTH ATLANTIC TREATY ORGANIZATION (NATO) which, in peacetime, is principally stationed in the designated combat zone of the NATO Command to which it is committed.

insertion 1. The placement of troops and equipment into an operational area in AIRMOBILE OPERATIONS. 2. The placement of OBSERVATION POSTS, PATROLS, or raiding parties either by helicopter or parachute.

insignia Distinctive devices worn on a military uniform to show grade (meaning rank), ORGANIZATION, RATING, and SERVICE. The United States Army, Air Force, and Marine Corps insignia for its officer ranks are as follows:

General	Four silver stars
Lieutenant General	Three silver stars
Major General	Two silver stars
Brigadier General	One silver star
Colonel	Silver eagle
Lieutenant Colonel	Silver oak leaf
Major	Gold oak leaf
Captain	Two silver bars
First lieutenant	One silver bar
Second lieutenant	One gold bar

inspection 1. A formal review of a military unit's preparedness. 2. A pro forma review of a military unit by a visiting dignitary. 3. In ARMS CONTROL, the physical process of determining compliance with arms-control measures. 4. The examination and testing of supplies and services (including, when appropriate, raw materials, components, and intermediate assemblies) to determine whether they conform to contract requirements.

inspection, challenge See CHALLENGE INSPECTION.

inspection arms 1. A position prescribed in the manuals for the RIFLE, AUTOMATIC rifle, CARBINE, and PISTOL, in which the weapon is held with the chamber open for inspection. 2. The command to take this position.

inspector general 1. A military branch outside the normal CHAIN OF COMMAND that monitors the efficiency of other units. 2. The commander of such a branch. 3. A job title (of military origin) for the administrative head of an inspection/investigative unit of a larger agency. See John W. Braden Jr., "The IG and the Staff Work for the Same Boss," *Army* 31, 8 (August 1981).

installation A grouping of facilities, located in the same vicinity, which support particular functions. Installations may be the elements of a BASE.

installation confinement facility A jail for prisoners awaiting COURT-MARTIAL, at an INSTALLATION which also serves as a transfer point for other prisoners pending their movement to an area confinement facility or correctional treatment facility.

instrument direction The recorded direction of a HIGH AIRBURST, as indicated on an instrument at an artillery battery position, which enables the correction of subsequent firing.

instrument landing system (ILS) A mechanical system used for landing aircraft in poor weather (or instrument) conditions. The pilot flies the aircraft by responding to an instrument that indicates whether the aircraft is on course, left or right of course, and whether the aircraft is on, above or below the correct angle of approach.

in support An expression used to denote the task of providing artillery supporting fire to a formation or unit. See also AT PRIORITY CALL; DIRECT SUPPORT.

in support of Assisting or protecting another formation, unit, or organization while remaining under original control.

insurgency An organization or movement whose purpose is the overthrow of a constituted government through the use of subversion and armed conflict. Compare to COUNTERINSURGENCY REVOLUTION. See Thomas J. Kuster Jr., "Dealing with the Insurgency Spectre," *Military Review* 67, 2 (February 1987).

integrated defense An AIR DEFENSE in which two or more vital areas are defended with a single overall defense.

integrated equipment Any equipment in which is embodied both a COMMUNICATIONS and an encoding (as well as decoding) capability.

integrated fire-control system A system that performs the functions of TARGET ACQUISITION, TRACKING, data computation, and engagement control, primarily using electronic means assisted by electromechanical devices.

integrated logistics support A composite of all the SUPPORT considerations necessary to assure the effective and economical support of a military system for its life cycle. It is an integral part of all other aspects of system acquisition and operation.

integrated staff An allied or joint STAFF in which one officer only is appointed to each POST on the establishment of the HEADQUARTERS, irrespective of nationality and Service. See also COMBINED STAFF; JOINT STAFF; PARALLEL STAFF.

integrated warfare The conduct of military operations in any combat environment wherein opposing forces employ non-conventional weapons.

integration 1. A stage in the INTELLIGENCE CYCLE in which a pattern is formed through the selection and combination of evaluated information. 2. The process of making differing military units or services more compatible so that they may better operate as an integrated force in time of war. See William T. Marriott III, "Force Integration's Next Big Challenge," *Military Review* 66, 4 (April 1986).

intelligence 1. An individual's ability to cope with his or her environment and deal with mental abstractions. 2. The military, as well as other organizations concerned with NATIONAL SECURITY, use the word "intelligence" in its original Latin sense—as information. Intelligence in this context also implies secret or protected information. It is the product resulting from the collection, evaluation, analysis, integration, and interpretation of all available information concerning an enemy force, foreign nation, or area of operations, and which is immediately or potentially significant to military planning and operations. The term is also applied to the activity that results in the product, and to the organizations engaged in such activity. Compare to: COMBAT INFORMATION. See Stansfield Turner and George Thibault, "Intelligence: The Right Rules," *Foreign Policy* No. 48 (Fall 1982); Michael I. Handel, "The Study of Intelligence," *Orbis* Vol. 26, No. 4 (Winter 1983); Viktor Suvorov, *Soviet Military Intelligence* (London: Hamish Hamilton, 1984); Jeffrey T. Richelson, *Sword and Shield: Soviet Intelligence and Security Apparatus* (Cambridge, MA: Ballinger, 1986).

intelligence, basic See BASIC INTELLIGENCE.

intelligence, combat See COMBAT INTELLIGENCE.

intelligence, communications See COMMUNICATIONS INTELLIGENCE.

intelligence, critical See CRITICAL INTELLIGENCE.

intelligence, current See CURRENT INTELLIGENCE.

intelligence, domestic See DOMESTIC INTELLIGENCE.

intelligence, electronic See ELECTRONIC INTELLIGENCE.

intelligence, human See HUMAN INTELLIGENCE.

intelligence, imagery See IMAGERY INTELLIGENCE.

intelligence, military See MILITARY INTELLIGENCE.

intelligence, nuclear See NUCLEAR INTELLIGENCE.

intelligence, operational See OPERATIONAL INTELLIGENCE.

intelligence, political See POLITICAL INTELLIGENCE.

intelligence, radar See RADAR INTELLIGENCE.

intelligence, security See SECURITY INTELLIGENCE.

intelligence, signals See SIGNALS INTELLIGENCE.

intelligence, strategic See STRATEGIC INTELLIGENCE.

intelligence, tactical See TACTICAL INTELLIGENCE.

intelligence, target See TARGET INTELLIGENCE.

intelligence, technical See TECHNICAL INTELLIGENCE.

intelligence, terrain See TERRAIN INTELLIGENCE.

Intelligence Agency, Central See CENTRAL INTELLIGENCE AGENCY.

intelligence annex A supporting document of an OPERATION PLAN or ORDER that provides detailed information on the enemy situation, assignment of intelligence tasks, and intelligence administrative procedures.

intelligence collection plan Generally, a plan for gathering information from all available sources to meet an intelligence requirement. Specifically, a logical plan for transforming the essential elements of information into orders or requests to sources within a required time limit.

intelligence collection requirement, specific See SPECIFIC INTELLIGENCE COLLECTION REQUIREMENT.

intelligence community 1. All of the spies in the world; the totality of the employees of the world's civilian and military INTELLIGENCE agencies. 2. All of a single nation's military and civilian intelligence-gathering and analysis agencies. Leading members of the American intelligence community include: the CENTRAL INTELLIGENCE AGENCY, the NATIONAL SECURITY AGENCY, and the DEFENSE INTELLIGENCE AGENCY. See Tyrus G. Fain, Katherine C. Plant and Ross Milloy, *The Intelligence Community: History, Organization and Issues* (New York: Bowker Co. 1977); William J. Casey, "The American Intelligence Community," *Presidential Studies Quarterly* Vol. 12, No. 2 (Spring 1982); Stephen J. Flanagan, "Managing the Intelligence Community," *International Security* Vol. 10, No. 1 (Summer 1985).

intelligence cycle The steps by which information is converted into INTELLIGENCE and made available to users. There are five steps in the cycle:

1. *Planning and direction*—The determination of intelligence requirements, preparation of a collection plan, issuance of orders and requests to infor-

mation-collection agencies, and a continuous check on the productivity of collection agencies.
2. *Collection*—The acquisition of information and the provision of this information to processing or production elements or both.
3. *Processing*—The conversion of collected information into a form suitable to the production of intelligence.
4. *Production*—The conversion of information into intelligence through the integration, analysis, evaluation, and interpretation of all source data and the preparation of intelligence products in support of known or anticipated user requirements.
5. *Dissemination*—The conveyance of intelligence to users in a suitable form.

intelligence estimate The appraisal of available intelligence relating to a specific situation or condition, with a view to determining the COURSES OF ACTION open to the enemy or potential enemy and the order of probability of their adoption.

intelligence estimate, national See NATIONAL INTELLIGENCE ESTIMATE.

intelligence journal A chronological log of INTELLIGENCE activities covering a stated period of time. It is an index of reports and messages that have been received and transmitted, and of important events that have occurred and actions that have been taken.

intelligence oversight The review of the policies and activities of intelligence agencies, such as the CENTRAL INTELLIGENCE AGENCY (CIA) by the appropriate committees of the U.S. Congress. This was not formally done by the Congress until the 1970s, when reports of FEDERAL BUREAU OF INVESTIGATION (FBI) and CIA abuses of these agencies' operating mandates encouraged Congress to watch over their activities carefully, systematically, and formally. Legislative bodies in other western countries like France and the United Kingdom have equivalent supervision. See Anne Karalekas, "Intelligence Oversight: Has Anything Changed?" *Washington Quarterly* Vol. 6, No. 3 (Summer 1983); Barry Goldwater, "Congress and Intelligence Oversight," *Washington Quarterly* Vol. 6, No. 3 (Summer 1983).

Intelligence Oversight Board (IOB) A permanent, non-partisan panel of three members from outside the government who are appointed by the president. Created in 1976, the IOB is responsible for discovering and reporting to the president any intelligence activities that raise questions of propriety or legality in terms of the Constitution, the laws of the United States or presidential executive order.

intelligence report A specific report of information, usually on a single item, made at any level of command in tactical operations and disseminated as rapidly as possible in keeping with the timeliness of the information. Also called INTREP.

intelligence summary, daily See DAILY INTELLIGENCE SUMMARY.

intelligence zone, tactical See TACTICAL INTELLIGENCE ZONE.

intensity factor A multiplying factor used in planning activities in order to evaluate the foreseeable intensity or the specific nature of an OPERATION in a given area for a given period of time. It is applied to the STANDARD DAY OF SUPPLY in order to calculate the COMBAT DAY OF SUPPLY.

intercardinal points The intermediate compass directions: northeast, southeast, southwest, northwest.

intercept 1. To make contact with the enemy; to contact a hostile force before it arrives at its planned destination. 2. The act of destroying a target. 3. To make use of enemy electronic signals.

intercepting search A type of search designed to intercept an enemy whose previous position is known, and the limits of whose subsequent course and speed can be assumed.

interception 1. The act of listening in on or recording communications intended for another party for the purpose of obtaining INTELLIGENCE. 2. Making contact with an element (friend or foe) that is being sought. 3. Engaging an enemy force.

interception, air See AIR INTERCEPTION.

interception, controlled See CONTROLLED INTERCEPTION.

interceptor A fighter aircraft designed to identify unknown airborne objects and to seek out and destroy long-range enemy bombers.

intercept point The point to which an airborne vehicle is vectored or guided to complete an interception.

interchangeability A condition that exists when two or more items possess such functional and physical characteristics as to be equivalent in performance and durability, and are capable of being exchanged one for the other without alteration of the items themselves, or of adjoining items, except for adjustment, and without selection for fit and performance. See also COMPATIBILITY.

interchangeability lists Lists of parts which are common to, and interchangeable among, various types of general purpose and COMBAT VEHICLES.

intercontinental ballistic missile (ICBM) A land-based BALLISTIC MISSILE with a range capability from about 3000 to 8000 nautical miles. Ballistic missiles that are a major element of the United States's strategic nuclear TRIAD. Soviet and American ICBMs can reach any target in the other's country from bases in their own heartlands. They were first developed after World War II, largely based on the technological breakthroughs of the German scientists who built the V2 rocket. The first generation of ICBMs had single warheads, often with huge megatonnage, and were highly inaccurate, with CIRCULAR ERROR PROBABLES (CEPs) as high as one kilometer, and could be used only against large urban targets. The next generation, represented by the U.S. MINUTEMAN class first deployed during the mid-1960s, are more accurate, carry MULTIPLE INDEPENDENTLY-TARGETABLE REENTRY VEHICLES (MIRVs) and at least in the case of American models are now entirely solid-fueled. Thus the American missiles are typically capable of delivering three warheads in the 150–350-kiloton range, with a CEP of about 250 meters. Because they are solid-fueled, they can be launched with very little warning. America has produced relatively few versions of the ICBM, and the third generation, the MX MISSILE, was only being deployed during the late 1980s. The Soviet Union keeps a far higher proportion of its warheads on ICBMs, rather than relying as the United States does on SUBMARINE-LAUNCHED BALLISTIC MISSILES (SLBMs) and has tended to experiment more. In 1985, for example, there were at least six different types of Soviet ICBMs deployed, with two new models, the SS-24 and SS-25, being tested. On the whole Soviet missiles are heavier, with more and much larger and more numerous warheads than American models. This has led to American fears of a Soviet drive for strategic superiority and the so-called window of vulnerability. The United States, meanwhile, has led the way in producing MIRV capacity, and has built larger and more accurate SLBMs. See Albert Carnesale and Charles Glaser, "ICBM Vulnerability: The Cures Are Worse Than the Disease," *International Security* Vol. 7, No. 1 (Summer 1982); David C. Morrison, "ICBM Vulnerability," *Bulletin of the Atomic Scientist* Vol. 40, No. 9 (November 1984).

interdict To isolate or seal off an area by any means; to deny the enemy use of a route or approach. Most usually, to carry out air or naval attacks on a supply route or other line of communication in order to hold up the enemy's resupply and reinforcement.

interdiction An action to divert, disrupt, delay or destroy the enemy's surface military potential before it can be used effectively against friendly forces; the use of military force of some kind to prevent transportation of supplies, equipment and troops past a particular point or along some route. In modern warfare, it typically will refer to the use of airpower to destroy bridges, major railway junctions, or other CHOKE POINTS well inside enemy territory, thus preventing not only supplies, but also reinforcements from reaching the battle area. During the Vietnam War, for example, a major role of American airpower was to bomb routes such as the Ho Chi Minh trail in an attempt to starve the Vietcong guerrillas inside South Vietnam of all their supplies. AIR INTERDICTION is a vital element of American and North Atlantic Treaty Organization (NATO) plans for a CENTRAL FRONT war, and is enshrined in official doctrines such as DEEP STRIKE and FOLLOW-ON FORCES ATTACK. Given the Soviet doctrine of attack in several waves that are spaced several days apart, it is hoped that interdiction would prevent the scheduled arrival of such ECHELONED ATTACKS. As EMERGING TECHNOLOGY weaponry seems to offer the hope of highly accurate conventional attacks hundreds of kilometers inside enemy territory, interdiction could be a decisive tactic in any future war between the superpowers. See AIR INTERDICTION.

interdiction, air See AIR INTERDICTION.

interdiction fire FIRE placed on an area or point to prevent the enemy from using or passing through that area or point.

interface A boundary or point common to two or more similar or dissimilar COMMAND AND CONTROL systems, subsystems, or other entities toward which or at which necessary information flow takes place. An interface can be, depending on circumstances, a person, an organization, or a particular job.

interference Any electrical disturbance that causes undesirable responses in electronic equipment.

interior guard A guard force to keep order within the limits of an INSTALLATION, enforce police regulations, and protect the property and personnel of the command.

interior lines See LINES OF OPERATION.

intermediate nuclear forces (INF) Those THEATRE NUCLEAR FORCES (TNF) that have been defined for the purposes of ARMS-CONTROL negotiations that took place in the mid-1980s as having missiles with ranges between 500 and 5,500 kilometers. The term covers those missiles that fall somewhere between strategic weapons, or intercontinental and submarine-launched ballistic missiles (ICBMs and SLBMs) owned by the superpowers, and very short-range BATTLEFIELD NUCLEAR WEAPONS. See INF TREATY.

intermediate obstacles Obstructions located between FORWARD OBSTACLES and REAR OBSTACLES; designed to assist in limiting enemy penetrations of a DEFENSE AREA, to CANALIZE enemy forces into selected target areas, and to impede the lateral movement of the enemy.

intermediate pack A wrap, box, or bundle that contains two or more UNIT PACKS of identical items.

intermediate-range ballistic missile (IRBM) A BALLISTIC MISSILE with a range capability from about 1500 to 3000 nautical miles; a missile that does not have the range to hit the Soviet Union from a base in mainland America, and vice versa. The term is not applied to weapons in the tactical battlefield and short-range theatre categories, so an IRBM would have a range of at least 1,000 kilometers. These missiles (the U.S. Pershing class among them) are the main subject of the INTERMEDIATE NUCLEAR FORCES (INF) Treaty between the United States and the Soviet Union.

intermediate-range bomber aircraft A BOMBER with a tactical operating radius of between 1000 and 2500 nautical miles at design gross weight and design bomb load.

interment flag A national flag used to drape the casket in a military funeral.

intermittent arming device A device included in a MINE so that it will be armed only at set times.

internal attack The full range of measures taken by organized insurgents to bring about the internal destruction and overthrow of a constituted government.

internal control 1. The plan of organization and all of the coordinate methods and measures adopted within a military entity to safeguard its assets, check the accuracy and reliability of its accounting data, promote operational efficiency, and encourage adherence to prescribed managerial policies. 2. The control exercised over the movement and discipline of a CONVOY, SERIAL, or MARCH UNIT, by its own officers.

internal defense The full range or measures taken by a government to free and protect its society from subversion, lawlessness, and insurgency.

internal-defense assistance operation Any operation undertaken by the military, paramilitary, police, or other security agencies of an outside power to strengthen the government of a HOST NATION politically, economically, psychosocially, or militarily.

internal development assistance operation Any organized action undertaken by government or nongovernment agencies of an outside power to support internal development efforts in a HOST NATION.

internal security 1. The state of law and order prevailing within a nation. 2. The prevention of action against resources, industries, and institutions, and the protection of life and property in the event of a domestic emergency, by the employment of all measures, in peace or war, other than military defense. 3. The condition resulting from the measures taken within a military command to safeguard defense information coming under its control, including the physical security of documents and materials.

international cooperative logistics Cooperation and mutual support in the field of LOGISTICS through the coordination of policies, plans, procedures, development activities, and the common supply and exchange of goods and services arranged on the basis of bilateral and multilateral agreements having appropriate cost-reimbursement provisions.

international identification code In railway terminology, a CODE that identifies a military train from point of origin to final destination. The code consists of a series of figures, letters, or symbols indicating the priority, country of origin, day of departure, national identification code number, and country of destination of the train.

international law The totality of treaties, customs, and agreements among sovereign states. Jeremy Bentham is credited with having coined the phrase in 1780 in his *Principles of Morals and Legislation*, but many others before and since have tried to give substance and theoretical cohesiveness to transnational practices, which are often chaotic and frequently break down into war. When the international concerns apply to individuals, it becomes a matter of the "conflict of laws." Hugo Grotius (1583–1645) is often called the "father" of international law because his 1625 *De Jure Belli Ac Pacis* (The Law of War and Peace), which asserted that it was possible to create a code of international law suitable for every time and place, has influenced all subsequent thinking on the subject. See also WORLD COURT. See F. S. Ruddy, "The Origin and Development of the Concept of International Law," *Columbia Journal of Transnational Law* Vol. 7, No. 2 (Fall 1968); Marian Nash Leich, "Contemporary Practice of the United States Relating to International Law," *American Journal of International Law* Vol. 79, No. 4 (October 1985); Anthony D'Amato, *International Law: Prospect and Process* (Ardsley-on-Hudson, NY: Transnational Publishers, 1986).

international military post An international POST (a temporary position in an international organization) authorized to be filled by a military person whose pay and allowances remain the responsibility of the parent nation.

international peace force An appropriately constituted organization established for the purpose of preserving peace between hostile parties. An example is the multinational UNITED NATIONS (U.N.) force in Cyprus, stationed between the Greek and Turkish sectors.

international relations 1. The academic field of study that examines the political, military, and economic interactions among nations. 2. The totality of private interactions among citizens of differing countries. 3. The practice of DIPLOMACY. See John A. Vasquez, editor, *Classics of International Relations* (Englewood Cliffs, NJ: Prentice-Hall, 1986).

internee A person who, during war, is kept within a particular country or is confined to a certain area. Protected persons, as defined in the GENEVA CONVENTIONS of 1949, may only be made internees in accordance with the requirements therein stated.

internee, civilian See CIVILIAN INTERNEE.

internment installation procedures Local regulations, standard operating procedures, or other instructions governing PRISONER-OF-WAR camp or CIVILIAN INTERNEE camp activities, or those persons interned therein.

interoperability 1. The ability of military systems, units, or forces to provide services to and accept services from other systems, units, or forces, and to use the services so exchanged to enable them to operate effectively together. 2. The condition achieved among communications-electronics systems or items of communications-electronics equipment when information or services can be exchanged directly and satisfactorily between them or their users. Also called INTERFACE. 3. The capability of two or more items or components of equipment to perform essentially the same function or to complement each other in a system, regardless of differences in technical characteristics and with negligible additional training of personnel. Interoperability has become a major concern for NORTH ATLANTIC TREATY ORGANIZATION (NATO) forces as they strive to transcend the conventional superiority of the WARSAW PACT. However, NATO is hardly one unified military force, numbering on the central front alone 10 separate forces, each with its own procurement policies and many with their own armaments industries to support. This means that incompatibilities between equipment and in operating procedures can very seriously reduce the efficiency level of the total NATO force to below that which its sheer numerical capacity would seem to provide. An oft quoted example of such difficulties is that of refueling ships of various NATO member navies. Though a frigate squadron of ships from the British Royal Navy, and from the navies of the Netherlands and of West Germany, may operate together, the bores of each navy's refueling pipes are of different diameters, making it impossible for the ships to aid each other, and indeed calling for several, rather than one, fleet auxiliary to support the squadron. Similar problems are legion—radio sets that cannot be inter-tuned, the need for many different calibers of ammunition for weapons of the same generic type, incompatible radar identification systems so that no one can be sure whether an aircraft is friendly or not, and so on. The cumulative effect is to increase the effective Warsaw Pact force ratio superiority considerably. Efforts are continually being made to remedy this problem, but it can only properly be dealt with at the political level. Real interoperability would require a substantial integration of the Western European and American armaments industries, and would necessitate the military procurement officials of NATO countries being allowed to standardize one type of each weapons system, regardless of where it is made. See Leo J. Pigaty, "Practicing Interoperability," *Army Logistician* 9, 5 (September–October 1977); John B. Walsh, "Initiatives in Standardization/Interoperability," *NATO Review* 26, 5 (October 1978).

interpretation A stage in the INTELLIGENCE CYCLE in which the significance of information is judged in relation to the current body of knowledge.

interrogator responder Those components of IDENTIFICATION, FRIEND OR FOE equipments, which challenge and receive replies.

interrupter A safety device in a FUZE that prevents it from acting until a PROJECTILE has left the MUZZLE of a gun.

intersection A method of locating a geographical point by plotting the AZIMUTH to that point from two or more known fixed points. The intersection of these azimuths indicates the location of the point.

intersection, graphic See GRAPHIC INTERSECTION.

interserviceable item An item that has been identified for use by more than one military service.

interservice rivalry Not a competition for glory with beneficial consequences; in the modern era, interservice rivalry denotes the intense maneuvering for finite defense procurement funds. Even under governments committed to high levels of defense spending, the costs of modern weapons systems are such that success by one service in winning the funds to buy a much-needed new weapon will inevitably cut down what the other services can gain. This rivalry tends to lead to a system where each service fights for its own preferred projects, and the central authorities are seldom in a position to create a rational overall policy based on independently assessed national needs. Partly as a consequence, the development of strategic doctrine itself becomes distorted; the doctrine tends to follow the procurement of various weapons systems rather than vice versa.

interservice support Action by one military service or element thereof to provide logistic and/or administrative support to another military service or element. Such action can be recurring or nonrecurring in character on an installation, area, or worldwide basis.

interservice support agreement A document wherein the participants, to preclude any misunderstanding, state clearly in writing their agreement to provide interservice support and especially to specify the obligations assumed by each and the rights granted to each.

interval 1. The space between adjacent groups of ships or boats measured in any direction between the corresponding ships or boats in each group. 2. The space between adjacent aircraft, measured from the front to rear in units of time or distance. 4. The time lapse between photographic exposures. 5. At BATTERY LEFT (RIGHT), a period ordered in seconds as the time between adjacent guns' firing. Five seconds is the standard interval. 6. At rounds of FIRE FOR EFFECT, the time in seconds between successive rounds from each gun. 7. As applied to two units of CIPHER TEXT (letter digraphs, code groups, etc.) or key sequence, the number of such units between them, counting either the first or last of the two units, but not both.

interval, close See CLOSE INTERVAL.

intervalometer 1. A device installed in an aircraft dropping air-delivery containers from its BOMB RACKS, to insure that the containers are released at regular, preplanned intervals. 2. An automatic timing device for regulating air camera exposures; also may cock and trip the shutter automatically. 3. A device installed in an armament control system to regulate the release of bombs or firing of rockets at preplanned intervals.

intervisibility The condition of being able to see one point from another. This condition may be altered or interrupted by weather, obscuration, or TERRAIN MASKING.

intransit strength A term that applies to PERSONNEL accountable to a reporting organization who are incoming or outgoing in a permanent change of station (a new assignment) status.

intra-war bargaining Negotiations during gaps or FIRE-BREAKS between rounds of a nuclear exchange. It is hoped that carefully developed strategic options and targeting policies will allow for bargaining and negotiation in such periods to terminate a war before it reaches MUTUAL ASSURED DESTRUCTION levels. Intra-war bargaining, and the allied concept of intra-war deterrence, hinge on the idea that each side would initially refrain from destroying targets of especial value to their enemy. The argument is that if, for example, the United States ensures that it does not hit any major civilian targets in the first round of CENTRAL STRATEGIC WARFARE, it can hope to deter the Soviet Union from retaliating, and persuade it to settle the war on relatively favorable terms because of the threat of COUNTERVALUE strikes in a further ESCALATION should the bargaining break down (see NUCLEAR HOSTAGE). Such a concept is indicative of the more general move toward new ideas of nuclear war-fighting, which are beginning to replace the older assumptions that any nuclear exchange would involve a once-and-for-all launch against every target the enemy holds dear. Even American advocates of intra-war bargaining or deterrence admit that there is a serious danger of destroying COMMAND, CONTROL, COMMUNICATIONS, AND INTELLIGENCE (C^3I) facilities. They specify that the United States must actually refrain from targeting the Soviet leadership, to ensure that there remains someone with sufficient national command authority with whom to negotiate and bargain. The critics of both theories of intra-war behavior suggest that expectations of rationality on the part of leaders, in the face of national fear, confusion and rage, are unreal. They doubt that nuclear war could be controlled in such a sophisticated way. It is noticeable that such conceptions play no part in public Soviet strategic thought about nuclear war. The declared policy of the Soviet Union is that, in such an eventuality, it will not spare any targets.

intra-war deterrence A strategy to convince the enemy that ESCALATION in an existing war is pointless, since the aggressor would merely be destroying himself by inviting retaliation. See ESCALATION CONTROL.

intruder An individual, UNIT, or WEAPONS SYSTEM in or near a military operational or exercise area, and which presents the threat of INTELLIGENCE gathering or disruptive activity.

intruder operation An air-offensive operation by day or night over enemy territory with the primary object of destroying enemy aircraft in the vicinity of their bases.

intrusion-resistant communications cable A cable designed to provide substantial physical protection and electrical isolation for the wire lines that make up its information-carrying core. The protective measures used are devices that detect slight changes in the physical or electrical state of the cable, and which provide visible or audible indications at a central control point of any attempted intrusion. Also known as an alarmed cable.

IOB See Intelligence Oversight Board.

IP See IMMEDIATE PERMANENT INCAPACITATION DOSES; INITIAL POINT.

Iron Curtain The political, social, and economic divisions between the countries of Eastern and Western Europe. It was popularized by Winston Churchill in a March 5, 1946, speech at Westminster College, Fulton, Missouri, in which he said: "From Stettin in the Baltic to Trieste in the Adriatic, an iron curtain has descended across the continent." Now the phrase is also used to refer to any hostile and seemingly permanent political division.

irredentism The long standing and frustrated desire of the people of one state to annex some area of an adjoining state that contain peoples of the same cultural or ethnic group. Examples of irredentism include France's attitude toward Alsace and Lorraine after 1870 and Germany's attitude toward the Sudetenland in Czechoslovakia after World War I.

irregular forces Armed individuals or groups who are not members of the regular armed forces, police, or other internal security forces. They often engage in GUERRILLA WARFARE.

irregular outer edge In LAND MINE WARFARE, short mine rows or strips laid in an irregular manner in front of a MINEFIELD facing the enemy, to deceive the enemy as to the type or extent of the minefield. Generally, an irregular outer edge will only be used in minefields with buried mines.

isolationism A nation's policy of curtailing international relationships as much as possible to avoid conflict and insure its peace. Isolationism is the name usually given to the foreign policy of the United States between the two World Wars. During that period, the United States would accept no obligation to either political or military alliances. Since 1945, the United States has officially eschewed isolationism, replacing it with a clear international commitment, enshrined in policies such as the TRUMAN DOCTRINE and formal alliances such as the NORTH ATLANTIC TREATY ORGANIZATION (NATO). From time to time, commentators believe they can detect a resurgence of American isolationist sentiment, particularly in relation to problems within NATO. There are some who believe that the Western European powers are selfish in not spending enough on their own defense, and that they are too weak to stand up to the Soviet Union or properly to support the United States in its international role (see BURDEN SHARING and GLOBALISM). Such sentiments oddly recall the attitudes of the 1930's—that the United States should not strain its own economy, or risk its own citizens, on behalf of "undeserving" foreign powers. See Thomas N. Guinsburg, *The Pursuit of Isolationism in the United States Senate from Versailles to Pearl Harbor* (New York: Garland, 1982); Charles Krauthammer, "Isolationism, Left and Right," *New Republic* 192 (March 4, 1985).

isolead curve A curved line, on a chart or diagram, used to show how far ahead of a moving TARGET a GUN must be aimed to allow for the time the PROJECTILE takes to reach the target.

IT See IMMEDIATE TRANSIENT INCAPACITATION DOSE.

J

jackboots Heavy CAVALRY boots; often used as a symbol for oppressive tyranny or a fascist regime.

jacket The metal casing that forms the outer covering of a lead bullet.

jamming The deliberate radiation, reradiation, or reflection of electromagnetic energy to prevent or degrade the receipt of information by a receiver. See ACOUSTIC JAMMING; BARRAGE JAMMING; ELECTRONIC COUNTERMEASURES; SELECTIVE JAMMING; SPOT JAMMING.

JAN grid Joint Army-Navy grid system; a GRID system covering the entire surface of the Earth, adopted to afford security in referring to geographical positions. It is set up by prescribing the location of the origin and size of the grid squares, both in terms of latitude and longitude. Joint Army-Navy grids are used generally on Mercator projections.

JATO unit Jet-assisted takeoff unit. A rocket motor unit normally used to assist the initial action of the main propulsion unit of an aircraft.

jeep 1. The U.S. Army's four-wheel drive general purpose (GP-jeep) vehicle dating from World War II. It is now being phased out and replaced by the HMMWV. 2. A smaller version of something; thus, an escort aircraft carrier is called a "jeep carrier." 3. A commercially sold four-wheel-drive light truck.

jihad See HOLY WAR.

jingoism Strong nationalist sentiment, characterized by a belligerent foreign policy. The term first became current in England in the mid 1870s, when the British seemed on the verge of war with Russia. A popular song went:

> We don't want to fight,
> But by Jingo, if we do,
> We've got the ships, we've got the men,
> We've got the money too.

Jingo was a euphemism for "by God" or "by Jesus." The term soon crossed the Atlantic and became increasingly popular in the United States.

joint A term that connotes activities, operations, organizations, and other functions, in which elements of more than one service of the same nation participate. See also COMBINED FORCE.

Joint Chiefs of Staff (JCS) The highest level of military command in the U.S. Department of Defense; its members function as the president's top military advisers. The JCS committee consists of five people: the senior naval officer, known as the Chief of Naval Operations, the Commandant of the U.S. Marine Corps, and the Chiefs of Staff of the Army and the Air Force, with the committee being headed by the fifth officer, the Chairman of the Joint Chiefs, a presidential appointee who can be drawn from any of the services. Together with the Secretary of Defense and the senior civil servants in the Office of the Secretary of Defense, the Joint Chiefs are unifying elements in the huge and divided American military machine. The Pentagon Reorganization Act of 1986 strengthened the role of the Chairman of the Joint Chiefs of Staff by making him the president's "principal military adviser," rather than simply the representative of the collective opinion of all the service chiefs. While an organization known as the Joint Chiefs of Staff operated during World War II, the present organization was created by the NATIONAL SECURITY ACT of 1947. See William J. Lynn and Barry R. Posen, "The Case for JCS Reform," *International Security* Vol. 10, No. 3 (Winter 1985–86); Mark Perry, *Four Stars* (Boston: Houghton Mifflin, 1989).

joint common-user item An item of an interchangeable nature that is in common use by two or more services of a nation. See INTERCHANGEABILITY.

joint doctrine Fundamental principles that guide the employment of FORCES of two or more SERVICES of the same nation in coordinated action toward a common OBJECTIVE. See also COMBINED DOCTRINE.

joint force A military command consisting of elements of two or more services, such as Army and Air Force. Compare to COMBINED FORCE; COMPONENT FORCE.

joint operation An operation carried on by two or more armed services of the same nation. See also COMBINED OPERATION.

joint operations center A jointly manned facility of a JOINT FORCE commander's headquarters established for planning, monitoring, and guiding the execution of the commander's decisions.

joint staff 1. A staff formed of two or more of the services of the same country. See also COMBINED STAFF; INTEGRATED STAFF; PARALLEL STAFF. 2. The Staff of the JOINT CHIEFS OF STAFF as provided for under the NATIONAL SECURITY ACT of 1947, as amended.

Joint STARS The joint surveillance and target attack radar system. It is "joint" because it is the United States Air Force that provides the aircraft and the radar to give information to army ground units. Overall Joint STARS is a battle management and targeting system that detects, locates, tracks, and assists in attacking enemy targets beyond the FORWARD LINE OF OWN TROOPS.

joint training procedures Those training procedures agreed to jointly by those service agencies charged with developing DOCTRINE involving more than one service. They may provide the basis for JOINT DOCTRINE.

join-up 1. To form separate aircraft or groups of aircraft into a specific FORMATION. See also RENDEZVOUS. 2. To enter military service.

Jomini, Antoine Henri (1779–1869) A general and military critic of the Napoleonic era whose systematic efforts to develop the principles of warfare made him a founder of modern military thought. He was the first to delineate the significant differences, between STRATEGY, TACTICS, and LOGISTICS. His theory of LINES OF OPERATION is extensively developed in his greatest work, *Precis on the Art of War* (1836). While Jomini and CLAUSEWITZ were contemporaries, it was Jomini who dominated the latter nineteenth-century thinking about the art of war. Clausewitz, in turn, has been far more influential in the twentieth century. To Jomini, war was a game to be played by the generals—not politicians. He divorced the concept of war from national survival, confining it instead to a theory of operations where universal conceptions and doctrines would provide the framework of victory regardless of the type of weapon or geography. In this approach he failed to take into account the irrational aspect of nations at war and the unpredictable acts of desperate men. He diverged from Clausewitz by holding that the fundamental objective of military operations was to occupy enemy territory rather than to annihilate the enemy's military system.

JSTARS See JOINT STARS.

judge advocate A staff officer who functions as the legal advisor to a command; often charged with the administration of military justice.

judge advocate general The chief legal officer of a military service.

jumpmaster The assigned, airborne-qualified individual (a combat parachuting expert) who controls parachutists from the time they enter an aircraft until they exit from it.

jump-off line See LINE OF DEPARTURE.

jump speed The airspeed at which parachute troops can jump with relative safety from an aircraft.

junior officer 1. A COMPANY grade officer; those officers with ranks below MAJOR (below lieutenant senior grade in the Navy). 2. Any officer who is junior in rank to another. If two officers are of the same rank, the junior officer is the one who has held the rank for the lesser period of time.

junta 1. The collective dictatorship of a group of senior officers. 2. A ruling board, council or committee that controls the government of a country, especially after a COUP D'ETAT.

just-war theory The realm of theory having its origins in medieval Catholic political theology, that considers when it is morally acceptable to go to war, and what forms of warlike activities are permissible once the justness of a general war effort has been established. The question of the acceptability of a war, *Jus ad Bellum* in the traditional terminology, has lately undergone the development of more restrictive criteria. At one time, for example, it was understood that one nation could make war on another to punish it for actions that had not even directly affected the first nation. Alternatively, it had also been considered acceptable for one country to attack another in retaliation for some hurt it had suffered which may have been, in itself, much less provocative than a full-scale attack. Most modern thinkers would limit the right to go to war to self-defense alone, but it is probable that an alliance war, in which war is made on a third party that has attacked an ally, would be understood in the general category of self-defense. Most of the other restrictions that the original just-war theory imposed, such as having "right intentions," and that the war must be declared by a legitimate authority, have little meaning in the modern context. See Anthony T. Bouscaren, "Just War, Nuclear Arms and the Catholic Bishops," *Strategic Review* 11, 3 (Summer 1983); Daniel F. Montaldi. "Toward a Human Rights Based Account of the Just War," *Social Theory and Practice* 11, 2 (Summer 1985); John Langan, "Just War Theory and Decisionmaking in a Democracy," *Naval War College Review* 38, 4 (July–August 1985).

K

Kahn, Herman (1922–1983) The nuclear strategist and futurologist whose pivotal analyses of defense strategy in a nuclear age helped the United States move away from its 1950s policy of massive retaliation to the present one of FLEXIBLE RESPONSE. Kahn was critical of those who failed to evaluate analytically the causes and effects of thermonuclear war, simply because such a war is too horrible to think about. Kahn argued that the moral questions should not inhibit the clear rational analyses of the many possibilities of nuclear war. See Herman Kahn, *On Thermonuclear War* (Princeton: Princeton University Press, 1960); Herman Kahn, *Thinking About the Unthinkable* (New York: Horizon Press: 1962); Herman Kahn, *Thinking About the Unthinkable in the 1980s* (New York: Simon and Schuster, 1984).

kamikaze 1. The Japanese suicide pilots (or their aircraft) who sought to crash into American ships during World War II. The word means "divine wind" in reference to the storm that destroyed a Chinese invasion fleet centuries before. 2. By analogy, any suicide mission. See Edwin Palmer Hoyt, *The Kamikazes* (New York: Arbor House, 1983).

key 1. In cartography, a synonym for legend. 2. In CRYPTOGRAPHY, a symbol or sequence of symbols (or electrical or mechanical correlates of symbols) that control the operations of encryption and decryption.

keyholding The tumbling of a BULLET in flight, caused by failure of the bullet to receive sufficient spin from the RIFLING in the barrel of the gun that fired it.

key list A publication containing the KEY for a CRYPTOSYSTEM in a given CRYPTOPERIOD.

key point A concentrated military SITE or INSTALLATION, the destruction or capture of which would seriously affect a war effort or the success of a military operation.

key symbol In psychological operations, a simple, suggestive, repetitive element (rhythm, sign, color, etc.) that has an immediate impact on a target audience and which creates a favorable environment for the acceptance of a psychological theme.

key terrain Any locality or area the seizure or retention of which affords a marked advantage to a combatant.

KGB Komitet Gosudarstuennoe Bezopasnosti; The Committee for State Security of the Soviet Union; the internal security police and international espionage organization of the Soviet Union. The KGB, which calls itself the "sword the Soviet Union, is far more than a Soviet version of the CIA; it also runs the internal police force, immigration, and prisons. It is as if the American CIA, FBI and all state and local police and prisons were run by one agency. See Amy W. Knight, "The KGB's Special Departments in the Soviet Armed Forces," *Orbis* Vol. 28, No. 2 (Summer 1984).

kill 1. To destroy an enemy soldier, vehicle, or other entity. 2. The consumption of SUPPLIES so that they no longer exist. Killing is the most important job of the soldier. Yet a basic problem with combat effectiveness is the reluctance of fully trained soldiers to kill other human beings. Compare to FIRE DISCIPLINE. 3. As applied to AIR DEFENSE, the term used to denote that a hostile airborne, ballistic, or orbiting object has been destroyed or rendered ineffective.

kill, functional See FUNCTIONAL KILL.

kill assessment 1. The process of determining the percentage of an enemy that has been nullified (destroyed). 2. The detection and assimilation of information indicating the destruction of an object under attack. Kill assessment is one of the many functions to be performed by the battle-management system of the United States's BALLISTIC MISSILE DEFENSE system.

killed in action (KIA) A battle CASUALTY who is killed outright or who dies as a result of wounds or other injuries before reaching a medical treatment facility.

killing zone An area in which the commander of a military force plans to force the enemy to concentrate so as to destroy him with conventional weapons or with tactically employed nuclear weapons.

kill probability A measure of the probability of destroying a target.

kill radius The maximum distance that a person may be from the center of a nuclear explosion and still have a 100 percent chance of being killed. Anyone standing outside the kill radius can have at least some chance (albeit slim) of surviving.

kiloton 1. One thousand tons. 2. In the context of nuclear weapons, one thousand tons of TNT.

kiloton weapon A NUCLEAR WEAPON whose yield is measured in terms of thousands of tons of trinitrotoluene (TNT) explosive equivalents, producing yields from 1 to 99 kilotons. See also MEGATON WEAPON; NOMINAL WEAPON; SUBKILOTON WEAPON.

kinetic energy ammunition AMMUNITION designed to inflict damage on FORTIFICATIONS, ARMOR, or ships by reason of its kinetic energy (body in motion) upon impact. The combination of weight and velocity often produces far more damage than the highest of explosives. Most anti-armor weaponry now follows this principle. Kinetic-energy weapons have become still more fashionable with the advent of serious research on the STRATEGIC DEFENSE INITIATIVE (SDI). The problem of destroying missile WARHEADS in space is peculiarly difficult because the near-vacuum environment experienced by RE-ENTRY VEHICLES makes all but the nearest of misses by even a nuclear warhead useless. However, the warheads, which are physically frail, are very vulnerable to direct impact with any hard object, however small, if it is traveling fast enough. One part of the SDI is based on this principle; large numbers of solid objects would be fired into the paths of incoming missiles. Sometimes called "smart rocks," such projectiles do not depend on complex and expensive warheads, and can be manufactured and fired in large numbers. These are kinetic energy weapons just as much as the heavy, solid anti-tank rounds, made of depleted uranium, with which modern armies are equipped.

kinetic lead The corrective allowance made for the relative motion of a target when computing the LEAD angle in gunnery.

KISS Keep It Simple, Stupid; slang concept which implies that complicated plans have a tendency to fail where a simplistic approach might have succeeded.

kit 1. A grouping of items to form a functional unit such as a repair kit. 2. A soldier's personal items.

kit, adaption See ADAPTION KIT.

kitchen police (KP) Military or civilian personnel detailed or hired to perform noncooking duties involving the preliminary preparation of fruits and vegetables and the sanitation and cleaning of dining facility buildings and equipment.

knife rest A portable wooden or metal frame strung with barbed wire, used as a barricade on roads and wherever else a readily removable barrier is needed. With a metal frame it can be used as an underwater obstacle in beach defenses.

knot A speed of one nautical mile per hour; approximately equal to 1.115 statute miles per hour. Knots are also used as a measure of aircraft speeds.

known datum point A clearly visible point to which the AZIMUTH and RANGE are known.

kopfring A metal ring welded to the nose of a BOMB to reduce its penetration in earth or water.

Korean War The war, that lasted from June 1950 to July 1953; the first major occasion in which the United States, with contingents from 14 other United Nations members, fought against an established Communist power bloc. It resulted from an invasion by the Communist North Korean regime south of the 38th parallel, which had been fixed as an arbitrary dividing line between North and South Korea at the end of World War II. UNITED NATIONS (UN) attempts to reunite Communist North and democratic South Korea had failed. Initially the landing at Inchon (on September 15, 1950) by a large American force, followed by an invasion of North Korea, drove the North Koreans back. However, the pursuing UN forces clashed with Communist Chinese forces on the Chinese border, and a very large Chinese army pushed the UN forces back into South Korea, with the southern capital, Seoul, being captured for the second time. Eventually the UN forces, predominantly American, were able to drive the Communists back across the 38th parallel, where the war fell into a stalemate. The cease-fire line agreed to in 1953 roughly follows the 38th parallel, and a demilitarized zone, supervised by UN forces, separates the two countries.

The Korean War had two main effects on American thinking. First, Americans regarded their war effort as a failure, and it represented a considerable shock to military morale which, perhaps, sowed the seeds of defeatism a decade later in Vietnam. Second, it was a deeply unpopular war, fought by reservists who had thought that their World War II experience was the last wartime experience they would have. Casualties were very high, with over 50,000 Americans (and 17,000 soldiers from other UN contingents) killed, compared with 58,000 in the Vietnam War.

These factors combined to encourage President Eisenhower, who came to office in the last months of the war, to enunciate his massive retaliation doctrine. One of the dilemmas of the Korean War was that while the United States enjoyed an almost complete nuclear monopoly at the time, it found no way of using the weapons, and was forced to fight a conventional war for which it was ill-prepared. It was also partly for this reason that massive retaliation, threatening nuclear war against civilian targets rather than against conventional forces, became the main line of American strategy for the rest of the 1950s. See Keith D. McFarland, *The Korean War: An Annotated Bibliography* (New York: Garland, 1986); Max Hastings, *The Korean War* (New York, NY: Simon & Schuster, 1987); Clay Blair, *The Forgotten War: America in Korea 1950–1953* (New York: Times Books, 1988).

KP See KITCHEN POLICE.

Kt A kiloton (see KILOTON WEAPON).

L

laager Defensive positions organized by a mobile force as it stops for a night or lesser period; a PERIMETER DEFENSE.

ladar A technique analogous to RADAR, but which uses laser light rather than radio or microwaves. The light is bounced off a target and then detected, with the return beam providing information about the distance and velocity of the target.

lag rate In manpower control usage, a percentage indicating the ratio of the shortage between the actual and AUTHORIZED STRENGTH of a military unit.

Lambert projection A conic map projection in which the meridians are straight lines converging in the direction of the poles, and the parallels are concentric circles intersecting the meridians at right angles.

lance corporal 1. The U.S. Marine Corps' and the British Army's rank for an enlisted man above a PRIVATE but below a CORPORAL. 2. A private who is temporarily assigned the duties of a corporal.

Lanchester equations The pioneering examples of operations research applied to military activities. Frederick William Lanchester (1868–1946) was an engineer who, during World War I, produced mathematical analyses of weapons systems and their relationship to manpower.

land-control operations The employment of ground forces, supported by naval and air forces as appropriate, to achieve military objectives in vital land areas. Such operations include the destruction of opposing ground forces, the securing of key terrain, the protection of vital land lines of communication, and the establishment of local military superiority in areas of land operations.

land diameter The diameter of the bore of a rifle measured from the top of the LANDS.

landing, diversionary See DIVERSIONARY LANDING.

landing aid Any illuminating light, radio beacon, radar device, communicating device, or any system of such devices for aiding aircraft to make an APPROACH and landing.

landing area 1. That part of an OBJECTIVE AREA within which are conducted the landing operations of an AMPHIBIOUS FORCE. It includes the beach, the approaches to the beach, TRANSPORT AREAS, FIRE-SUPPORT areas, the air occupied by close supporting aircraft, and the land included in the advance inland to the initial objective. 2. The general area used for landing troops and materiel, either by airdrop or air landing. This area includes one or more DROP ZONES or landing strips. 3. Any specially prepared or selected surface of land, water, or a ship's deck designated or used for the take-off and landing of aircraft.

landing attack An attack against enemy defenses by troops landed from ships, aircraft, boats, or amphibious vehicles. See also ASSAULT.

landing craft A craft employed in AMPHIBIOUS OPERATIONS, specifically designed for carrying troops and equipment and for beaching, unloading, and retracting. Also used for logistic cargo resupply operations.

landing force A TASK ORGANIZATION of aviation and ground troop units assigned to an amphibious assault. It is the highest troop echelon in the amphibious operation. See also AMPHIBIOUS FORCE.

landing group A subordinate organization of a LANDING FORCE, capable of conducting landing operations, under a single tactical command, against an enemy position or group of positions.

landing mat A prefabricated, portable mat so designed that any number of planks (sections) of it may be rapidly fastened together to form surfacing for emergency runways, landing beaches, and other temporary approaches for an assault.

landing plan An AIRBORNE or AIRMOBILE plan prescribing the sequence, place of arrival, and method of entry into an OBJECTIVE AREA. The purpose of the plan is to get the proper units to the correct place in the correct order to properly execute the ground TACTICAL PLAN. Compare to: GROUND TACTICAL PLAN, MARSHALLING PLAN.

landing point A point within a LANDING SITE where a helicopter or vertical take-off and landing aircraft can land.

landing roll The movement of an aircraft from touchdown through deceleration of taxi speed or a full stop.

landing schedule In AMPHIBIOUS OPERATIONS, a schedule that shows the beach, hour, and priorities of landing of assault units, and which coordinates the movements of LANDING CRAFT from transport ships to the beach in order to execute the SCHEME OF MANEUVER ASHORE.

landing ship An assault ship designed for long sea voyages and for rapid unloading over and onto a beach.

landing ship, dock (LSD) A ship designed to transport and launch loaded amphibious craft or AMPHIBIOUS VEHICLES with their crews and embarked personnel or equipment, and to render limited docking and repair services to small ships and craft. It is also capable of acting as a control ship in an AMPHIBIOUS OPERATION.

landing ship, tank See TANK LANDING SHIP.

landing site 1. A site within a LANDING ZONE containing one or more LANDING POINTS. 2. In AMPHIBIOUS OPERATIONS, a continuous segment of coastline over which troops, equipment, and supplies can be landed by surface means.

landing site, helicopter See HELICOPTER LANDING SITE.

landing team, brigade See BRIGADE LANDING TEAM.

landing threshold The beginning of the portion of a runway usable for landing.

landing zone Any specified zone used for the landing of aircraft.

landing zone, helicopter See HELICOPTER LANDING ZONE.

landing-zone control party A group specially trained and equipped to establish and operate COMMUNICATIONS and SIGNAL devices from the ground for the traffic control of aircraft or helicopters for a specific LANDING ZONE. Compare to PATHFINDERS.

land mine warfare See MINE WARFARE.

lands A raised portion between grooves in the BORE of a gun or rifle. Spiral channels cut in the bore of a gun are called grooves.

land search The search of TERRAIN by earth-bound personnel.

land tail That part of an AIRBORNE or air-transported unit that will not travel to combat by air; and travel by land to rejoin the rest of its air-transported unit in the combat zone.

lane 1. A clear route through an OBSTACLE. A single lane is normally 8 meters wide. A lane for foot troops is a minimum of 2 meters in width. 2. A strip of roadway intended to accommodate the forward movement of a single line of vehicles, generally 8 feet to 13 feet in width.

lane marker In LAND MINE WARFARE, sign used to mark a minefield lane.

laser Originally an acronym for Light Amplification by Stimulated Emission of Radiation; a device that produces a narrow beam of coherent radiation through a physical process known as stimulated emission. Lasers are able to focus large quantities of energy at great distances, and are among the leading candidates for BALLISTIC MISSILE DEFENSE weapons.

laser, gamma-ray See GAMMA-RAY LASER.

laser designator The use of a low-power laser to illuminate a target so that a weapon equipped with a special tracker can home in on the designated target.

laser-guided weapon A weapon that utilizes a seeker to detect laser energy reflected from a laser-marked or laser-designated target, and which through signal processing provides guidance commands to a control system that guides the weapon to the point from which the laser energy is being reflected.

laser rangefinder A device that uses laser energy for determining the distance from the device to a place or object.

laser ranging The use of laser transmissions to determine the RANGE to a target, normally as an input to the delivery computer for a weapon.

laser seeker A device based on a direction-sensitive receiver which detects the energy reflected from a laser-designated target and defines the direction of the target relative to the receiver. See also LASER-GUIDED WEAPON.

laser target-designating system A system used to direct (aim or point) laser energy at a target. The system consists of a LASER DESIGNATOR or laser target marker, with the display and control components necessary to acquire the target and direct the beam of laser energy onto it. See TARGET ACQUISITION.

laser tracker A device that locks on to the energy reflected from a laser-marked or laser-designated target and defines the direction of the target relative to itself.

laser weapons Devices that produce tightly-focused beams of very high energy ELECTROMAGNETIC RADIATION. The radiation can, in principle, be of any wavelength, although the radiations with which we are more familiar from civilian research emit radiation in the infra-red, visible, and ultra-violet regions of the electromagnetic spectrum, and are usually referred to as optical radiations. One technique on which much research is being done is an X-ray laser that could pump out an incredibly powerful beam. This, however, would require a nuclear explosion to trigger it. However, all the power of that explosion would effectively be focused into a tight beam, rather than being diffused in all directions, as in a standard nuclear detonation. The main interest in laser weapons comes from the research and planning for the STRATEGIC DEFENSE INITIATIVE. While laser-beam-producing technology with enough power to be of military use has been too bulky to be of much battlefield use, lightweight models are presently being developed. See Keith B. Payne, *Laser Weapons in Space: Policy Issue* (Boulder, CO: Westview Press, 1982); Ellis M. Madsen, "Defending Against Battlefield Laser Weapons," *Military Review* 67, 5 (May 1987).

last-ditch The final line of defense, assuming that the defensive position consists of a series of EARTHWORKS. Many a king proclaimed that he would die in the last-ditch on the battlefield rather than surrender. But from the last-ditch, surrender looks increasingly like a viable option.

last stand A final, "to the death" defense against an overwhelmingly superior force because surrender is not an option allowed by the enemy (as was the case concerning General George Armstrong Custer at the Little Big Horn) or because of the suicidal inclinations of the defenders (as was true of the Japanese in many World War II battles); or because it is the only honorable action the defenders can take under the circumstances (as with the 300 Spartans at Thermopylae). It was often observed in the United States during World War II that the Japanese who fought to the death were called fanatics, while the Americans who died in last-stand battles were called heroes.

latent lethality dose (LL) That RADIATION DOSE (determined to be 650 rads) which causes functional impairment within 2 hours of exposure. Persons so exposed may respond to medical treatment and survive; however, the majority of exposed persons will remain functionally impaired until death, which will probably occur within several weeks.

lateral 1. An underground gallery (small tunnel) that is constructed parallel to the FRONT LINE, and from which other parallel galleries for attack, defense, and listening are projected toward the enemy. A lateral differs from a FISHBONE MINE SYSTEM, which is a series of independent galleries cut in the direction of the enemy. 2. To one side of a line, such as the observer target line in FIRE CONTROL.

lateral spread An artillery technique used to place the mean POINT OF IMPACT of two or more units 100 meters apart on a line perpendicular to the gun-target line.

latest arrival date The latest date a unit should arrive "in-theater" in support of a specific OPERATION PLAN.

launch The transition from static repose to dynamic flight of a MISSILE.

launch, cold See COLD LAUNCH.

launch, precautionary See PRECAUTIONARY LAUNCH.

launcher 1. A structure designed to support and hold a MISSILE in position for firing; situated on a LAUNCH PAD. 2. A ship's catapult for launching aircraft.

launching, zero-length See ZERO-LENGTH LAUNCHING.

launching site Any site or installation with the capability of launching SURFACE-TO-AIR or SURFACE-TO-SURFACE MISSILES.

launch on warning A nuclear retaliatory attack undertaken as soon as it is believed that enemy missiles had been fired; as soon as there was warning of an attack of incoming missiles. Such a "launch on warning" strategic policy is sometimes suggested as a way to avoid the vulnerability of land-based missiles to a FIRST-STRIKE attack. Many fear that an enemy could gain overwhelming strategic superiority after having destroyed on the ground the bulk of a country's land-based missiles and many of its bombers in a surprise nuclear attack. It is assumed that no country would ever fire a retaliatory strike before it had actually received at least one nuclear detonation on its soil (see LAUNCH UNDER ATTACK); yet the time required to give a launch order after the first explosion over a silo, and have it confirmed and executed would take sufficiently long that many of the defender's weapons would be destroyed by the time the counterattack was to begin. Launch on warning would effectively require a bypass of the complex confirmation procedures, such as checks and codes necessary to ensure that accidental or unofficial launches do not occur. Communications are a particular problem, because if launch were delayed until at least some warheads had exploded, the ELECTROMAGNETIC PULSE would be liable to seriously hamper rapid communication, putting the remaining missiles further at risk. Even if a sizeable number of the defender's missiles were launched during an attack, it has to be remembered that in the first few minutes of their trajectory, missiles are actually more vulnerable to nuclear explosions going off around them than they would be sitting in their silos. There seems to be very little possibility of a safe "launch on warning" policy operating successfully, and it is believed that both the American and Soviet military leaderships have accepted that it cannot be relied on, even if it is still frequently suggested as a policy. See John D. Williams, "Launch on Warning in Soviet Nuclear Strategy," *Air University Review* 38, 1 (November–December 1986).

launch pad A concrete or other hard surface area on which a missile launcher is positioned.

launch time The time at which an aircraft or missile is scheduled to be airborne.

launch tube Any one of a variety of structures that positions a MISSILE for launching from a submarine.

launch under attack A policy of waiting until enemy strategic nuclear missiles actually detonate on home soil before retaliating. Accepting a launch-under-attack policy necessitates that a country's strategic forces be either very large or very well protected, in order to ensure a secure SECOND-STRIKE capacity. Certain American analysts fear that the Soviet Union could gain strategic superiority by being in a position to destroy most of America's land-based missiles in their silos in a surprise attack. Because the United States rejects the LAUNCH ON WARNING policy, it does keep the bulk of its warheads in submarines (SSBNs), while the major part of the Soviet Union's strategic force consists of multi-warhead intercontinental ballistic missiles (ICBMs) in heavily-defended SILOS. The distinction between the policies of "launch on warning" and "launch under attack" involves only those missiles which, not being capable of recall, cannot be committed without certainty that an attack is actually under way. In contrast, the STRATEGIC AIR COMMAND would order its bombers to take off as soon as there was warning of an attack. This presents a major strategic problem to the Soviet Union. Even if it managed to destroy most of America's airbases (by firing submarine-launched missiles from near the coast of the United States) and from ICBM silos (with their own long-range ballistic missiles) simultaneously, the United States would still retain its assured destruction capacity in the form of the SSBNs.

law, higher See HIGHER LAW.

law, international See INTERNATIONAL LAW.

law, martial See MARTIAL LAW.

law, military See MILITARY LAW.

law of war That part of INTERNATIONAL LAW that regulates the conduct of armed hostilities. It is often termed the law of armed conflict. See also RULES OF ENGAGEMENT.

lay 1. To direct or adjust the aim of a weapon. 2. The setting of a given range, given direction, or both for a weapon. 3. To drop one or more aerial bombs or aerial mines onto the ground or water surface from an aircraft. 4. To spread a smoke screen on the ground from an aircraft. 5. To calculate or project a course.

laydown bombing A very low-level bombing technique wherein DELAY FUZES or devices are used to allow the attacker to escape the effects of his bomb.

layered defense A term usually used in strategic and naval contexts; a defensive posture consisting of multiple defense measures, each of which is designed to ward off an attack at a specific distance from the defended area. If one level is penetrated, others remain effective. Increasingly, defense systems depend on multiple layers of detection and destruction. In the naval context, for example, an aircraft-carrier BATTLE GROUP will have at least three layers of defense. The first will be FIGHTER aircraft from the carriers, which hope to intercept attacking BOMBERS before they can launch their AIR-TO-SURFACE MISSILES (ASMs). However, the detection range of the fighters will probably not greatly exceed the range at which the bombers can launch their ASM salvos. As the attacking missiles approach the fleet, specialized area defense ships, such as the United States's AEGIS cruisers, will attempt to destroy the incoming missiles with SURFACE-TO-AIR MISSILES (SAMs) of their own. At still closer ranges the POINT DEFENSE systems on individual ships will attempt to destroy remaining enemy weapons that are targeted on them. Even this last stage may have two layers, with point defense SAM systems first and then, at very close range, rapid-firing MACHINE GUNS with automatic radar control. Modern defenses against air attack have to cope with a large number of MISSILES with very sophisticate target-seeking WARHEADS that can be launched from ranges of well over a hundred miles. As a result, it is inevitable that at least some attacking weapons will "leak through" any one defensive shield. Thus, the same concept of layered defense is crucial to the BALLISTIC MISSILE DEFENSE measures currently being designed as part of the STRATEGIC DEFENSE INITIATIVE. See Jerome Bracken, "Layered Defense of Deceptively Based ICBMs," *Naval Research Logistics Quarterly* 31, 4 (December 1984).

laying-up position Any suitable position where naval units can berth, camouflage, and replenish in preparation for forthcoming operations.

lay on 1. To execute a BOMBER strike. 2. To set up a MISSION.

lead 1. The distance ahead of a moving target that a gun must be aimed in order to hit the target. 2. The vertical and lateral angles between a gun target line and the axis of the BORE at the moment of firing at a moving target. 3. To aim a gun ahead of a moving target. 4. One target length, as it appears to the gunner, is used as a unit for measuring lead. 5. In military highway operations, the linear spacing between the heads of successive vehicles, SERIALS, MARCH UNITS, or COLUMNS. 6. An ice crack too wide for men, sledges, and dogs to cross easily; i.e., any crack wider than 3 to 5 feet.

lead aircraft 1. The aircraft designated to exercise command of other aircraft within the flight. 2. An aircraft in the van (the front) of two or more aircraft.

lead azide A colorless and poisonous chemical compound that is used as a DETONATOR for explosives.

lead curve A line on a chart, recording in graphic form the LEAD necessary to aim a gun at a moving target.

leader's rule A method of determining the safe RANGE for MACHINE GUNS firing over the heads of friendly troops when the range to the target is more than 900 meters. Compare to GUNNER'S RULE.

lead in An EXPLOSIVE TRAIN that conducts a detonating impulse into an explosive-filled cavity.

lead styphnate A primary HIGH EXPLOSIVE consisting of reddish-brown rhombic crystals, used extensively in some EXPLOSIVE TRAINS.

leaflet bomb A device used for dropping large quantities of PROPAGANDA leaflets from high-flying aircraft, to insure their reaching their targets with a minimum of drift caused by air currents.

leaflet projectile A standard ejection projectile especially designed for leaflet dissemination; when loaded with PROPAGANDA leaflets, the projectile matches the ballistic characteristics of a high explosive projectile.

leakage The percentage of WARHEADS that penetrate a defensive system in intact and operational form.

leaker 1. Someone who deliberately discloses confidential or classified information in order to advance the public interest, embarrass a bureaucratic or military rival, or help a reporter disclose incompetence or skulduggery to the public. *New York Times* columnist James Reston has written, "The government is the only known vessel that leaks from the top." 2. Someone who inadvertently discloses secret information. 3. A term for a BOMB or PROJECTILE filled with a CHEMICAL AGENT, which is leaking its contents and contaminating the area.

leapfrog 1. A form of movement in which like supporting elements of a military force are moved successively through or by one another along the axis of movement of the supported forces. 2. A technique in medical tactics for maintaining continuous medical support for maneuvering forces and forces in combat by alternately displacing medical units performing the same or similar functions along a common axis of movement.

least-developed countries A list of nations established by the United Nations (UN) GENERAL ASSEMBLY in 1971 and consisting of those developing countries without significant economic growth, with very low per-capita income, and with low literacy rates. This list, plus a later UN list of countries most seriously affected by the oil price increases of 1973 and 1974, comprises the FOURTH WORLD.

least separation distance (LSD) The minimum distance in meters that a designated GROUND ZERO must be separated from an object to preclude, with 90 percent assurance, damage to the object or OBSTACLES.

leatherneck A U.S. Marine; early Marine uniforms had leather neckbands.

leave 1. Permission. 2. An authorized absence from duty. Winston Churchill wrote during World War II that: "A certain amount of leave, although not in any contract of service, is a recognized part of a soldier's life." The Duke of Wellington is famous, in part, for having granted an officer 48 hours leave on the grounds that 48 hours "is as long as any reasonable man can wish to stay in bed with the same woman." In peacetime, American armed forces personnel are allowed 30 days of leave per year.

leave en route Ordinary LEAVE granted to military personnel when traveling to a new STATION in connection with temporary duty or a permanent change of station.

leaver An independent merchant ship which breaks off from a main CONVOY.

leaver convoy A CONVOY that has broken off from a main convoy and is proceeding to a different destination.

leaver section A group of ships forming part of a main convoy, which will subsequently break off to become LEAVERS or a LEAVER CONVOY.

left face 1. In DISMOUNTED drill, a movement from the position of ATTENTION by which a person turns on the heel of the left foot and the ball of the right foot so as to face 90 degrees to the left of the original position. 2. The command to execute this movement.

left flank, MARCH See RIGHT (LEFT) FLANK, MARCH.

left (or right) 1. Terms used to establish the relative position of a body of TROOPS. The person using the terms "left" or "right" is assumed to be facing in the direction of the enemy regardless of whether the troops are advancing toward or withdrawing from the enemy. 2. A correction used in the ADJUSTMENT of fire to indicate that a lateral shift of the MEAN POINT OF IMPACT perpendicular to the REFERENCE LINE or SPOTTING LINE is desired.

left (or right) bank That bank of a stream or river on the left (right) of an observer when he is facing in the direction of flow or downstream.

leg infantry See STRAIGHT LEG INFANTRY.

legion 1. The basic military organization of Ancient Rome; comparable to a modern DIVISION in function, but smaller in size. 2. A special-purpose military organization such as a FOREIGN LEGION. 3. Many; a large group, such as the American Legion, an organization for American veterans.

length of column The length of roadway occupied by a COLUMN in movement, including the gaps inside the column in movement from the front of the leading vehicle to the rear of the last vehicle.

lensatic compass A compass equipped with a magnifying glass for reading the scale, used in making accurate measurements such as AZIMUTHS, for FIRE CONTROL.

lethal dose, median See MEDIAN LETHAL DOSE.

lethality A measure of the amount of damage that a weapon can be expected to inflict on a specific type of target. The concept can be applied to all weapons—there is no reason why the lethality of an infantry rifle should not be measured—but it is usually restricted either to NUCLEAR WEAPONS or to the more technologically-sophisticated among conventional armaments. The lethality coefficient is mainly a combination of two factors, accuracy and YIELD, or, as with weapons using KINETIC ENERGY AMMUNITION, some other measure of the force applied. Accuracy is normally measured in terms of CIRCULAR ERROR PROBABLE (CEP), while yield is calculated according to the EQUIVALENT MEGATONNAGE (MTE) of a WARHEAD. The formula for the latter is MTE = nominal yield, which effectively increases the equivalent yield of a small weapon of perhaps a few hundred KILOTONS, in relation to the impact of a megaton-level warhead. As lethality is measured for nuclear missiles, it is inversely proportional to the square of the accuracy of a weapon, but directly proportional only to the ⅔ power of the yield: this explains the vital role of low-CEP weapons in modern nuclear arsenals. For example, a relatively small warhead, such as that of a TRIDENT missile of perhaps 200 kilotons and a CEP of 100 meters, may well be more effective in destroying a HARD TARGET than a megaton-level weapon with a much higher CEP. The advent of PRECISION-GUIDED MUNITIONS has led to very much higher lethality in modern nuclear weapons, even though the individual warheads are, in general, much smaller than in the past.

lethality dose, latent See LATENT LETHALITY DOSE.

letter of instructions A form of order by which a superior commander gives information about broad aims, policies, and strategic plans for military OPERATIONS in a large area over a considerable period of time. It is issued to large units of a command and has the same authority as an OPERATION ORDER.

leveling, differential See DIFFERENTIAL LEVELING.

level of strength The personnel and personnel authorizations, in accordance with capabilities, mission, or both, as approved in a TABLE OF ORGANIZATION AND EQUIPMENT.

level point The point on the descending branch of the TRAJECTORY of a projectile or other device that is at the same altitude as the origin; the point at which the trajectory is level with the muzzle of the gun; sometimes referred to as the POINT OF FALL.

leverage In the context of the STRATEGIC DEFENSE INITIATIVE (SDI) a term that refers to the advantage gained by BOOST-PHASE interception, when a single booster kill may eliminate many RE-ENTRY VEHICLES and DECOYS before they are deployed. This could provide a favorable cost-exchange ratio for the defense, and would reduce stress on later layers of the defense system. See also LAYERED DEFENSE.

levy 1. To conscript (draft) troops. 2. The troops so conscripted. 3. The mandatory reassignment of ENLISTED PERSONS in specified military occupational specialties or grades.

liaison That contact or intercommunication maintained between elements of military forces to insure mutual understanding and unity of purpose and action.

liaison, combat See COMBAT LIAISON.

liaison, command See COMMAND LIAISON.

liaison officer, fighter See FIGHTER LIAISON OFFICER.

liaison officer, ground See GROUND LIAISON OFFICER.

liberated territory Any area, domestic, neutral, or friendly, which, having been occupied by an enemy, is retaken by friendly forces.

LIC 1. LOW INTENSITY CONFLICT. 2. Legally correct interpretation; the philosophic disposition to concede as little as possible to the other sides in an arms-control agreement—to agree to do what is legally correct according to the letter of the agreement but go no further in its spirit.

Liddell Hart, Basil H. (1895–1970) An English army captain who, after he was forced out of a career as an officer because of World War I wounds, became a military historian and military journalist and wrote some of this century's most insightful critiques of STRATEGY and TACTICS. Liddell Hart believed that military strategy should be based on the principle that one pursues war to create a future state of peace. He reintroduced the idea (from CLAUSEWITZ) that destruction of the enemy's will to fight (either populace or military) provides the basis for true victory. A strong advocate of the INDIRECT APPROACH in fighting a war, he maintained that the objective of battle is to create a situation so disadvantageous to the enemy that he will not fight, or if he does fight, will lose quickly and decisively. Liddell Hart's ideas were already adopted by the German military before World War II, and formed the basis for the BLITZKREIG. See Basil Henry Liddell Hart, *Strategy*, (New York: Praeger, 1967); Ronald Lewin, "Sir Basil Liddell Hart: The Captain Who Taught Generals," *International Affairs (Great Britain)* Vol. 47, No. 1 (January 1971); Jerry D. Morelock, "The Legacy of Liddell Hart," *Military Review* 66, 5 (May 1986); John J. Mearsheimer, ed. *Liddell Hart and the Weight of History* (Ithaca, NY: Cornell University Press, 1988).

lieutenant 1. The lowest grade of COMMISSIONED OFFICER; divided into the ranks of FIRST (higher) and SECOND (lower) LIEUTENANT. According to General Dwight David Eisenhower, "the most terrible job in warfare is to be a second lieutenant leading a platoon when you are on the battlefield." Statistically speaking, JUNIOR OFFICERS in combat are twice as likely to become casualties as the enlisted men they lead. 2. In the U.S. Navy, an ENSIGN is the lowest grade of commissioned officer; then a lieutenant, junior grade (which corresponds to the Army's first lieutenant) and lieutenant (which corresponds to the Army's CAPTAIN). 3. Any assistant or second in command.

lieutenant, first See FIRST LIEUTENANT.

lieutenant colonel A FIELD-GRADE officer ranking below a COLONEL and above a MAJOR. The INSIGNIA is a silver leaf.

lieutenant commander. A naval officer above lieutenant and below commander, equivalent to a major in the Army.

lieutenant general A three-star general ranking below a full, (or four star) GENERAL and above a MAJOR (or two star) GENERAL.

light Less powerful as opposed to the more powerful "heavy" versions of some MATERIEL. Thus there exist light and heavy tanks, machine guns, divisions, and so on.

light, carry See CARRY LIGHT.

light at the end of the tunnel A metaphor for the promise of a safe exit from a desperate situation that became the standard hopeful response of many American officials in the 1960s about when the war in Vietnam would be successfully concluded. They were so consistently and embarrassingly wrong that the phrase is now associated with stupid or false predictions of military success.

light colonel A LIEUTENANT COLONEL.

light discipline A BLACKOUT so that the enemy cannot detect positions.

light division A traditional INFANTRY division, with only a small amount of light SUPPORTING ARTILLERY weapons, and no armored component other than light reconnaissance vehicles. It is also mainly "foot-mobile," with very little in the way of motorized troop carriers. The essence is that it should be easily and rapidly deployable, entirely by air, to fight anywhere in the world. In contrast, a heavy division, as its name suggests, has a considerable armored element; its infantry is equipped with ARMORED FIGHTING VEHICLES, its artillery and anti-aircraft elements are very strong, and it has sizeable engineering and other support elements, all well equipped. Such a division cannot be made AIRMOBILE, and consequently takes a very long time to deploy and is dependent on major logistics support. The only way to use such divisions in a hurry, unless they are actually stationed in the theatre of war during a crisis, is to have all their equipment pre-positioned, and to fly in only the soldiers themselves. In order to have an adequate number of divisions in Europe should a war break out without months of warning, the United States has developed such a system, known as POMCUS. See also RAPID DEPLOYMENT FORCE. See Wayne A. Downing, "Light Infantry Integration in Central Europe," *Military Review* 66, 9 (September 1986); Robert S. Rush, "Comparing Light Divisions," *Military Review* 67, 1 (January 1987); James D. Marett, "The Republic of Korea Army: The 'Light,' Light Infantry Division," *Military Review,* 17, 11 (November 1987).

light elephant-steel shelter A splinter-proof shelter for PERSONNEL and MATERIEL made from steel arch sections of medium size and weight.

lighter The portage means by which personnel and cargo discharged from ships off shore are carried to, and in some cases beyond, the shoreline.

light level of operations Operations involving less than 30 percent of all echelons in a maneuver force and less than 50 percent of FIRE-SUPPORT means engaged in sporadic combat over a period during which the employment of next higher echelon resources, to assure accomplishment of the force mission, will not be required.

light shelter A shelter that can protect against continuous bombardment from 8-inch shells. A light shelter is not to be confused with a light shellproof shelter, which is built to withstand 6-inch shells.

limited distribution messages Those messages that must receive limited distribution but may be handled by regular communications personnel within the normal handling precaution afforded by the SECURITY CLASSIFICATION of messages.

limited nuclear options (LNOs) Targeting plans for missile, or possibly bomber, attacks on specific targets. They range from single SHOT ACROSS THE BOW demonstration strikes, through smaller or larger COUNTERFORCE measures and attacks on other military targets to, presumably, a restricted intimidatory attack on one or two cities. The idea is to have a whole range of carefully prepared responses to any form of Soviet nuclear use (see FLEXIBLE RESPONSE). Limited nuclear options first came to public knowledge in 1974, when President Nixon's Secretary of Defense, James Schlesinger, announced a new strategy that moved America away from simple assured-destruction targeting (see SCHLESINGER DOCTRINE). See Ian Clark, *Limited Nuclear War: Political Theory and War Conventions* (Princeton, NJ: Princeton University Press, 1982).

limited special inspection An INSPECTION, other than an annual general inspection, which is limited to specific subjects of inquiry.

limited standard article An item that is not as satisfactory as a standard type, but is a usable substitute therefor, and is either in use or available for issue to meet a demand.

Limited Test Ban Treaty See TEST-BAN TREATY.

limited visibility operations Operations conducted at night and during other periods of reduced visibility.

limited war 1. A war in which at least one of the parties refrains from using all of its resources in order to defeat the enemy; a war that is limited by technology, objectives or weapons. It is used in strategic discussions to mean virtually anything short of CENTRAL STRATEGIC WARFARE. War can be seen as limited in two dimensions: the area and number of participants, or the means used to wage it. A third variable that once would have been important refers to the goal of the opposing sides. In the modern period it is more likely that "limited war" will refer to a confrontation that does not involve the United States and the Soviet Union or, if it does, one that is fought away from the CENTRAL FRONT and which does not involve NUCLEAR WEAPONS. The problem with the concept is that war is only ever "limited" from an external perspective. So, for example, many would regard the Arab–Israeli wars as "limited." They only involved second- and third-rank powers, did not involve nuclear or chemical weapons, and were geographically restricted and short in duration. However, the Israeli high command would not have viewed such conflicts as limited, given that the very existence of the state of Israel was at stake. Perhaps the emptiness of the term "limited war" is best demonstrated by recent discussions of "limited nuclear war," which is meant to describe a central front war restricting nuclear exchanges to the European continent west of the Soviet border. Such a war would, of course, be Armageddon for Europeans. Similarly, even a purely conventional war, given the effect of modern FIREPOWER, would be so devastating inside East and West Germany that it could not be, in any meaningful way, regarded as limited. (See also LIMITED NUCLEAR OPTIONS.) The only useful application of the concept of limited war is for military planning by the superpowers, who have to be ready, at least in the United States's perspective, to fight what is sometimes described as "one-and-a-half wars"—a major central front war and a smaller war, limited geographically and in weapons usage, elsewhere in the world. See William Vincent O'Brien, *The Conduct of Just and Limited War* (New York: Praeger, 1981); Stephen Peter Rosen, "Vietnam and the American Theory of Limited War," *International Security* Vol. 7, No. 2 (Fall 1982).

limit of advance An easily recognized terrain feature beyond which attacking elements should not advance.

limit of fire 1. The boundary marking off the area on which gunfire can be delivered. 2. The safe angular limit for firing at aerial targets.

line 1. COMBAT as opposed to SUPPORT troops; combat arms. 2. Any FORMATION in which the units (men, tanks, trucks, etc.) are positioned side by side. 3. A queue. 4. The terrain separating hostile forces; thus, being "behind the lines" means being in enemy held territory.

line, handover See HANDOVER LINE.

line, holding See HOLDING LINE.

line, no-fire See NO-FIRE LINE.

line, orienting See ORIENTING LINE.

line, reference See REFERENCE LINE.

linear defense The even distribution of defensive forces along an entire front. However, if an attacker manages to breakthrough at any point, he may easily move at will behind the line; thus, a DEFENSE IN DEPTH is generally preferable. See also FORWARD DEFENSE, MAGINOT MENTALITY.

linear obstacle spacing The distance between obstacles that cross an entire terrain unit and have a somewhat regular pattern, such as row crops or ricefield dikes.

linear speed method A method of calculating FIRING DATA in which the future position of a moving target is determined by finding the direction of its flight and the GROUND SPEED of the target. By multiplying the ground speed by the time of flight of the projectile being fired, the future position of the target is determined. The linear speed method and the ANGULAR TRAVEL METHOD are two methods of computing firing data.

linear tactics Any arrangement that uses successive lines of soldiers to attack an enemy; as the first line fires, it moves to the rear to reload and wait its turn to advance and fire again. This system was introduced by GUSTAVUS ADOLPHUS in 1630.

line haul In highway transportation, a type of haul involving long trips over the road and in which the proportion of running time is high in relation to time consumed in loading and unloading. Line hauls usually are evaluated on the basis of ton-miles-forward per day. In rail transportation, this term applies to the movement or carriage of material over tracks of a carrier from one point to another, but excluding switching service.

line of aim A direct line from a weapon's sight to the AIMING POINT.

line of battle 1. The classic linear deployment of opposing forces facing one another. 2. A formation of infantry on a FRONT. 3. Any FRONT LINE.

line of collimation A line that passes through the optical center of the objective lens of an instrument and the point of reference at the point of principal focus.

line of constant bearing A line from a fixed or moving point to a moving object or fixed point that retains a constant angular value with respect to a reference line.

line of contact A general trace delineating the location where two opposing forces are engaged.

line of departure 1. A line designated to coordinate the departure of attacking units or scouting elements at a specified time of attack. Also called the jump-off line. 2. A suitably marked offshore coordinating line to assist ASSAULT CRAFT to land on designated beaches at scheduled times. 3. In ground operations, a line ordinarily located on or behind the last available natural cover, which can be reached without exposure to hostile observation and small arms fire; suitable, clearly defined terrain features such as roads, the edges of woods, and friendly front lines may be used. 4. A line tangent to the TRAJECTORY of a projectile at the instant of the projectile's departure from the origin. It is displaced vertically from the line of elevation by the amount of the vertical jump.

line of departure is line of contact The designation of forward friendly positions as the LINE OF DEPARTURE when opposing forces are in contact.

line of deployment, probable See PROBABLE LINE OF DEPLOYMENT.

line of drift 1. A natural route along which wounded men may be expected to go back for medical aid from a combat position. 2. A route along which stragglers may be expected to go from the ZONE OF ACTION to rear areas.

line of duty 1. Authorized duty in military service. 2. The classification of all sickness, injury, or death suffered by personnel in active military service, unless caused by individual fault or neglect, and unless the disease, injury, or condition existed prior to service and was not aggravated by service.

line of elevation The axis of the bore prolonged when a piece is laid for firing.

line of fall A line tangent to the TRAJECTORY of a projectile at the LEVEL POINT.

line officer An officer belonging to a combatant branch of an army; also called an officer of the line.

line of fire The flight path of a bullet or other projectile. Compare to TRAJECTORY.

line of impact A line tangent to the TRAJECTORY of a projectile at the point of impact or burst.

line of march 1. The route selected for a military force to move along. 2. The disposition of a military unit's various elements during the course of a march.

line of own troops, forward See FORWARD LINE OF OWN TROOPS.

line of sight 1. The line between the target and the aiming references of a weapon. 2. The straight line between two points. This line is in the plane of the great circle, but does not follow the curvature of the Earth.

line-of-sight radar A RADAR system that can detect only objects within an unobstructed path of view from the radar; not over the horizon.

line of skirmishers A line of DISMOUNTED soldiers in staggered formation at extended intervals.

line overlap, oblique See OBLIQUE LINE OVERLAP.

liner 1. A helmet liner. This plastic underhelmet is often variously painted and used to identify the function of troops in training at military bases. 2. The rifled inside part of a gun barrel that may be replaced when worn out.

line replaceable unit A composite group of modules or subassemblies performing one or more discrete functions in an electronic communications system, constructed as an independently packaged unit for direct installation in the equipment.

lines, interior See INTERIOR LINES.

line search RECONNAISSANCE along a specific line of communications, such as a road, railway, or waterway, to detect fleeting targets and activities in general.

line shot 1. A projectile that strikes on the line from OBSERVER to target. 2. A PROJECTILE that passes through any part of the cone of sight formed by an air target.

lines of communication All the routes (land, water, and air) that connect an operating military force with a BASE OF OPERATIONS, and along which supplies and military forces move.

lines of operation JOMINI's concept that the commander who has the more advantageous line of communication and supply is better able to dominate the battlefield. A commander is said to have the more advantageous or "INTERIOR" LINES when he can concentrate his forces on a given point more rapidly than the enemy. A disadvantaged commander has "EXTERIOR" LINES relative to the enemy force when he is unable to concentrate his forces on a given point faster than the enemy. Prussian Field Marshall Helmuth von MOLTKE (1800–1891) wrote that: "The unquestionable advantages of the interior line of operations are valid only as long as you retain enough space to advance against one enemy . . . gaining time to beat and pursue him, and then to turn against the other. . . . If this space, however, is narrowed down to the extent that you cannot attack one enemy without running the risk of meeting the other who attacks you from the flank or rear, then the strategic advantage of interior lines turns into the tactical disadvantage of encirclement."

link 1. In COMMUNICATIONS, a general term used to indicate the existence of communications facilities between two points. 2. A maritime route, other than a coastal or transit route, that links any two or more routes. 3. The metal unit that connects the CARTRIDGES for an AUTOMATIC weapon and, with them, forms a FEED BELT. 4. A Link Trainer, a flight simulator used by pilots for practice.

linkage An international political strategy relating two or more issues in negotiations, and then using them as tradeoffs or pressure points. Compare to BARGAINING CHIP. See Arthur A. Stein, "The Politics of Linkage," *World Politics* Vol. 33, No. 1 (October 1980).

linkup A meeting of friendly ground forces (e.g., when an advancing force reaches an OBJECTIVE AREA previously seized by an AIRBORNE or AIRMOBILE force), when an encircled element breaks out to rejoin friendly forces, or when converging COMBAT MANEUVER FORCES meet.

linkup point An easily identifiable point on the ground where two forces conduct a LINKUP. When one force is stationary, linkup points are normally established where the moving force's routes of advance intersect the stationary force's SECURITY elements. Linkup points for two moving forces are established on boundaries where the two forces are expected to converge.

listening silence A period of time specified by a commander during which the transmitters of all radio sets used for signal communication within the command will be completely shut down and will not be operated except during emergencies specifically described in orders. All of the receivers must remain in operation on net frequencies unless special orders are issued to the contrary.

listening watch A continuous radio receiver watch established for the reception of traffic addressed to, or of interest to, the unit maintaining the watch.

litter-relay point The place where a new litter team takes over further movement of a CASUALTY and the first team returns for another casualty. The object is to provide short litter hauls for the bearers. The casualty may or may not be placed on a wheeled litter at this point.

live ammunition Real as opposed to blank cartridges.

live-fire exercise Training that uses real bullets to give the troops some sense of what it is like to be under fire. The problem with this is that occasionally a trainee is actually hit by a live round.

LNO See LIMITED NUCLEAR OPTIONS.

loading tray 1. A trough-shaped carrier on which heavy PROJECTILES are placed so that they can be more easily and safely slipped into the BREECH of a gun. 2. A hollowed slide that guides shells into the breech of some types of AUTOMATIC weapons.

local parole The status of prisoners in confinement who have freedom of movement outside their enclosure but within a defined area.

local security SECURITY elements established in the proximity of a unit to prevent surprise by the enemy.

local war See LIMITED WAR.

lock 1. A safety device on a firearm; when it is "on" the weapon cannot be fired. 2. To make a loaded firearm safe by setting its safety device or lock. 3. The firing mechanism of a gun.

lock and load A command to prepare to FIRE.

lock on A term signifying that a TRACKING or target-seeking system is continuously and automatically tracking a target in one or more coordinates (e.g., range, bearing, elevation).

lodgment 1. Quarters for soldiers. 2. Enemy ground occupied and held by friendly forces. 3. The consolidation of two or more adjoining BEACHHEADS or AIRHEADS.

loft bombing A method of bombing in which the delivery plane approaches the target at a very low altitude, makes a definite pull-up at a given point, releases the bomb at a predetermined point during the pull-up, and tosses the bomb onto the target. See also LOW ALTITUDE BOMBING; OVER-THE-SHOULDER BOMBING; TOSS BOMBING.

logarithmic range scale A type of scale used on the RANGE DISKS of guns when the disks are graduated for the combination of power charge and projectile. It is so named because it follows a logarithmic rather than an arithmetic curve.

logistic constraint 1. A constraint in terms of numbers of standard obstacles by type, given to all tactical commanders developing OBSTACLE PLANS. The availability of materials, transportation, and construction effort in specific SECTORS dictate the constraint. Its purpose is to force the development of obstacle plans within limits that can reasonably be accomplished. 2. Any other LOGISTIC shortage that impacts on tactical operations.

logistic estimate of the situation An appraisal resulting from an orderly examination of the LOGISTIC factors influencing contemplated courses of action to provide conclusions about the degree and manner of that influence. See also ESTIMATE OF THE SITUATION.

logistician A logistics officer.

logistic implications test An analysis of the major LOGISTIC aspects of a joint strategic war plan and the consideration of its logistic implications as they may limit the acceptability of the plan. The logistic analysis and consideration are conducted concurrently with the development of the STRATEGIC PLAN. The objective is to establish whether the logistic requirements generated by the plan are in balance with availabilities.

logistic movement, nuclear See NUCLEAR LOGISTIC MOVEMENT.

logistics The art and science of moving military forces and keeping them supplied; those inventory, production, and traffic-management activities that seek the timely placement of MATERIEL and PERSONNEL at the proper time and in the appropriate quantities. While unglamorous, logistics is often the key to victory in battle, and even more frequently the essential skill of long-term STRATEGIC success in a campaign or war. Many of the great military leaders in history have been noted as much for the care they took over supply and transport as for their tactical skills. NAPOLEON and Wellington are famous as a pair of opponents who had this quality in common. JOMINI wrote in his *Precis of the Act of War* (1838) that: "Logistics comprises the means and arrangements which work out the plans of strategy and tactics. Strategy decides where to act; logistics brings the troops to this point." As military technology developed, and as campaigning came to be a matter of huge armies fighting protracted battles, the capacity of soldiers to live off the land and carry most of their munitions with them disappeared. For example in both Field Marshal Erwin Rommel's desert campaigns in North Africa and General George S. Patton's advance through France during World War II, the supply of fuel for tank armies was as decisive as any traditional "military" factor. Today, the standard main battle tank in the U.S. Army has a fuel consumption of one gallon per 5 miles; an infantry soldier is able to carry perhaps 60 rounds of rifle ammunition, while modern rifles are capable of being fired at a rate of several hundred rounds a minute. There is no doubt that logistical competence and preparation will be of vast importance in any NORTH ATLANTIC TREATY ORGANIZATION (NATO)–WARSAW PACT conflict in Europe, where ammunition is as much of a worry to military planners as is the general force imbalance between the two sides (see POMCUS). With the pace, mobility, and firepower of modern warfare, no army unit can hope to carry fuel, ammunition, medical supplies, or even food for more than two or three days' consumption. Indeed, it is estimated that the ratio of logistics personnel to front-line combat soldiers may be as high as ten to one. See Martin Van Creveld, *Supplying War: Logistics From Wallenstein to Patton* (Cambridge: Cambridge University Press, 1977); Bruce P. Schoch, "Soviet Army Logistics-Mobile Tail of the Armored Fist," *Army Logistician* 13, 2 (March–April 1981); D. J. Wilson, "Canadian Forces Logistics," *Army Logistician* 15, 3 (May–June 1983); ALOG Staff Feature, "German Army Logistics," *Army Logistician* 15, 6 (November–December 1983); Septime Marie d'Humieres, "Logistics of the French Army," *Army Logistician* 18, 1 (January–February 1986); ALOG Staff Feature, "Chinese Army Logistics," *Army Logistician* 18, 2 (March–April 1986); Warren L. Kempf, "British Army Logistics," *Army Logistician* 18, 4 (July–August 1986).

logistics, consumer See CONSUMER LOGISTICS.

logistics, cooperative See COOPERATIVE LOGISTICS.

logistics, strategic See STRATEGIC LOGISTICS.

logistics, tactical See TACTICAL LOGISTICS.

logistic support The provision of adequate MATERIEL and SERVICES to a military force to assure the successful accomplishment of its assigned MISSIONS.

logistic support, unified See UNIFIED LOGISTIC SUPPORT.

long gray line An affectionate description for the corps of cadets at West Point, because of their gray uniforms.

long-range bomber aircraft A BOMBER designed for a tactical operating radius over 2,500 nautical miles at design gross weight and design bomb load.

long-range radar 1. RADAR equipment whose maximum range on a reflecting target of 1 square meter, and normal to the signal path, exceeds 300 miles but is less than 800 miles, provided a LINE OF SIGHT exists between the target and the radar. Compare to VERY-LONG-RANGE RADAR.

long-range theatre nuclear forces INTERMEDIATE NUCLEAR FORCES; the name change has come about as a result of ARMS-CONTROL negotiations between the United States and the Soviet Union since 1977. A variety of names have come into circulation for such forces because of an essential ambiguity about exactly what it was that was sought to be controlled through negotiations. It is clear that the superpowers' INTERCONTINENTAL BALLISTIC MISSILES (ICBMs) and SUBMARINE-LAUNCHED BALLISTIC MISSILES (SLBMs) belong in one class of weapons. The category of TACTICAL or BATTLEFIELD NUCLEAR WEAPONS is clearly another class. There is a fundamental difference between these two groups: an ICBM or SLBM owned by one superpower can hit the homeland of another, whereas tactical missiles, with ranges up to perhaps 500 kilometers, could not reach far beyond the battlefield area. Theatre nuclear forces cover weapons intermediate to these, with ranges varying from 500 to 5,500 kilometers. "Long-range theatre" weapons is probably an entirely accurate description of these weapons. The PERSHING II missile, for example, carries a rather small warhead, of not more than 50 kilotons, but with very high accuracy, enabling it to be used against super-hard targets, such as command bunkers. The Soviet-made SS-20 is very much less accurate, but almost certainly was intended for use against NORTH ATLANTIC TREATY ORGANIZATION (NATO) airbases and similar military targets. Neither side has the slightest need to use these theater forces if it wishes to destroy European cities or targets inside the Soviet Union. The phrase "intermediate nuclear forces" has meaning mainly in the context of the INF TREATY which calls for the destruction of both the American Pershing II and the Soviet SS-20.

long supply The situation in which the total quantity of an item of MATERIEL on hand within a military service exceeds the service's mobilized materiel requirement for the item. This situation, when it occurs, requires a further determination about that portion of the quantity in long supply which is to be retained (either as economic retention stock or contingency retention stock), and that portion of the quantity in long supply which is not to be retained (excess stock).

long thrust A BAYONET thrust made with arms extended and the body thrown forward, with the weight shifted to the leading foot.

loose issue stock Military supplies removed from their original containers for issue in small quantities.

loran A long-range radio navigation system that uses the time difference in the reception of pulse-type transmissions from two or more fixed stations for position fixing. This term is derived from the words *lo*ng-*ra*nge electronic *n*avigation.

loss replacement PERSONNEL added to a unit to fill a vacancy due to loss. See also replacement.

lost In ARTILLERY and NAVAL GUNFIRE SUPPORT, a spotting or observation used to indicate that artillery ROUNDS fired were not observed.

lot A quantity of supplies of the same general classification, such as subsistence supplies, clothing, or equipage received and stored at any one time.

lot integrity The perpetual segregation of AMMUNITION by lot number, whether in a storage environment or at the firing site, throughout the life cycle of the ammunition.

lot number The identification number assigned to a particular quantity or lot of material, such as ammunition, from a single manufacturer.

low airburst The FALLOUT-SAFE HEIGHT OF BURST for a NUCLEAR WEAPON that maximizes damage to or casualties at surface target.

low-altitude bombing HORIZONTAL BOMBING with the height of release between 900 and 8000 feet.

low-altitude bombing system In a flight-control system, a control mode in which the low-altitude bombing maneuver of an aircraft is controlled automatically.

low-altitude defense system A generic term for anti-missile defense systems that wait until the last possible moment before destroying incoming missiles; thus, the explosions occur at low altitudes. Also called LoADS.

low-angle fire FIRE delivered at angles of elevation below the elevation that corresponds to the maximum range of the gun and ammunition concerned.

low-angle loft bombing A type of LOFT BOMBING, using free-fall bombs, in which weapon release occurs at an angle less than 35 degrees above the horizontal.

low explosive An explosive that undergoes a relatively slow chemical transformation, thereby producing a deflagration or auto-combustion at rates that vary from a few centimeters per minute to approximately 400 meters per second. It is suitable for use in IGNITER TRAINS and certain types of PROPELLANTS.

low-grade cryptosystem A CRYPTOGRAPHIC system designed to provide temporary security, such as with e.g., combat or operational codes.

low-intensity conflict 1. A small war that does not directly involve the superpowers. 2. A PROXY WAR. 3. INSURGENCY or COUNTERINSURGENCY operations limited to relatively inexpensive conventional WEAPONS. 4. A show of force whereby military forces are dispatched with no real intention to engage them in combat; this is as much a diplomatic as it is a military action. Compare to LIMITED WAR. See Rod Paschall, "Low-Intensity Conflict Doctrine: Who Needs It?" *Parameters: Journal of the US Army War College* Vol. 15, No. 3 (Autumn 1985); John S. Fulton, "The Debate about Low-Intensity Conflict," *Military Review* 66, 2 (February 1986).

low-level flight operations The operation of aircraft at optimum altitudes which afford cover and concealment from ground visual and electronic detection, in order to exploit surprise to the fullest.

low-level navigation The technique of directing an aircraft along a desired course at low altitudes (generally below 500 feet absolute altitude) by using pilotage, dead reckoning, and electronic navigational aids in such a manner that the position of the aircraft is known at any time.

low-level signaling The use of low levels of voltage and current on signal lines, e.g., positive or negative 6 volts plus or minus 1 volt at 1 milliampere or less.

low-order burst 1. The breaking of a PROJECTILE into a few large fragments instead of a large number of smaller fragments, as a result of a LOW-ORDER DETONATION. 2. A malfunctioning artillery SHELL. 3. Anything that does not function as planned.

low-order detonation The incomplete detonation of the explosive charge in a BOMB, PROJECTILE, or other similar HIGH EXPLOSIVE.

low velocity A MUZZLE VELOCITY of an artillery projectile that is 2,499 feet per second or less.

low-visibility operations Sensitive operations wherein the political or military restrictions inherent in COVERT and CLANDESTINE operations are either not necessary or not feasible; actions are taken as required to limit exposure of those involved and their activities. The execution of these operations is undertaken with the knowledge that the action or sponsorship of the operation may preclude plausible denial by the initiating power.

loyalty 1. Allegiance. A loyalty oath is an affirmation of allegiance. Allegiance may be given to a state or to any organization or religious group that may expect special commitment by its members to the interests of the organization or the group. When the interests of such groups are in conflict with one another, individuals who hold membership in more than one of the groups may find themselves, or be considered by others to be, conflicted in their loyalties; hence they may be thought inherently or potentially disloyal to one or the other. 2. A conspiracy for mutual inefficiency. This is B. H. LIDDELL HART's explanation for what an artificial loyalty to one's military unit often amounts to.

LPD See AMPHIBIOUS TRANSPORT DOCK.

LSD See LANDING SHIP, DOCK; also LEAST SEPARATION DISTANCE.

LST See TANK LANDING SHIP.

Lt. An abbreviation for LIEUTENANT.

Lt. Col. An abbreviation for LIEUTENANT COLONEL.

Lt. Gen. An abbreviation for LIEUTENANT GENERAL.

lucrative target Any target that is important enough to either side as to not be a waste of time and materiel to destroy.

lunette 1. A defensive FIELDWORK shaped like a half-moon and open at the rear. 2. The towing ring (or hole) at the trail of a gun carriage.

lyddite A powerful explosive containing picric acid and used in artillery SHELLS. Sometimes called melinite.

LZ An abbreviation for LANDING ZONE.

M

MAAG Military Assistance Advisory Group; an American military mission to a foreign country. It is usually a joint service group representing the Secretary of Defense.

MAC See MILITARY AIRLIFT COMMAND.

machine, blasting See BLASTING MACHINE.

machine, doomsday See DOOMSDAY MACHINE.

machine-gun A fully automatic, rapid-fire SMALL ARM. While the Gatling gun of 1862 was the forerunner of modern versions of the machine gun, it was not fully automatic. The modern machine gun dates from the Maxim gun of 1894. Invented by an American, Hiram Maxim (1840–1916), it and its later versions would have a revolutionary effect upon war. The machine-guns that both sides had during World War I forced millions of men into stalemated trench warfare. A stalemate that was finally broken by the tank, a weapon specifically invented to counter the machine-gun. See John Ellis, *The Social History of the Machine Gun* (New York: Pantheon, 1975).

machine-gun, coaxial See COAXIAL MACHINE-GUN.

machine-gun, heavy See HEAVY MACHINE-GUN.

machmeter An instrument that measures the speed of an aircraft in relation to the speed of sound.

Mach number A number indicating the ratio of the airspeed of an aircraft to the SPEED OF SOUND. Mach 1 is flying at the speed of sound. Mach .9 is ninety percent of the speed of sound. Named after the Austrian physicist Ernst Mach (1838–1916).

MAD Mutual Assured Destruction; a strategic policy of the United States. In 1965, American defense officials concluded that it would be impossible to create a nuclear strategic superiority capable of preventing irreparable damage to the United States in case of enemy attack; therefore the United States would no longer target only enemy military installations, but would also target enemy population centers to maintain a "convincing capability to inflict unacceptable damage on an attacker." See ROBERT S. MCNAMARA. See Samuel F. Wells, Jr., "America and the 'MAD' World," *Wilson Quarterly* (Autumn 1977); Robert Jervis, "MAD is the Best Possible Deterrence," *Bulletin of the Atomic Scientists* Vol. 41, No. 3 (March 1985).

Mae West A brightly colored inflatable life jacket first worn for emergency use by American pilots during World War II flights over water. It was informally named for Mae West (1892–1980), the American film star noted for her full figure.

magazine 1. A detachable part of a FIREARM that holds BULLETS. 2. A place for the storage of AMMUNITION or SUPPLIES.

magazine, box See BOX MAGAZINE.

magazine area A place specifically designed and set aside for the storage of EXPLOSIVES or AMMUNITION.

magazine space The area in a covered structure, above or below ground, constructed for the storage of AMMUNITION and EXPLOSIVES.

Maginot mentality A dogmatic belief in a static defense. The Maginot Line (named for the French Minister of War for most of the 1920s, Andre Maginot) was the fortified defensive line built by the French between the two World Wars to defend against German invasion across their eastern border. When originally built, it was thought to be impregnable. The rapid defeat of the French army in May 1940 by the numerically inferior invading German army has often been blamed on the weakness of the Maginot strategy. Not only did the Maginot Line itself fail but, more generally, the entire attitude of static defense is seen as having made the Allied forces quite unable to cope with a rapidly maneuvering army which exploited the advantages of surprise, initiative, and shock. Many critics of modern NORTH ATLANTIC TREATY ORGANIZATION (NATO) strategy have accused it of suffering from a similar doctrinal malaise, called the "Maginot mentality." Particularly disliked by these critics is the doctrine insisted on by the West Germans, that invading WARSAW PACT forces must be held at the frontier. This, usually called a fixed forward defense, is seen as repeating the French mistake by ceding the initiative to the Warsaw Pact. They would be able to assemble their forces where they liked, break through, and maneuver behind the forward defensive lines, cutting off communications and penetrating deeply into the NATO rear area. Because these tactics worked so well for the Germans in 1940, it is thought that NATO is in danger of re-enacting a basic mistake in the art of war. The criticism is, however, vastly oversimplified, and ignores several major points. First, the Maginot Line itself did not actually fail—no German units even attempted to penetrate it; they simply went around the ends of it or parachuted over it. Second, fixed defense has worked many times in history, and is practiced with great success today by the Israeli army on large parts of their borders. Finally, the demand for forward linear defense is a condition for guaranteeing West German faith in its membership of NATO. The political impossibility of the West German army accepting a doctrine of maneuver warfare that would turn large parts of the Federal Republic of Germany into a battlefield is not difficult to understand. See Judith Hughes, *To the Maginot Line: The Politics of French Military Preparation in the 1920's* (Cambridge, MA: Harvard University Press, 1971); Anthony Kemp, *The Maginot Line: Myth and Reality* (New York: Stein and Day, 1982); Sidney Lens, *The Maginot Line Syndrome: America's Hopeless Foreign Policy* (Cambridge, MA: Ballinger, 1982).

magnetic contour matching A GUIDANCE SYSTEM for missiles that makes use of the Earth's magnetic field for steering toward a target.

magnetic mine A MINE that responds to the magnetic field of a target.

magnetic north The direction indicated by the north-seeking pole of a freely suspended magnetic needle, influenced only by the earth's magnetic field.

Mahan, Alfred Thayer (1840–1914) The single most influential American naval officer in world history—not because of his fighting; but because of his writing. Mahan, who rose to admiral over a 30-year-career, used history to show the relationship between the development of sea power and the relative success of European nations. He used this historical evidence to synthesize general principles relating to geography, territorial configurations, and the nature of government that can be applied regardless of technical changes in weaponry. His major works, *The Influence of Sea Power Upon History 1660–1763* (1890) and *The Influence of Sea Power Upon the French Revolution and Empire, 1793–1812* (1892), became classics and greatly influenced the worldwide naval buildup prior to World War I. Mahan's theories were also a major influence on American desires to fight the 1898 Spanish-American War and to build the Panama Canal. See William E. Livezey, *Mahan on Sea Power* (Norman, OK: University of Oklahoma Press, 1981); Thomas R. Pollock, "The Historical Elements of Mahanian Doctrine," *Naval War College Review* 35, 4 (July–August 1982).

main armament The largest guns installed on a WARSHIP.

main attack The principal ATTACK or effort into which a military commander throws the full weight of the offensive power at his disposal. An attack directed against the chief objective of a CAMPAIGN or BATTLE.

main battle area (MBA) That portion of a BATTLEFIELD in which the decisive part of the battle is fought to defeat an enemy attack.

main battle tank See TANK, MAIN BATTLE.

main body 1. The principal part of a TACTICAL COMMAND or FORMATION. It does not include detached elements of the command such as ADVANCE GUARDS, FLANK GUARDS, or covering forces. 2. In a motor column, all vehicles exclusive of the column head, trail, and control vehicles, which consist primarily of the vehicles carrying the bulk of the cargo or troops with the column.

main defense area An area in which a main defensive battle is fought. For any particular command, this area extends from the FORWARD EDGE OF THE BATTLE AREA to the rear boundaries of those units comprising the main defensive forces of the command.

main guard The regular INTERIOR GUARD of a POST or UNIT, whose principal duties are to patrol the area and protect the personnel, buildings, and equipment. A main guard is a subdivision of the interior guard of a command, the other subdivision being the special guard, such as an ESCORT GUARD or guard of honor.

main line of resistance A line at the FORWARD EDGE OF THE BATTLE AREA, designated for the purpose of coordinating the fire of all units and supporting weapons, including air and naval gunfire. It defines the forward limits of a series of mutually supporting defensive areas, but it does not include the areas occupied or used by COVERING or SCREENING forces.

main road A road capable of serving as the principal ground line of communication to an area or locality. Usually it is wide enough and suitable for two-way, all-weather traffic at high speeds.

main stage That stage of a ROCKET which develops the greatest amount of thrust.

main supply route The route or routes designated within an AREA OF OPERATIONS over which the bulk of traffic flows in support of military operations.

maintain watch To listen continuously on a given radio frequency.

maintenance 1. All action taken to retain MATERIEL in or to restore it to a specified condition. It includes inspection, testing, servicing, classification as to serviceability, repair, rebuilding, and reclamation. 2. All supply and repair action taken to keep a force in condition to carry out its mission. 3. The routine recurring work required to keep a facility (plant, building, structure, ground facility, utility system, or other real property) in such condition that it may be continuously utilized, in its original or designed capacity and efficiency, for its intended purpose. See Kenneth L. Privratsky, "Comparing U.S. and Soviet Maintenance Practices," *Army Logistician* 18, 5 (September–October 1986); James A. Dunn, Jr., "Measuring Maintenance Capability," *Army Logistician* 18, 6 (November–December 1986).

maintenance status 1. A deliberately imposed inactive condition with adequate personnel to maintain and preserve INSTALLATIONS, MATERIEL, and facilities in such a state that they may be readily restored to operable condition in a minimum time by the assignment of additional personnel and without extensive repair or overhaul. 2. That condition of MATERIEL which is in fact, or is administratively classified as, unserviceable, pending the completion of required servicing or repairs.

major The lowest ranking field-grade officer ranking below a LIEUTENANT COLONEL but above a captain; the INSIGNIA is a gold leaf. Typically a major, if a line officer, would be the executive officer to a lieutenant colonel commanding a battalion.

major assembly Any self-contained unit of individual identity. A completed assembly of component parts ready for operation, but utilized as a portion of and intended for further installation in an end item or major item.

major general The two-star general ranking below a LIEUTENANT GENERAL and above a BRIDADIER GENERAL. Because the title was originally sergeant-major general, a modern major general ranks below a LIEUTENANT GENERAL.

major nuclear power Any nation that possesses a nuclear striking force capable of posing a serious threat to every other nation.

make safe One or more actions needed to prevent or interrupt the function of a WEAPONS SYSTEM (traditionally synonymous with "dearm," "disarm," and "disable"). Among these actions are: (1) installation (e.g., safety devices such as pins or locks); (2) disconnection (e.g., hoses, linkages, batteries); (3) removal (e.g., explosive devices such as INITIATORS, FUZES, DETONATORS); (4) intervention (e.g., welding, lockwiring).

management A process of establishing and attaining objectives. Military management consists of those continuing actions of planning, organizing, directing, coordinating, controlling, and evaluating the use of men, money, materials, and facilities to accomplish MISSIONS and tasks. Management is inherent in COMMAND, but it does not include as extensive authority and responsibility as command. The post-World War II U.S. Army has been severely criticized for employing management at the expense of leadership. According to Richard A. Gabriel and Paul L. Savage in *Crisis in Command: Mismanagement in the Army* (New York: Hill and Wang, 1978) this led to the poor performance of American forces in Vietnam because "managerial efficiency instead of 'honor' becomes the standard of performance. The managerial disposition undermines, it seems to us, the sense of military honor commonly associated with unit integrity. Inasmuch as the latter is involved with 'profitless' personal sacrifice, a managerially oriented commander may come to see his troops as a resource to be used to advance his own career rather than as a moral charge, placed upon his honor and rested ultimately in reciprocal trust and self-sacrifice." See also CAREERISM.

management headquarters A HEADQUARTERS primarily concerned with long-range planning, programming, and budgeting of resources; the development of policy and procedures; coordination of effort, and evaluation, as opposed to the planning for and direct control of operations.

management system An integrated group of procedures, methods, policies, practices, and personnel used by a commander or other supervisor in planning, organizing, directing, coordinating, and controlling an organization.

mandrel 1. A mold used for shaping CARTRIDGE CASES. 2. The inner configuration of solid propellants used in ROCKET MOTORS.

maneuver 1. A movement to put ships or aircraft in a position of advantage over the enemy. 2. A tactical exercise carried out at sea, in the air, on the ground, or on a map in imitation of war. While maneuvers are critically important elements of training, it is important to remember Barbara Tuchman's observation in *The Guns of August* (1962): "Human beings, like plans, prove fallible in the presence of those ingredients that are missing in maneuvers—danger, death, and live ammunition." 3. The operation of a ship, aircraft, or vehicle, to cause it to perform desired movements. 4. The employment of forces on the battlefield through movement in combination with FIRE, or fire potential, to achieve a position of advantage over the enemy in order to accomplish a MISSION (compare to PRINCIPLES OF WAR). Maneuver warfare must be compared to ATTRITION warfare, which relies on numerical and materiel superiority to batter through the enemy lines by killing troops and destroying equipment in a more or less head-on attack. In contrast, maneuver warfare depends on the skill of commanders rather than the sheer force of the UNITS under them. As the name suggests, it depends on outmaneuvering the enemy: moving rapidly to exploit temporary weak points, pouring troops through gaps in a line, attacking from the flank or from behind, and continually seeking to surprise the enemy. Rather than physically destroying the enemy, the idea is to throw him off balance, attacking where least expected, and cutting lines of communication, information, and supply. Thus the enemy is prevented from using his forces in a decisive way. A model for such tactics is the German BLITZKRIEG, which proved so successful in the battle for France in 1940. The German armored divisions advanced far more rapidly than the Allies expected, avoiding the French and British strong points and penetrating deep behind the "official lines." Allied armies were cut off from each other and isolated, with the Allied high command being unable even to know quite where the Germans were, still less to launch effective COUNTERATTACKS. See John J. Mearsheimer, "Maneuver, Mobile Defense, and the NATO Central Front," *International Security* 6 (Winter 1981–1982); William S. Lind, *Maneuver Warfare Handbook* (Boulder, CO: Westview Press, 1985); David A. Fastabend and Ralph H. Graves, "Maneuver, Synchronization and Obstacle Operations," *Military Review* 66, 2 (February 1986).

maneuver, forms of See FORMS OF MANEUVER.

maneuver, free See FREE MANEUVER.

maneuver forces, combat See COMBAT MANEUVER FORCES.

maneuverable re-entry vehicle (MARV) The next stage in sophisticated warhead design after MULTIPLE INDEPENDENTLY TARGETABLE RE-ENTRY VEHICLES (MIRVS). While a MIRVed missile can deliver several WARHEADS against separate targets, once ejected from the missile's BUS (or front end), they follow predetermined and predictable courses. A MARVed missile, on the other hand, would eject as many warheads, each also destined for a separate pre-selected target, but which would follow a variable and changing course, making them much harder for BALLISTIC MISSILE DEFENSES to destroy.

maneuvering area That part of an airfield used for takeoffs, landings, and associated maneuvers.

maneuvering force An element of a COMBAT UNIT that seeks to seize an attack OBJECTIVE through movement to a more advantageous position with respect to the enemy.

Manhattan Project The code name for a federally financed research project during World War II that resulted in the development of the ATOMIC BOMB. This project is generally referred to as the first major involvement with science by the federal government in a policymaking role. The Manhattan Project cost about two billion dollars then; equivalent to about ten billion dollars in today's money. See Leslie R. Groves, *Now It Can Be Told: The Story of the Manhattan Project* (New York: Harper & Row, 1962); Richard Rhodes, *The Making of the Atomic Bomb* (New York: Simon & Schuster, 1988).

manipulative communications cover Those measures taken to alter or conceal the characteristics of COMMUNICATIONS so as to deny to an enemy (or potential enemy) the means to identify them.

manipulative communications deception The alteration or simulation of friendly telecommunications for the purpose of DECEPTION.

manipulative electronic deception The manipulation of friendly electromagnetic radiations by measures such as traffic leveling, providing false traffic peaks, and padding traffic to deceive the enemy as to the intentions of friendly units.

manning level A PERSONNEL ceiling imposed against normally authorized troop strengths owing to the reduced availability of personnel because of limited procurement, funding, or other factors.

manning table A chart that gives a list of the PERSONNEL of an organization and shows the duties to which each person is assigned.

man on horseback 1. A military figure, such as NAPOLEON, who becomes a dictator when a civilian regime falters. 2. Any former military figure (usually a general) who aspires to civilian leadership. In American politics, the phrase was first applied to Ulysses S. Grant during his 1868 presidential campaign. See Samuel E. Finer, *The Man on Horseback: The Role of the Military in Politics* (New York: Praeger, 1962).

man-portable Capable of being carried by one man. In land warfare, equipment that can be carried by one man over a long distance without seriously interrupting the performance of his usual duties.

man space The space and weight factor used to determine the capacity of vehicles, craft, and transport aircraft, based on the requirements of one person with individual equipment. The person is assumed to weigh between 222 and 250 pounds, and to occupy 13.5 cubic feet of space.

man transportable Items that are usually transported on wheeled, tracked, or air vehicles, but can be handled by one or more individuals for limited distances (100–500 meters). That upper weight limit is approximately 65 pounds per individual.

manual 1. Something to be done by hand as opposed to electronically. 2. A formally prescribed drill for handling something. For example, the "manual of arms" prescribes the formal movements for the rifle at drills and ceremonies. There are also manuals for the color, the guidon and the sword.

manufacturer, military See MILITARY MANUFACTURER.

map A graphic representation, usually on a plane surface, and at an established scale, of natural or artificial features on the surface of a part or the whole of the earth or another planetary body. The features are positioned relative to a coordinate reference system.

map, administrative See ADMINISTRATIVE MAP.

map, contour See CONTOUR MAP.

map, hypsographic See HYPSOGRAPHIC MAP.

map, provisional See PROVISIONAL MAP.

map, situation See SITUATION MAP.

map chart A representation of a land-sea area, using the characteristics of a map to represent the land area and the characteristics of a chart to represent the sea area, with such special characteristics as to make the map-chart most useful in military operations, particularly amphibious operations.

map compilation 1. The preparation of a new map, less final drafting and reproduction, from air photography, existing maps, charts, and other source materials used either singly or in combination. 2. The map drawing containing all information to be shown on the finished map ready for final drafting for reproduction.

map convergence The angle at which one meridian is inclined to another on a map or chart.

map exercise A MAP MANEUVER.

map K A proportional correction for the discrepancy between the scale of a FIRING CHART and that of the plotting scale being used.

map maneuver An exercise in which military OPERATIONS with opposing sides are conducted on a map, the troops and the military establishments being represented by markers or symbols that are moved to represent the maneuvering of the troops on the ground.

map orientation The act of placing a map so that its north lines point to the corresponding true north direction.

map reconnaissance The study of ground features on a map, such as roads, woods, and waterways, to obtain the information needed in preparing a TACTICAL PLAN or MANEUVER.

map reference A means of identifying a point on the surface of the Earth by relating it to information appearing on a map, generally the graticule or grid.

map scale The relationship between distance on a map and distance on the ground expressed as a ratio; e.g., 1:25,000 or 1/25,000 means that one inch on the map equals 25,000 inches on the ground. A large-scale map covers a lesser area (a 1/25,000 map is a larger scale map than a 1/100,000 map).

map series A group of maps or charts usually having the same scale and cartographic specifications, and with each sheet appropriately identified by the producing agency as belonging to the same series.

map sheet An individual map or chart either complete in itself or part of a series. British General Sir William Slim wrote in his *Unofficial History* (1962): "While the battles the British fight may differ in the widest possible ways, they have invariably two common characteristics—they are always fought uphill and always at the junction of two or more map sheets."

march, forced See FORCED MARCH.

march-collecting post A location on a route of march at which CASUALTIES who cannot continue to march are given medical treatment and are moved to medical stations in the rear.

march column All elements of a FORCE using the same route for a single movement under the control of a single commander. Whenever possible, a force marches over multiple routes to reduce CLOSURE TIME. A large column may be composed of a number of subdivisions, each under the control of a subordinate commander. March columns, regardless of size, are composed of three elements: a head (the first vehicle of the column, which normally sets the pace), a main body (made up of the major elements of column SERIALS and march UNITS), and a trail element (including personnel and the equipment necessary for emergency vehicle repair and recovery, medical aid, evacuation, and so on).

marching fire An INFANTRY tactic that calls for a unit to advance in one thin SKIRMISH LINE while firing at anything that could be a potential danger.

march outpost Either an OBSERVATION POST or PATROL established for the protection of a command during a halt in a march.

march unit A subdivision of a SERIAL; normally a SQUAD, SECTION, PLATOON, COMPANY, TROOP, or BATTERY. It moves and halts under the control of a single commander using voice commands, visual signals, or radio signals when no other means or communication can be used.

marginal weather Weather that is sufficiently adverse to a military operation as to require the imposition of procedural limitations.

marines 1. Troops used for service at sea and for AMPHIBIOUS OPERATIONS. 2. The U.S. Marine Corps, part of the U.S. Navy. See George Custance, "A New Role for Soviet Marines," *Marine Corps Gazette* (December 1981); Karl C. Lippard, *The Warriors: The United States Marines* (Novato, CA: Presidio Press, 1983).

marines, tell that to the A phrase indicating that something is unbelievable;

that marines are too smart or experienced to believe it. Yet the phrase, which originated in the British navy centuries ago, originally meant just the opposite; that the marines were so stupid that they would believe almost anything. In the days of press gangs, when marines functioned much like prison guards, sailors tended to grow contemptuous of them and, right or wrong, spread stories of the low intellectual capacity of the British marines.

maritime strategy 1. A force-PROJECTION capability based on naval power and sea-lift capability. A maritime strategy is necessarily adopted by any power that has a substantial navy and widely scattered military and commercial interests. 2. A master plan for the use of sea power in peace and war. See Robert W. Komer, "Maritime Strategy vs. Coalition Defense," *Foreign Affairs,* 60, 5 (Summer 1982); Norman Polar and Scott C. Truver, "The Maritime Strategy," *Air Force,* 70, 11 (November 1987).

mark 1. In ARTILLERY AND NAVAL GUNFIRE SUPPORT, a call for fire on a specified location to orient the OBSERVER or SPOTTER or to indicate targets. 2. In CLOSE AIR SUPPORT and AIR INTERDICTION, an air-control-agency term used to indicate the point of WEAPON release. It is usually preceded by the word "stand by" as a preparatory command.

marker circle A circular band marking the approximate center of a LANDING AREA or the intersection of the principal landing strips in an airport or landing field.

marker ship In an AMPHIBIOUS OPERATION, a ship that takes accurate STATION on a designated CONTROL POINT. It may fly identifying flags by day and show lights to seaward by night.

marking fire FIRE placed on a target for the purpose of identification.

marking panel A sheet of material displayed for visual communication, usually between friendly units. See also PANEL CODE.

marking team A group of personnel landed in a LANDING AREA with the task of establishing navigational aids. See also PATHFINDERS.

mark mark A command from a ground controller for aircraft to release their bombs; it may indicate electronic ground-controlled release or a voice command to the aircrew.

marksman 1. A soldier who can accurately use FIREARMS. 2. The lowest level of attained proficiency with a RIFLE; the next levels are SHARPSHOOTER and EXPERT.

mark target An order to a marker, in target practice, to mark the location of a shot on a target.

mark time A command that requires the feet to be moved as in marching, but without advancing.

marshal 1. The highest ranking officer in some armies; sometimes comparable to a GENERAL with equivalent graduations, sometimes equivalent to a five-star general in the United States Army. A field marshal is usually the highest grade of marshal except in France where it is second to a Marshal of France. 2. See MARSHALLING.

marshal, air See AIR MARSHAL.

marshalling 1. The process by which units participating in an AMPHIBIOUS or AIRBORNE OPERATION group together, assemble when feasible, or move to temporary camps in the vicinity of embarkation points, complete preparations for combat, or prepare for loading. 2. The process of assembling, holding, and organizing supplies, equipment, or both, especially transportation vehicles for onward movement. See also STAGING AREA.

marshalling area 1. The general area in which unit-preparation areas and departure airfields may be located and from which air movement of a military force is initiated. 2. In AMPHIBIOUS OPERATIONS, the designated area in which, as part of the MOUNTING phase, units are reorganized for embarkation; vehicles and equipment are prepared to move directly to embarkation areas; and housekeeping facilities are provided for troops by other units.

marshalling plan An airborne operational plan detailing the process by which units of the force complete final preparations for combat, move to departure airfields, and load for takeoff. It begins when elements of the force are literally "sealed" in marshalling camps, and terminates at loading.

Marshall Plan The economic aid program for post-World War II Europe proposed by George Catlett Marshall (1880–1959), U. S. Army Chief of Staff during World War II. In 1947, President Harry S. Truman made him Secretary of State. In June of that year he proposed the European Recovery Program, a massive aid program that became known as the Marshall Plan. The plan worked so well and became so well known that the phrase entered the language, meaning any massive use of federal funds to solve a major social problem. See Donald Hester, "The Marshall Plan: A Study of U. S. Interests, Values and Institutions," *Orbis* Vol. XVIII, No. 4 (Winter 1974); Michael J. Hogan, *The Marshall Plan: America, Britain, and the Reconstruction of Western Europe* (New York: Cambridge University Press, 1987).

Marshall, S. L. A. See FIRE DISCIPLINE.

martial law 1. The law exercised over an OCCUPIED TERRITORY by military units of the occupying power. 2. The exercise of partial or complete military control over police powers in domestic territory in time of emergency because of public necessity. In the United States, it is usually authorized by the President, but may be imposed by a military commander in the interests of public safety. 3. Arbitrary military rule imposed not by constitutional means, but by force. See Stanley Robert Rankin, *When Civil Law Fails: Martial Law and Its Legal Basis in the United States* (New York: AMS Press, 1965).

martinet A strict disciplinarian. The word comes from an inspector general in the army of France's Louis XIV, Jean Martinet, who was so despised for his spit-and-polish discipline that he was "accidentally" killed by his own soldiers while leading an assault in 1672.

MARV See maneuverable RE-ENTRY vehicle

Marxism The doctrine of REVOLUTION based on the writings of Karl Marx (1818–1883) and Friedrich Engels (1820–1895), which maintains that human history is a history of struggle between the exploiting and exploited classes. They wrote the *Communist Manifesto* (1848) "to do for history what Darwin's theory has done for biology." The basic theme of Marxism holds that the proletariat will suffer so from alienation that they will rise up against the bourgeoisie who own the means of production, or overthrow the system of capitalism. After a brief period of rule by the DICTATORSHIP OF THE PROLETARIAT, the classless society of communism would be forthcoming. As far as the actual doctrines of communist societies go, it is probably better to talk of Marxist-Leninism, because Lenin had such a major impact in the process of turning a general theory into a practical doctrine for revolutionaries and subsequent post-revolutionary governments. While Marxism currently has a strong influence on the economies of the SECOND, THIRD, and FOURTH WORLDS, its intent has never been fully achieved. Indeed, because Marx's writings are so vast and often contradictory, serious Marxists spend considerable time arguing about just what Marx "really" meant. Marx's magnum opus, *Das Kapital* (1867), is frequently referred to as the "bible of socialism." See Alfred Meyer, *Marxism: The Unity of Theory and Practice—A Critical Essay* (Cambridge, MA: Harvard University Press, 1959); Sholomo Avineri, *The Social and Political Thought of Karl Marx* (Cambridge, England: Cambridge University Press, 1968); Tom Bottomore, editor, *Dictionary of Marxism* (Cambridge, MA: Harvard University Press, 1985).

mask See PROTECTIVE MASK.

mask clearance 1. The absence of any obstruction in the path of a TRAJECTORY. 2. The amount of clearance by which a PROJECTILE passes over any object between the weapon that fires it and its target.

masking In ELECTRONIC WARFARE, the use of additional transmitters to hide the location or source of a particular electromagnetic radiation, or purpose of the radiation.

mass 1. The concentration of COMBAT POWER. 2. The military FORMATION in which units are spaced at less than the normal distances and intervals. 3. To concentrate or bring together, as to mass the FIRE of multiple weapons or units. Also see PRINCIPLES OF WAR.

mass casualties 1. Any large numbers of CASUALTIES produced in a relatively short period of time and which far exceed local LOGISTIC SUPPORT capabilities. 2. Those unreplaced soldiers who are killed, wounded, or ill; and, because of their numbers or duties, reduce the strength and effectiveness of a military unit to a point at which the success of its mission is doubtful.

massed fire 1. The fire of the BATTERIES of two or more ships directed against a single target. 2. Fire from a number of WEAPONS directed at a single point or small area. NAPOLEON wrote that: "In a battle like in a siege, skill consists in converging a mass of fire on a single point: once the combat is opened, the commander who is adroit will suddenly and unexpectedly open fire with a surprising mass of artillery on one of these points, and is sure to seize it." See also CONCENTRATED FIRE.

mass formation A formation of a COMPANY or any larger unit in which the SQUADS standing in COLUMN are abreast of one another.

massive retaliation The strategic doctrine adopted and publicly announced by the Eisenhower Administration in 1954 as the basis for America's support not only for the North Atlantic Treaty Organization (NATO) but for all allies the United States agreed to defend. It effectively replaced the conventional concept of defense with deterrence and retribution. Instead of promising a commitment of troops to defend an American ally invaded by the Soviet Union or one of its client states, the new doctrine relied entirely on the threat of strategic nuclear force. In theory, any real incursion over an ally's border would be met immediately with a major and devastating nuclear strike not on the enemy's troops, but on the Soviet homeland. There were two main problems with the massive retaliation doctrine. First, it lacked credibility. Quite apart from any Soviet capacity to retaliate, it seemed to many highly improbable that the United States would kill millions of Soviet civilians to retaliate against a possibly minor border incursion by a minor ally of the Soviet Union. Such a nuclear strike was not even considered very likely if the Soviet army were to expel the West from Berlin. Second, the doctrine could only work as long as the United States had a near monopoly in nuclear INTERCONTINENTAL BALLISTIC MISSILES. As soon as the Soviet Union could develop the capacity to destroy a few American cities, the threat of massive retaliation would weaken, and once the Soviet Union gained anything like nuclear parity with the West it would be totally self-defeating. Massive retaliation was abandoned overtly in 1967 when the United States persuaded NATO to accept the FLEXIBLE RESPONSE doctrine. See Samuel F. Wells, Jr., "The Origins of Massive Retaliation," *Political Science Quarterly* Vol. 96, No. 1 (Spring 1981).

master depot A depot on a military base that is delegated responsibility for accounting for and controlling the distribution of all supplies of the class or type assigned to it for an entire theater of operations or major section thereof. A master depot becomes the theater stock control point for the designated items.

master sergeant A high-ranking NONCOMISSIONED OFFICER.

master station That station in a given system of transmitting stations that controls the transmissions of the other stations (the slave stations) and maintains the time relationship between the pulses sent by these stations.

material, active See ACTIVE MATERIAL.

material, bleaching See BLEACHING MATERIAL.

materiel All items (including ships, tanks, self-propelled weapons, aircraft, etc., and related spare parts, repair parts, and support equipment, but excluding real property, installations, and utilities) necessary to equip, operate, maintain, and support military activities without distinction as to its application for administrative or combat purposes.

materiel, on-equipment See ON-EQUIPMENT MATERIEL.

materiel and powder report A report on the performance of WEAPONS and AMMUNITION.

materiel damage The damage to MATERIEL caused by hostile acts. It is classified by degrees: light, moderate, or severe.

materiel pipeline The quantity of an item required in the worldwide supply system to maintain an uninterrupted replacement flow of it to military forces.

materiel readiness The availability of MATERIEL required by a military organization to support its wartime activities or contingencies (disaster relief, flood, earthquake, etc.), or other emergencies.

materiel requirements Those quantities of items of equipment and supplies necessary to equip, provide a MATERIEL pipeline, and sustain a SERVICE, FORMATION, ORGANIZATION, or UNIT in the fulfillment of its purposes or tasks during a specific period.

maximum effective range The maximum distance at which a WEAPON may be expected to be accurate and achieve its desired result.

maximum gradeability The steepest slope that a vehicle can negotiate in low gear. This is usually expressed in terms of the percentage of slope: namely, the ratio between the vertical rise and the horizontal distance traveled. It is sometimes expressed by the degree of angle between the slope and the horizontal.

maximum-issue quantity The maximum quantity of an item approved for issue per requisition; quantities above this maximum must be reviewed by supply personnel.

maximum obstacle elevation figure A figure shown in designated areas on aeronautical charts to indicate the minimum altitude needed for an aircraft to clear any possible vertical obstructions.

maximum ordinate In ARTILLERY AND NAVAL GUNFIRE SUPPORT: 1. the highest point along the TRAJECTORY of a projectile; 2. the difference in altitude (vertical interval) between the origin and the summit of the trajectory of a projectile.

maximum permissible dose That RADIATION DOSE which a military commander or other appropriate authority may prescribe as the limiting cumulative radiation dose to be received over a specific period of time by members of his command, consistent with current operational military considerations.

maximum range The greatest distance a weapon can fire with accuracy without consideration of DISPERSION.

maximum security institution The designation given to DISCIPLINARY BARRACKS in which more serious offenders are usually confined, and so constructed as to reduce the possibility of escape of prisoners to a minimum.

maximum sustained speed The highest speed at which a transport vehicle, with its rated payload, can be driven for an extended period on a level first-class highway without sustaining damage.

maximum thermometer A thermometer in which the mercury, or the indicator used for registering temperature, remains at the highest point it has reached since its last setting.

mayday An international distress call for aircraft and ships.

MBA 1. MAIN BATTLE AREA. 2. Master of Business Administration.

McNamara, Robert S. The Secretary of Defense under presidents John F. Kennedy and Lyndon B. Johnson, from 1961 to 1968. It was McNamara who pushed the PENTAGON into developing the strategy of MUTUAL ASSURED DESTRUCTION, and abandon its goal to always maintain more nuclear power than the Soviet Union. Due to his efforts, a new emphasis was placed on building and retaining a secure SECOND-STRIKE capacity at a level that would guarantee the destruction of the enemy if he were to initiate any nuclear action. McNamara's other major change, although it took much longer to enforce, was to shift the NORTH ATLANTIC TREATY ORGANIZATION (NATO) from reliance on MASSIVE RETALIATION, under which NATO conventional forces were regarded as TRIPWIRES. Although his new doctrine of flexible response was first suggested in 1962, he did not manage to persuade NATO formally to accept it until 1967, and he was never truly successful in pressuring European NATO members to increase their conventional force strength to a level at which a strategic reliance on the early use of nuclear weapons could be abandoned. He was equally famous in American defense circles for importing a new management style into the Pentagon, often referred to as systems analysis. Weapons programs and even military tactics were subjected to a combination of cost analysis and operations research more familiar in the manufacturing industry (see PLANNING PROGRAM BUDGETING SYSTEMS). In recent years, McNamara has again come to the fore in nuclear debate by joining with other leading strategists of the 1960s in calling for NATO to abandon its doctrine of first use of nuclear weapons. See James M. Roherty, *Decisions of Robert S. McNamara: A Study of the Role of Secretary of Defense* (Coral Cables, Florida: University of Miami Press, 1970); Clark A. Murdock, *Defense Policy Formation: A Comparative Analysis of the McNamara Era* (Albany: State University of New York Press, 1974).

meaconing Transmission by the enemy of false navigational signals to confuse or hinder the navigation of aircraft and ships and to confuse ground stations.

meal basis of issue Edible items issued for a given day containing the quantities of food needed for the number of individuals expected to consume each meal.

meal ticket A government voucher authorizing a public eating place to furnish meals up to a certain price limit to the person or persons named, and to charge the cost to the government.

mean point of impact The point whose coordinates are the arithmetic means of the coordinates of the separate POINTS OF IMPACT or burst of a finite number of artillery PROJECTILES fired or released at the same AIMING POINT under a given set of conditions.

mean sea level The average height of the surface of the sea for all stages of the tide, used as a reference for the elevations of geographic features.

means of communications A medium by which a message is actually conveyed from one person or place to another. It includes radios of all types, wire lines, messengers, mail, and visual or sound-signaling devices.

measurement and signature intelligence Information obtained by the quantitative and qualitative analysis of data (metric, angle, spatial, wavelength, time-dependence, modulation, plasma, and hydromagnetic) derived from specific technical sensors for the purpose of identifying any distinctive features associated with the source, emitter, or sender of the data, and to facilitate subsequent identification or measurement of the data. See also ELINT.

mechanical sweep In naval mine warfare, any SWEEP used with the object of physically contacting a mine or its appendages.

mechanical time fuze A FUZE with a clocklike mechanism that controls the time at which it will go off.

mechanized A description of any ground unit that has considerable use of tanks, armored fighting vehicles, armored personnel carriers, trucks, and other means of transport. Mechanized is in contrast to motorized, which has motor transportation but no armor.

mechanized division A ground force in which all elements travel in and fight from TANKS or ARMORED FIGHTING VEHICLES.

mechanized infantry division A mechanized DIVISION that has a greater proportion of INFANTRY compared to other elements such as ARMOR.

Medal of Honor The United States's highest decoration, awarded in the name of Congress for conspicuous heroism in combat at the risk of life. The heroic act must be such that its omission would not have justly subjected the recipient to censure for failure to perform his duty. Often incorrectly referred to as the Congressional Medal of Honor. See also ROSETTE. See Joseph A. Blake, "The Congressional Medal of Honor in Three Wars," *Pacific Sociological Review* 16, 2 (April 1973).

median incapacitating dose The amount or quantity of a CHEMICAL AGENT which when introduced into the body will incapacitate 50 percent of exposed, unprotected personnel.

median lethal dose 1. The amount of RADIATION over the whole body that would be fatal to 50 percent of the exposed personnel in a given period of time. 2. The dose of a CHEMICAL AGENT that would kill 50 percent of exposed, unprotected, and untreated personnel. It is expressed in milligram minutes per cubic meter.

medic Any member of a military medical corps. Thus a medic could be a physician or, more commonly, an enlisted man trained in first aid who may accompany troops in battle. The shout of "medic" is a call for first aid. Compare to CORPSMAN.

medical support The provision of health services in support of an army in the field.

medical treatment, emergency See EMERGENCY MEDICAL TREATMENT.

medical warning tag A tag that serves as a means of rapid recognition of selected health problems when military personnel records are not immediately available and the individual requiring treatment is unable to give a medical history.

medium-altitude bombing Horizontal bombing with the height of bomb release between 8,000 and 15,000 feet.

medium-angle loft bombing A type of LOFT BOMBING wherein bomb release occurs at an angle between 35 and 75 degrees above the horizontal.

medium atomic-demolition munition A low-YIELD, team-portable, ATOMIC DEMOLITION MUNITION that can be detonated either by remote control or by a timer device.

medium-range ballistic missile A BALLISTIC MISSILE with a range capability from about 600 to 1,500 nautical miles.

medium-range bomber aircraft A BOMBER designed for a tactical operating radius of under 1,000 nautical miles at design gross weight and design bomb load.

medium-range radar Radar equipment whose maximum range on a reflecting target of 1 square meter normal to the signal path exceeds 240 kilometers but is less than 480 kilometers, provided a LINE OF SIGHT exists between the target and the radar.

medium-security institution The designation given to DISCIPLINARY BARRACKS or rehabilitation centers in which less serious offenders may be committed for confinement, and usually enclosed with a fence rather than a wall. The prisoners are normally housed in barracks or dormitories. Compare to a MAXIMUM SECURITY INSTITUTION.

meeting engagement A COMBAT action that occurs when a moving FORCE, completely deployed for battle, engages an enemy at an unexpected time and place.

meet your waterloo To suffer a final defeat. The last battle of NAPOLEON was at Waterloo in 1815, where he lost to the British Duke of Wellington and the Prussian Marshal Blucher.

mega A prefix meaning one million.

megaton The amount of a conventional explosive, such as TNT, that would have to be exploded to produce the same energy release as the YIELD of a particular NUCLEAR WEAPON. A megaton-level detonation is equivalent to exploding one million tons of TNT, a kiloton being equivalent to one thousand tons. When it is remembered that the typical terrorist car bomb contains at most a few hundred pounds of conventional explosive, some idea may be gained of the magnitude of these figures. Extensive calculations have been made about the likely impact of megaton-level weapons, although there is obviously little direct experience. The atomic bombs dropped on Hiroshima and Nagasaki in 1945 were in the range of 10–20 kilotons, whereas the typical reentry vehicle in a missile equipped with MULTIPLE INDEPENDENTLY TARGETABLE REENTRY VEHICLES (MIRVS) carries at least 150 kilotons.

megatonnage, equivalent See EQUIVALENT MEGATONNAGE.

megaton weapon A NUCLEAR WEAPON whose yield is measured in terms of millions of tons of trinitrotoluene (TNT) explosive equivalents. See also KILOTON WEAPON; NOMINAL WEAPON; SUBKILOTON WEAPON.

melinite A powerful explosive for shells similar to LYDDITE, made by combining picric acid with guncotton.

Mercator chart A map based upon a Mercator projection and widely used for surface and air navigation. The Mercator projection, named for Gerhardus Mercator, the 16th century Flemish cartographer, tends to distort (by increasing) the size of land areas in the southernmost and northernmost latitudes.

mercenary A professional soldier who will fight for whoever pays him, without any particular regard to ideological concerns. Modern mercenaries range from hired killers who dub themselves with the name because it makes what they do sound more legitimate, to former members of some of the world's best armies, who tend to work for the revolutionary or counterrevolutionary movements of the THIRD WORLD. See Grant E. Courtney, "American Mercenaries and the Neutrality Act: Shortening the Leash on the Dogs of War," *Journal of Legislation* Vol. 12, No. 2 (Summer 1985); Anthony Mockler, *The New Mercenaries* (New York, NY: The Paragon House Publications, 1987).

merchant of death 1. A private business that SELLS ARMS and MUNITIONS in quantities suitable for military use. The term was widely used in congressional investigations after World War I. 2. A government that sells military supplies and equipment. See Robert H. Ferrell, "The Merchants of Death, Then and Now," *Journal of International Affairs* Vol. 26, No. 1 (1972).

mercuric fulminate (fulminate of mercury) An high-explosive initiator that is detonated by heat, impact, or friction. (To a large extent, being replaced by LEAD AZIDE.)

mess A place where food is prepared and served for military personnel; originally a group who cooked and ate their rations together.

message indicator A group of symbols usually put at the beginning of the text or an encrypted message or transmission, which identifies or governs the arrangement of the CRYPTOVARIABLES applicable to the message or transmission. See also CRYPTOGRAPHY.

message parts The result of the division of a long message into several shorter messages of different lengths as a transmission security measure, or to comply with communications requirements. Message parts must be prepared in such a manner as to appear unrelated externally; statements identifying the parts are encrypted in the texts. See ENCRYPTION/ENCRYPTED TEXT.

mess dress Formal wear for officers.

messkit Eating equipment that a soldier uses in the field; MESS gear. It includes a knife, fork, spoon, cup, and meat can.

metalled road A road constructed of gravel, crushed stone, slag, or similar material with a binder of fine aggregate tar or cement.

metascope A hand-carried device designed for locating a source of infrared rays; used for night vision.

meters error An error in FIRING DATA expressed in meters; a linear deviation from the target at a POINT OF IMPACT, measured in terms of meters.

method of resupply The means by which a UNIT makes its requirements for supply known to the issuing installation. Examples are requisitions, automatic, on call, status report, and expenditure report.

MIA See MISSING IN ACTION.

midcourse guidance The guidance applied to a MISSILE between termination of the launching phase and the start of the terminal phase of the missile's flight.

midcourse phase That portion of the trajectory of a BALLISTIC MISSILE between the BOOST PHASE and the RE-ENTRY phase. In the first or boost phase the ROCKET engines of the missile are burning and the missile is propelled by powered flight. As the engines cut out, the missile enters the longest period of its flight, the midcourse phase, during which it is coasting on a predetermined BALLISTIC TRAJEC-

TORY outside the Earth's atmosphere. If the points of launch and impact were in the United States and the Soviet Union, respectively, or vice versa, this phase would last about 20 minutes. The missile is effectively "in orbit" at this stage. This midcourse phase, before a missile with MULTIPLE INDEPENDENTLY TARGETABLE REENTRY VEHICLES (MIRVS) ejects its WARHEADS for their descent during the terminal phase, is the prime opportunity for BALLISTIC MISSILE DEFENSES (BMDs) to operate. A single hit will destroy the entire warhead load, and the missile is in a predictable path for a long enough time to allow for the tracking and aiming of interceptor devices. At the same time, because it is at its maximum height, hitting the missile requires the development of new defensive technology such as the STRATEGIC DEFENSE INITIATIVE. See also TERMINAL PHASE.

Midgetman The popular name given to an American plan to build what is more formally referred to as a Small Intercontinental Ballistic Missile (SICBM). The Scowcroft Commission recommended it as the next generation of American missiles, after the deployment of the MX. The MIDGETMAN is designed both to be invulnerable to a FIRST STRIKE, but also not to give the Soviet Union any reason to fear that it is itself intended for a first strike. The MX in contrast looks very much like a first-strike weapon but fails, given the problem of basing modes, to be invulnerable. Midgetman would achieve invulnerability by being mobile enough to be capable of being driven all around North America. Missiles would be so widely dispersed that the SICBM batteries would not constitute a reliable target for a Soviet first strike. The small size of the missile, which enables this mobility, gives it its nickname. Its size is also meant to reassure the Soviet Union that it is purely defensive: the SICBM is intended to be a single WARHEAD missile. See Jeffrey R. Smith, "Midgetman Missile Plans Generate Political Debate," *Science,* 232, 2 (June 6, 1986); Albert Gore, Jr., "Let's Develop the Midgetman Now" *Technology Review,* 89, 3 (November–December 1986).

midpoint In gunnery, the point on an air target's course that is at a minimum SLANT RANGE from the gun position.

midshipman 1. A student training for a commission at the United States Naval Academy. Comparable to a cadet at West Point. 2. A student training for a commission at the United States Coast Guard Academy. 3. A student in a Navy ROTC program at a civilian university. 4. In the Royal Navy and others that follow the British practice, a noncommissioned officer. 5. In the last century, a boy serving an apprenticeship aboard a warship; their battle station was "amidships," thus the name.

mil 1. A unit of measurement for angles based on the angle subtended by 1/6400 of the circumference of a circle. A mil is the angle subtended by one unit at one thousand units. 2. 1/1000 of an inch (wire measurement).

mile, nautical See NAUTICAL MILE.

mil formula A mil relation used in gunnery; expressed by M W/R, where M is the angular measurement in mils between two points, W is the lateral distance in meters between the points, and R is the mean distance to the points in thousands of meters. The mil relation is approximately true for angular measurements of less than 600 mils.

mil-gridded oblique An air photograph taken with a camera tilted below the horizontal, on which has been printed an angle-measuring GRID.

militarily significant fallout Radioactive contamination capable of inflicting RADIATION DOSES that may result in a reduction of the combat effectiveness of military personnel.

militarism 1. A term used to denote a state's policy of gaining its ends by overt or threatened use of military force. As such, in the heavily ideologically loaded language in which such concepts are couched, the nation would be a threat to world peace, and any armaments increase it carried out would be evidence of its "militaristic" tendencies. President Woodrow Wilson told the 1916 graduating class at West Point that: "Militarism does not consist in the existence of any army, nor even in the existence of a very great army. Militarism is a spirit. It is a point of view. It is system. It is a purpose. The purpose of militarism is to use armies for aggression." 2. A political culture or ideology's exaltation of military values, patriotism, and the associated group behavior or symbols. Obviously, states that are militaristic are likely to exhibit militarism in much of their social symbolism, such as with the general love of uniforms in Nazi Germany. 3. A government and its society dominated by a military class and its ideals. See Volker Rolf Berghahn, *Militarism: The History of an International Debate, 1861–1979* (New York: Cambridge University Press, 1984).

military 1. Of or pertaining to war, or the affairs of war, whether on land, sea, or in the air. 2. The whole organization of defensive and offensive arms and armed forces in a society; the armed forces and the civil service and political direction of them. 3. Land as opposed to naval or air forces. See Susan K. Kinnell, *Military History of the United States: An Annotated Bibliography* (Santa Barbara, CA: ABC-Clio, Inc., 1986).

military academy 1. An institution of higher education that trains new officers; for example, the United States Military Academy at West Point, the Virginia Military Institute, and The Citadel in South Carolina. 2. A private residential school where life for the students follows a modified military regimen.

Military Airlift Command (MAC) The U. S. Air Force's strategic airlift organization, which has the capability to transport strategically significant amounts of personnel and MATERIEL to a theater of operations.

Military Assistance Advisory Group See MAAG.

military authority, national See NATIONAL MILITARY AUTHORITY.

military capability The ability to achieve a specified wartime objective (e.g., win a war or battle, destroy a target set). It includes four major components: force structure, modernization, readiness, and sustainability.

1. *Force Structure*—The numbers, size, and composition of the units that comprise a defense force; e.g., divisions, ships, airwings.
2. *Modernization*—The technical sophistication of forces, units, weapon systems, and equipments.
3. *Readiness*—The ability of forces, units, weapon systems, or equipments to deliver the outputs for which they were designed (including the ability to deploy and employ without unacceptable delays).
4. *Sustainability*—The "staying power" of forces, units, weapon systems, and equipments, often measured in numbers of days.

military characteristics Those characteristics of equipment upon which depends its ability to perform desired military functions. Military characteristics include physical and operational characteristics, but not technical characteristics.

military civic action The use of preponderantly indigenous military forces on projects useful to a local population at all levels in such fields as education, training, public works, agriculture, transportation, communications, health, sanitation, and other areas contributing to economic and social development, which would also serve to improve the standing of the military forces with the population. Also called military civil action.

military climb corridor Controlled airspaces of defined vertical and horizontal dimensions extending from a military airfield.

military commission A court convened by military authority for the trial of persons not usually subject to military law who are charged with violations of the LAWS OF WAR; and in places subject to MILITARY GOVERNMENT or MARTIAL LAW, for the trial of such persons when charged with violations of proclamations, ordinances, and valid domestic civil and criminal laws of the territory concerned.

military convoy A land or maritime CONVOY that is controlled and reported as a military UNIT. A maritime convoy can consist of any combination of merchant ships, auxiliaries, or other units.

military courtesy Rules of conduct that are required, either by regulation or by tradition, for military personnel.

military crest An area on the forward slope of a hill or ridge from which maximum observation can be obtained, covering the slope down to the base of the hill or ridge.

military currency Currency prepared by a power and declared by its military commander to be legal tender for use by civilian or military personnel as prescribed in the areas occupied by its forces. Also called military payment currency.

military damage assessment An appraisal of the effects of an attack on a nation's military forces, to determine residual military capability and to support planning for recovery and reconstitution. See also DAMAGE ASSESSMENT.

military deception Actions executed to mislead foreign decisionmakers, causing them to derive and accept desired APPRECIATIONS of military capabilities, intentions, operations, or other activities that evoke foreign actions that contribute to the originator's objectives. There are three categories of military deception:

1. *Strategic military deception*—Military deception planned and executed to result in foreign national policies and actions that support the originator's national objectives, policies, and strategic military plans.
2. *Tactical military deception*—Military deception planned and executed by and in support of operational commanders against a pertinent threat, to result in opposing operational actions favorable to the originator's plans and operations.
3. *Department/Service military deception*—Military deception planned and executed by military services about military systems, doctrine, tactics, techniques, personnel, or service operations, or other activities to result in foreign actions that increase or maintain the originator's capabilities relative to those of adversaries.

See also DECEPTION and STRATEGIC DECEPTION.

military defense The activities and measures designed, either in whole or in part, to prevent the successful completion of any organized enemy military action.

military department One of the departments within the U.S. DEPARTMENT OF DEFENSE, created by the NATIONAL SECURITY ACT of 1947, as amended; the Department of the Army; the Department of the Navy; and the Department of the Air Force.

military-designed vehicle A vehicle having military characteristics resulting from military research and development processes, designed primarily for use by forces in the field in direct connection with, or support of, COMBAT or tactical operations.

military dynamite A blasting explosive in cartridges especially suitable for use in military construction, quarrying, and service demolition work, which is less sensitive than commercial dynamite and has good storage characteristics.

military education The systematic instruction of individuals in subjects that will enhance their knowledge of the science and ART OF WAR. See also MILITARY TRAINING.

military geography The specialized field of geography dealing with natural and man-made physical features that may affect the planning and conduct of military operations.

military government See CIVIL AFFAIRS.

military government court A court established by a commander having the responsibility for MILITARY GOVERNMENT, with jurisdiction over cases arising under enactments of military government or indigenous law over all persons in an occupied territory, except members of the occupying and allied forces, who are subject to MILITARY LAW.

MILITARY GOVERNMENT ORDINANCE An enactment, on the authority of a MILITARY GOVERNOR, promulgating laws or rules regulating an OCCUPIED TERRITORY under his control.

military governor The military commander or other designated person who, in an OCCUPIED TERRITORY, exercises supreme authority over the civil population, subject to the laws and usages of war and to any directive received from his government or his superiors.

military grid Two sets of parallel lines intersecting at right angles and forming squares; the grid is superimposed on maps, charts, and other similar representations of the surface of the Earth in an accurate and consistent manner, to permit the identification of ground locations with respect to other locations, and the computation of direction and distance to other points.

military grid reference system A system that uses a standard-scaled grid square, based on a point of origin on a map projection of the surface of the Earth in an accurate and consistent manner to permit either position referencing or the computation of direction and distance between grid positions.

military-industrial complex A vague term for a nation's ARMED FORCES and their industrial suppliers. During his farewell address in 1961, President Dwight D. Eisenhower warned that "in the councils of government we must guard against the acquisition of unwarranted influence, whether sought or unsought, by the military-industrial complex. The potential for the disastrous rise of misplaced power exists and will persist." This is a warning that has been well heeded, at least in political analysis. Senator Barry Goldwater had a slightly different attitude: "Thank heaven for the military-industrial complex. Its ultimate aim is peace in our time." See Jerome Slater and Terry Nardin, "The 'Military-Industrial Complex' Muddle," *Yale Review* Vol. LXV, No. 1 (October 1975); Barry S. Rundquist, "On Testing a Military Industrial Complex Theory," *American Politics Quarterly* Vol. 6, No. 1, (January 1978); Paul A. C. Koistinen, *The Military Industrial Complex: A Historical Perspective* (New York: Praeger, 1980).

military intelligence INTELLIGENCE on any foreign military or military-related situation or activity that is significant to military policymaking or the planning and conduct of military operations and activities.

military intervention The deliberate act of a nation or a group of nations to introduce its military forces into the course of an existing controversy.

military jurisdiction 1. The power and authority to impose MILITARY LAW. 2. The administration of military law, usually exercised by military courts.

military justice The application of MILITARY LAW to persons subject thereto and accused of the commission of offenses under the UNIFORM CODE OF MILITARY JUSTICE.

military law 1. MARTIAL LAW. 2. The laws governing military organizations.

military manufacturer A military arsenal, factory, manufacturing depot, or fabricating activity producing items of a purely military nature.

military necessity The principle whereby a belligerent has the right to apply any measures required to bring about the successful conclusion of a military operation and which are not forbidden by the LAWS OF WAR.

military nuclear power A nation that has NUCLEAR WEAPONS and the capacity for their employment.

military occupation A condition in which territory is under the effective control of a foreign armed force. See also OCCUPIED TERRITORY.

military occupational specialty A term used to identify a grouping of DUTY POSITIONS having such a close occupational or functional relationship that persons within each specialty are interchangeable at any given level of skill.

military occupational specialty code A fixed number which indicates a given MILITARY OCCUPATIONAL SPECIALTY. Also known as military occupational specialty number and specification serial number. In the Air Force this is called the Air Force specialty code.

military operations on urbanized terrain (MOUT) All military actions planned and conducted on a topographical complex and its adjacent natural terrain where man-made construction is the dominant feature. It includes combat in cities, which is that portion of MOUT involving house-to-house and street-by-street fighting in towns and cities.

military packaging The materials and methods or procedures prescribed in military specifications, standards, drawings, etc., designed to provide the degree of packaging protection necessary to prevent damage and deterioration during the worldwide distribution of MATERIEL.

military police (MP) Officers or enlisted men charged with controlling the conduct of service personnel through the promotion of compliance with and the enforcement of MILITARY LAW, orders, and regulations; traffic control; crime prevention, investigation and reporting; the apprehension of military absentees and escaped military prisoners; the custody, administration, and treatment of military prisoners; providing security for military supplies, equipment and materiel; and other duties. Military police provide such support to COMBAT ZONES; exercise prescribed control over PRISONERS OF WAR and indigenous civilians; and fight as INFANTRY when the situation requires. The military poilice are often called by their initials, MP. The MPs of the Air Force are the air police; in the Navy, they are the shore patrol.

military policy A broad principle or course of action in respect to military affairs, adopted at an appropriate level within a military organization and made applicable to actions that fall under such authority. Military policy is derived from, and is an integral part of, national policy in either war or peace.

military post 1. A place of assignment such as a given fort or air base. 2. A specific duty assignment such as guard, cook, etc.

military post, international See INTERNATIONAL MILITARY POST.

military posture The military disposition, strength, and condition of readiness of a force as they affect the force's capabilities.

military reform caucus A bipartisan group of members of the U.S. Senate and House of Representatives who are alarmed at the escalation of defense expenditure in the face of what they see as a continued weakening in United States strategic capacity. The members are drawn from the more conservative elements of the Democratic party and from the whole spectrum of the Republicans. They are not typical among defense critics because all stand for a strong American MILITARY POSTURE, and do not object to high defense expenditure as such. Their concern is, rather, for military efficiency. In particular they tend to focus on the overreliance on NUCLEAR WEAPONS resulting from American and European NORTH ATLANTIC TREATY ORGANIZATION (NATO) weakness in conventional forces and weapons, as well as on escalation in defense procurement costs. See Jeffrey Record, "The Military Reform Caucus," *Washington Quarterly* 6, 2 (Spring 1983).

military reform movement A loosely-organized body of American defense analysts who since the mid 1970s have, in general, been asserting that:

1. The American military establishment has been woefully inadequate and unsuccessful since 1945.
2. A major reason for such inadequacy is the American preference for ATTRITION warfare rather than MANEUVER warfare.
3. A second reason is the overly bureaucratic nature of the military establishment, and the careerist and managerial ambitions of the officer corps, which has replaced a "service and leadership" attitude to soldiering.
4. At the same time, and partly for reasons connected with the above, defense procurement has gone on the wrong track, focusing on buying more and more technologically advanced WEAPONS SYSTEMS, in smaller and smaller quantities but at greater and greater prices, so that the defense budget, although ever-increasing, actually buys less and less real security.
5. Consequently, the whole strategy of the United States has outgrown existing military capacity and, in particular, the NORTH ATLANTIC TREATY ORGANIZATION (NATO) commitment is both wrongly handled in tactical terms and distorting the overall defense posture, so that other vital American interests cannot be protected. See Edward Luttwak, *The Pentagon and the Art of War: The Question of Military Reform* (New York: Simon & Schuster, 1984).

military regime An autocratic government under which the military controls a nation's political system—usually following a COUP D'ETAT. In military regimes, the civil liberties of the subjects and normal political and constitutional arrangements may be suspended. Thus it is unlikely that opposition parties will be allowed to operate freely. Although military regimes are frequently dictatorial, it is not necessarily the case that they will be totalitarian. If they occur because of a national crisis or political emergency, such regimes may have a degree of political legitimacy. And in some cases the leaders of the regime may intend to restore the democratic system of government as soon as it is deemed safe to do so. Military coups and military regimes are most often associated with THIRD WORLD countries, though both Greece and Turkey in contemporary Europe have experienced them.

military satellite (MILSAT) A SATELLITE used for military purposes, such as navigation, intelligence gathering, or communications.

military science 1. The study of the ways and means, as well as the hows and whys, of military affairs. Dame Rebecca West (1892–1983), the English novelist and journalist, once wrote that: "Before a war, military science seems a real science, like astronomy, but after a war it seems more like astrology." 2. The technology of war. 3. RESERVE OFFICERS' TRAINING CORPS course of instruction conducted at colleges or universities. Air Force ROTC coursework is aerospace studies. Navy ROTC coursework is naval science.

military science, professor of 1. A senior officer detailed by the Department of the Army for duty with a college-level civilian educational institution for the purpose of supervising instruction in authorized military subjects, usually in conjunction with ROTC programs. 2. A senior military instructor provided by an educational institution and approved by the Department of the Army for duty with the civilian educational institutions sponsoring NATIONAL DEFENSE CADET CORPS units for the purpose of supervising instructions in authorized military subjects. 3. A loose term for any full-time academic who teaches military and national security-related subjects.

military specialist 1. An ENLISTED PERSON who has been rated, on the basis of training or experience, as qualified to perform a specified military duty. 2. An expert in military affairs.

military strategist 1. Any commander of a large military force preparing for combat. 2. An individual, whether a serving officer or an academic in a university or think tank, who understands the inter-relationships of the international environment, national power, national resources, national security, and military and national strategies; is knowledgeable in the role of military forces in support of national objectives and policies; and understands the process of strategy formulation used both by his country and its potential adversaries.

military strategy The art and science of employing the armed forces of a nation to secure the objectives of national policy by the application of force or the threat of force. See also STRATEGY.

military strength 1. All PERSONNEL in active military service who are assigned to a TABLE OF ORGANIZATION AND EQUIPMENT unit, or who are in the pipeline. Personnel of the civilian components of military services who are on active duty for training purposes are not included. 2. A measure of a military unit's strength in terms of personnel or equipment.

military targets, other See OTHER MILITARY TARGETS.

military training 1. The instruction of military PERSONNEL to enhance their capacity to perform specific military functions and tasks. 2. The exercise of one or more military UNITS, conducted to enhance their combat readiness. See also MILITARY EDUCATION.

military training company A UNIT established at a DISCIPLINARY BARRACKS to train prisoners with a view of enabling them to demonstrate their worthiness for restoration to honorable duty status.

militia Part-time citizen soldiers as opposed to full-time regular soldiers. See NATIONAL GUARD.

milk run An easy, routine MISSION; presumably as safe as delivering milk.

Mil rule A method by which the hand, held a known distance from the eye, enables an observer to make rough estimates of angular distance.

mils error An error in FIRING DATA expressed in terms of MILS of angular distance. A mils error differs from a meters error, which is an expression of error in meters of linear distance. The mils error is often referred to by a forward artillery observer as "x" number of "clicks" right, left, short, or long in adjusting indirect fire.

mine 1. In land mine warfare, an explosive or other material, normally encased, designed to destroy or damage ground vehicles, boats, or aircraft, or designed to wound, kill, or otherwise incapacitate personnel. It may be detonated by the action of its victim, by the passage of time, or by controlled means. 2. In naval mine warfare, an explosive device laid in the water with the intention of damaging or sinking ships or of deterring shipping from entering an area. The term does not include devices attached to the bottoms of ships or to harbor installations by personnel operating underwater, nor does it include devices that explode immediately upon the expiration of a predetermined time after laying. 3. To install mines on land or water. 4. A subterranean passage under or toward an enemy position. See FISHBONE, LATERAL.

mine, acoustic See ACCOUSTIC MINE.

mine, activated See ACTIVATED MINE.

mine, actuated See ACTUATED MINE.

mine, air See AIR MINE.

mine, antenna See ANTENNA MINE.

mine, antipersonnel See ANTIPERSONNEL MINE.

mine, antitank See ANTITANK MINE.

mine, boobytrapped See BOOBYTRAPPED MINE.

mine, bounding See BOUNDING MINE.

mine, bouquet See BOUQUET MINE.

mine, chemical See CHEMICAL MINE.

mine, contact See CONTACT MINE.

mine, controlled See CONTROLLED MINE.

mine, conventional See CONVENTIONAL MINE.

mine, creeping See CREEPING MINE.

mine, delayed action See DELAYED ACTION MINE.

mine, homing See HOMING MINE.

mine, independent See INDEPENDENT MINE.

mine, inert See INERT MINE.

mine, influence See INFLUENCE MINE.

mine, magnetic See MAGNETIC MINE.

mine, oscillating See OSCILLATING MINE.

mine, passive See PASSIVE MINE.

mine, pressure See PRESSURE MINE.

mine, scatterable See SCATTERABLE MINE.

mine, unconventional See UNCONVENTIONAL MINE.

mine defense The defense of a POSITION or AREA by land or underwater MINES. A mine defense system includes the personnel and equipment needed to plant, operate, maintain, and protect the MINEFIELDS that are laid.

mine detector 1. An electromagnetic instrument for locating buried or other concealed land mines. 2. A soldier ordered to probe for mines without the benefit of sophisticated electronic equipment.

mine disposal The OPERATION, by suitably qualified personnel, designed to render safe, neutralize, recover, remove, or destroy mines.

minefield 1. In land warfare, an area of ground containing MINES laid with or without a pattern. See also DEFENSIVE MINEFIELD; MIXED MINEFIELD; NUISANCE MINEFIELD; PROTECTIVE MINEFIELD. 2. In naval warfare, an area of water containing mines laid with or without a pattern.

minefield, antisubmarine See ANTISUBMARINE MINEFIELD.

minefield, closure See CLOSURE MINEFIELD.

minefield, deep See DEEP MINEFIELD.

minefield, defensive See DEFENSIVE MINEFIELD.

minefield, dummy See DUMMY MINEFIELD.

minefield, mixed See MIXED MINEFIELD.

minefield, nuisance See NUISANCE MINEFIELD.

minefield, offensive See OFFENSIVE MINEFIELD.

minefield, protective See PROTECTIVE MINEFIELD.

minefield, tactical See TACTICAL MINEFIELD.

minefield lane A marked lane, unmined or cleared of mines, leading through a MINEFIELD.

minefield marking The visible marking of all points required in laying and indicating the extent of a minefield

minefield record A complete written record of all pertinent information concerning a MINEFIELD, submitted on a standard form by the officer in charge of the minelaying operation.

minehunting The employment of ships, airborne equipment, or divers to locate and dispose of individual MINES.

minehunting, acoustic See ACOUSTIC MINEHUNTING.

mine strip In LAND MINE WARFARE, two parallel mine rows laid simultaneously six meters or six paces apart.

minesweeper 1. A specially designed ship used for the removal and destruction of naval mines. 2. A heavy road roller pushed in front of a tank that is used to destroy land mines by exploding them.

minesweeping The technique (on land or water) of searching for or clearing MINES, using mechanical or explosion gear, which physically removes or destroys a mine, or produces, in the area, the influence fields necessary to actuate the mine. See also INFLUENCE SWEEP.

mine system, fishbone See FISHBONE MINE SYSTEM.

mine system, lateral See LATERAL.

mine warfare The strategic and tactical use of MINES and their countermeasures.

mine warfare, land See LAND MINE WARFARE.

minimax A technical term within GAME THEORY, the mathematical analysis of conflict situations that has been found to have a certain utility in STRATEGIC reasoning. In a minimax strategy, attempts are made to minimize the worst eventuality, rather than to maximize the best possible outcome. For example, a nuclear deterrence posture that maximized DAMAGE LIMITATION and rested on a secure SECOND-STRIKE capacity could be said to minimize the damage to a country if its enemy launched an unprovoked attack. Such a strategy would not be the same as one intended to give a chance of total victory with a first strike. The latter might leave the aggressor open to much greater damage if a miscalculation had been made, and is thus not a minimax strategy. Most simple game-theory analyses end up predicting minimax as the strategy likely to be adopted, independently, by both sides, as long as they are unable to trust each other.

minimum-altitude bombing HORIZONTAL or GLIDE BOMBING with the height of bomb release under 900 feet.

minimum attack altitude The lowest altitude, that permits the safe conduct of an air attack or minimizes effective enemy counteraction, as determined by the tactical use of weapons, terrain consideration, and weapons effects.

minimum clearance The vertical distance by which the CONE OF FIRE must clear friendly troops when delivering overhead fire.

minimum deterrence The extent of strategic nuclear force, most probably of a secure SECOND-STRIKE nature, such as a SUBMARINE-LAUNCHED BALLISTIC MISSILE (SLBM), that is thought to be just sufficient to deter an enemy from mounting a first strike. This is also known as finite deterrence. The British, and sometimes the French, strategic nuclear forces are regarded as minimum deterrents. Although very weak compared with those of the United States and Soviet Union, it is thought that the British nuclear force could inflict enough damage on the Soviet Union to make it very improbable that the Soviet government would risk making a nuclear attack on the United Kingdom. Obviously there is no technical answer to the question of how much damage must be threatened to deter an aggressor. Deterrence is inherently a psychological phenomenon. The British government has always used the argument that its strategic forces are at the minimum deterrence level to justify their contention that they have no obligation to offer to reduce them as part of any ARMS-CONTROL agreement. See Vincent Ferraro and Kathleen FitzGerald, "The End of a Strategic Era: Proposal for Minimal Deterrence," *World Policy Journal* 1, 2 (Winter 1984).

minimum essential equipment That part of authorized allowances of army equipment, clothing, and supplies needed to preserve the integrity of a UNIT during movement, without regard to the performance of its COMBAT or SERVICE mission. Items common within this category will normally be carried by or accompany troops to a port of embarkation, and will be placed aboard the same ships with the troops. As used in MOVEMENT DIRECTIVES, minimum essential equipment refers to specific items of both organizational and individual clothing and equipment.

minimum normal burst altitude The altitude below which nuclear WARHEADS used in AIR-DEFENSE are not normally detonated.

minimum quadrant elevation The lowest QUADRANT ELEVATION of a weapon at which the PROJECTILE it fires will safely clear an obstacle between the weapon and its target.

minimum range 1. The least range setting of a GUN at which the PROJECTILE it fires will clear an obstacle or friendly troops between the gun and its target. 2. The shortest distance to which a gun can fire from a given position. 3. The range at which a projectile or FUZE will be armed.

minimum - residual - radioactivity weapon A NUCLEAR WEAPON designed to have the fewest unwanted effects from FALLOUT, RAINOUT, and burst site radioactivity. See also SALTED WEAPON.

minimum safe distance The minimum distance in meters from a desired GROUND ZERO at which a specific degree of risk and vulnerability will not be exceeded with a 99 percent assurance.

mining effect Destruction or damage caused by the force of an explosion below the surface of the ground or water.

mining system A series of underground passages through which enemy fortified positions can be reached secretly and be blown up. See also FISHBONE, LATERAL.

minor repair Repair which, in general, permits the quick return to serviceability of a piece of equipment without extensive disassembly; it can be accomplished with few tools and little or no equipment, and normally does not require EVACUATION of the equipment to a rear ECHELON.

minute gun A gun fired at regular intervals as a signal or a mark of respect to a deceased individual. When fired on the latter occasion, the time interval of the rounds will be one minute, and the number of rounds to be expended will be in strict accordance with current regulations on personal salutes.

Minuteman A family of American missiles, named after the early American colonial GUERRILLA fighters against the British in the American War of Independence. Minuteman I, deployed from 1960 onward, was the first solid-fuel missile with intercontinental range and, like its successors, was housed in concrete SILOS in the American Midwest. The advantage of solid-fuel missiles is that they can be fired very rapidly. Liquid-fuelled missiles, which continue to be prominent in the Soviet INTERCONTINENTAL BALLISTIC MISSILE (ICBM) force, have a lengthy preparation time and cannot be kept at permanent readiness. Two subsequent generations of Minuteman have been deployed (all Minuteman I missiles have been withdrawn from service). Minuteman II entered service in 1966, and still accounts for much of the American force of approximately 1,000 ICBMs. The third generation, Minuteman III, are equipped with multiple independently targetable reentry vehicles (MIRVS) and carry three independently targetable warheads, variously rated at YIELDS between 170 and 335 kilotons. They are not particularly accurate, which makes them only marginally useful against HARD targets. The older Minuteman II missiles have not been equipped with MIRVs, and are reported to have single warheads of slightly over one MEGATON yield. See Roy Neal, *Ace in the Hole: The Story of the Minuteman Missile* (Garden City, New York: Doubleday, 1962).

MIRV A modern nuclear missile that breaks into as many as a dozen separate missiles before they come down to Earth to hit as many separate targets. Weight limitations on the PAYLOAD capacity of the missile mean that each of the MIRVs it carries has to be less powerful than the single WARHEAD that earlier generations of missiles carried. Most MIRVs have a destructive power of under 300 KILOTONS, compared with earlier single warheads rated at or above one MEGATON. However, the phenomenon of EQUIVALENT MEGATONNAGE, and the fact that most targets, except for HARD targets, would be destroyed by yields considerably less than a megaton, usually make this restriction unimportant. The number of MIRVs in a missile's front end varies considerably. Until recently, the usual payload for an American missile, such as an ICBM like the Minuteman III or a SUBMARINE-LAUNCHED BALLISTIC MISSILE (SLBM) such as the Poseidon, was only three warheads, each of between 150 and 350 kilotons. The Soviet Union's large ICBMs, such as the SS-18, can carry up to 10 MIRV warheads of around 900 kilotons. The newest generation of American missiles, the MX ICBM and the TRIDENT II SLBM, have the capacity to carry as many as 14–17 warheads, each likely to be of about 350 kilotons. The new Trident system, due for deployment by the Royal Navy in the 1990s, will probably have eight MIRV warheads, and the French are also developing MIRV capacity for the generation of SLBMs that they will deploy at about the same time. There have been several serious consequences of MIRV capacity for arms control, the most obvious being a problem of comparability. When missiles had single warheads it was relatively easy to know what the respective size of the two superpowers' nuclear arsenals was. Now, however, the number of missiles may be a very poor indicator of actual strength. In addition, the problems of designing BALLISTIC MISSILE DEFENSES have been vastly increased because a small number of missiles carried by a few submarines, for example, can launch hundreds of warheads. As they will be accompanied by DECOYS, this presents the enemy with a virtually insoluble problem. Compare to MANEUVERABLE REENTRY VEHICLE. See Ronald L. Tammen, *MIRV and the Arms Race* (New York: Praeger, 1973); Ted Greenwood, *Making the MIRV: A Study of Defense Decision Making* (Cambridge, MA: Ballinger, 1975).

misfire 1. The failure of a WEAPON or explosive device to fire or explode properly. 2. The failure of a PRIMER or the PROPELLING CHARGE or a ROUND or PROJECTILE to function wholly or in part.

missile 1. Any PROJECTILE. 2. A weapon or object to which propulsive energy is applied or continues to be applied after it is launched.

missile, aerodynamic See AERODYNAMIC MISSILE.

missile, antiballistic See ABM.

missile, ballistic See BALLISTIC MISSILE.

missile, cruise See CRUISE MISSILE.

missile, guided See GUIDED MISSILE.

missile, heavy See HEAVY MISSILE.

missile, nontactical See NONTACTICAL MISSILE.

missile, operational See OPERATIONAL MISSILE.

missile, production See PRODUCTION MISSILE.

missile, surface-to-surface See SURFACE-TO-SURFACE MISSILE.

missile, tactical See TACTICAL MISSILE.

missile-assembly-checkout facility A building, van, or other type of structure located near an operational missile-launching location and designated for the final assembly and checkout of the missile system.

missile base, hard See HARD MISSILE BASE.

missile base, soft See SOFT MISSILE BASE.

missile crisis, Cuban See CUBAN MISSILE CRISIS.

missile-effective rate The percentage of tactical MISSILE LAUNCHERS that are either prepared or ready to launch missiles.

Missile Experimental See MX.

missile gap Presidential candidate John F. Kennedy's 1960 charge that the United States was behind the Soviet Union in nuclear missile production. After the election, the administration of President Kennedy "discovered" that there was no such gap after all. The belief in the gap was sincere, but by the time of the biggest crisis in United States-Soviet relations, the CUBAN MISSILE CRISIS of 1962, the United States had learned from Soviet defectors that the gap was illusory. In fact, some estimates suggest that by 1963, when the United States had several hundred MINUTEMAN I INTERCONTINENTAL BALLISTIC MISSILES (ICBMs) the Soviet Union had deployed only eight missiles capable of reaching North America. See Edgar M. Bottome, *The Missile Gap: A Study of the Formulation of Military and Political Policy* (Rutherford, New Jersey: Fairleigh Dickinson University, 1971); James C. Dick, "The Strategic Arms Race, 1957–61: Who Opened a Missile Gap?" *Journal of Politics* Vol. 34, No. 4 (November 1972).

missile-intercept zone That geographical division of the destruction area where SURFACE-TO-AIR MISSILES have primary responsibility for destroying airborne objects.

missile launcher, operational See OPERATIONAL MISSILE LAUNCHER.

missile master A complete electronic fire-distribution system designed for use in coordinating all elements of AIR DEFENSE from target detection to target destruction. Through the use of electronic computers, memory devices, and communications and display equipment, it automatically collects, displays, and disseminates all information describing the tactical air situation almost instantaneously to the battery commanders in the defense, in order to enable them to employ their weapons with maximum effectiveness.

missile round A MISSILE with its WARHEAD, ready for launching.

missile section A portion of a MISSILE, individually packaged, which when assembled to other portions constitutes a whole missile; examples of these portions are the fore section; nose section; warhead section; guidance section; and propulsion section.

missing A term referring to a nonbattle CASUALTY whose whereabouts and status are unknown, provided the absence appears to be involuntary and the individual is not known to be in a status of unauthorized absence.

missing in action (MIA) A term used to describe a battle CASUALTY whose whereabouts and status are unknown, provided the absence appears to be involuntary and the individual is not known to be in a status of unauthorized absence.

mission 1. The task, together with the purpose, that clearly indicates the action to be taken (but not necessarily how) in a military operation, and the reason therefor. 2. In common usage, especially when applied to lower military units, a duty assigned to an individual or unit; a task. 3. The dispatching of one or more aircraft to accomplish a particular task. See David R. Segal and Young Hee Yoon, "Institutional and Occupational Models of the Army in the Career Force: Implications for Definition of Mission and Perceptions of Combat Readiness," *Journal of Political and Military Sociology,* 12, 2 (Fall 1984).

mission, blunting See BLUNTING MISSION.

mission, call See CALL MISSION.

mission, close support See CLOSE-SUPPORT MISSION.

mission, suppression See SUPPRESSION MISSION.

mission capable A condition status of a item of equipment or system, meaning that it is either fully or partially capable of performing its function.

mission-capable, fully See FULLY MISSION CAPABLE.

mission essential-support item A secondary item, not otherwise authorized for stockage, but required to insure the continued operation of an essential major item, system, or facility that is determined to be vital to an essential defense mission, and the unserviceability or failure of which would jeopardize a basic defense assignment or objective.

mission item An item that is authorized to be stocked by a depot, and for which a stock level has been established.

mission load That quantity of SUPPLIES authorized to be on hand in support units, or stored in depots for them, which will permit such units to accomplish their peacetime and COMBAT-SUPPORT missions until resupply can be effected. The mission load is generally computed in 15-day increments, and is basically designed to satisfy combat requirements.

mission-oriented protective posture (MOPP) A flexible system for protection against a chemical attack, devised to maximize a unit's ability to accomplish its mission in a toxic environment. This posture requires personnel to wear individual protective clothing and equipment consistent with the chemical threat, work rate imposed by their mission, temperature, and humidity without excessive mission degradation.

mission request, immediate See IMMEDIATE MISSION REQUEST.

mission review report An INTELLIGENCE report containing information on all of the targets covered by one aerial photographic sortie.

mission-support site In UNCONVENTIONAL WARFARE, a relatively secure site, utilized by a force as a temporary storage site or stopover point during the conduct of operations.

mission-type order 1. An order issued by a higher to a lower unit that includes the accomplishment of the total mission assigned to the higher headquarters. 2. An order to a unit to perform a mission without specifying how it is to be accomplished.

Mitchell, William "Billy" (1879–1936) The American World War I air corps general who was so insistent that the United States develop viable air power that he was court-martialed in 1935 for his persistent and insubordinate criticism. Mitchell was the first to demonstrate that airplanes could sink a battleship. In 1924 he predicted the exact time of day (7:30 A.M.) and location at which Japanese air power would one day attack Pearl Harbor. In 1942 he was posthumously promoted to major general. In 1946 he was awarded a special medal by the U.S. Congress; and in 1947 his dream of an independent air force became a reality. See Isaac D. Levine, *Mitchell, Pioneer of Air Power* (New York: Arno, 1972); Alfred F. Hurley, *Billy Mitchell: Crusader for Air Power* (Bloomington, Indiana: Indiana University, 1975).

mixed air In ARTILLERY AND NAVAL GUNFIRE SUPPORT, a spotting or observation to indicate that the rounds fired resulted in both air and impact bursts, with the majority of the bursts being airbursts.

mixed force A military FORCE that includes several different BRANCHES (such as artillery and infantry) or SERVICES (such as army and navy). Compare to JOINT FORCE.

mixed graze In ARTILLERY AND NAVAL GUNFIRE SUPPORT, a spotting or observation to indicate that the rounds fired resulted in both air and impact bursts with a majority of the bursts being impact bursts.

mixed minefield A MINEFIELD containing both ANTITANK and ANTIPERSONNEL MINES.

mixed salvo A series of SHOTS in which some fall short of the target and some hit beyond it. A mixed salvo differs from a BRACKETING SALVO during which the number of shots going over the target equals the number falling short of it.

MLRS See MULTIPLE-LAUNCH ROCKET SYSTEM.

mobile army surgical hospital (MASH) 1. A nonfixed medical treatment facility. 2. The movie and TV show named after the above.

mobile defense 1. A defense of an area or position in which MANEUVER is used along with organization of fire and utilization of terrain to seize the initiative from the enemy. 2. A type of defense that consists of canalizing or disrupting the enemy by means of delaying action, and thereafter destroying him by OFFENSIVE action.

mobile employment The use of air-defense artillery in the defense of ground combat forces in a moving situation.

mobile equipment pool A group of weapons, special munitions, and special equipment that may be assigned at the BATTALION or COMPANY, PLATOON, and SQUAD levels, for example, for use as required by a tactical situation and unit mission. The term should be preceded by the appropriate organizational designation whenever a specific level of organization (e.g., squad, platoon, company, or battalion mobile equipment pool) is intended.

mobile reserves 1. Troops held ready in favorable positions for probable reinforcement or COUNTERATTACK. 2. Reserve supplies loaded on trucks or cars for prompt movement to the FRONT.

mobile striking force That portion of a GENERAL RESERVE, including combat and support elements from all components, that is available for immediate employment in any area on, or immediately after, D-DAY.

mobile supply point That location at which equipment, supplies, and ammunition are placed on motor vehicles or railcars and are readily available for rapid displacement in support of a designated COMBAT force. Usually to support a fast-moving situation.

mobile unit Any unit equipped with a sufficient number of ORGANIC vehicles for the purpose of transporting all assigned personnel and equipment from one location to another at one time.

mobile warfare Warfare in which the opposing sides seek to seize and hold the initiative by the use of MANEUVER, organization of fire, and utilization of terrain. Also called war of movement.

mobility A quality or capability of military forces which permits them to move from place to place while retaining the ability to fulfill their primary mission.

mobility operations Work done by engineer units to reduce or negate the effects of existing or reinforcing OBSTACLES. The objectives are to improve the movement of MANEUVER and WEAPON SYSTEMS and critical supplies, and to construct covered and concealed routes to and from BATTLE POSITIONS.

mobilization 1. Preparing for war or other emergencies through assembling and organizing national resources. 2. The process by which ARMED FORCES are brought to a state of immediate readiness for war or other national emergency. 3. The positioning of troops and other forces in a way that they function as a diplomatic warning to a potential aggressor. Sometimes a total mobilization will be seen by a potential enemy as a de facto declaration of war. See John D. Stuckey and Joseph H. Pistorius, "Mobilization for the Vietnam War: A Political and Military Catastrophe," *Parameters: Journal of the US Army War College* Vol. 15, No. 1 (Spring 1985); Gregory D. Foster and Karen A. McPherson, "Mobilization for Low Intensity Conflict," *Naval War College Review* 38, 3 (May–June 1985).

mobilization, industrial See INDUSTRIAL MOBILIZATION.

mobilization plan A plan for assembling and placing, in a state of readiness for war, the manpower and material resources of a nation.

mode, basing See BASING MODE.

moderate level of operations Operations involving 30 to 60 percent of all echelons in a maneuver force and over 50 percent of all FIRE-SUPPORT means engaged in continuous combat over a period of time, during which the employment of next higher echelon resources, to assure accomplishment of the force mission, is not anticipated.

moderate risk (nuclear) The measure of troop safety involving the medium degree of risk to friendly troops used in the computation of MINIMUM SAFE DISTANCE from a NUCLEAR DETONATION. It is associated with maximum 2.5 percent incidence of casualties or a 5 percent incidence of nuisance effects. Moderate risk should not be exceeded if troops are expected to operate at their full efficiency after a friendly burst.

modify In ARTILLERY, an order by the authorized person to make modifications to a FIRE PLAN.

Molotov cocktail A "home made" incendiary grenade consisting of a bottle filled with gasoline or oil and ignited by a cloth "fuse" that seals and extends from the bottle. The cloth wick is ignited before the "cocktail" is thrown at a target. First used widely during the Spanish Civil War of the 1930s and named after V. M. Molotov (1890–1986), then the Foreign Minister of the Soviet Union.

Moltke, Helmuth Karl Bernhard von (1800–1891) The Prussian general who is generally credited with perfecting the modern concept of a GENERAL STAFF capable of devising strategic and tactical plays for mass armies on broad fronts. He should not be confused with his less talented nephew Helmuth Johannes Ludwig von Moltke (1848–1916), who was chief of the German general staff at the beginning of World War I. See Helmuth Karl Bernhard von Moltke, *Strategy: Its Theory and Application: The Wars for German Unification, 1866–1871* (Westport, Conn.: Greenwood Press, 1971); Jack D. Hoschouer, "Vol Moltke and the General Staff," *Military Review* 67, 3 (March 1987).

monostatic radar See MULTISTATIC RADAR.

Monroe Doctrine The assertion by President James Monroe during his 1823 State of the Union message that the Western Hemisphere was closed to colonization and aggressive actions by European powers. In return the United States promised not "to interfere in the internal concerns" of Europe. Backed by the British fleet, the United States was able to maintain this policy of nonintervention, noninterference, and noncolonization throughout the 19th century and beyond. The Monroe Doctrine has had strong rhetorical and political usage in the twentieth century, but its relevance is declining. See Ernest R. King, *The Making of the Monroe Doctrine* (Cambridge, MA: Belknap Press of Harvard University Press, 1975); Jerald A. Combs, "The Origins of the Monroe Doctrine: A Survey of Interpretations by United States Historians," *Australian Journal of Politics and History* Vol. 27, No. 2 (1981); David F. Ronfeldt, "Rethinking the Monroe Doctrine," *Orbis* Vol. 28, No. 4 (Winter 1985).

MOOSEMUSS See PRINCIPLES OF WAR.

mopping up The liquidation of remnants of enemy resistance in an area that has been surrounded or isolated, or through which units of a force have previously passed without eliminating all active resistance.

morale 1. The level of psychological and emotional functioning of an individual or group with respect to sense of purpose, confidence, loyalty and ability to accomplish tasks. 2. The intangible quality of enthusiasm or eagerness to fight that all commanders try to instill in their troops; a military unit's attitude toward its current tasks and military life generally. Throughout the ages, commanders have tried to raise their troops' morale by speeches before battle. The classic morale-raising speech in literature comes from Shakespeare's *Henry V*, when (in Act IV, Scene 2) the King seeks to raise the spirits of his outnumbered soldiers just before the Battle of Agincourt in 1415:

> This day is call'd the feast of Crispian:
> He that outlives this day, and comes safe home,
> Will stand a tip-toe when this day is named,
> And rouse him at the name of Crispian.
> He that shall live this day, and see old age,
> Will yearly on the vigil feast his neighbors,
> And say 'To-morrow is Saint Crispian:'
> Then will he strip his sleeve and show his scars,
> And say 'These wounds I had on Crispin's day.'
> Old men forget; yet all shall be forgot,
> But he'll remember with advantages
> What feats he did that day: then shall our names,
> Familiar in his mouth as household words,
> Harry the king, Bedford and Exeter,
> Warwick and Talbot, Salisbury and Gloucester.
> Be in their flowing cups freshly remember'd.
> This story shall the good man teach his son;
> And Crispin Crispian shall ne'er go by,
> From this day to the ending of the world,
> But we in it shall be remembered:

> We few, we happy few, we band of brothers;
> For he to-day that sheds his blood with me
> Shall be my brother; be he ne'er go by,
> This day shall gentle his condition:
> And gentlemen in England now a-bed
> Shall think themselves accursed they were not here,
> And hold their manhoods cheap whiles any speaks
> That fought with us upon Saint Crispin's day.

Compare this heroic style to General George S. Patton's parallel statement to American troops just before D-day in 1944: "There's one great thing you men can say when it's all over and you're home once more. You can thank God that twenty years from now, when you're sitting around the fireside with your grandson on your knee and he asks you what you did in the war, you won't have to shift him to the other knee, cough, and say, 'I shovelled shit in Louisiana.' " See John Baynes, *Morale: A Study of Men and Courage* (London: International Institute for Strategic Studies, 1982).

morale-support activities A term used to denote a functional grouping of military morale-support services, which normally include arts, crafts, recreation centers, automotive repair, music, theater, youth activities, information, touring and travel, sports, outdoor recreation, and libraries.

morale support company Technical troops provided for special-services operations. The companies are separate, semimobile organizations trained and equipped to produce and provide entertainment and recreational facilities and activities for troops, including motion pictures, libraries, sports, live entertainment shows and novelty acts, musical entertainment, and crafts.

morale support officer A COMMISSIONED OFFICER trained in military recreation who is responsible for the development and operation of voluntary free-time activities or facilities for military personnel, such as crafts, libraries, soldier music, soldier shows, service clubs, sports, and motion pictures. Sometimes this function is performed by a CHAPLAIN.

morning report A military unit's daily record in which the number and status of all assigned personnel are reported.

mortality factor A numerical factor used to determine the quantity of replacement parts to be allowed any given item of equipment. It is based on the durability of the part relative to the durability of the entire item.

mortality rate The number of deaths occurring in a military force during a given period, per 1,000 strength. The rate is calculated by dividing the number of deaths that occur during a given period by the average strength of the force during the same period, and multiplying by 1,000. The period must be expressed, and is usually 1 year.

mortar A muzzle-loading, INDIRECT-FIRE weapon with either a rifled or SMOOTH BORE. It is usually carried by infantry, has a shorter range than a HOWITZER, employs a higher angle of fire, and has a tube length of 10 to 20 CALIBERS.

mortar, base See BASE MORTAR.

mortar deflection board A control instrument used for finding the corrections for wind, drift, and other factors, and the adjustment that must be applied to the AZIMUTH settings of a MORTAR.

MOS See MILITARY OCCUPATION SPECIALTY.

mosaic mountant Paper of a sticky base to which aerial photographs are attached to make a mosaic illustration of a ground location.

motor convoy Two or more vehicles under single control, with or without an escort, used in the transportation of military PERSONNEL or MATERIEL.

motor march A controlled movement of troops in which all ELEMENTS move by motor.

motor pool 1. A group of motor vehicles for use as needed by different organizations or individuals. 2. A place where motor vehicles are stored and serviced.

motor transport Motor vehicles used for transporting military personnel, weapons, equipment, and supplies, excluding combat vehicles such as tanks, scout cars, and armored cars.

motorized unit A unit equipped with complete motor transportation that enables all of its personnel, weapons, and equipment to be moved at the same time without assistance from other sources.

mount 1. A carriage or stand for a gun. 2. To put guns into position for firing. 3. A horse. Compare to MOUNTING.

mount adapter A device to make a gun fit properly into a mount.

mountain troops Soldiers equipped and trained in mountain warfare, including skiing and mountain climbing.

mounted Traditionally this referred to troops on horseback; now it is far more likely to mean troops riding in vehicles. The troops are DISMOUNTED once they leave their vehicles.

mounting 1. All preparations made in designated areas in anticipation of an OPERATION. It includes the assembly in the MOUNTING AREA, preparation and maintenance within the mounting area, movement to loading points, and subsequent EMBARKATION into ships, craft, or aircraft if applicable. 2. A carriage or stand upon which a WEAPON is placed.

mounting area A general locality where assigned forces of an AMPHIBIOUS or AIRBORNE OPERATION, with their equipment, are assembled, prepared, and loaded in ships or aircraft preparatory to an ASSAULT. See also EMBARKATION AREA.

mounting phase The first part of a major military operation, such as an AIRBORNE OPERATION, when men, equipment and supplies are assembled prior to movement into combat. See also ASSAULT PHASE.

mount up An order (of CAVALRY origin) for troops to get into machines (e.g., trucks, tanks) in preparation for moving out.

movement 1. A march. 2. A component part of a MANEUVER. 3. The repositioning of forces for tactical or strategic reasons.

movement control 1. The planning, routing, scheduling and control of personnel and freight movements over LINES OF COMMUNICATION. 2. An organization responsible for these functions.

movement credit The allocation of time granted to one or more vehicles in order to move over a controlled route in a fixed time according to MOVEMENT INSTRUCTIONS.

movement directive The basic document, published by a competent authority, that authorizes a command to take action to move a designated unit from one location to another.

movement instructions Detailed instructions for the execution of a MOVEMENT. They are issued by a transportation officer as an implementation of the movement program, and represent accepted procedure to be followed by the shipper or receiver and transport services involved in the movement.

movement order An order issued by a commander, covering the details for a move of his command.

movement plan 1. A naval plan providing for the movement of an AMPHIBIOUS TASK FORCE to an OBJECTIVE AREA. It includes information and instructions concerning the departure of ships, loading points, and the passage at sea; and the approach to and arrival in assigned positions in the objective area. 2. Up-to-date LOGISTICS data reflecting a summary of transportation requirements, priorities, and limiting factors incident to the MOVEMENT of one or more units or other special groupings of personnel by highway, marine, rail, or air transportation.

movement priority The relative precedence given to each MOVEMENT requirement.

movement techniques Various means of traversing terrain, depending on the likelihood of enemy contact. Major movement techniques include:

1. *Traveling*—Used when speed is necessary and contact with enemy forces is not likely. All elements of the moving UNIT move simultaneously, with the unit leader located where he can best exercise control.
2. *Traveling overwatch*—Used when contact with enemy forces is possible. The lead element and trailing element of the unit are separated by a short distance which varies with the terrain. The trailing element moves at variable speeds and may pause for short periods to overwatch the lead element; it keys its movement to terrain and the lead element. It OVERWATCHES at a distance such that enemy engagement of the lead element will not prevent the trailing element from firing or moving to support the lead element.
3. *Bounding overwatch*—Used when contact with enemy forces is expected. The unit moves by bounds. One element is always halted in position to overwatch another element while it moves. The overwatching element is positioned to support the moving unit by FIRE or fire and MANEUVER.

movement to contact An OFFENSIVE operation designed to gain initial ground contact with the enemy or to regain lost contact.

move out A command that follows instructions in spoken field orders. It indicates that the men addressed are to leave and carry out orders.

moving map display A display in which a symbol, representing a vehicle, remains fixed on a screen while a map or chart image moves beneath the symbol, so that the display simulates the movement of the vehicle.

moving pivot 1. A person who acts as the turning point or pivot for a line of troops when they change their direction of march. 2. The arc of a circle about which a COLUMN turns when it changes its direction of march.

moving screen A system of PATROLS, often motorized or mechanized detachments, used to keep enemy scouting parties at a distance and deny hostile observation of troop movement.

MP MILITARY POLICE.

M-16 The standard U.S. infantry rifle in Vietnam.

mudcapping A method of breaking up large rocks without drilling. A charge of HIGH EXPLOSIVE is laid on a rock and covered with a shovelful of mud, then set off.

multi-modal In transport operations, a term applied to the movement of passengers and cargo by more than one method of transport.

multiple employment A concept whereby an attack helicopter or air cavalry unit, because of its mobility, can be assigned more than one MISSION during a single OPERATION. For example, when employed in reserve, an attack helicopter battalion can simultaneously reinforce ground units with specific missions. It can then be quickly reconstituted to execute a contingency mission elsewhere in the battlefield.

multiple gun A group of guns emplaced and adjusted for firing as a unit, or any group of guns mounted in one position and fired as a unit.

multiple independently targetable re-entry vehicle. See **MIRV.**

multiple-launch rocket system (MLRS) The descendant of the multi-barrel rocket projectors used during World War II; modern versions just coming into service with NORTH ATLANTIC TREATY ORGANIZATION (NATO) armies are capable of firing their 12 rockets within a few seconds. They are used for two distinct purposes. The first is in support of conventional artillery tasks, along with the usual HOWITZERS and CANNONS. Far more important is the second role, in which they are in the vanguard of the new generation of EMERGING TECHNOLOGY weaponry. In this role they are capable of bombarding a target far beyond normal artillery range with about 7,000 grenade-size submunitions spread over an area equivalent to six football fields. Thus, any personnel and equipment not in a heavily-armored cover will be destroyed by a single launch. In the near future, MLRS will become even more deadly, since the United States is co-operating with European NATO members to develop "terminally guided" or SMART BOMBS to be fitted to these rockets, giving the overall weapon a very long-range anti-armor capacity in addition to its current, predominantly antipersonnel function.

multiple phenomenology A system using repeated observations of potential targets by means of different physical principles and different SENSOR systems. In the case of sensor systems, the use of multiple phenomenology makes it more difficult for an adversary to deceive them.

multiple re-entry vehicle (MRV) The RE-ENTRY VEHICLE of a ballistic missile delivery system which places more than one warhead over an individual target. There are no MRV systems left in the American or Soviet strategic nuclear forces because of the development from MRV to MULTIPLE INDEPENDENTLY-TARGETABLE RE-ENTRY VEHICLE (MIRV) technology. The British and French SUBMARINE-LAUNCHED BALLISTIC MISSILE (SLBM) forces, however, still rely on this generation of weapons. In the MRV system a missile carries several warheads, and releases them while outside the Earth's atmosphere so that they may individually complete their BALLISTIC TRAJECTORY to permit their warheads to hit the target area separately. However, they cannot, unlike MIRV warheads, be aimed at separate targets, and instead detonate in a scattered pattern within a single area. The principal advantage of an MRV BUS over the older generation of single-warhead missiles is that the several incoming warheads present a multiplicity of targets for any ANTI-BALLISTIC MISSILE system to cope with. Furthermore, for certain targets, such as very large cities, the quite widely separated detonation of three or more relatively small warheads can be far more destructive than the detonation of a single MEGATON-level warhead.

multipolarity An international situation in which there exist several nations of roughly equal power, rather than the United States–Soviet Union bipolarity of today. It is, in fact, the basis of classic BALANCE-OF-POWER theories, in which individual states are protected by an automatic tendency for alliances to be formed against any one nation that seems to be getting relatively more powerful than its partners in a multipolar system.

multisection charge A PROPELLING CHARGE in SEPARATE LOADING or SEMI-FIXED AMMUNITION that is loaded into a number of powder bags. Range adjustments can be made by increasing or reducing the number of bags used, as contrasted with a single-section charge, in which the size of the charge cannot be changed. Three types of multisection charges are equal-section charges, base-and-increment charges, and unequal-section charges.

multi-service doctrine A group of fundamental principles that guide the employment of forces of two or three SERVICES of the same nation in coordinated action toward a common OBJECTIVE. Compare to JOINT DOCTRINE.

multispectral sensing The use of many different bands of the electromagnetic spectrum (e.g., visible and infrared light) to sense a target. If several bands are used, it is much more difficult for deceptive measures to be effective.

multistatic radar A RADAR system that has transmitters and receivers stationed at multiple locations; typically, a radar system with a transmitter and several receivers, all of which are geographically separated. An advantage of multistatic radar over monostatic radar (a radar system in which the receiver and transmitter are co-located) is that even if its transmitters—which might be detected by the enemy when operating—are attacked, its receivers, in other locations, might not be noticed and might thereby escape attack.

munitions Supplies and equipment of all kinds, including weapons and ammunition, needed by the military SERVICES for direct military purposes.

mustang 1. An officer who has risen from the enlisted ranks; of naval origin. 2. The P-51, a World War II long-range fighter aircraft.

mustard Dichlorodiethyl sulfide; the chemical warfare gaseous blistering agent first introduced by the Germans in World War I, which damaged lungs and often caused blindness and death. So called because of its mustard-like smell.

mustard, cut the To meet a minimal standard of performance; to do well.

mustard, distilled See DISTILLED MUSTARD.

mustard H A dark, oily, liquid CHEMICAL AGENT that injures the eyes and lungs and blisters the skin; it has a garlic- or horseradish-like odor when impure. See also DISTILLED MUSTARD.

mutiny The simultaneous active or passive resistance by multiple subordinates to lawful military authority. Mutiny is an inherent group effort. A single soldier cannot mutiny, he can only be insubordinate. The most famous mutinies are the 1789 mutiny on the HMS *Bounty*, which actually happened, and the World War II *Caine Mutiny*, which was a novel and play (and later made into a movie) by fiction writer Herman Wouk.

mutual aid Arrangements made at government level between one nation and one or more other nations to assist each other.

Mutual and Balanced Force Reductions (MBFR) Talks conducted between the NORTH ATLANTIC TREATY ORGANIZATION (NATO) and the WARSAW PACT from 1973 to 1989, when they were superseded by the CFE (Conventional Forces in Europe) talks. The Vienna-based negotiations are unlike the SALT or START talks, for example, in that they directly involve most members of the two alliances rather than simply the United States and the Soviet Union. All of the Warsaw Pact nations participate, as do all NATO members, with the exceptions of France, Iceland, Portugal, and Spain. This in itself leads to problems, because the NATO Council, in constructing the negotiating position for its diplomats to present, has to satisfy too many divergent internal views, while the Soviet Union completely dominates the position adopted by the Warsaw Pact. The aim of the talks is to contribute to stability and reduce armament expenses by limiting land-based troops on the CENTRAL FRONT. Although the process has been accused of being political time-wasting by both sides, it may in fact become of more urgent importance for the West. There are two serious problems facing NATO. One is the likelihood that the United States will need to reduce its forward based troops because of budgetary pressure. The other is that West Germany, with the biggest single permanently deployed army, is facing a severe demographic problem and will not be able, even with conscription, to maintain its troop numbers over the next decade. Therefore, any agreement that can be gained from the talks to reduce Warsaw Pact troops, even if the Eastern bloc retains an advantage, would be of great benefit to NATO. See John G. Keliber, *The Negotiations on Mutual and Balanced Force Reductions: The Search for Arms Control in Central Europe* (Elmsford, New York: Pergamon, 1980); Richard F. Staar, "The MBFR Process and Its Prospects," *Orbis* (Winter 1984).

mutual assured destruction (MAD) See MAD.

mutual support 1. That SUPPORT which military units render to each other against an enemy, because of their assigned tasks, their position relative to each other and to the enemy, and their inherent capabilities. 2. A condition which exists when positions are able to support each other by DIRECT FIRE, thus preventing the enemy from attacking one position without being subjected to direct fire from one or more other positions.

muzzle The end part of a gun barrel; that part of the barrel from which a fired projectile exits.

muzzle blast The violent surge of hot air and powder gases that escapes from the barrel of a gun as it is fired.

muzzle bell A bell-shaped, built-up section at the MUZZLE of some types of CANNON.

muzzle boresight A disk with crosshairs or other marking that is fitted into a gun at the MUZZLE. The crosshairs show the exact center of the BORE, along which a soldier sights through another disk, set into the BREECH RECESS of the gun, to line the bore of the gun upon a fixed AIMING POINT in order to make the axis of the bore correspond with the axis of the GUNSIGHTS. The attachment at the breech is called the breech boresight.

muzzle brake A device attached to the MUZZLE of a weapon that utilizes escaping gas to reduce RECOIL. Also called a muzzle compensator.

muzzle velocity The velocity of a PROJECTILE with respect to the MUZZLE of the weapon that fires it, at the instant the projectile leaves the weapon.

muzzle-velocity error The numerical difference between the corrections determined by application of meteorological data and other known variations and those determined at approximately the same time by registration, expressed in meters per second variations from standard muzzle velocity. It is assumed to represent the difference between firing table muzzle velocity and developed muzzle velocity.

muzzle wave A compression wave or reaction of the air in front of the MUZZLE of a weapon immediately after firing.

mx/missile experimental The American INTERCONTINENTAL BALLISTIC MISSILE (ICBM) officially called the "peacekeeper." While it is one missile, it has 10 individually targeted warheads that have a combined destructive power 350 times that of the first American nuclear bomb, which destroyed Hiroshima, Japan, in 1945. The MX missile is designed to close a gap between Soviet and American missile forces that has led to the fear of a WINDOW of vulnerability. This fear has come about because the Soviet Union has deployed a large force of very heavy and fairly-accurate ICBMs capable of destroying HARD targets in America, and possibly capable of carrying out a first strike. The existing American ICBM force of MINUTEMAN missiles are neither accurate enough, nor carry a sufficient number or weight of WARHEADS, to have an equivalent capacity. The MX has a much higher throw-weight, of about 3,600 kilograms, than Minuteman III, and a similar range, at 13,000 kilometers. Most importantly, it can carry from 8 to 14 MIRV warheads against Minuteman III's three, each with the same yield of some 335 kilotons, but with far greater accuracy. Clearly MX is a major enhancement of the United States's hard-target strike capability. However, if deployment never exceeds the currently-planned 50 missiles, it will not constitute a first-strike weapon, because of the large number of targets it would leave undestroyed. The major disadvantage of the MX is that, since no one has yet solved the BASING-MODE problem, it will remain as prone to problems of vulnerability as the Minuteman, into the SILOS of which it is being placed. See Herbert Scoville, Jr., *MX* (Cambridge, MA: M.I.T. Press, 1981); Lauren H. Holland and Robert A. Hoover, *The MX Decision: A New Direction in U. S. Weapons Procurement Policy?* (Boulder, CO: Westview Press, 1985).

N

napalm 1. Powdered aluminum soap or a similar compound, often a mixture of naphthenate and palmitate (thus the name), used to gelatinize oil or gasoline for use in INCENDIARY bombs or FLAME THROWERS. 2. The resultant gelatinized substance.

Napoleon Bonaparte (1769–1821) The Corsican-born French general and dictator who dominated the political and military affairs of Europe from the time he seized control of France in a coup d'etat in 1799 until his final defeat at the Battle of Waterloo in 1815. For a short period this ambitious general, later emperor of France, conquered most of continental Europe. But while a brilliant tactician, he was a failure as a grand strategist because he grossly overextended his power and failed to disrupt the coalitions that formed in his wake. But as a campaign strategist and innovator, he has had few peers in history. As a battle tactician, his greatest skill was in the mass use of artillery and the application of the principle of shock in his handling of infantry and cavalry. He was one of the first successful practitioners of combined arms operations. He was also unique in grasping the vital importance of logistics, which allowed him to field the largest armies Europe had seen to date. He also espoused the doctrine of defeating an enemy in detail by rapid movement of forces to exploit the situation. Jomini in his *Precis on the Art of War* (1838) wrote that "[Napoleon] was sent into the world to teach generals and statesmen what they should avoid. His victories teach what may be accomplished by activity, boldness, and skill; his disasters what might have been avoided by prudence."

nation, host See HOST NATION.

national command A military COMMAND that is organized by and functions under the authority of a specific nation. It may or may not be placed under a NORTH ATLANTIC TREATY ORGANIZATION (NATO) or other alliance commander.

National Command Authorities (NCA) The president and the secretary of defense of the United States, or their duly deputized alternates or successors. A major problem in designing a strategic nuclear system, especially one based on the FAIL-SAFE concept of protection against accidental war, is that of establishing who can give the order to launch a nuclear attack and under what conditions. Ideally, the authority to order the firing of nuclear weapons is restricted to the highest political level possible, in all countries. However, a surprise attack by an enemy nation could easily kill such individuals, or cut them off from communication, regardless of any precautions that might be taken. The phrase "National Command Authorities" thus becomes the inevitably vague way of describing the list of those on whom such authority might devolve under conditions of nuclear war. Although the first few steps of this chain are political in all countries, it is less widely realized that military officers, possibly not of the highest seniority, could quite conceivably end up with national command authority. If the president of the United States and his cabinet are killed or prevented from exercising their authority, responsibility could be transferred to the commander of the STRATEGIC AIR COMMAND (SAC). But he could also be killed in a major attack aimed at destroying the command and control bunkers of the SAC. The standby position in the United States, and almost certainly elsewhere, is to have an officer some ranks lower permanently airborne or on alert for immediate takeoff. This officer would be empowered, under specific conditions, to authorize nuclear attacks; and many credible scenarios involve such military personnel being the only surviving authorities in a major attack on the United States. See also FAIL-SAFE.

national commander A commander who controls only forces of his nation and is normally not in an allied chain of command.

national component Any national forces of one or more SERVICES under the command of a single NATIONAL COMMANDER.

National Defense Cadet Corps Secondary school students that receive military training under the general guidance of the Department of Defense.

National Defense University (NDU) A joint-service educational institution established by the U.S. Department of Defense in 1976. It comprises two divisions: the NATIONAL WAR COLLEGE (NWC) and the INDUSTRIAL COLLEGE OF THE ARMED FORCES (ICAF). The University is located at Fort Lesley J. McNair, in Washington, D.C.

national ensign The flag of a nation flown from the staff at the stern of its ships.

national force commanders Commanders of NATIONAL FORCES assigned as separate elements of subordinate allied commands.

National Guard The military forces of the states; often used for civil emergencies such as major fires or floods. Normally under the command of each state's governor, any or all of a state's individual Guard units may be called into federal service at any time. Once a guard unit is called into federal service, it is no longer subject to state control. The National Guard was organized in 1916. Until that time each state had a volunteer militia. See Martha Derthick, *The National Guard in Politics* (Cambridge, Mass.: Harvard University Press, 1965); John K. Mahon, *History of the Militia and the National Guard* (New York: Macmillan, 1983).

national industrial base A nation's MILITARY-INDUSTRIAL COMPLEX; the total capability of a state to produce its own military supplies and equipment. The bigger the base, the better able a state is to defend itself or wage war. See F. Michael Rogers, "The Impact of Foreign Military Sales on the National Industrial Base," *Strategic Review* 5, 2 (Spring 1977).

national infrastructure A military INFRASTRUCTURE provided and financed by a NORTH ATLANTIC TREATY ORGANIZATION (NATO) member in its own territory and solely for its own forces.

national intelligence estimate A STRATEGIC estimate of the capabilities, vulnerabilities, and probable COURSES OF ACTION of foreign nations, produced at the national level as a composite of the views of the INTELLIGENCE COMMUNITY. See Avi Shlaim, "Failure in National Intelligence Estimates: The Case of the Yom Kippur War," *World Politics* 28, 3 (April 1976).

national interest 1. Those policy aims and objectives that are identified as the special concerns of a given nation. Violation of them either in the setting of domestic policy or in international negotiations would be perceived as damaging to the nation's future, both in domestic development and in international competition. The four classic national interests are peace, prosperity, prestige, and power. 2. In the context of foreign policy, the security of the state. In this context a national interest would be called a vital interest. See Fred A. Sondermann, "The Concept of the National Interest," *Orbis* Vol. 21, No. 1 (Spring 1977); Alan Tonelson, "The Real National Interest, *Foreign Policy* No. 16 (Winter 1985).

nationalism The development of a national consciousness; the totality of the cultural, historical, linguistic, psychological, and social forces that pull a people together with a sense of belonging and shared values. This tends to lead to the political belief that this "national" community of people and interests should have their own political order that is independent from, and equal to, all of the other political communities in the world. The modern nation-state was forged from nationalistic sentiment and most of the wars of the last two centuries have been efforts to find relief for a frustrated nationalism. Compare to JINGOIZM. See Gale Stokes, "The Undeveloped Theory of Nationalism," *World Politics,* Vol. 31, No. 1 (October 1978).

national military authority The government agency, such as a ministry of defense, empowered to make decisions on military matters on behalf of its country.

National Military Establishment The formal name for the United States cabinet level department containing the Army, Navy, and Air Force from 1947 to 1949. In 1949 the name was changed to the present Department of Defense.

national objectives Those fundamental aims, goals, or purposes of a nation, as opposed to the means for seeking these ends, toward which a policy is directed and the efforts and resources of the nation are applied.

national of the United States 1. A citizen of the United States. 2. A person who, though not a citizen of the United States, owes permanent allegiance to it.

national policy A broad course of action or statements of guidance adopted by a government at the national level in pursuit of NATIONAL OBJECTIVES.

national salute 1. A SALUTE of 21 guns in honor of a national flag, the ruler of a foreign country, and under certain circumstances, the present or former president of the United States. 2. A salute of 50 guns, one for each State in the Union, fired at noon on July 4 each year, to commemorate the signing of the Declaration of Independence. In this meaning, usually called the Salute to the Union.

national security 1. A condition of military or defense advantage. 2. A favorable foreign-relations position. 3. A phrase used to justify hiding from the public embarrassing or illegal activities on the part of a national government. 4. A defense posture capable of successfully resisting hostile or destructive action from within or without, overt or covert. Compare to SECURITY. See Richard D. Cotter, "Notes Toward a Definition of National Security," *Washington Monthly* Vol. 7, No. 10 (December 1975); C. Maxwell Stanley, "New Definition for National Security," *Bulletin of the Atomic Scientists* Vol. 37, No. 3 (March 1981).

National Security Act The 1947 law that combined the U.S. Army, Navy, and Air Force into one National Military Establishment. It also created the NATIONAL SECURITY COUNCIL and the CENTRAL INTELLIGENCE AGENCY. Amendments to the Act in 1949 replaced the National Military Establishment with the present Department of Defense and placed the National SECURITY COUNCIL in the Executive Office of the President. See Frank N. Trager, "The National Security Act of 1947: Its Thirtieth Anniversary," *Air University Review* 29, 1 (November–December 1977).

National Security Adviser The Assistant to the President for National Security Affairs, who directs the staff of the NATIONAL SECURITY COUNCIL within the Executive Office of the President. Since the 1960s there has been a large degree of institutional competitiveness between the National Security Adviser and the Secretary of State over control of foreign policymaking. See Thomas M. Franck, "The Constitutional and Legal Position of the National Security Adviser and Deputy Adviser," *American Journal of International Law* Vol. 74, No. 3 (July 1980).

National Security Agency (NSA) The agency that handles the interception, decoding, and interpretation of virtually all SIGNALS INTELLIGENCE, and most of the other forms of electronic intelligence for the United States. Unlike the CENTRAL INTELLIGENCE AGENCY (CIA) it is not fully independent, but comes loosely under the control of the Department of Defense, although most of its employees are civilians.

National Security Council (NSC) The organization within the Executive Office of the President whose statutory function is to advise the president with respect to the integration of domestic, foreign, and military policies relating to NATIONAL SECURITY. Members of the Council include the president, vice president, and the secretaries of State and Defense. The Council's staff is directed by the Assistant to the president for National Security Affairs. Unlike the Department of Defense or the Department of State, which are independent government departments represented by cabinet secretaries fighting for their own departmental interests, the NSC is directly under presidential control. As a source of INTELLIGENCE evaluation, as well as representing the president's own think-tank for developing policy in all military, strategic and foreign affairs matters, the NSC is a natural rival to such huge bureaucratic institutions as the Department of State and the PENTAGON. Very few policies have any chance of presidential support unless the National Security Adviser and the NSC staff can be persuaded to agree to them. In late 1986, revelations that the NSC was heavily involved with covert operations and functioning like a "little" CIA suggested that some NSC staff members went beyond their statutory authority. The ensuing scandal forced the Reagan administration to reorganize the NSC internally and substantially change its staff. See also NSC-68.

National Security Organization The overall organization of the United States for NATIONAL SECURITY, under the president as COMMANDER-IN-CHIEF. It consists of the NATIONAL SECURITY COUNCIL, the Office of Emergency Planning, the CENTRAL INTELLIGENCE AGENCY, and the Department of Defense.

national service 1. The concept that a nation's youth should serve the state for a set time period in a military or civilian capacity prior to completing higher education and starting a career. 2. A euphemism for CONSCRIPTION. National Service was the official name of the military conscription system operated in Britain from early in 1939, when war was already feared, until the early 1960s. See Richard Danzig and Peter Szanton, *National Service: What Would It Mean?* (Lexington, MA: Lexington Books, 1986).

national strategy The art and science of developing and using the political, economic, and psychological powers of a nation, together with its armed forces, during peace and war, to secure NATIONAL OBJECTIVES. See also STRATEGY.

national technical means Methods of verification in ARMS CONTROL agreements: principally, SATELLITES that can relay sufficiently detailed photographs to reveal whether a nation is building missile

SILOS, concentrating forces, or engaging in any other activity in contravention of an arms control agreement. However, developments in WEAPONS SYSTEMS are increasingly making such methods less effective. A major example is the introduction by both the Soviet Union and the United States of mobile INTERCONTINENTAL BALLISTIC MISSILES (ICBMS) which, because they can be hidden or camouflaged, are less easily verifiable. Some crucial aspects of desired arms control agreements cannot, even in principle, be verified by satellite; for example, what is going on inside a civilian chemical plant being used to illegally make chemical warfare weapons.

National War College (NWC) A division of the National Defense University chartered to conduct senior-level courses of study and research in the development and implementation of NATIONAL SECURITY policy and STRATEGY (and the application of military power in support thereof). Attendees are usually FIELD-GRADE officers and high-level civilians in defense-related positions. Compare to NATIONAL DEFENSE UNIVERSITY; INDUSTRIAL COLLEGE OF THE ARMED FORCES.

NATO (North Atlantic Treaty Organization) An organization also known as the Atlantic Alliance, consisting of the signatories of the 1949 North Atlantic Treaty, which unites Western Europe and North America in a commitment of mutual security; Article 5 states that "the parties agree that an armed attack against one or more of them in Europe or North America shall be considered an attack against them all." The creation of NATO took place at the height of the COLD WAR in order to contain Soviet expansionist tendencies. It would seem to have worked, in that no member of NATO has ever been turned into a Soviet SATELLITE. Another vital clause, which has come to be more politically sensitive than was originally expected, is Article 6, which limits the geographical area in which treaty support of a member state is required. The simplest definition of this area is that it applies to attacks on the territories of the member states in Europe or North America, and their ships and aircraft when north of the Tropic of Cancer. Thus it excludes areas such as the Indian Ocean and the Persian Gulf, where the United States nowadays tends to argue that vital NATO interests lie, and where it would like European NATO members to assist it in its defense activities. The most significant event in the development of NATO, without which it would probably not have been an important check on Soviet expansionism, was the entry of West Germany in 1955. In many ways NATO was, and remains, a pact with the primary aim of defending West Germany, not only out of consideration for that country, but because West Germany is a barrier to any Soviet incursion into Western Europe in general. It made no sense to establish a huge, expensive, multi-nation defense alliance without including the country whose defence was central to the whole enterprise. Furthermore, the troop numbers required could not be raised without the use of West Germany's population. The internal history of NATO has been one of continued tension, particularly between the United States and the European members. Europe tends to resent American leadership of the alliance, and America resents what it sees as European reluctance to pay properly for its own defense. (See BURDEN SHARING.) There is validity in both arguments, but little chance of change. On the one hand, the United States can hardly fail to lead when the official policy of NATO, FLEXIBLE RESPONSE, requires the use of NUCLEAR WEAPONS, in which the United States has a near monopoly. On the other hand, although the United States does spend about 60 percent of its defense budget on NATO, this is roughly its proportional share of total NATO wealth, in terms of Gross National Product (GNP) per capita, rather than just total GNP. Despite the tensions, NATO is a robust organization that seems capable of endless readjustment, and has weathered all problems so far. No other defensive alliance of its complexity has lasted so long, at such a state of readiness, during peacetime, and remained, at least in broad terms, a democratic partnership of equals. The structure of NATO is complex because it is simultaneously a high-technology military organization and a pluralistic political alliance. The political control and decision-making of the alliance is in the hands of the North Atlantic Council (NAC), where Ministers of Foreign Affairs from each of the 16 nations consult at least twice a year and reach binding decisions. Meetings may also be held at the heads of state or government level, and the NAC also holds sessions at least weekly at the permanent representative level. NATO's Defense Planning Committee (DPC) contains all the members except France and exists in permanent session with twice yearly meetings of defense ministers. The DPC deals with the more strictly military and technical decisions that have to be made. Specifically nuclear matters are dealt with by the same people as the DPC, minus the Icelandic representative, at both levels, under the title of the NUCLEAR PLANNING GROUP. Technical military advice is given to these committees by the NATO MILITARY COMMITTEE, assisted by an integrated International Military Staff.

The most important routine work of these committees and staffs is the preparation of force plans on a rolling two-year basis. A "ministerial guidance" document is passed by the DPC, consisting of guidelines and political targets for defense planning both by NATO as such and by the individual members. On the basis of this a set of "force goals" is prepared, delineating specific planning targets for each country for the next six years. Against the background of these documents there is an annual review of individual national defense policy actions over the previous year, which allows, finally, the writing of a comprehensive NATO force plan for the next five years. This allows SACEUR to do strategic and tactical planning over a five-year period. Compare to TRIPWIRE, WARSAW PACT; See William H. Park, *Defending the West: A History of NATO* (Boulder, CO: Westview Press, 1986); Augustus Richard Norton, *NATO: A Bibliography and Resource Guide* (New York: Garland, 1985).

The North Atlantic Treaty Organization

Member (with year of accession)
Belgium (1949)
Canada (1949)
Denmark (1949)
France (1949)
Germany, Federal Republic of (1955)
Greece (1952)
Iceland (1949)
Italy (1949)
Luxembourg (1949)
Netherlands (1949)
Norway (1949)
Portugal (1949)
Spain (1982)
Turkey (1952)
United Kingdom (1949)
United States (1949)

NATO Military Committee One of the organs of the NATO policymaking structure. It is the highest purely military body in NATO, although it has no command functions, these being handled by SACEUR and the equivalent supreme commanders for the naval roles. The committee consists formally of the chiefs of staff of all the member countries (except France, which, despite not having belonged to the integrated military structure of NATO since 1966, is represented by a Chief of Mission, and Iceland, which has no military forces of its own). The committee has the responsibility for making recommendations on defense policy to the higher civilian organs, and of giving guidance to the major NATO commanders. If, for example, SACEUR wants to adopt a new tactical doctrine, such as FOLLOW ON FORCES ATTACK, or a new weapons system, it would be necessary to persuade this military committee to make an official recommendation to the

NORTH ATLANTIC COUNCIL. The regular work of the committee is done mainly by the national permanent military representatives, although the chiefs of staff do meet at least twice a year. The presidency of the committee rotates annually among the member nationals in alphabetic order.

nautical mile A measure of distance equal to one minute of arc at the Earth's equator. The United States has adopted the international nautical mile, equal to 1,852 meters or 6,076.1 feet.

naval beach group A permanently organized naval command, within an amphibious force, comprised of a commander, his staff, a BEACHMASTER unit, an amphibious construction battalion, and an ASSAULT CRAFT unit, designed to provide an administrative group from which required naval tactical components may be made available to the attack force commander and to the amphibious landing force commander to support the landing of one DIVISION (reinforced). See also SHORE PARTY.

naval campaign An OPERATION or connected series of operations conducted essentially by naval forces, including surface, subsurface, air, and amphibious troops for the purpose of gaining, extending, or maintaining control of the sea.

naval district A geographically defined area in which one naval officer, designated commandant, is the direct representative of the Secretary of the Navy and the Chief of Naval Operations; for example, the 14th Naval District in Hawaii. The commandant has the responsibility for local naval defense and security, and for the coordination of naval activities in the area.

naval gunfire liaison team Personnel and equipment required to coordinate and advise ground or landing forces, or both, on the employment of naval gunfire.

naval gunfire operations center The agency established in a ship to control the execution of plans for the employment of naval gunfire, process requests for naval-gunfire support, and to allot ships to forward OBSERVERS. See Carl White, "Naval Gunfire Support: Coming Back Slowly," *Sea Power* (June 1983).

naval gunfire spotting team The unit of a SHORE FIRE CONTROL PARTY that designates targets for naval gunfire; controls the commencement, cessation, rate, and types of fire; and and spots fire on the target. See also FIELD ARTILLERY OBSERVER; SPOTTER.

navigation, pursuit See PURSUIT NAVIGATION.

navigation guidance, radio See RADIO NAVIGATION GUIDANCE.

navigation system, inertial See INERTIAL NAVIGATION SYSTEM.

NBC warfare Nuclear, biological, and CHEMICAL WARFARE, collectively; the three forms of war that fall outside the definition of CONVENTIONAL WAR. Of the three, biological warfare is the one least understood, most feared, and least clearly part of the arsenals of modern states. The grouping together of these three forms of attack has come about because, while scientifically very different, the three threats require essentially the same defensive responses. While there is clearly no defense against the direct blast and heat of a nuclear explosion, the problems of RADIATION and FALLOUT necessitate the construction of airtight compartments in ships, aircraft and tanks, and the wearing of protective suiting by soldiers exposed to nuclear warfare. Exactly the same sort of protection, and the provision of a filtered air supply, are needed to protect against the short- and possibly longer-term effects of both biological and chemical warfare. Regular EXERCISES are carried out in all branches of the military to practice both the cleansing of personnel and equipment after a nuclear, biological or chemical attack, and the carrying out of routine military tasks while under such an attack. It makes relatively little difference to these exercises which of the three forms of NBC warfare is being considered. See C. N. Donnelly, "Winning the NBC War: Soviet Army Theory and Practice," *International Defense Review* (August 1981); David Rosser-Owen, "NBC Warfare and Anti-NBC Protection," *Armada International* (January 1984); Helmut Stelzmuller, "NBC Defense—NATO Needs New Devices," *Military Technology* (February 1983).

NCA See NATIONAL COMMAND AUTHORITIES.

NCO See NONCOMMISSIONED OFFICER.

near miss Any circumstance in flight in which the degree of separation between two aircraft is considered by either pilot to have constituted a hazardous situation involving the potential risk of a collision.

near real time A delay caused by the automated processing and display of data between the occurrence of an event and reception of the data at some other location. See also REAL TIME; REPORTING TIME INTERVAL.

neatlines The lines that bound the body of a map, usually parallels and meridians.

need to know A criterion used in SECURITY procedures which requires the custodians of CLASSIFIED INFORMATION to establish, prior to disclosure of the information, that the intended recipient must have access to the information to perform official duties. While the system is a sensible one, it has drawbacks, principally in overregimenting the organization of data so that items of information that might be very useful, although apparently unconnected, to a particular officer's task will be withheld. Like all security systems, it often fails because of an unavoidable problem: the sheer mechanical workings of the most highly confidential material have to be carried out by those who have no "need to know" at all. Thus, many ESPIONAGE successes have come about by corrupting messengers and clerks who have an adequately-high general SECURITY CLEARANCE, combined with an unlimited access to all of the information in their department that would be denied to senior officers on the need-to-know basis.

negative A term used in AIR INTERCEPTION that means "cancel" or "no" in response to an operation or question; as opposed to affirmative, which means "yes."

negligible risk (nuclear) The measure of troop safety providing the least risk to friendly troops used in computating the MINIMUM SAFE DISTANCE from a NUCLEAR DETONATION. It is associated with a maximum of 1 percent incidence of CASUALTIES or 2.5 percent incidence of NUISANCE EFFECTS. Negligible risk should not be exceeded unless significant tactical advantage will be gained from taking such risk. Compare to MODERATE RISK.

nerve agent A lethal AGENT that causes paralysis by interfering with the transmission of nerve impulses.

net An organization of COMMUNICATIONS stations capable of direct communications on a common channel or frequency. See also SPOT NET.

net, chain, cell systems Patterns of clandestine organization, especially for operational purposes. A net is the broadest of the three systems, and usually consists of a succession of ECHELONS; and such functional specialists as may be required to accomplish the clandestine organization's mission. When a net consists largely or entirely of nonstaff employees, it may be called an agent net. A chain is commonly defined as a series of agents and informants who receive instructions from and pass information to a principal agent by means of cutouts (go-betweens) and couriers. The cell system emphasizes a variant of the echelon element of a net; its distinctive feature is the grouping of personnel into small units that are relatively isolated and self-contained. In the interest of maximum security for the organization as a whole, each cell has contact with the rest of the organization only through an agent of the organization and a single member of the cell. Others in the cell do not know the agent, and nobody in the cell knows the identities or activities of members of other cells.

net assessment An objective, realistic comparison of the military forces of two rival nations (such as the United States and Soviet Union) or two alliances (such as THE NORTH ATLANTIC TREATY ORGANIZATION (NATO) and the WARSAW PACT).

net weight 1. Weight of a ground vehicle without fuel, engine oil, coolant, on-vehicle materiel, cargo, or operating personnel. 2. Weight of a vehicle, fully equipped and serviced for operation, including the weight of the fuel, lubricants, coolant, vehicle tools, and spares, but not including the weight of the crew, personal equipment, and load. 3. Weight of a container or pallet without freight and binding. See also GROSS WEIGHT.

neutralism A political ideology that manifests itself in a rejection of commitment to the political ideologies or foreign policies of other states or groups of states. The term may sometimes be used to connote a positive attitude toward assuming an obligation to help reduce tension between groups of states, most particularly to prevent the outbreak of war.

neutrality In INTERNATIONAL LAW, the attitude of impartiality during periods of war, adopted by third states toward belligerents (the warring countries) and recognized by the belligerents. If a state wishes to adopt a position of neutrality between two or more others who are at war, it has an obligation under international law to refrain from aiding any war-making party, or from allowing them to use its territory for any warlike purpose at all. In return for this it is to be allowed to continue trading with any or all of the war-making powers, except that the latter have the right to blockade and prevent any prohibited trading, although they must exercise care to protect the nationals and ships of the neutral country. To remain within international law, no war-making party may attack the neutral state. Legally, not all nations even have the right to announce a general neutrality. All members of the UNITED NATIONS, for example, share a common duty to defend each other and to assist in the punishment of an aggressor under certain conditions, and could not claim that their neutrality required or allowed them to be impartial between two parties if one had the sanction of the United Nations. In practice, the only effective neutrality is what has come to be known as "armed neutrality." This state of affairs, and modern Sweden may be the best example, involves not just the general intention not to be involved in any war, but a manifest ability to defend its own frontiers effectively. Being able to defend oneself actually comes close to a legal definition of neutrality, because it is always open to a combatant nation to claim the need to occupy a neutral state to prevent its enemy from so doing, if it cannot trust the neutral state itself to be able to honor its legal obligation not to allow any other party to benefit from its weakness. Considering the readiness of aggressors to invade neutral countries in the major wars of this century, the notion of neutrality in any third world war is largely imaginary. Not only is a CENTRAL-FRONT war inherently likely to be nuclear, but the strategic position of a country such as Sweden would make it extremely difficult for either the WARSAW PACT or the NORTH ATLANTIC TREATY ORGANIZATION (NATO) to respect the neutrality of at least that country's airspace. Neutrality is, of course, entirely possible in limited and small wars not involving the SUPERPOWERS or the major alliances, but this is largely the neutrality of those who do not care to be involved, rather than that of a small nation which fears to be involved. See Walter L. Williams, Jr., "Neutrality in Modern Armed Conflicts: A Survey of the Developing Law," *Military Law Review* Vol. 90 (Fall 1980); Marek Thee, *Towards a New Conceptualization of Neutrality: A Strategy for Conflict Resolution in Asia* (Los Angeles: California State University Press, 1982).

neutralization In MINE WARFARE, the condition in which a mine is rendered, by external means, incapable of firing upon the passage of a target, although it may remain dangerous to handle.

neutralization fire FIRE that is delivered to temporarily eliminate or interrupt the movement or the firing of WEAPONS.

neutralize 1. To render ineffective or unusable for military purposes. 2. To render enemy PERSONNEL or MATERIEL incapable of interfering with a particular operation. 3. To render safe MINES, BOMBS, MISSILES, and BOOBY TRAPS.

neutralized area An area whose independence and integrity (inviolability) have been conferred and guaranteed by treaty, either voluntarily or involuntarily.

neutral state In INTERNATIONAL LAW, a state that pursues a policy of NEUTRALITY (non-involvement) during war.

neutron bomb See ENHANCED RADIATION WEAPON.

nickname Two short, separate words that may be formally or informally assigned by any appropriate authority to an event, project, activity, placename, topographical feature, or item of equipment for convenience of reference but not for the SECURITY of information.

night cap Night combat air patrol (written NCAP).

nitrocellulose A chemical substance formed by the action of a mixture of nitric and sulfuric acids on cotton or some other form of cellulose. Guncotton, an explosive, is a nitrocellulose that has very high nitrogen content.

nitrogen mustards A class of poisonous blistering compounds similar to mustard gas but containing nitrogen instead of sulfur.

Nixon Doctrine The American foreign policy enunciated by President Richard M. Nixon at a press conference on Guam on July 25, 1969, that sought to minimize the role of the United States as world policeman. The central thesis of the doctrine is that "America cannot—and will not—conceive all the plans, design all the programs, execute all the decisions and undertake all the defense of the free nations of the world. We will help where it makes a real difference and is considered in our interest." See Stephen P. Gibert, "Implications of the Nixon Doctrine for military aid policy," *Orbis 16* (Fall 1972); Werner Kaltefleiter, "Europe and the Nixon Doctrine: A German Point of View," *Orbis* 17 (Spring 1973).

no-cities doctrine A policy of attacking military as opposed to civilian targets in a nuclear war. When ROBERT MCNAMARA became President John F. Kennedy's Secretary of Defense in 1961, he inherited President Dwight D. Eisenhower's MASSIVE RETALIATION doctrine for the employment of American strategic nuclear weapons. Initially, in an effort to increase the credibility of this threat by making it more likely that it would in fact be carried out, he adopted the "no cities doctrine." This was essentially a statement that nuclear weapons would be used, as far as possible, in the traditional way of all military power. In effect, he was abandoning the threat to destroy the civilian population of the Soviet Union for what would now be called a COUNTERFORCE strategy. Opposition to the policy was so vigorous that McNamara had to withdraw it and replace it with the idea of MUTUAL ASSURED DESTRUCTION, which remains the basis of the NORTH ATLANTIC

TREATY ORGANIZATION'S (NATO) FLEXIBLE RESPONSE strategy.

no-fire area An area in which no fire or effects of fire are allowed except for specifically approved missions or in response to enemy action.

no-fire line A line short of which artillery or ships do not fire except on request or approval of the supported commander, but beyond which they may fire at any time without danger to friendly troops.

no first use A STRATEGIC policy, publicly proclaimed, of not being the first to use nuclear weapons in a potential war. The United States has never proclaimed this as a policy. Demands that the NORTH ATLANTIC TREATY ORGANIZATION (NATO) in general, or the United States in particular, should adopt such a policy have been made with increasing frequency, and from increasingly different circles. A well-publicized call for no first use was published in *Foreign Affairs* in 1983, in an article signed by, among others, ROBERT MCNAMARA and McGeorge Bundy. Far from being well-known peace activists, these men had been of vital importance in the development of the American nuclear arsenal. Their call was mainly addressed to the first use of BATTLEFIELD NUCLEAR WEAPONS or INTERMEDIATE NUCLEAR FORCES in the European theater. Their argument was that such weapons serve no useful military function whatsoever and possession of them acts only to deter the Soviet Union from making a theater nuclear attack themselves. If NATO were publicly to promise never to initiate such an attack, it was felt, ARMS CONTROL progress would be facilitated. However, any such no first-use declaration would contradict the official NATO policy of FLEXIBLE RESPONSE, which is deliberately ambivalent about the particular level of ESCALATION NATO members would choose in response to a Soviet attack. The Soviet Union has itself promulgated a no-first-use declaration, but this is not taken very seriously by Western military planners, and there is no reason to assume that the WARSAW PACT nations would be any more impressed by a similar NATO statement. The Soviet no-first-use declaration can be seen from one perspective as a reflection of the Soviet Union's numerical superiority in conventional forces. See John D. Steinbruner & Leon V. Sigal, editors, *Alliance Security: NATO and the No-First-Use Question* (Washington, D.C.: The Brookings Institution, 1983); Lawrence D. Weiler, "No First Use: A History," *Bulletin of the Atomic Scientists* Vol. 39, No. 2 (February 1983).

no man's land 1. The area between two opposing armies or between two fortified frontiers. 2. The World War I term for the area between opposing trenches.

noise 1. An unwanted RECEIVER response, other than another SIGNAL (interference). Noise may be audible in voice communication equipment, or visible in equipment such as RADAR. In the latter case it is also known as snow. 2. An intelligence concept referring to all the incoming data that makes it difficult to pick out the truly important data that should be acted upon.

noise discipline A blackout of sound.

nomenclature, standard See STANDARD NOMENCLATURE.

nominal weapon A NUCLEAR WEAPON producing a YIELD of approximately 20 kilotons. See also KILOTON WEAPON; MEGATON WEAPON; SUBKILOTON WEAPON.

nonaligned countries Nations that have deliberately chosen not to be politically or militarily associated with either the West or the Soviet bloc. This word has lost much of its meaning because some of the self-professed "nonaligned" nations are aligned with the Soviet Union.

nonbattle casualty A person who is not a BATTLE CASUALTY, but who is lost to his organization by reason of disease or injury, or by reason of being missing where the absence does not appear to be voluntary or due to enemy action or to being interned.

nonboresafe fuze A FUZE that does not include a safety device to make impossible the explosion of the main charge of a PROJECTILE prematurely while it is still in the BORE of a gun.

noncom A NONCOMMISSIONED OFFICER.

noncommissioned officer (NCO) An ENLISTED PERSON in any of the graduations of corporal or sergeant (in pay grade E-4 or higher, excluding SPECIALIST) normally appointed to fill positions wherein some qualities of leadership are required. Baron von Steuben (1730–1794), the officer who did so much to train the infant United States Army, wrote in his 1779 *Regulations:* "The choice of non-commissioned officers is an object of the greatest importance: the order and discipline of a regiment depends so much upon their behavior, that too much care cannot be taken in preferring none to that trust but those who by their merit and good conduct are entitled to it. Honesty, sobriety, and a remarkable attention to every point of duty, with a neatness in their dress, are indispensable requisites." Rudyard Kipling (1865–1936) was right when he wrote that "the backbone of the army is the noncommissioned man."

nondefense information Information that does not require safeguarding in the interest of national defense.

nondelay fuze A FUZE that functions as a result of inertia of the FIRING PIN (or PRIMER) as a MISSILE is retarded during penetration of its target. The inertia causes the firing pin to strike the primer (or primer the firing pin), initiating action of the fuze. This type of fuze is inherently slower in action than the SUPERQUICK (or instantaneous) FUZE, since its action depends upon deceleration (retardation) of the missile during penetration of the target.

nonduty status The status of an OFFICER or ENLISTED PERSON who, for any reason, such as arrest, leave, sick confinement, or being ABSENT WITHOUT LEAVE, is not available for duty with the organization to which he or she belongs, other than an absence pursuant to a PASS.

noneffective rate The measure of the effect on the strength of a command of personnel who are excused from duty because of illness, injury, or BATTLE CASUALTY, expressed as the ratio of the number of such noneffective personnel at a particular time or an average day in a given period to the strength of the command over the same period in thousands. The noneffective rate may be based on all personnel excused from duty, or it may be specifically due to a particular disease or injury group or other entity. The rate may be computed for a particular area or command or for an entire army.

noneffective sortie Any aircraft that is dispatched and which for any reason fails to carry out the purpose of its MISSION. Abortive SORTIES are included.

nonexpendable supplies and material Supplies that are not consumed by use and which retain their original identity during the period of use, such as weapons, machines, tools, and equipment.

nonintervention The state policy of not intervening in the domestic life or internal affairs of other countries. This is an especially touchy international concern between two nations when the stronger of them (such as the United States) has had a history of interfering in the domestic policies of nations in a particular region (such as Latin America). See Yale H. Ferguson, "Reflections on the Inter-American Principle of Nonintervention: A Search for Meaning in Ambiguity," *Journal of Politics* Vol. 32, No. 3 (August 1970).

nonjudicial punishment Light punishments and other corrective measures imposed by a commanding officer upon any military person who does not demand trial by COURT-MATERIAL. Article 15 of the UNIFORM CODE OF MILITARY JUSTICE provides for these measures. This is known as captain's mast in the Navy. Compare to COMPANY PUNISHMENT.

nonpay status The status of an OFFICER or ENLISTED PERSON who is not entitled to receive pay while in a NONDUTY STATUS; that is, when not available for duty with one's own organization due to one's own fault or neglect. Being ABSENT WITHOUT LEAVE and time lost from duty because of illness, due to the soldier's fault, are cases in which the soldier is placed in a nonpay status.

nonproliferation, nuclear See NUCLEAR NONPROLIFERATION.

nonscheduled units Units of a LANDING FORCE that are held in readiness for landing during the initial unloading period, but not included in either scheduled or on-call WAVES.

nonsparking tools A term referring to nonferrous tools used in AMMUNITION maintenance and EXPLOSIVE disposal operations.

nontactical missile A standard production MISSILE made and allocated for nontactical use such as training, engineering, weapons development testing and evaluation, target work, or for modification to other programs. It may be inert (without a PROPELLANT or EXPLOSIVE components).

nontactical wheeled-vehicle fleet Motor vehicles used in support of general transportation services and facility and equipment-maintenance functions, and which are not directly connected with combat or tactical operations.

nonvisible path segment A portion of the path of a moving TARGET along which the target is continuously nonvisible to a particular SENSOR.

nonvisible time segment The length of time during which a target is on a NONVISIBLE PATH SEGMENT.

normal charge An EXPLOSIVE CHARGE employing a standard amount of PROPELLANT to fire a gun under ordinary conditions, as compared with a REDUCED CHARGE.

normal impact The striking of a PROJECTILE against a surface that is perpendicular to the line of flight of the projectile.

normal interval The space between individual soldiers standing side by side. It is obtained by extending the left arm sideways at shoulder height so that the fingertips touch the shoulder of the soldier next in line.

normal operations Generally and collectively, the broad functions that the commander of a unified combatant command undertakes when he is assigned responsibility for a given geographic or functional area.

North American Air Defense Command (NORAD) The combined American and Canadian headquarters responsible for aerospace SURVEILLANCE and the defense of North America against air and ballistic missile attack.

North Atlantic Treaty Organization See NATO.

northerly turning error The error caused in a magnetic compass by the vertical component of the Earth's magnetic field, which is at its maximum when an aircraft on a northerly or southerly heading banks to turn off that heading.

northern flank Usually taken to mean Norway in terms of the NORTH ATLANTIC TREATY ORGANIZATION (NATO) nations, and particularly northern Norway, where Soviet territory and Soviet troops are nearer to vital NATO interests than anywhere except on the CENTRAL FRONT. The particular importance of northern Norway is that aircraft based there could command the sea approaches to the Kola Inlet, which is vital for Soviet naval deployment. While there is no other intrinsic value to this largely desolate area, and even though it does not provide a useful avenue of invasion to the rest of Western Europe, northern Norway is likely to be an immediate target for Soviet forces should a central-front war start.

nose spray Fragments of a bursting SHELL that are thrown forward in the line of flight. See also BASE SPRAY; SIDE SPRAY.

notional ship A theoretical or average ship of any one category used in transportation planning (e.g., a Liberty ship for dry cargo; a T-2 tanker for bulk petroleum, oils, and lubricants; a personnel transport of 2,400 troop spaces).

NRC See NUCLEAR REGULATORY COMMISSION.

NSC See NATIONAL SECURITY COUNCIL.

NSC-68 NATIONAL SECURITY COUNCIL memorandum Number 68, approved by President Harry S. Truman in 1950. This was the first formal decision that the post World War II strategy of the United States would be based on NUCLEAR WEAPONS. The memorandum did not argue that the United States had a special advantage in nuclear weapons. Indeed, it predicted a major Soviet atomic bomber force by 1954. It stressed that the natural way to wage a nuclear war was by surprise attack, and therefore that the United States would always be more vulnerable than the Soviet Union to such a strategy. Nevertheless, NSC-68 saw no alternative to American reliance on nuclear force until the American weakness in CONVENTIONAL weapons compared to the Soviet Union, could be rectified. As it never was rectified, the memorandum was effectively the first step to a permanent reliance on possible first use of nuclear weapons.

Nth country A reference to any country which has been added to the group of powers possessing NUCLEAR WEAPONS—the next nation of a series to acquire nuclear capabilities. The Nth-country problem was a fashionable way of talking about what is now known as the problem of nuclear PROLIFERATION, and was in common usage during the late 1950s and early 1960s. The real concern of those, mainly Americans, who worried about the problem was their distaste for the British and French independent deterrents. The use of the term "Nth country" was an attempt to disguise this fact. There were two distinct aspects of the problem. One was a purely American concern that the NORTH ATLANTIC TREATY ORGANIZATION'S (NATO) power, particularly its nuclear power, should be as much under American control as possible. The second subject of concern was more genuinely indifferent to which country might become "the Nth" to join the NUCLEAR CLUB. This, a common worry on the part of the more academic strategic theorists, was largely the fear of escalation into a CATALYTIC WAR between the SUPERPOWERS as a result of minor nuclear powers using their weapons against other countries which were linked to a superpower by alliance obligations. See Robert L. Bledsoe, "Laser Isotope Enrichment: A New Dimension to the Nth Country Problem?" *Air University Review* 29, 3 (March–April 1978).

NUCINT See NUCLEAR INTELLIGENCE.

nuclear, biological, and chemical element That part of an operations center that performs the primary functions of coordinating chemical operations and biological defensive actions with other SUPPORT operations, predicting FALLOUT resulting from the employment of NUCLEAR WEAPONS by friendly and enemy forces, and evaluating chemical, biological, and radiological contamination.

nuclear, biological, and chemical warfare See NBC WARFARE.

nuclear airburst The explosion of a NUCLEAR WEAPON in the air, at a height greater than the maximum radius of the FIREBALL.

nuclear artillery See ARTILLERY, NUCLEAR.

nuclear bonus effects Desirable damage or casualties produced by the effects of friendly NUCLEAR WEAPONS that cannot be accurately calculated in targeting, because the uncertainties involved preclude depending on them for a militarily significant result. A "bonus" might be the destruction of a hitherto secret enemy weapons storage depot.

nuclear cloud An all-inclusive term for the volume of hot gases, smoke, dust, and other particulate matter from a NUCLEAR WEAPON itself and from its environment, which is carried aloft in conjunction with the rise of the FIREBALL produced by the detonation of the weapon.

nuclear club Those states that had gained NUCLEAR WEAPONS by the early 1960s and which, openly admitting to owning them, base their defense policy to a greater or lesser extent around the use of such weapons; the traditional five major powers of the UNITED NATIONS (UN) Security Council—the United States, Soviet Union, France, United Kingdom, and the People's Republic of China. Any more recent nuclear powers are probably not to be considered members of the nuclear club for several reasons. First, there is no other power that is absolutely open about owning nuclear weapons. The only other country that is generally acknowledged to have nuclear weapons beyond any doubt is Israel, even though the Israeli government still does not publicly admit this. India is known to have tested, but not known to have built, a nuclear weapon. Among a further moderate-sized group of countries, evidence mainly points to a capacity to rapidly test and build nuclear weapons, South Africa being a prime example, but their nuclear status remains one of potential rather than actuality. Second, even if several THIRD WORLD countries did build nuclear weapons, it is improbable that their defense strategies would ever be based on these rather than on conventional forces, in part because they are not locked into a complex deterrence system with other nuclear powers. An exception here might be the potential for India and Pakistan both to develop such weapons and end up in a nuclear-deterrence deadlock.

nuclear collateral damage See COLLATERAL DAMAGE.

nuclear column A hollow cylinder of water and spray thrown up from an underwater burst of a NUCLEAR WEAPON, through which the hot, high-pressure gases formed in the explosion are vented to the atmosphere. A somewhat similar column of dirt is formed in an underground explosion.

nuclear damage assessment The determination of the damage effect to the population, forces, and resources of a nation resulting from a nuclear attack. It is performed during and after the attack. It does not include the function of evaluating the operational significance of nuclear damage.

nuclear detonation An explosion resulting from FISSION, FUSION, or both types of reactions in nuclear materials, such as that from a NUCLEAR WEAPON.

nuclear dud A NUCLEAR WEAPON which, when launched at or emplaced on a target, fails to provide any explosion of that part of the weapon designed to produce the nuclear YIELD.

nuclear exchange A nuclear war where the combatants are both able to explode nuclear weapons on the opposition. A "full" nuclear exchange means that all possible nuclear weapons on each side are exploded.

nuclear fission The physical process harnessed to produce ATOMIC BOMBS. It depends on the instability of atoms of some of the heavy elements, which release very large amounts of energy when split by bombardment with subatomic particles. A heavy element is one with a large number of neutrons and protons in its atomic nucleus. The neutrons, being electronically neutral, can be detached from the nucleus when struck by a free neutron, which acts as a submicroscopic bullet. With naturally unstable elements there is a continual process of particular loss which, over time, leads to the transformation of the element from one variant (isotope) to another. This process was first understood during the years between World Wars I and II, and it became apparent that there was a theoretical possibility of a chain reaction being artificially created. Such a chain reaction would occur if the impact of one neutron "bullet" was to force the expulsion from a nucleus of two or more further neutrons, which would split more nuclei, and so on, with increasing rapidity. Some heavy isotopes (such as uranium 233 or 235, and plutonium 239) were known to be particularly prone to this natural neutron release. Amounts of one of these heavy elements too small to give way to a chain reaction are referred to as "subcritical." Bringing together two samples of a suitable element which together formed a CRITICAL MASS would cause an immediate and spontaneous release of huge amounts of energy. The chain reaction requires containment of the free-flying neutrons so that they impact on other atoms of uranium, or those of other radioactive elements, rather than escaping into the atmosphere. The technical problem is that the neutron-emission/nucleus-splitting process is very fast indeed. If the two subcritical masses were brought together too slowly they would produce enough energy to throw themselves apart, or to melt their container, before the full chain reaction could start. It was the solution to this problem of keeping the two radioactive samples sufficiently apart to prevent premature energy release, and then bringing them together fast enough to force the chain reaction, that was finally solved by engineers working on the MANHATTAN PROJECT. See also NUCLEAR FUSION. See Robert Chadwell Williams and Philip Louis Cantelon, *The American Atom: A Documentary History of Nuclear Policies from the Discovery of Fission to the Present, 1939–1984* (Philadelphia: University of Pennsylvania Press, 1984).

nuclear forces, theater See THEATER NUCLEAR FORCES.

nuclear freeze A policy of mutually stopping the testing, production and deployment of NUCLEAR WEAPONS by all sides. The U.S. House of Representatives approved a resolution calling for an "immediate, mutual, and verifiable freeze" on such weapons in 1983; but all nuclear-freeze motions have been defeated in the Senate. See Howard Stoertz, Jr., "Monitoring a Nuclear Freeze," *International Security* Vol. 8, No. 4 (Spring 1984); James M. McCormick, "Congressional Voting on the Nuclear Freeze Resolutions," *American Politics Quarterly* Vol. 13, No. 1 (January 1985).

nuclear-free zone (NFZ) 1. A nation or area in which there is no effort to store, deploy, or transport nuclear weapons, as established by a local government in Western Europe or by the United States. Some authorities, particularly in the United Kingdom, have taken this a step further by refusing even to cooperate with CIVIL-DEFENSE planning for nuclear war, either on the grounds that such preparations are futile or that they actually encourage war. 2. An area of the globe in which NUCLEAR-WEAPONS deployments are banned by treaty. Examples include the New Zealand governments's demand for a Pacific NFZ and the suggestions of a nuclear-free corridor on either side of the border between East and West Germany. Local nuclear-free zones mean nothing at all in terms of national and international politics. Even the more general demands for treaty-based denuclearization of whole areas are less powerful than they may seem. Given the range of most missile- or aircraft-delivered nuclear weapons, both THE NORTH ATLANTIC TREATY ORGANIZATION (NATO) and the WARSAW PACT could accept a broad range of zones in which such weapons could not be deployed. Targets within a continental European NFZ, for example, could easily be hit by missiles based outside the zone. Treaties forbidding the use of nuclear weapons in specified zones would not, in the appalling context of a nuclear or potentially nuclear, war, be worth the paper on which they were written. 3. An area that is free of nuclear weapons by treaty or other international understandings; Antarctica for example. See Robert K. German, "Nuclear-Free Zones: Norwegian Interest, Soviet Encouragement," *Orbis* Vol. 26, No. 2 (Summer 1982).

nuclear fuel, enriched See ENRICHED NUCLEAR FUEL.

nuclear fusion The physical process generated in THERMONUCLEAR weapons, by which huge amounts of energy are released when two hydrogen atoms combine to form a heavier element. The lightest elements that can be used for this process within the capacity of modern physics are isotopes of hydrogen. (An isotope is a variant of an element. Each element has a fixed number of protons in its nucleus, but can have more than one form, depending on the number of neutrons. The isotope number is the sum of protons and neutrons.) The two isotopes of hydrogen that are usable for nuclear weapons manufacture are deuterium and tritium. The problem with nuclear transformations of a light element into a heavier one is that while an enormous amount of energy is released, very high temperatures are required to start the process. A vivid illustration of this can be seen if it is remembered that the energy-release phenomenon that powers a star involves the nuclear fusion of helium and hydrogen atoms to yield heavier elements. It is the extremely high gravitational pressure generated in the core of such immense objects as stars that creates the temperature to sustain energy release by fusion reactions. The only way in which it is possible to create such temperatures by human design is to use the heat of a nuclear fission chain reaction. Thus, the trigger for a hydrogen bomb is itself an atomic device. As a result, a thermonuclear explosion is an explosion generated by the heat of a nuclear explosion. Unlike atomic bombs, in which considerations of CRITICAL MASS limit the possible yield, there is no theoretical limit to the size of a hydrogen bomb. With the provision of a fission trigger, the amount of deuterium or tritium present is the size-determining factor. See Edwar Teller, *Fusion* (New York: Academic Press, 1981).

nuclear hostage A target of presumedly great importance to an enemy that is not attacked in a FIRST (or subsequent) STRIKE, but which an assailant clearly retains the capacity to destroy. The usual version is that by attacking the enemy's nuclear forces or other military targets, but leaving its cities undamaged, the latter are effectively held hostage. The enemy is forced to consider whether it wishes to retaliate to the limited strike, knowing that doing so carries a great risk of incurring a SECOND STRIKE that would do very much more damage to the social and economic structure of the nation than the inevitable collateral damage suffered in the original COUNTERFORCE attack. The enemy's civilians are thus held hostage against the moderation of their own leaders. Setting aside the macabre precision and rationality of the theory, which is found in all nuclear thinking, the assumption that the Soviet Union shares the West's sense of nuclear policy does present a serious problem in considering the concept of nuclear hostages. While there is good evidence that Soviet targeting is, if anything, less likely to involve the deliberate destruction of urban targets than is that of the Western nuclear powers, it is also true that the Soviet Union has much less faith in either limited nuclear options or ESCALATION CONTROL. A Soviet counter strike after a "non-city" strike by the United States, which avoided hitting American cities, would probably depend much more on whether the Soviets saw any point in targeting American population centers than on the fact that their own had been left alone. A further problem is that the doctrine is, in any case, dependent on the viability of SURGICAL STRIKES that could avoid killing millions of the enemy's citizens. No one but the country attacked can determine what citizen death rate is to be regarded as "merely" COLLATERAL DAMAGE, and when a nuclear strike would be taken as virtually a deliberate COUNTERVALUE strike. Consequently it would be very difficult to be sure that the attack planned was, indeed, going to be seen by the enemy as having left its vital concerns safe and available to be considered as nuclear hostages. See Wolfgang K. H. Panofsky, "The Mutual Hostage Relationship between America and Russia," *Foreign Affairs* 52 (October 1973).

nuclear incident An unexpected event involving a nuclear weapon, facility, or component, but not constituting a NUCLEAR WEAPON(S) ACCIDENT.

nuclear intelligence Intelligence information derived from the collection and analysis of RADIATION and other effects resulting from radioactive sources. Also called NUCINT.

nuclearism According to Robert J. Lifton, in *Boundaries: Psychological Man in Revolution* (New York: Random House, 1970), "The passionate embrace of nuclear weapons as a solution to our anxieties (especially our anxieties concerning the weapons themselves). That is, one turns to the weapons, and to their power, as means of restoring boundaries. Nuclearism, then, is a secular religion, a total ideology in which grace, mastery of death, is achieved by means of a new technological deity."

nuclear load, prescribed See PRESCRIBED NUCLEAR LOAD.

nuclear logistic movement The transport of NUCLEAR WEAPONS or components of nuclear weapons in connection with supply or maintenance operations.

nuclear nation 1. Military nuclear powers. 2. Civil nuclear powers.

nuclear nonproliferation The policy of stopping the spread of NUCLEAR WEAPONS to non-nuclear states. The United States, since 1945, has followed a policy of nonproliferation, seeking to reduce the incentives for non-nuclear states to go nuclear and to control the flow of weapons-grade nuclear materials to other nations. See David Dewitt, *Nuclear Non-Proliferation and Global Security* (New York: St. Martin's Press, 1987); John Simpson, *Nuclear Non-Proliferation: An Agenda for the 1990s* (New York: Cambridge University Press, 1987).

Nuclear Non Proliferation Treaty The 1968 treaty on the non proliferation of NUCLEAR WEAPONS which was signed by 115 nations; it called for those nations having nuclear weapons not to transfer them to other nations, and for non-nuclear states not to become nuclear. The treaty is of little use. The three nuclear powers who signed it (the United States, United Kingdom, and Soviet Union) had not the slightest intention of increasing world instability by letting other, smaller powers acquire such weapons. The non-nuclear powers of the day who had any desire to build such weapons simply refused to sign the treaty, for example, Canada is a signatory, since it had long been Canadian policy to renounce the ownership of nuclear weapons, while India, South Africa, and Israel, all of whom are now known or, believed to have such weapons, or at least the capacity to build them, all refused to sign. The more curious omissions from the list of signatories were the other two countries which did have nuclear weapons in 1968, France and the People's Republic of China. In both cases it was more a matter of status than of politics. France was still in its mood of Gaullism, in which it resented America's nuclear leadership, and China was at the height of its ideological split with the Soviet Union. The uses of nuclear energy for peaceful purposes are not covered by the treaty. See D. M. Edwards, "International Legal Aspects of Safeguards and the Non-Proliferation of Nuclear Weapons," *International and Comparative Law Quarterly* Vol. 33, No 1 (January 1984); Ian Bellany, Coit D. Blacker and Joseph Gallacher, *The Nuclear Non-Proliferation Treaty* (Totowa, NJ: F. Cass, 1985).

nuclear options, limited See LIMITED NUCLEAR OPTIONS.

nuclear ordnance items Assemblies, equipment, components, and parts that are peculiar in design to NUCLEAR WEAPONS programs.

nuclear parity A condition at a given point in time when opposing forces possess nuclear offensive and defensive systems approximately equal in overall combat effectiveness.

nuclear power, major See MAJOR NUCLEAR POWER.

nuclear power, military See MILITARY NUCLEAR POWER.

nuclear-powered submarine See SSN.

nuclear precursor A nuclear explosion near an adversary's sensors or weapons shortly before the arrival of a large number of nuclear warheads on nearby targets. The aim is to prevent the adversary from launching his own weapons or from using his sensors because of the background or debris produced by the precursor.

nuclear proliferation See PROLIFERATION.

nuclear rainfall See RAINFALL, NUCLEAR.

nuclear reactor A facility in which fissile material is used in a self-supporting chain reaction (NUCLEAR FISSION) to produce heat, radiation, or both for both practical application and research and development.

Nuclear Regulatory Commission (NRC) The federal agency of the United States that licenses and regulates the uses of nuclear energy to protect the public health and safety and the environment. It does this by licensing persons and companies to build and operate NUCLEAR REACTORS and to own and use nuclear materials. The NRC makes rules and sets standards for these types of licenses, and also inspects the activities of the persons and companies licensed, to ensure that they do not violate the safety rules of the commission. The NRC was created in 1975 under provisions of the Energy Reorganization Act of 1974 and supplanted the Atomic Energy Commission (AEC), which had performed similar functions since 1946. The AEC was created in the first place in an effort to separate civilian and military uses of nuclear energy. But this "battle" has effectively been lost now that the NRC has responsibilities for both the military and civilian use of nuclear power. See Jeffrey S. Klein, "The Nuclear Regulatory Bureaucracy," *Society* Vol. 18, No. 5 (July–August 1981); George T. Mazuzan and J. Samuel Walker, *Controlling the Atom: The Beginnings of Nuclear Regulation, 1946–1962* (Berkeley: University of California Press, 1985).

nuclear safety Those design features, procedures, and actions that protect against intentional and unintentional acts that could lead to a NUCLEAR INCIDENT or accident.

nuclear safety line A demarcation line selected, if possible, to follow well-defined topographical features and used to delineate levels of protective measures, degrees of damage or risk to friendly troops, and to prescribe limits to which the effects of nuclear weapons used by friendly forces may be permitted to extend.

nuclear sanctuary See SANCTUARIZATION.

nuclear stalemate A concept which postulates a situation in which the relative strength of opposing nuclear forces results in mutual deterrence against the employment of NUCLEAR WEAPONS. Compare to BALANCE OF TERROR. See Michael Krepon, *Strategic Stalemate: Nuclear Weapons and Arms Control in American Politics* (New York: St. Martin's Press, 1984).

nuclear stockage, prescribed See PRESCRIBED NUCLEAR STOCKAGE.

nuclear strike warning A warning of impending FRIENDLY or suspected enemy nuclear attack.

nuclear sufficiency See EFFECTIVE PARITY.

nuclear support The use of NUCLEAR WEAPONS against hostile forces in support of friendly air, land, and naval operations.

nuclear support, preplanned See PREPLANNED NUCLEAR SUPPORT.

nuclear surface burst An explosion of a NUCLEAR WEAPON at the surface of land or water, or above the surface, at a height less than the maximum radius of the FIREBALL.

nuclear test-ban treaty See TEST-BAN TREATY.

nuclear threshold That point during a war at which NUCLEAR WEAPONS of any sort are introduced, in addition to or instead of CONVENTIONAL WEAPONS. As such, it is one of the FIRE-BREAKS discussed in the theory of ESCALATION, and is of particular importance to NORTH ATLANTIC TREATY ORGANIZATION (NATO) planners. The most common use of the concept is in the context of NATO's FLEXIBLE RESPONSE strategy for defeating a WARSAW PACT attack. It is generally assumed that NATO would be able to sustain a purely conventional defense effort for only a limited time, often put at as little as 10 days, and certainly no longer than a few weeks. If the Soviet Union was still able to mount a forceful conventional attack after this time, NATO would be forced to "go nuclear," to cross the threshold and use at least BATTLEFIELD NUCLEAR WEAPONS. Thus, political initiatives to improve NATO's conventional defense strength are described as raising the nuclear threshold because they will defer the stage at which NATO might have to use nuclear weapons. See Benjamin S. Lambeth, "On Thresholds in Soviet Military Thought," *Washington Quarterly* (Spring 1984); Bernard W. Rogers, "Raising the Nuclear Threshold," *Defense* (June 1984); Roman Kolkowicz and Ellen Propper Mickiewicz, *The Soviet Calculus of Nuclear War* (Lexington, Mass.: Lexington Books, 1986).

nuclear umbrella An informal term describing the protection extended to Western Europe and Japan by American extended deterrence. This nuclear guarantee has underpinned NORTH ATLANTIC TREATY ORGANIZATION (NATO) policy from the days of MASSIVE RETALIATION to the current doctrine of FLEXIBLE RESPONSE. The "umbrella" is provided, at its strongest by American INTERCONTINENTAL BALLISTIC MISSILES (ICBMs). If necessary, the United States would initiate CENTRAL STRATEGIC WARFARE against the Soviet Union to prevent the latter from capturing a significant part of Western Europe, or as retaliation for a nuclear strike in Europe. This umbrella has become increasingly leaky for two reasons. The first, relatively technical reason, is that it is unclear what sort of nuclear response would be made to a particular WARSAW PACT incursion. Would American presidential permission for SACEUR to use battlefield nuclear weapons be enough to fulfill the nuclear guarantee? Would a Soviet nuclear strike against a French town with, for example, an SS-20 demand that an American ICBM be launched against a Soviet city? The second, and more important reason for doubting the American nuclear umbrella is the essential lack of credibility that it carries in a period of NUCLEAR PARITY. When the doctrine was developed, and even until the early 1970s, the United States had such superiority in intercontinental nuclear weapons that it could be plausible for an American president to threaten to strike at the heartland of the Soviet Union to prevent even a conventional defeat of NATO. However, in a situation where a broad nuclear parity exists, it is inevitable that the question is raised of whether an American President would, for example, risk Detroit to avenge Dusseldorf. A major reason why European governments fear the withdrawal of American troops from Western Europe is that a substantial American presence makes it more likely that, if only to protect them from being overrun by Soviet armies, the United States would be forced to use its nuclear arsenal. This argument is, to some extent, a re-creation of the TRIPWIRE thesis.

nuclear vulnerability assessment The estimation of the probable effect on a nation's population, military forces, and resources of a hypothetical nuclear attack. It is performed predominantly in the preattack period; however, it may be extended to the periods during or after the attack.

nuclear warfare Warfare involving the employment of NUCLEAR WEAPONS.

nuclear warning message A warning message disseminated to all affected friendly forces any time a NUCLEAR WEAPON is to be detonated if the effects of the weapon will have an impact upon those forces.

nuclear weapon A complete assembly in its intended ultimate configuration which, upon completion of the prescribed arming, fusing, and firing sequence is capable of producing the intended nuclear reaction and release of energy.

nuclear weapon, clean A nuclear weapon that causes damage by means of explosive force only, with radioactive byproducts kept to a minimum. It is intended to do more harm to inanimate objects than flesh. Compare to a DIRTY NUCLEAR WEAPON.

nuclear weapon, dirty See DIRTY NUCLEAR WEAPON.

nuclear weapon degradation The degeneration of a nuclear WARHEAD to such an extent that the anticipated nuclear YIELD is lessened.

nuclear weapon employment time The time required for the arrival of a NUCLEAR WEAPON at its target after the decision to fire it has been made.

nuclear weapon exercise An operation not directly related to the immediate operational readiness to use a nuclear weapon. It includes the removal of a nuclear weapon from its usual storage location, preparing it for use, delivery of it to an employment unit, the movement in a ground training exercise to include loading of the weapon aboard an aircraft or missile, and return of the weapon to storage. It may include any or all of the operations listed above, but does not include launching or flying operations.

nuclear weapon maneuver A NUCLEAR WEAPON exercise that is extended to include flyaway of a nuclear weapon in a combat aircraft, but does not include the actual use of the weapon.

nuclear weapon package A discrete grouping of NUCLEAR WEAPONS according to their specific YIELDS, planned for employment in a specific area for a designated time frame. It is employed at CORPS level.

nuclear weapons, battlefield See BATTLEFIELD NUCLEAR WEAPONS.

nuclear weapons, deployed See DEPLOYED NUCLEAR WEAPONS.

nuclear weapon(s) accident An unexpected event involving NUCLEAR WEAPONS or radiological nuclear-weapon components that results in the accidental or unauthorized launching, firing, or use of a nuclear-capable WEAPON SYSTEM which could create the risk of war; NUCLEAR DETONATION; nonnuclear detonation; contamination by radioactivity; seizure, theft, loss or destruction of a nuclear weapon; or a public hazard from such a weapon, whether real or implied.

nuclear weapons surety The totality of MATERIEL, PERSONNEL, and procedures that contribute to the security, safety, and reliability of NUCLEAR WEAPONS.

nuclear winter A theoretical climatic change caused by the effects of smoke from a full nuclear exchange among the superpowers. Because the smoke would blot out most of the Earth's sunlight for weeks or months after a nuclear war, most plant and animal life that survived the initial blast would be subsequently destroyed. Theories such as this depend on precise assumptions fed into computer models of nuclear weapons effects on climate. No one doubts that some smoke would reach the upper atmosphere or that very large amounts of smoke would produce some temperature drop, or that large temperature reductions would have some impact on agriculture. Whether an agreed-upon model can be produced is improbable, because there can be little solid evidence for any set of assumptions on something as untested as CENTRAL STRATEGIC WARFARE. At the moment, apparently equally valid models suggest a temperature drop of only a few degrees, producing September temperatures during July in the northern hemisphere, with little attendant long-term danger, or a drop of as much as 20 degrees centigrade, with potential near extinction of the human race. All that can be said with certainty is that the nuclear-winter hypothesis does raise a problem that strategists had never considered for the first three decades of the nuclear age. See Thomas F. Malone, "International Scientists on Nuclear Winter," *Bulletin of the Atomic Scientists* Vol. 41, No. 11 (December 1985); Paul R. Ehrlich, Carl Sagan, Donald Kennedy and Walter Orr Roberts, *The Cold and the Dark: The World after Nuclear War* (New York: Norton, 1985).

nuclear yields The energy released in the detonation of a NUCLEAR WEAPON, measured in terms of the KILOTONS or megatons of trinitrotoluene (TNT) required to produce the same energy release. Yields are categorized as:

Very low—Less than 1 kiloton.
Low—1 kiloton to 10 kilotons.
Medium—Over 10 kilotons to 50 kilotons.
High—Over 50 kilotons to 500 kilotons.
Very high—Over 500 kilotons.

nuisance effects The annoying but not incapacitating results of exposing troops to light fallout from nuclear weapons.

nuisance minefield A MINEFIELD laid to delay and disorganize the enemy and to hinder his use of an area or route.

(number of) rounds In ARTILLERY AND NAVAL GUNFIRE SUPPORT, a command or request used to indicate the number of PROJECTILES per gun to be fired on a specific target.

O

objective 1. An end goal or aim to be attained by the employment of military force. See also TARGET. 2. The physical object of a military action, e.g., a definite tactical feature, the seizure or holding of which is essential to a commander's plan. The objective may be the most critical of the PRINCIPLES OF WAR because it gives coherence to all of the others; it provides the "what" so that the others can provide the "how."

objective area 1. A defined geographical area within which is located an OBJECTIVE to be captured or reached by military forces. 2. In AIRBORNE, AIRMOBILE, and AMPHIBIOUS OPERATIONS, it is the proposed area of operations and includes the AIRHEAD or BEACHHEAD.

objective force A force that can meet a projected threat and carry out a NATIONAL STRATEGY at a level of prudent risk and in consideration of the reasonable attainability of its objective.

objective-force level The level of military force that must be attained within a finite time frame and resource level to accomplish approved military objectives, missions, or tasks.

objective plane A plane tangent to the ground or coinciding with the surface of a TARGET, especially such a plane at the POINT OF IMPACT of a BOMB or PROJECTILE.

obligated stocks SUPPLIES reserved for issue only for a specifically designated purpose, such as special project, mobilization of reserves, etc.

obligated tour The duration of an initial tour of active duty.

oblique compartment A COMPARTMENT OF TERRAIN whose long axis is diagonal to the direction of march or to the FRONT.

oblique line overlap A succession of overlapping oblique photographs taken in a straight line. They cannot be pieced together, as can a vertical line overlap, but they give a series of useful perspective views. The overlap between successive photographs is usually 50 percent.

oblique order 1. An ECHELONED ATTACK formation calling for an advance by a reinforced WING to be followed by subsequent attacks on an adjacent LINE, thus preventing the enemy from shifting forces for fear of exposing a FLANK. 2. A FORMATION that has troops positioned at a 45-degree angle from the supposed FRONT.

obscurant A material (e.g., smoke or chaff) used to conceal an object from observation by a radio or optical SENSOR. Smoke may be used to conceal an object from observation by an optical sensor, and chaff may be used to conceal an object from observation by a radio sensor (e.g., RADAR).

obscuration The effects of weather, battlefield dust and debris, or the use of smoke MUNITIONS to hamper observation and TARGET ACQUISITION capability or to conceal the activities or movement of a military force.

obscuration fire A category of FIRE using smoke of other OBSCURANT directly on or near the enemy, with the primary purpose of suppressing OBSERVERS and minimizing the enemy's vision both within and beyond his position area.

obscuration smoke Smoke used on or near the enemy with the primary purpose of minimizing his vision both within and beyond his position area, or to cause an enemy force to vary its speed, inadvertently change direction, deploy prematurely, or rely on nonoptical means of communication. See also IDENTIFICATION SMOKE, SCREENING SMOKE, SMOKE.

observation post A position, possibly airborne, from which military observations are made or FIRE is directed and adjusted, and which has appropriate communications. See also FORWARD OBSERVER.

observed fire FIRE for which the POINT OF IMPACT or burst can be seen by an observer. The fire can be controlled and adjusted on the basis of observation. Compare to INDIRECT FIRE.

observed fire chart A chart, usually a gridded sheet, on which the relative locations of BATTERIES of a BATTALION and its TARGET are plotted from data obtained as a result of firing.

observer A member of a fire support unit (artillery or naval) who is positioned forward enough to the line of fire that he can watch rounds hit and report back (via radio) information to the gunners that will allow them to deliver more accurate fire. Compare to FORWARD OBSERVER.

observer-target distance The distance along an imaginary straight line from an OBSERVER to a TARGET.

observer-target line (OT line) An imaginary straight line from an OBSERVER or SPOTTER to a TARGET.

observers Representatives from other allied or neutral nations, usually military officers, who are invited to watch military exercises or, during wartime, actual combat operations.

observing angle The angle at the target between a line to an artillery or naval gunfire OBSERVER and a line to the GUN or BATTERY; the angular distance of an observer from a gun or battery; ANGLE T.

observing interval The time between two successive observations made to secure FIRING DATA on a moving target.

observing line See OBSERVER TARGET LINE.

observing point That point on a target on which an OBSERVER sights to secure FIRING DATA. See also ADJUSTING POINT.

observing sector 1. An area visible from a point of observation. 2. An area assigned to a given POST for observation.

obstacle Any natural or man-made obstruction that canalizes, delays, restricts, or diverts the movement of a military force. The effectiveness of an obstacle is considerably enhanced when it is covered by FIRE. Obstacles can include: ABATIS, antitank ditches, blown bridges, built-up areas, MINEFIELDS, rivers, road craters, terrain, and wire. Types of obstacles include:

1. *Existing obstacles*—Either natural or cultural obstacles already in existence when battle planning begins.
2. *Cultural obstacles*—A man-made feature or series of connected man-made features that disrupt or impede the movement of a combat force.
3. *Natural obstacle*—Any existing obstacle or area created by nature that disrupts or impedes the movement of a combat force.
4. *Reinforcing obstacles*—Obstacles specifically constructed, emplaced, or detonated to assist an anticipated military action or one already in progress by canalizing, delaying, or disorganizing enemy movement.
5. *Standard obstacles*—A guide normally prepared by a division engineer which lists all the types of obstacles that will be employed by the division's units.

obstacle, flanking See FLANKING OBSTACLE.

obstacle-approach angles The angles formed by the inclines at the base of a positive (e.g., a hill) or top of a negative (e.g., a decline) vertical OBSTACLE that a vehicle must negotiate in surmounting the obstacle.

obstacle base width The distance across the bottom of an OBSTACLE.

obstacle course An area filled with hurdles, fences, ditches, and other OBSTACLES. It is used to train soldiers in overcoming similar obstacles in the field, and to develop their quickness, endurance, and agility.

obstacle length The length of the long axis of an obstacle.

obstacle plan That part of an OPERATION PLAN (or order) concerned with the use of OBSTACLES to enhance friendly firing or to canalize, direct, restrict, delay, or stop the movement of an opposing force. Obstacle plans are used at CORPS level and below.

obstacles, forward See FORWARD OBSTACLES.

obstacles, intermediate See INTERMEDIATE OBSTACLES.

obstacles, rear See REAR OBSTACLES.

obstacle spacing The horizontal distance between the edges of vertical OBSTACLES.

obstacle-spacing type The pattern of location of a series of OBSTACLES (linear or random).

obstacle system A coordinated series of related OBSTACLES designed to canalize and disorganize enemy forces, to delay or stop enemy movement, and to otherwise aid in the accomplishment of a UNIT mission.

obstacle vertical magnitude The vertical distance from the base of a vertical OBSTACLE to the crest of the obstacle; how high it is.

occulter A shutter for closing off the beam of a searchlight when it is not being used, so that it cannot be seen and located by the enemy.

occupation of position Movement into and proper organization of an area to be used as a battle position.

occupied territory Territory under the authority and effective control of a belligerent armed force. The term is not applicable to territory being administered pursuant to peace terms, a treaty, or other agreement, express or implied, with the civil authority of the territory. See also CIVIL AFFAIRS AGREEMENT.

offense 1. An attack or assault on an enemy. 2. A combat operation designed to carry the fight to the enemy. Offensive operations are undertaken to destroy enemy forces, secure key terrain, deprive the enemy of resources, deceive or divert the enemy, develop intelligence, and destroy the enemy's will to continue a battle. Offensive operations include DELIBERATE ATTACK, HASTY ATTACK, MOVEMENT TO CONTACT, EXPLOITATION , PURSUIT, and other limited-objective operations.

offense, preferential See PREFERENTIAL OFFENSE.

offensive 1. The condition of a force when it is attacking. 2. Attacking; ready to attack. 3. Suitable for attack; used for attack. Guns and tanks are often offensive weapons. 4. An attack, especially one on a large scale. NAPOLEON advised that: "When you have once undertaken the offensive, it should be maintained to the last extremity. A retreat, however skillful the maneuvers may be, will always produce an injurious moral effect on the army, since by losing the chances of success yourself you throw them into the hands of the enemy. Besides, retreats cost far more, both in men and materiel, than the most bloody engagements; with this difference, that in a battle the enemy loses nearly as much as you, while in a retreat the loss is all on your side." Compare to PRINCIPLES OF WAR. See Phillip A. Petersen and John G. Hines, "The Conventional Offensive in Soviet Theater Strategy," *Orbis* 27, 3 (Fall 1983); Gary L. Guertner, "Offensive Doctrine in a Defense-Dominant World," *Air University Review* 37, 1 (November–December 1985).

offensive, cult of See CULT OF THE OFFENSIVE.

offensive, strategic See STRATEGIC OFFENSIVE.

offensive air support TACTICAL AIR OPERATIONS that directly support a land battle.

offensive counter-air operation An operation designed to destroy enemy air power as close to its source as possible.

offensive grenade A high-explosive HAND GRENADE designed for use by troops advancing in the open. The body of the grenade is made of fiber so that metal fragments are not thrown back at the attackers.

offensive minefield In naval mine warfare, a MINEFIELD laid in enemy territorial water or waters under enemy control.

officer A person holding a commission or warrant in one of the armed forces signifying his or her position of command or authority. See also COMMISSIONED OFFICER, NONCOMMISSIONED OFFICER FIELD GRADE, company grade.

officer, advance See ADVANCE OFFICER.

officer, commissioned See COMMISSIONED OFFICER.

officer, confinement See CONFINEMENT OFFICER.

officer, control See CONTROL OFFICER.

officer, duty See DUTY OFFICER.

officer, executive See EXECUTIVE OFFICER.

officer, fighter liaison See FIGHTER LIAISON OFFICER.

officer, fire support See FIRE SUPPORT OFFICER.

officer, flag See FLAG OFFICER.

officer, general See GENERAL OFFICER.

officer, ground liaison See GROUND-LIAISON OFFICER.

officer, junior See JUNIOR OFFICER.

officer, line See LINE OFFICER.

officer, morale-support See MORALE-SUPPORT OFFICER.

officer, noncommissioned See NONCOMMISSIONED OFFICER.

officer, operations See OPERATIONS OFFICER.

officer, ordnance See ORDNANCE OFFICER.

officer, range See RANGE OFFICER.

officer, releasing See RELEASING OFFICER.

officer, reserve See RESERVE OFFICER.

officer, responsible See RESPONSIBLE OFFICER.

officer, safety See SAFETY OFFICER.

officer, warrant See WARRANT OFFICER.

officer in charge 1. Any officer in a position of command during the absence of the commanding officer. 2. An officer given specific responsibility for something.

officer of the day The officer on duty, usually for a 24-hour period, who represents the commander of a unit on all routine matters. The naval equivalent is "officer of the deck." In the Air Force the equivalent is "air officer of the day."

officer of the guard The officer in charge of a guard detail or unit.

offset bombing Any bombing procedure that employs a REFERENCE or AIMING POINT other than the actual target.

offset method A way of describing locations on a map by giving the distance from the bottom of the map and to the left or right of a secretly designated north-and-south line.

offset plotting A method of plotting FIRING DATA when different RANGES and AZIMUTHS must be sent to each gun of a BATTERY.

offset point In AIR INTERCEPTION, a point in space relative to a target's flight path, toward which an interceptor is vectored and from which the final or a preliminary turn to an attack heading is made.

offset post A POST identified for elimination or disestablishment when establishing a newly authorized post.

offset registration fire In FIELD-ARTILLERY operations, REGISTRATION FIRE from a supplementary position.

ogive The curved pointed nose (front) of a bullet, rocket or other projectile.

on An element of a tank fire command, directing the gunner to halt the traverse of the TURRET; usually preceded by steady. See also STEADY ON.

on call 1. A standby arrangement for materiel or personnel. 2. A term used to signify that a prearranged concentration, air strike, or final protective fire may be called for. See also CALL FOR FIRE; CALL MISSION.

on-call target 1. In ARTILLERY AND NAVAL GUNFIRE SUPPORT, a planned target other than a scheduled target on which FIRE is delivered when requested. The purpose of designating an on-call target is to reduce the reaction time required to initiate firing on a target. The degree of prearrangement for an on-call target will influence the reaction time from request to execution (the greater the prearrangement, the less the reaction time). 2. A planned NUCLEAR WEAPONS target (other than a scheduled nuclear target) for which a strike can be anticipated but which will be destroyed upon request rather than at a specific time.

one-day's supply A unit or quantity of supplies adopted as a standard of measurement, used in estimating the average daily expenditure of the same supplies under stated conditions. It may also be expressed in terms of a factor, e.g., rounds of ammunition per weapon per day. See also STANDARD DAY OF SUPPLY; COMBAT DAY OF SUPPLY.

one-hundred percent rectangle An area that includes practically all of the shots fired by an ARTILLERY gun or BATTERY at a target.

on-equipment materièl Items of supply which, although not part of the equipment proper, are issued with and accompany equipment. They are required for equipment maintenance, operation, armament, fire protection, communications, etc., Examples are gun mounts, guns, radios, flashlights, fire extinguishers, sighting and fire control equipment, specified equipment (spare) parts, and tools for maintenance of the equipment.

one-time cryptosystem 1. A CRYPTOSYSTEM employing a one-time key (a code used only once). 2. A cryptosystem in which a CIPHER alphabet is used only once.

one-time pad A one-time key for a CRYPTOSYSTEM, printed on the pages of a pad and designed to permit the destruction of each page as soon as it has been used.

one-time tape A tape used as the keying element in a ONE-TIME CRYPTOSYSTEM.

on guard 1. Ready to defend or protect. 2. Watching, as a GUARD or a member of a guard. 3. The first position of readiness in BAYONET exercises.

on-launcher reliability The percentage of TACTICAL MISSILES, of those loaded on LAUNCHERS for firing, that are fired within the required time limits.

online crypto-operation The use of CRYPTOEQUIPMENT that is directly connected to a signal line, making ENCRYPTION and transmission, or reception and decryption, or both together, a single, continuous process. See also OFFLINE CRYPTO-OPERATION.

on-site inspection 1. A procurement inspection performed at a contractor's facilities for the purpose of determining that military supplies and services conform to the specifications cited in the contractual documents. 2. Allowing OBSERVERS from another country with which one has entered into an ARMS CONTROL agreement check one's own military bases or arms-manufacturing and supply points. The INF TREATY provides for mutual on-site inspection in both the United States and the Soviet Union to insure that certain classes of nuclear missiles are being destroyed according to the terms of the treaty. See Richard L. Shearer Jr., *On-Site Inspection for Arms Control: Breaking the Verification Barrier* (Washington, D.C.: National Defense University Press, 1984).

on station 1. In AIR INTERCEPTION, a code meaning, "I have reached my assigned station." 2. In CLOSE AIR SUPPORT and AIR INTERDICTION, a code meaning "airborne aircraft are in position to attack targets or to perform the mission designated by a control agency." 3. A plane orbiting on patrol.

on the double A command meaning twice as fast as normal; as fast as humanly possible. This phrase is often appended to an order to indicate how rapidly it should be carried out.

open city A concept in INTERNATIONAL LAW that allows an occupying army to withdraw its troops from a city and declare it an "open city" so that incoming opposition forces know that they do not have to fight or bomb their way in and thus destroy much of it and its people. For example, when defeat by the Germans was inevitable in 1940, France declared Paris an "open city." Under international law, further defense of the city by French troops would have constituted a crime.

open code A cryptosystem using external text which has meaning to disguise a hidden meaning.

open column A motor column in which the distance between vehicles is increased to accomplish greater dispersion.

open order 1. A spread-out, irregular FORMATION of troops as opposed to a close order. Troops expecting contact with the enemy would be in open order to make it more difficult for the enemy to target many of them at once. 2. An order allowing a member of the military to do anything so long as it pertains to his or her basic job.

open ranks 1. An arrangement of ranks in a CLOSE ORDER DRILL in which the normal distance between ranks is increased by the length of a full step. 2. The preparatory command to take this position.

open route A roadway over which a minimum of supervision is exercised. Supervision ordinarily is limited to the control of traffic at intersections, and is analogous to civilian control over rural roads carrying a low volume of traffic.

open sheaf 1. The distribution of the FIRE of two or more PIECES so that adjoining POINTS OF IMPACT or points of burst are separated by the maximum effective width of burst of the type of SHELL being used. 2. A term used in a CALL FOR FIRE to indicate that the observer desires a wider SHEAF than the one being employed. See also CONVERGED SHEAF; PARALLEL SHEAF; SPECIAL SHEAF.

open source information Information of potential INTELLIGENCE value which is available to the general public.

operating forces Those forces whose primary missions are to participate in and support COMBAT. See also COMBAT SERVICE SUPPORT; COMBAT SUPPORT.

operating handle The handle or bar with which the operating lever of a GUN is operated to open and close the BREECH of the gun.

operating-level factor A factor used to identify the days of supply needed for a military operation. This factor, when divided into the total quantity demanded results in the operating level quantity.

operating level of supply The quantities of MATERIEL required to sustain military operations in the interval between requisitions or the arrival of successive shipments. These quantities should be based on the established replenishment period (monthly, quarterly, etc.).

operating program The plan prepared by a command, agency, or installation that lists the annual OBJECTIVES to be attained by a military force by relating the objectives to available resources.

operating schedule 1. A schedule that indicates the required time phasing of accomplishments. 2. A detailed schedule required in programming and budgeting that sets forth the time phasing of a particular military objective.

operating slide A mechanism in a Browning MACHINE GUN that permits opening the BREECH for loading, unloading, and cleaning out STOPPAGE, and closing the breech for firing.

operating strength The present and absent strength of an organization; it does not include INTRANSIT STRENGTH.

operation The carrying out of a strategic, tactical, service, training, or administrative military mission; COMBAT, including any movements or maneuvers needed to gain the objectives of a battle or campaign.

operation annexes Amplifying instructions which are so voluminous or technical that their inclusion in the body of a plan or order is undersirable.

operation, autonomous See AUTONOMOUS OPERATION.

operation, biological See BIOLOGICAL OPERATION.

operation, combined See COMBINED OPERATION.

operation, deception See DECEPTION OPERATION.

operation, defoliant See DEFOLIANT OPERATION.

operation, delay See DELAY OPERATION.

operation, delaying See DELAYING OPERATION.

operation, denial See DENIAL OPERATION.

operation, intruder See INTRUDER OPERATION.

operation, joint See JOINT OPERATION.

operation, radiological See RADIOLOGICAL OPERATION.

operation, withdrawal See WITHDRAWAL OPERATION.

operation exposure guide The maximum amount of nuclear radiation exposure that a commander considers permissible for his unit while performing a particular mission or missions.

operation map A map showing the location and strength of friendly forces involved in an operation. It may also indicate the predicted movement and location of enemy forces.

operation order A formal directive issued by a commander to subordinate commanders for the purpose of effecting the coordinated execution of an operation.

operation plan A plan for a single or series of connected operations to be carried out simultaneously or in succession. It is usually based on stated assumptions, and is the form of directive used by higher authorities to permit subordinate commanders to prepare supporting plans and orders. The designation "plan" is usually used instead of "order" in preparing for operations well in advance. An operation plan may be put into effect at a prescribed time, or on signal, and then becomes the operation order. It basically explains the who, what, when, where, and how of a mission.

operational chain of command The CHAIN OF COMMAND established for a particular operation or series of continuing operations. The operational chain of command in the U.S. goes from the president to the secretary of defense to the joint chiefs of staff to unified commanders. This is in contrast to the administrative chain of command, which goes from the president to the service secretaries.

operational characteristics The specific military qualities required of an item of equipment to enable it to meet its intended operational function.

operational command 1. The authority granted to a commander to assign missions or tasks to subordinate commanders, to deploy units, to reassign forces, and to retain or delegate OPERATIONAL or TACTICAL CONTROL as may be deemed necessary. It does not of itself include responsibility for the administration or LOGISTICS. See also ADMINISTRATIVE CONTROL. 2. The forces assigned to a commander.

operational control The authority delegated to a commander to direct forces provided to him so that he may accomplish specific missions or tasks that are usually limited by function, time, or location; and to deploy the necessary UNITS and retain or assign TACTICAL CONTROL of those units. It does not of itself include administrative or logistic responsibility, discipline, internal organization, or UNIT TRAINING.

operational environment A composite of the conditions, circumstances, and influences that affect the employment of military forces and bear on the decisions of the unit commanders.

operational evaluation The test and analysis of a specific end item or system, insofar as practicable under actual operating conditions, to determine if its production or purchase in quantity is warranted. See also ACCEPTANCE TRIAL.

operational exposure guide A method used to specify the acceptable level of radiation exposure for friendly troops during a tactical operation. It is used for operational planning and requires records of radiation exposure. See also RADIATION STATUS.

operational headquarters A HEADQUARTERS primarily concerned with COMMAND AND CONTROL of the execution of operational missions.

operational intelligence INTELLIGENCE required for planning and executing all types of operations.

operational interchangeability The ability to substitute one item for another of different composition or origin without loss in effectiveness, accuracy, and safety of performance.

operational level A mid-range between tactical and strategic levels of operations. Traditionally, military activities have been divided into two levels, STRATEGY and TACTICS. "Strategy" tended to cover the planning and movements of whole armies, while "tactics" was understood to deal with factors at the immediate combat level. The latter covered procedures for maneuvering units up to the level of a DIVISION, but also as small as an infantry rifle group. In slightly different ways, during World War II, both the Germans and the Soviets began to think about an intermediate range of activities. This went beyond the relatively mechanical business of bringing small units into contact with the enemy, but did not touch on the major, often political, questions of where a nation's overall effort should be directed, and what its war aims should be. This intermediate range of activities has subsequently come to be called the "operational level." It forms part of the DOCTRINE and war planning of many armies, including not only its originators in Germany and the Soviet Union, but also armies as different as those of Israel and the United States. No two armies define the operational level (the technique is sometimes called "operational art") in quite the same way. The easiest way of thinking about it is to see operational activities as covering the whole of a THEATRE OF WAR, for example the CENTRAL FRONT in Europe, and necessarily involving all services and arms, in the pursuit of very broad definitions of victory in that area. It differs from strategy in that decisions on major resource allocation and general war objectives, such as whether to aim for unconditional surrender, armistice, or some other final outcome, will already have been made. However, commanders will retain a considerable degree of discretion within the operational level, and their task will be much greater and more complex than simply to bring the enemy to battle in favorable circumstances. See Edward N. Luttwak, "The Operational Level of War," *International Security* 5 (Winter 1980–1981); William J. Bolt and David Jablonsky, "Tactics and the Operational Level of War," *Military Review* 67, 2 (February 1987); Mark H. Gerner, "Leadership at the Operational Level," *Military Review* 67, 6 (June 1987); Gordon R. Sullivan, "Learning to Decide at the Operational Level of War," *Military Review* 17, 10 (October 1987).

operational maneuver groups Mobile military formations of WARSAW PACT forces to be used in a CENTRAL FRONT war. Because of the fear of BATTLEFIELD NUCLEAR WEAPONS, a wide distribution of troop formations is essential to avoid presenting easy targets for the NORTH ATLANTIC TREATY ORGANIZATION'S (NATO) nuclear forces. As a consequence, the Warsaw Pact is committed to a plan in which its armies would approach the battle area in a series of waves (see ECHELONED ATTACK.) This means that the whole military force is not assembled for a BREAKTHROUGH at any one or two points, and does not, therefore, present such a suitable target for nuclear attack. The main disadvantage to this deployment is that it deprives the commanders of the weight of troops with which to exploit any breakthrough that does occur (see CONCENTRATION OF FORCE). To overcome this difficulty, the Warsaw Pact plans to combine its echeloned forces with a series of compact, mobile formations called operational maneuver groups. These would be of roughly divisional size (or perhaps 10,000 troops, in armored and armored infantry units), too small and fast-moving to make a particularly attractive target, but capable of exploiting weak spots exposed by the linear attack of the first or second echelon. They would not fight PITCHED BATTLES against strong reserves, but bypass any fortified areas and penetrate as far as possible behind NATO front lines in order to cut communications and paralyse efforts to reorganize. See C. N. Donnelly, "The Soviet Operational Maneuver Group: A New Challenge for NATO," *International Defense Review* (September 1982); Michael Ruehle, "The Soviet Operational Maneuver Group: Is the Threat Lost in a Terminological Quarrel?" *Armed Forces Journal* (August 1984); Chris Bellamy, "Antecedents of the Modern Soviet Operational Manoeuvre Group (OMG), *Journal of the Royal United Services Institute (RUSI)* (September 1984).

operational missile A missile that has been accepted for tactical or strategic use.

operational missile launcher A LAUNCHER that has been accepted by the using service and has been issued to its units for tactical or strategic use.

operational ration A specially designed RATION of food normally composed of nonperishable items, for use under actual or simulated combat conditions. It is used in peacetime for emergencies, contingencies, travel, and training.

operational readiness The capability of a UNIT or WEAPON SYSTEM to perform the missions or functions for which it was organized or designed; to be combat ready.

operational readiness, immediate See IMMEDIATE OPERATIONAL READINESS.

operational readiness evaluation An evaluation of the operational capability and effectiveness of a unit.

operational reserve An emergency reserve of PERSONNEL or MATERIEL or both established for the support of a specific operation. (See also RESERVE SUPPLIES.) Any sound military plan will retain an operational reserve under the overall commander's control, in order to reinforce active units either in terms of exploiting their strength or to shoring up weak points. Typically the reserve will be one that is organized one level below the level of the unit involved. Thus, if an attack is being carried out by a DIVISION, there should be a BRIGADE in reserve, if by a PLATOON, a SQUAD.

operational stocks Supplies held to meet possible operational requirements over and above minimum allowances. See WAR RESERVE.

operational supplies Those supplies, over and above the normal allowances of an overseas theater of operations, which are required to support the logistic and operational plans of the theater.

operational test An OPERATIONAL EVALUATION.

operational training Training that develops, maintains, or improves the OPERATIONAL READINESS of individuals or units.

operationally ready Capable of performing assigned missions or functions; available and qualified to perform assigned missions or functions.

operationally ready missile An OPERATIONAL MISSILE on a serviceable LAUNCHER, connected to serviceable FIRE CONTROL EQUIPMENT.

operations A command or planning center.

operations, area of See AREA OF OPERATIONS.

operations, base of See BASE OF OPERATIONS.

operations, commando See COMMANDO OPERATIONS.

operations, continuous See CONTINUOUS OPERATIONS.

operations, counterbarrier See COUNTERBARRIER OPERATIONS.

operations, countermobility See COUNTERMOBILITY OPERATIONS.

operations, covert See COVERT OPERATIONS.

operations, normal See NORMAL OPERATIONS.

operations, overt See OVERT OPERATIONS.

operations, psychological See PSYCHOLOGICAL OPERATIONS.

operations, supporting See SUPPORTING OPERATIONS.

operations center, battery See BATTERY OPERATIONS CENTER.

operations center, joint See JOINT OPERATIONS CENTER.

operations code A code capable of being used for general communications. It is composed largely, though not exclusively, of single words and phrases and permits spelling.

operations officer An officer at the REGIMENT or SQUADRON level designated to assist the unit commander. He will be charged with the detailed planning of combat operations decided on by the commander and his staff. For example, an air force operations officer will pick crews and allot them targets and routes to accomplish the general mission selected for them by the squadron commander.

operations security (OPSEC) The process of denying adversaries information about friendly capabilities and intentions by identifying, controlling, and protecting indicators of future military operations.

opportunity target See TARGET OF OPPORTUNITY.

opposing-force program A training program that focuses peacetime preparedness training on the tactical vulnerabilities of potential adversaries. It is designed to emphasize the competition inherent in battle by providing a credibly realistic opposing force in training, which utilizes the DOCTRINE, TACTICS, and WEAPONS SYSTEMS of actual potential adversaries.

oppositive numbers Officers having corresponding (parallel) duty assignments within their respective military establishments.

OPSEC See OPERATIONS SECURITY.

optimum height The height of an explosion that will produce the maximum effect against a given target.

orbit, geostationary See GEOSTATIONARY ORBIT.

order A communication, written, oral, or by signal, which conveys instructions from a superior to a subordinate. In a broad sense, the terms "order" and "command" are synonymous. However, an order implies discretion as to the details of execution, whereas a command does not. It is critical that orders be clear and unambiguous. HELMUTH VON MOLTKE (1800–1891), the founder of the German general staff in the last century, was fond of saying: "Remember, gentlemen, an order that can be misunderstood will be misunderstood."

order arms 1. In close order drill the holding of a rifle at the right side with the butt resting on the ground. 2. The command to move rifles to the above position. 3. In a British context, to bring a rifle to a salute position.

order of battle The identification, strength, command structure, and disposition of the personnel, units, and equipment of any military force. An order of battle is simply a list of the units in a military force deployed and intended to take part in any imminent or ongoing conflict. To discover the enemy's order of battle is a prime task of MILITARY INTELLIGENCE, because knowing exactly which units, with their special capabilities, are scheduled for combat in any theater can tell an intelligence analyst a good deal about the tactics and strategy that the enemy has planned. In past wars the order of battle was a highly prized secret, but the development of SIGNALS INTELLIGENCE and other technical means of intelligence gathering, such as SATELLITE observation, has made it much less possible to disguise the order of battle.

order of battle, electronic See ELECTRONIC ORDER OF BATTLE.

order of battle card A single or master standardized card containing basic information on enemy ground forces, units, and formations, providing all pertinent information about the enemy's order of battle.

order of march The sequence by which subunits of a larger military force travel to a new location.

orderly An enlisted rank assigned to cleaning tasks (such as a latrine orderly) or to wait upon officers.

orderly book A record book for a military unit.

orderly room The office of a COMPANY, in which the business of the company is done.

orders, general See GENERAL ORDERS.

orders group A standing group of key personnel that a commander at any level desires to be present when he issues his concept of an operation and his order to execute it. For example, a company commander will expect his PLATOON commanders, and often their senior NONCOMMISSIONED OFFICERS, as well as his immediate staff to be present. In the United Kingdom this is abbreviated to "0 group."

ordinary leave An authorized absence from assigned duty.

ordinary priority A category of IMMEDIATE MISSION REQUEST that is lower than URGENT PRIORITY but takes precedence over SEARCH AND ATTACK priority, e.g., a TARGET that is delaying a unit's advance but which is not causing casualties.

ordnance 1. Military weapons in general along with all of the associated ammunition and equipment. 2. Just artillery. 3. That brand of a military organization that deals with the development, procurement, issuance, and storage of weapons.

ordnance officer 1. An officer who is a member of an ORDNANCE corps. 2. A special staff officer who advises commanders on technical matters of ordnance. In this meaning, also called an ordnance staff officer. 3. An officer having responsibilities for dealing with ordnance maintenance, ammunition, and general supply, including that of placing captured enemy MATERIEL into usable condition.

ordnance service All activities necessary to maintain the ORDNANCE equipment of a command in usable condition, and such other equipment as directed by proper authority.

ordnance troops Technically trained troops assigned or attached to a TACTICAL UNIT to provide ORDNANCE maintenance, supply, or technical service. They also give instruction in the use, maintenance, and adjustment of ordnance MATERIEL.

organic Assigned to and forming an essential part of a military organization.

organization 1. Any military unit or larger COMMAND composed of two or more smaller units. In this meaning, a military ELEMENT of a command is an organization in relation to its components, and a UNIT in relation to higher commands. 2. The definite structure of a military element prescribed by a component authority such as a table of organization.

organization for embarkation The administrative grouping of a LANDING FORCE for movement overseas. It includes, in any ship or EMBARKATION group, the TASK ORGANIZATION that is established for landing, as well as additional forces embarked for purposes of transport, labor, or security.

organization for landing The specific tactical grouping of a LANDING FORCE for an ASSAULT.

organization of the ground The development of a defensive position by strengthening the natural defenses of the terrain and by assignment of the occupying troops to specific localities.

organizational equipment That equipment, other than individual equipment, that is used in furthering the common mission of a unit.

organizational repair Equipment parts that are authorized to an organization for its own use.

organize for combat To develop an organization in such a way that the unique capabilities of different-type forces complement each other.

organized position An area in which TROOPS and WEAPONS have been put in position for future action, and in which FIELD FORTIFICATIONS have been constructed.

organized strength The actual, authorized, or programmed strength of all TABLE OF ORGANIZATION AND EQUIPMENT units of an army, or subdivisions thereof.

orient 1. To place in the right position; to place a map so that the arrow on it that shows direction points in the proper direction, or so that the meridian lines of a map point north. 2. To set the correct angular readings for a weapon or instrument so that they read correctly for the location and for the direction in which the weapon or instrument is pointing. For some weapons and instruments, the ELEVATION adjustment is included in this category.

orienting angle A horizontal clockwise angle from the LINE OF FIRE to the ORIENTING LINE.

orienting line A line of known direction, established on the ground, that is used as a reference line in aiming ARTILLERY or other weapons.

orienting point A distant object used in aligning a DIRECTOR or other instrument with a gun.

orienting station A point on the ORIENTING LINE near the gun position from which the gun BATTERY may be oriented.

origin 1. A fixed point of reference on a graph, map, or chart. 2. The center of the MUZZLE of a gun at the instant of its firing.

origin of the trajectory The center of the MUZZLE of a gun at the instant when a PROJECTILE leaves it.

oscillating mine A hydrostatically controlled MINE that maintains a pre-set depth below the surface of the water independently of the rise and fall of the tide.

OT line See OBSERVER-TARGET LINE.

other military targets Military targets, including armaments depots, logistic bottlenecks, troop concentrations, and airbases and ports, not directly part of nuclear attack forces. While these are not HARD targets, they still require a precision of attack, particularly if COLLATERAL DAMAGE to civilians is to be minimized, that makes them unsuitable targets for many NUCLEAR WEAPONS.

Outer Space Treaty of 1967 A multilateral treaty signed and ratified by both the United States and the Soviet Union. Article IV of the Outer Space treaty forbids basing NUCLEAR WEAPONS or other weapons of mass destruction in space.

outflank To gain a tactical advantage by maneuvering troops to the side and rear of an enemy; thus taking the enemy by surprise and forcing him to weaken his front line in response.

over In ARTILLERY AND NAVAL GUNFIRE SUPPORT, an observation by a SPOTTER to indicate that a BURST(s) occurred beyond the TARGET.

overkill A concept found more in the language of nuclear disarmament than strategic theory, which has no very precise meaning. In general, those who believe that overkill has been reached in NUCLEAR WEAPONS capability argue that the total of the nuclear arsenals of members of the NUCLEAR CLUB is so high that enough destructive power has been accumulated to kill every inhabitant of the world many times over with power to spare. An alternative and weaker version simply holds that the United States and Soviet Union each have more weapons than they need to destroy the other nation completely. See David Alan Rosenberg, "The Origins of Overkill: Nuclear Weapons and American Strategy," *International Security* 7, 4 (Spring 1983).

overlay A printing or drawing on a transparent or semi-transparent medium at the same scale as a map, chart, or other cartographic device to show details not appearing or requiring special emphasis on the original.

overpackaging The use of more (quantitative or qualitative) preservation, packaging, or packing materials than is necessary to protect an item adequately. This term should not be confused with OVERPACKING.

overpacking The repacking of containers or items into more substantial and suitable containers, to withstand handling and transportation hazards, or the addition of packaging materials such as steel, stripping, waterproof caseliners into fiberboard sleeves or boxes, for example, to render the existing containers less susceptible to damage or pilferage during handling, transportation, and storage.

overpressure The pressure resulting from the BLAST WAVE of an explosion. It is referred to as "positive" when it exceeds atmospheric pressure, and "negative" when the pressures resulting from the wave are less than atmospheric pressure.

overshipment Freight received in excess of that listed or manifested.

over-the-horizon backscatter radar This radar uses high-frequency radio waves that refract in the ionosphere (a layer of charged particles 50 to 250 miles about the Earth) and return to Earth at over the horizon ranges (500 to 1,800 miles from the radar). This is the same principle that allows amateur band (ham) radio operators to talk to each other around the world. The radio energy reflects off aircraft and returns to the radar via the same ionospheric path as the transmitted energy. The radar avoids other users of the high-frequency band by using a very narrow range of frequencies at any time.

over-the-horizon radar A RADAR system that makes use of the atmospheric reflection and refraction phenomena to extend its range of detection beyond line of sight.

over-the-shoulder bombing A special case of LOFT BOMBING in which the bomb is released so that it is thrown back to the target. See also TOSS BOMBING.

overt operations The collection of INTELLIGENCE information openly, without concealment. Compare to COVERT OPERATIONS.

overwatch 1. The readiness of a military unit or part of a unit to FIRE or MANEUVER as a result of enemy actions against another friendly unit. 2. The tactical role of an ELEMENT positioned to observe the movement of another element and to support it with fire.

oxidizer That portion of the materials in an EXPLOSIVE that oxidizes or causes combustion in another substance. Such oxidizing agents include liquid oxygen and nitric acid.

oxime A class of drugs that is used in the pre-treatment of NERVE AGENT poisoning.

P

P The letter designation for a pursuit tactical aircraft, such as the World War II P-38 or P-51; no longer in current use.

pace The speed of a COLUMN regulated to maintain a prescribed average speed.

pace setter An individual, selected by a column commander, who travels in the lead vehicle or element of a column to regulate the column speed and establish the required PACE.

pacification 1. A military MISSION to gain control of an area so that no hostile activities can occur there. Pacification was the primary objective of the United States in Vietnam. It was officially defined by the U.S. Military Assistance Command Vietnam in November 1967 as "the military, political, economic, and social process of establishing or re-establishing local government responsive to and involving the participation of the people. It includes the provision of sustained, credible territorial security, the destruction of the enemy's underground government, the assertion or reassertion of political control and involvement of the people in government, and the initiation of economic and social activity capable of self-sustenance and expansion. The economic element of pacification includes the opening of roads and waterways and the maintenance of lines of communication important to economic and military activity." 2. The role a modern military force plays when directed against a society with only primitive weapons.

package (nuclear) A grouping of NUCLEAR WEAPONS, planned for employment in a specified area during a short time period according to their specific YIELDS.

packaged forces Forces of varying size and composition preselected for specific MISSIONS in order to facilitate planning, training and loading on transports.

packboard A lightweight, rectangularly shaped frame, fitted with shoulder straps and bindings. It facilitates carrying loads on a person's back by properly distributing their weight.

padding Extraneous text added to a coded message for the purpose of concealing its beginning, ending, or length. See also CRYPTOGRAPHY.

padre A military chaplain of any Christian denomination.

pancake 1. In AIR INTERCEPTION, a code meaning, "Land," or, "I wish to land." 2. To crash-land a plane.

panel 1. A specifically shaped or colored cloth or other material displayed in accordance with a prearranged CODE to convey messages. 2. An electrical switchboard or instrument board.

panel code A prearranged CODE designed for visual communications, usually between friendly units, by making use of MARKING PANELS.

paper tiger A Chinese expression for someone or some institution that is not as strong or powerful as its appearance or reputation would suggest. China's Chairman Mao was famous for calling the atomic bomb a "paper tiger."

parabomb A specially prepared equipment container with a parachute that is capable of opening automatically after a delay drop (a period of free fall).

parabundle Packed supplies to be dropped by air.

paracaisson A small, two-wheeled, hand-drawn vehicle, the body of which forms an air-delivery container for artillery ammunition and which, upon being assembled, becomes a utility cart.

parachute, free type See FREE TYPE PARACHUTE.

parachute tower, controlled See CONTROLLED PARACHUTE TOWER.

parade 1. An end of day ceremony in which troops assemble in formation and salute the flag as it is lowered. 2. A review of troops assembled for inspection. 3. A place where troops regularly assemble for inspection. 4. A large level area within a fort.

parade rest 1. A standing position in which the left foot is 12 inches to the left of the right foot, the legs are straight, and the hands are clasped behind the back. While at parade rest, a soldier remains motionless and silent. When a soldier has a rifle, parade rest is taken with the feet in this position, with the butt of the rifle on the ground, with the trigger to the front and the muzzle of the rifle in his right hand, extended forward, and with his left hand behind his back. 2. The command to take this position. The British equivalent order is "stand at ease."

paradrop Delivery by PARACHUTE of personnel or cargo from an aircraft in flight.

parallax 1. The seeming change of position of an observed object caused by a change in the observer's position. 2. The angular difference, because of displacement, between the direction from gun to target and the direction from a DIRECTING POINT to the same target.

parallax, false See FALSE PARALLAX.

parallax correction 1. The allowance to be made for the difference in position of a target as measured from a GUN and from the observer's position. 2. In AIR DEFENSE weapon systems, the correction that must be made to compensate for the displacement between remotely located equipment, such as RADAR units and LAUNCHERS, from a battery-directing or reference point such as TARGET TRACKING RADAR.

parallax error 1. The error in an observation caused by making the observation from a position different from the usual one or from the one where FIRING DATA are used. 2. An error made by reading the dial of an instrument from a slant rather than directly from the front.

parallel A trench dug in front of a besieged fort that is parallel to the fort's defenses.

parallel order The classic military formation on the line of battle that has opposing forces lined up opposite each other. Compare to OBLIQUE ORDER.

parallel sheaf In ARTILLERY and NAVAL GUNFIRE SUPPORT, a SHEAF in which the planes (lines) of fire of all PIECES are parallel. See also CONVERGED SHEAF; OPEN SHEAF.

parallel staff An allied or joint STAFF in which one officer from each nation, or service, is appointed to work side by side in each port. Compare to INTEGRATED STAFF.

parallel training A method of instruction in which an individual is given technical training, either basic or advanced, by another individual who is an expert; apprentice training.

paramilitary forces 1. Forces or groups that are distinct from the regular armed forces of any country, but resemble them in organization, equipment, training, or mission. In Europe these are police forces like the French Gendarmerie National, which nevertheless comes under the French Department of Defense, not the Ministry of Justice, as do the other armed forces of France. 2. Fascist private armies, such as the "Brown Shirts" of Nazi Germany. 3. State or local police forces.

paratrooper A soldier trained to jump from an aircraft into a combat area using a parachute. All paratroopers are AIRBORNE troops but not all airborne forces are paratroopers.

parent station An organization or INSTALLATION designated to furnish all or a portion of the common SUPPORT requirements of another installation or separate organization.

parent unit 1. A military unit that commands a variety of subunits. 2. Units, regardless of size, that have a numerical designation and unit identification code.

parity, effective See EFFECTIVE PARITY.

parity, nuclear A vague term for the rough equivalency in the strength of opposing nuclear forces. Parity is an idea of increasing importance to ARMS CONTROL analysis. It is the basic goal of both the United States and Soviet Union in arms-control negotiations, because neither can expect the other to accept a situation of strategic inferiority as legally acceptable in a treaty, although they might in practice settle for such a situation as a *de facto* condition. Equality of nuclear weapons as such is a largely meaningless concept because of the large number of variables involved, and the different needs each side may have both because of its STRATEGIC posture and because of the nature of the opponent as a target. Parity tends to be defined in terms of a situation in which neither side would believe itself to have an adequate superiority to risk an attack on the other. However, this essentially subjective element leads to problems in which one superpower insists that it is in an inferior position, while its opponent has a quite different interpretation. See also ESSENTIAL EQUIVALENCE. See Michael Don Ward, "Differential Paths to Parity: A Study of the Contemporary Arms Race," *American Political Science Review* Vol. 78, No. 2 (June 1984).

parlimentaire An agent employed by a commander of belligerent forces in the field to go in person within the enemy lines for the purpose of communicating or negotiating openly and directly with the enemy commander.

parole 1. A PASSWORD. 2. The practice of releasing prisoners of war upon their swearing not to take up arms again during the war. This was common practice during the early part of the American Civil War.

partial storage monitoring A periodic inspection of major assemblies or components for NUCLEAR WEAPONS, consisting mainly of external observations of humidity, temperatures, and visible damage or deterioration during storage. This type of inspection is also conducted prior to and upon the completion of a movement.

particle beam weapons Weapons similar in general principle to LASER WEAPONS. Like lasers, they involve the output of a narrowly focused beam of very high-energy radiation. But while the energy output of a laser is electromagnetic, particle beams, as the name implies, consist of high-energy nuclear particles. These can be electrically charged particles, protons or electrons, or neutral particles like neutrons. They would be produced in a machine like the accelerators used in laboratories to study subatomic matter: to be able to design such a machine that is also mobile and flexible enough to serve any military purpose seems highly unlikely. Nevertheless, particle-beam projectors are among the many weapons under consideration in the United States' STRATEGIC DEFENSE INITIATIVE. See Kurt Arbenze, "Particle Beam Weapons," *Armada International* (July–August 1982).

partisan warfare See GUERRILLA WARFARE.

party 1. A work detail, such as a GI party to clean barracks. 2. A small group of detached troops, a DETACHMENT. 3. A forthcoming military MISSION.

party commander, beach See BEACH PARTY COMMANDER.

pass 1. A short tactical run or dive by an aircraft at a target; a single sweep through or within the firing range of an enemy air formation. 2. Officially authorized LEAVE for a short period such as 48 or 72 hours. 3. A document authorizing leave.

passage lanes Areas along which a passing UNIT moves to avoid stationary units and OBSTACLES.

passage of command 1. The exchange of responsibility for a SECTOR or ZONE between the commanders of two military units. The time when the command is to pass is determined by mutual agreement between the two unit commanders unless directed by higher HEADQUARTERS. See also: HANDOFF, RELIEF IN PLACE. 2. A formal change of command when a new officer officially assumes control of a unit.

passage of lines A military operation in which a UNIT attacks through another unit which is in contact with the enemy, as when elements of a COVERING FORCE withdraw through the FORWARD EDGE OF THE BATTLE AREA, or when an EXPLOITING FORCE moves through elements of the force that conducted the initial attack. A passage of lines may be designated as a forward or rearward passage.

passage point A place where units will pass through one another either in an advance or a withdrawal. It is located where the commander desires subordinate units to physically execute a PASSAGE OF LINES.

pass in review 1. A march in front of a reviewing officer, during a CEREMONY. 2. The command given to start this movement.

passive A characteristic of SURVEILLANCE actions or equipment that emit no energy capable of being detected; for example, passive sonar is used to hunt for submarines.

passive air defense All measures, other than ACTIVE AIR DEFENSE, taken to minimize the effects of hostile air action. These include the use of COVER, CONCEALMENT, CAMOUFLAGE, DISPERSION, and protective construction. See also air defense.

passive defense Measures taken to reduce the probability of and to minimize the effects of damage caused by hostile action, without the intention of taking the initiative. The use of offensive WEAPONS SYSTEMS to "keep the peace" is the only difference between active defense and passive defense.

passive detection The detection of any object, airborne or otherwise, via observation of the light or energy that it gives off. It differs from active detection in that the latter involves the aiming of a light or energy beam in the vicinity of an object by an observer who then detects the reflected light or energy.

passive homing guidance A system of HOMING GUIDANCE in the receiver in a missile which utilizes radiation from a target. Often this will involve a missile homing in on aircraft's RADAR or on a ground-based radar unit. See also GUIDANCE.

passive mine 1. A mine that will not explode usually for a comparatively short time. 2. A mine that does not emit a signal to detect the presence of a target, in contrast to an active mine. See also ACTIVE MINE.

passive satellite defense All defensive measures, other than ACTIVE DEFENSE, taken to minimize the capability of enemy terrestrial SATELLITES to perform their assigned missions. These measures include the use of COVER, CONCEALMENT, CAMOUFLAGE, DISPERSAL, protective construction, planned movements, DECOYS, and other measures.

passive sensor A SENSOR that detects naturally occurring emissions from a target for tracking or identification purposes.

passover An officer formally considered for promotion but passed over; one who is not promoted.

pass time The time that elapses between the moment when the leading vehicle of a COLUMN passes a given point and the moment when the last vehicle passes the same point.

password A secret word or distinctive sound used to reply to a CHALLENGE. See also COUNTERSIGN.

pathfinder beacon A transmitting device utilizing ELECTROMAGNETIC RADIATION, such a visible light or infrared, ultraviolet, radar, or radio waves, and which provides an identifiable point to assist in the GUIDANCE of aircraft and assembly of ground units.

pathfinders 1. Experienced aircraft crews who lead a formation to the DROP ZONE, BOMB-RELEASE POINT, or TARGET. 2. Teams dropped or air-landed at an OBJECTIVE to determine the best approaches and landing sites for aircraft and helicopters and to establish and operate navigational aids for the purpose of guiding aircraft and helicopter-borne forces to DROP and LANDING ZONES. 3. A RADAR device used for navigating or homing to an objective when visibility precludes accurate visual navigation.

patrol 1. A DETACHMENT of ground, sea, or air forces sent out for the purpose of gathering information or carrying out a destructive, harassing, MOPPING-UP, or SECURITY mission. 2. A MILITARY POLICE patrol, normally consisting of two military policemen, performing enforcement activities in an assigned area during a specific time. See also COMBAT AIR PATROL; COMBAT PATROL; RECONNAISSANCE PATROL; STANDING PATROL.

patrol, combat See COMBAT PATROL.

pattern bombing The systematic covering of a target area with bombs uniformly distributed according to a plan.

payload 1. The sum of the weight of passengers and cargo that a transport aircraft carries. 2. The WARHEAD, its container, and activating devices in a military MISSILE. The payload of a missile is the amount of ORDNANCE that it can carry. The term does not necessarily have the same meaning as "warhead," because the payload may also include PENETRATION AIDS (see DECOY). An alternative use of the term includes the whole of the FRONT END of a missile. In the context of the THROW-WEIGHT of a missile, which refers to the capacity to carry a load into the atmosphere, "payload" usually has this alternative, broader meaning. 3. The SATELLITE or research vehicle of a space probe or research missile. 4. The load (expressed in tons of cargo or equipment, gallons of liquid, or numbers of passengers) that a military motor vehicle is designed to transport under specified conditions of operation, in addition to its unladen weight.

P-day That point at which the rate of production of an item available for military consumption equals the rate at which the item is required by military forces.

P-day concept The procurement and storage of the minimum MATERIEL needed to equip, train, and sustain forces after D-DAY until the production of materiel is expanded and capable of continually meeting its consumption. The P-day for a given item varies because of the variation in the consumption and production capability for each item.

PD-59 See PRESIDENTIAL DIRECTIVE 59.

peacekeeping force A military unit deployed by the United Nations (but comprised of soldiers from neutral nations) to separate hostile parties. Soldiers in such units typically have small arms for personal protection but do not engage in combat; their role is limited to maintaining a buffer between two belligerents. In 1988 the United Nations' peacekeeping forces were awarded the Nobel Peace Prize because they "represent the manifest will of the community of nations to achieve peace through negotiations and the forces have, by their presence, made a decisive contribution toward the initiation of actual peace negotiations." Thousands of peacekeeping forces have long been stationed along the India-Pakistan border, in the Sinai, on the Golan Heights, in Lebanon, and in Cyprus. In 1988 peacekeeping forces were sent to monitor the cease-fires in Afghanistan and between Iran and Iraq. See John Theodorides, "The United Nations Peace Keeping Force in Cyprus (UNFICYP)," *International and Comparative Law Quarterly* 31, 4 (October 1982); Joachim Hutter, "United Nations Peace-keeping Operations," *Aussenpolitik* 36, 3 (1985).

peacetime establishment A table setting out the authorized peacetime manpower requirement for military UNIT, FORMATION, or HEADQUARTERS. Also called "peacetime complement."

peak overpressure The maximum value of OVERPRESSURE from an explosion at a given location; generally experienced at the instant the shock (or blast) wave reaches the location. See also SHOCK WAVE.

peak strength The highest strength of an army, or subdivision thereof, attained during the specific time period.

pelorus A simple mechanical or optical instrument used to obtain the positions of objects on the Earth or of celestial bodies.

penetrability The ability of NUCLEAR WEAPONS to penetrate defensive systems.

penetration 1. In land operations, a form of OFFENSIVE that seeks to break through the enemy's defense and disrupt the defensive system. 2. The recruitment of AGENTS within, or the infiltration of agents or technical monitoring devices into an organization or group for the purpose of acquiring INFORMATION or of influencing its activities. 3. A projectile passing through a target.

penetration, ballistics or See BALLISTICS OR PENETRATION.

penetration, complete See COMPLETE PENETRATION.

penetration aid 1. A method to defeat a military defense, using camouflage, deception, decoys, and countermeasures. 2. A device, mounted on a post-boost stage ROCKET with RE-ENTRY VEHICLES, that is used to confuse defenses. It may be a DECOY or anything else that renders more difficult the defense's job of detecting and killing the re-entry vehicles.

penetration aid, active Any device that helps a nuclear WARHEAD to reach its target by destroying parts of an enemy's defense system.

penetration aid, passive Any device that helps a nuclear WARHEAD to reach its target by confusing enemy defense systems; a DECOY (see entry) is a common example.

Pentagon, The 1. The building that has become the symbol and the headquarters of the Department of Defense. In 1941 it was proposed by the Army as an alternative to the erection of a variety of temporary buildings. In less than four days, plans were made for a mammoth three-story facility to house 40,000 people. The pentagonal design derived from the fact that the original construction site was bounded by five existing roads. Immediately after the Japanese attack on Pearl Harbor, a fourth floor was added to the plan, and later a fifth. While parts of the building were occupied as early as eight months after groundbreaking, it wasn't finished until January 15, 1943. The five-story Pentagon building has five concentric rings connected by ten spoke-like corridors ranging out from the inner, or "A," ring. The combined length of the corridors is seventeen and a half miles, and the total gross floor space is more than 6.5-million square feet. The length of each of the five outer walls is 921 feet, and the structure is just over 71 feet high. Because of the unique design of the building, it should take no more than 10 minutes to walk between any two extremities. About 28,000 people, both civilians and military, work at the Pentagon. 2. Either the High Command of America's military forces, especially the Joint Chiefs of Staff, the civilian authorities in the DOD, or both. In this sense the use is figurative, because important parts of both of these insitutions are not located in the Pentagon building at all.

peptized fuel Thickened flame-thrower fuel to which water or other chemicals are added before mixing, in order to reduce mixing time and increase storage stability.

percentage corrector A mechanical device for correcting the RANGE of a gun and for determining the CORRECTED ELEVATION.

percussion 1. A sharp, light blow, especially one for setting off an EXPLOSIVE. 2. A command to set the TIME FUZE of a PROJECTILE or BOMB in a nonoperating position, to allow the projectile or bomb to be set off by the force of impact.

percussion charge A small HIGH-EXPLOSIVE charge that is set off by the blow of the FIRING PIN. A percussion charge is used to ignite the PRIMER charge in order to fire the PROPELLING CHARGE in a gun.

percussion detonator An item consisting of a BLASTING CAP and EXPLOSIVE elements, designed to detonate an explosive charge.

percussion mechanism A device that contains a FIRING PIN assembly; it slides in the center BORE of the BREECHBLOCK of a gun.

percussion primer A cap or cylinder containing a small charge of HIGH EXPLOSIVE that may be set off by a blow; used in all FIXED and SEMI-FIXED AMMUNITION and in certain types of SEPARATE-LOADING AMMUNITION to ignite the main PROPELLING CHARGE.

perimeter defense A defense without an exposed FLANK, consisting of forces deployed along the perimeter of a defended area.

peripheral war 1. A LIMITED WAR. 2. A secondary war of limited importance to a simultaneous major conflict.

permanent emplacement A fixed setting for a GUN. A permanent EMPLACEMENT is usually made of reinforced concrete, with the BASE PLATE and BASE RING set in the concrete and bolted down.

permanently operating factors The phrase originally used by Stalin to organize a Marxist-Leninist theory of war. In fact, the permanently operating factors have little to do with Marxism, and resemble many military historians' lists of the PRINCIPLES OF WAR (because they dealt with concepts such as the stability of the rear, morale, and the quality and quantity of divisions). It is unclear how significant they have been to the development of Soviet military DOCTRINE. Much lip service was, and some still is, paid to them, but it is characteristic of the presentation of Soviet thinking to use approved concepts and theories, even when the meaning is quite novel. Nevertheless, there seems to be a historical continuity of Soviet military planning which is at least compatible with, even if it does not derive from, Stalin's "factors." There are three characteristics that continue to dominate WARSAW PACT strategy. The first is the enormous importance attached in Soviet strategic thought to surprise, at least tactical surprise, in military operations. The second and third characteristics tend to mitigate against surprise; they are the paramount importance put on massive concentration of force, and the crucial role of ARTILLERY in winning battles.

permissive action link A device included in or attached to a NUCLEAR WEAPON system to preclude arming or launching until the insertion of a prescribed discrete CODE (released by the president of the United States in the event of a war).

Pershing II One of the two intermediate nuclear WEAPONS SYSTEMS deployed by the NORTH ATLANTIC TREATY ORGANIZATION (NATO) in response to the Soviet deployment of the SS-20 missile during the Euromissile crisis from 1977 onwards. The INTERMEDIATE NUCLEAR FORCES (INF) treaty calls for the destruction of all Pershing II missiles.

person, displaced See DISPLACED PERSON.

personal salute A cannon SALUTE given to distinguished visitors at a military establishment.

personal staff Those staff members that a commander elects to coordinate and administer operations directly, instead of through a CHIEF OF STAFF. See also GENERAL STAFF.

personnel 1. People. 2. Those individuals required in either a military or civilian capacity to accomplish an assigned MISSION. 3. All of the employees of an organization.

personnel, filler See FILLER PERSONNEL.

personnel, indigenous See INDIGENOUS PERSONNEL.

personnel, retrograde See RETROGRADE PERSONNEL.

personnel reaction time (nuclear) The time required by personnel to take prescribed protective measures after receipt of a NUCLEAR STRIKE WARNING.

perspective spatial model An optical reconstruction of an area of terrain showing depth. It entails viewing a pair of air photographs through a stereoscope.

PFC Private first class.

PGMs See PRECISION-GUIDED MUNITIONS.

phantom order A draft contract with an industrial establishment for wartime production of a specific product, with provisions for necessary preplanning in time of peace and for immediate execution of the contract upon receipt of proper authorization.

phase A specific part of an OPERATION that is different from those that precede or follow it. Phasing assists in planning and controlling, and may be indicated in terms of time (PREPARATORY FIRE phase), distance (intermediate objective or REPORT LINE phase), terrain (crossing of an OBSTACLE), or occurrence of an event (commitment of a RESERVE). It is normally associated with operations of larger units and with SPECIAL OPERATIONS (e.g., river-crossing and airborne operations).

phased-array radar A type of RADAR that depends on complex signal processing and electronics. A phased-array radar station is immobile and relies on a very large number of transmitting and receiving circuits with overlapping arcs of search. By very rapid switching between arrays of receptors, the same area of coverage is achieved as with a moving radar bowl, with no or only very short interruptions. Traditional radar systems rely on the familiar rotating radar disc to cover the whole sky, or any large portion of it beyond the arc that can be covered by a stationary aerial. Traditional radar systems, while adequate for most purposes, are far too slow for some modern military needs. Obviously the rotating aerial cannot receive signals from an object from which it is turned away. Some military tasks, especially those dealing with BALLISTIC MISSILE DEFENSE, cannot tolerate the delay involved in waiting for the aerial to come round for a second position check on an incoming object. The speed with which incoming RE-ENTRY VEHICLES move, and the tremendous pressure to observe, calculate, and respond to incoming objects in fractions of a second, have created a requirement for rapid and continuous radar monitoring. Phased-array radars provide the technology to meet this demand. Both the United States and the Soviet Union have built major phased-array radar stations.

phase line A line utilized for the control and coordination of military operations, usually a terrain feature extending across the ZONE OF ACTION. Units should always report crossing phase lines, but do not halt unless specifically directed. Phase lines are often used to prescribe the timing of DELAY OPERATIONS. See also REPORT LINE.

phases of training The five formal phases of military training: basic combat training, advanced individual training, basic unit training, advanced unit training, and training in field exercises and maneuvers. A sixth phase, OPERATIONAL READINESS training, is entered into as determined by major commanders.

phonetic alphabet A list of standard words used to identify letters in a message transmitted by radio or telephone. The following are the authorized words, listed in order, for each letter of the alphabet: Alpha, Bravo, Charlie, Delta, Echo, Foxtrot, Golf, Hotel, India, Juliet, Kilo, Lima, Mike, November, Oscar, Papa, Quebec, Romeo, Sierra, Tango, Uniform, Victor, Whiskey, X-ray, Yankee and Zulu.

phosphate finish The black finish applied to SMALL ARMS, ARTILLERY, or automotive components to provide resistance to erosion.

phosphorous A chemical contained in grenades, shells and bombs that, because it continues to burn after an initial explosion, is used for marking positions as well as causing casualties.

photocharting The process of making photocharts or photomaps from aerial photographs.

photographic dosimetry The determination of personal radiation dosage by use of photographic film.

photographic flight line The prescribed path in space along which an airborne vehicle moves during the execution of a photographic mission.

photographic intelligence (PHOTINT) The collected products of photographic interpretation, classified, and evaluated for intelligence use.

photographic interpretation See IMAGERY INTERPRETATION.

photographic interpretation report Any report issued as a result of the interpretation of photography.

photographic panorama A continuous assemblage of overlapping oblique or ground photographs that have been matched and joined together to form a continuous photographic representation of the area.

photographic reading The simple recognition of natural or man-made features from photographs not involving imagery interpretation techniques.

photographic reconnaissance Military aerial photography to obtain information about the results of bombings or about enemy movements, concentrations, activities, and forces.

phototopography The process of making topographic maps, charts, photomaps, or mosaics from air photographs.

physical security That part of security concerned with physical measures designed to safeguard personnel, to prevent unauthorized access to equipment, installations, material, and documents, and to safeguard them against espionage, sabotage, damage, and theft. See also COMMUNICATIONS SECURITY; PROTECTIVE SECURITY; SECURITY.

picket A guard detailed to watch for the approach of the enemy.

picket, air See AIR PICKET.

pickup field An open area where aircraft in flight may approach the ground to snatch messages, other aircraft, personnel, or supplies into the air.

pickup message A message picked up from the ground by a cable trailing from a low-flying aircraft.

pickup point A point on a TRAJECTORY visible to a RADAR and from which data has been obtained by computation or radar observation.

pickup zone A geographical area used to pick up troops or equipment by helicopter. Compare to LANDING ZONE.

piece 1. A CANNON; any item of ARTILLERY. 2. A SMALL ARM such as a rifle or pistol.

piecemeal attack An offensive action in which the various units are employed as they become available, or wherein the timing of a planned action breaks down and the action is reduced to uncoordinated phases. Piecemeal attacks should only be ordered in desperation or when the attackers retain superiority in fire power throughout the piecemeal commitment of troops.

piezolectric crystal The initiating element in many FUZES. When mechanically bent or stressed, it generates a voltage proportional to the stress.

pillaring The rapid vertical movement of smoke that sometimes results from the explosion of a white PHOSPHORUS bomb or projectile. The effect is undesirable because it does not produce OBSCURATION over a large area.

pillbox A small, low FORTIFICATION that houses machine guns, antitank weapons, etc; usually made of concrete, steel, or filled sandbags. Fields and beaches all over Europe are still dotted with the remains of pillboxes from World War II.

pincer movement A DOUBLE ENVELOPMENT.

pineapple A FRAGMENTATION GRENADE, so nicknamed because its surface resembles that of a pineapple.

pinpoint 1. A precisely identified point, especially on the ground, that locates a very small target; a reference point for rendezvous or for other purposes; the coordinates that define this point. 2. The position of an aircraft relative to the ground as determined by direct observation of the ground.

pinpoint target In ARTILLERY and NAVAL GUNFIRE SUPPORT, a target less than 50 meters in diameter.

pintle 1. The vertical bearing about which a GUN CARRIAGE revolves; the pin used as a hinge or axis for this revolution. 2. A hook, with a latch, on the rear of a towing vehicle to which a gun or trailer is attached by means of the LUNETTE.

pintle center The assumed center of a WEAPON, on which all FIRING-DATA computations are based.

pip The figure displayed on a RADAR indicator, caused by the ECHO from an aircraft or other reflecting object.

pipehead The downstream end of a PIPELINE, where products are received for storage, distribution, or forwarding by another means of transportation.

pipeline In LOGISTICS, the channel of support or a specific portion thereof by means of which MATERIEL or PERSONNEL flow from sources of procurement to their point of use.

pitched battle A major combat action between relatively evenly matched and well-prepared enemies. Pitch once referred to an orderly set of things, such as soldiers lined up opposite each other.

plaindress A type of message in which the originator and addressee designations are indicated externally of the text. See also CODRESS.

plain text Intelligible text or signals that have meaning and can be read or acted upon without the application of any DECRYPTION.

plan, administrative See ADMINISTRATIVE PLAN.

plan, campaign See CAMPAIGN PLAN.

plan, contingency See CONTINGENCY PLAN.

plan, defense See DEFENSE PLAN.

plan, landing See LANDING PLAN.

plan, marshalling See MARSHALLING PLAN.

plan, mobilization See MOBILIZATION PLAN.

plan, tactical See TACTICAL PLAN.

plane of departure The vertical plane containing the path of a PROJECTILE as it leaves the MUZZLE of a weapon.

plane of fire The vertical plane containing the axis of the BORE of a gun when it is ready to be fired.

plane of position The vertical plane containing a gun and its target; a vertical plane containing a line of sight.

plane of site A plane made by two lines, one from the MUZZLE of a gun to the target, the other line horizontal but perpendicular to the first line at the muzzle of the gun.

planned resupply The shipping of supplies in a regular flow as envisaged in preplanned schedules and organizations, and which will usually include some form of planned procurement.

planned target 1. In ARTILLERY AND NAVAL GUNFIRE SUPPORT, a target on which fire is prearranged. Individually planned targets may be further subdivided into either SCHEDULED or ON-CALL TARGETS. 2. A NUCLEAR WEAPONS target planned in an area or on a point in which a need for nuclear attack is anticipated; it may be scheduled or on call.

planning The formal process of making decisions for the future of individuals and organizations. Military planning is usually done by a GENERAL STAFF or UNIT staff.

planning, force See FORCE PLANNING.

planning horizon The time limit of organizational planning beyond which the future is considered too uncertain or unimportant to waste time on.

planning, strategic Long-range planning that has three dimensions: the identification and examination of future opportunities, threats, and consequences; the process of analyzing the military's environment and developing mutually compatible objectives along with the appropriate strategies and policies capable of achieving those objectives; and the integration of the various elements of the plan into an overall structure of plans so that each military unit would know in advance what had to be done, when, and by whom. See Joseph D. Douglass, Jr., "Strategic Planning and Nuclear Insecurity," *Orbis* 27, 3 (Fall 1983); Colin S. Gray, *Nuclear Strategy and Strategic Planning* (Philadelphia: Foreign Policy Research Institute, 1984).

planning, tactical Divisional planning concerned with the implementation of the larger goals and strategies that have been determined by STRATEGIC PLANNING; improving current operations; and the allocation of resources.

planning chart A chart of world scope, usually with a scale of 1:5,000,000, used for route planning and the control of TACTICAL MOVEMENTS and developments.

planning factor A multiplier, used in planning, to estimate the amount and type of effort involved in a contemplated OPERATION. Planning factors are often expressed as rates, ratios, or lengths of time.

planning-program budgeting systems (PPBS) A budgeting system that requires agency directors to identify program objectives, develop methods of measuring program output, calculate total program costs over the long run, prepare detailed multi-year program and financial plans, and analyze the costs and benefits of alternative program designs. Planning-program budgeting systems were first developed in the U.S. Department of Defense during the late 1950s. In the 1960s, PPBS took the budgeting world by storm. Its proponents insisted that it could interrelate and coordinate the three management processes constituting its title. Planning would be related to programs that would be keyed to budgeting. To further emphasize the planning dimension, PPBS pushed the budgeting time horizon out to half a decade, requiring five-year forecasts for program plans and cost estimates. Additionally, PPBS put a completely new emphasis on program objectives, outputs, and alternatives, and stressed the new watchword of evaluation—the "effectiveness" criterion. Finally, PPBS required the use of analytical techniques from STRATEGIC PLANNING, operations research, SYSTEMS ANALYSIS, and cost-benefit analysis to make governmental decision-making more systematic and rational. President Lyndon Johnson made PPBS mandatory for all federal agencies of the United States in 1965. Greatly concerned about the lack of objectives being formulated in the federal government, the nonconsideration of ends and preoccupation with means, the woeful lack of analysis and planning, and the ever-present dilemma of the lack of viable budget alternatives, Johnson embraced PPBS as the method and system that would ensure the success of his new programs. Within a short time, PPBS became the preeminent budgeting system in the federal government. According to PPBS, an armed service would have been required to identify a specific military need, compare the efficiency of various weapon designs, and carry out the long-term budgeting to obtain the desired weapons. While PPBS was formally abandoned in the federal government when the Nixon Administration discontinued it in 1971, many aspects of it are still in use. See David Novick, "The Origin and History of Programming Budgeting," *California Management Review* Vol. XI, No. 1 (Fall 1968); Allen Schick, "A Death in the Bureaucracy: The Demise of Federal PPB," *Public Administration Review* Vol. 33, No. 2 (March–April 1973); Gregory Palmer, *The McNamara Strategy and the Vietnam War: Program Budgeting in the Pentagon, 1960–1968* (Westport, Connecticut: Greenwood Press, 1978).

plastic explosive An EXPLOSIVE material that is malleable at normal temperatures and used mainly for demolition or by terrorists to produce crude bombs.

plastic range The stress range in which a material will not fail when subjected to the action of a force, but will not recover completely, with the result that a permanent deformation will be left when the force is removed.

plastic zone The ground region beyond the RUPTURE ZONE associated with crater formation resulting from an explosion in which there is no visible rupture, but in which the soil is permanently deformed and compressed to a high density.

platform drop The airdropping of loaded platforms from rear-loading aircraft with roller conveyors. See also AIRDROP.

platoon Usually a grouping of four SQUADS under a lieutenant. Two platoons often make up a COMPANY, which is commanded by a captain. See James R. McDonough, *Platoon Leader* (Novato, CA: Presidio Press, 1985).

plot 1. A map, chart, or graph representing data of any sort. 2. A representation on a diagram or chart of the position or course of a TARGET in terms of angles and distances from positions; the location of a position on a map or chart. 3. The visual display of a single location of an airborne object at a particular instant of time. 4. A portion of a map or overlay on which are drawn the outlines of the areas covered by one or more photographs.

plotting and relocating board A board on which the FIELD OF FIRE of a BATTERY is represented to scale. On a plotting board, the observation stations, the BASE LINE, and the DIRECTING POINT or BASE PIECE are shown to scale, and are located in their proper relation to one another. The board is used to locate the observed positions of a TARGET and to make predictions so that necessary FIRING DATA can be computed.

plunging fire Gunfire that strikes the Earth's surface at a high angle.

pneumatic deception device A dummy tank, vehicle, or weapon, made of inflatable material. It is used for deceiving enemy INTELLIGENCE about the location of friendly INSTALLATIONS.

point See POINT MAN.

point-blank A position so close to a target that taking aim is almost unnecessary; so close that it is almost impossible to miss.

point defense 1. A concept applying to anti-aircraft or ANTIBALLISTIC MISSILE (ABM) systems. It is different from, but often complementary to, AREA DEFENSE, which consists of rapid-firing guns or missiles, which track and attack incoming threats at some distance and over a broad area. Thus, some specialist ships in any modern fleet will have the duty of trying to protect the entire fleet from airborne attack. (See LAYERED DEFENSE.) Point defense is a capacity to protect only one specified target at relatively close range. An example of the contrast between point defense and area defense can be seen in various ANTIBALLISTIC MISSILE (ABM) defenses, particularly as incorporated in the United States STRATEGIC DEFENSE INITIATIVE (SDI). Although there are different versions of SDI plans, the one that caught President Reagan's attention was a general area defense over the United States. However, a more likely and technically feasible option could simply involve providing point defenses around INTERCONTINENTAL BALLISTIC MISSILE (ICBM) sites. 2. A special effort to defend the most critical points of a position, such as a hill or a headquarters.

point detonating fuze A FUZE located in the nose of a PROJECTILE, which is initiated upon impact.

point feature An object whose location can be described by a single set of coordinates.

point man The advance man of an ADVANCE GUARD; the single soldier who is most exposed to possible enemy fire.

point of departure In night attacks, a specific place on the LINE OF DEPARTURE where a unit will cross.

point of fall The point in the curved path of a falling PROJECTILE that is level with the MUZZLE of the gun that fired the projectile. Also called the LEVEL POINT.

point of impact 1. The point in a DROP ZONE where the first parachutist or air-dropped cargo item lands or is expected to land. 2. The point at which a PROJECTILE, BOMB, or RE-ENTRY VEHICLE impacts or is expected to impact.

point of no return 1. That point in a voyage when it is just as far to go on to the destination as it is to return to the place of origin. For example, if a plane were flying from San Francisco to Honolulu, a distance of 2500 miles; the point of no return would be just over halfway, or 1251 miles out over the Pacific Ocean. 2. In policy analysis, the point at which it is just as costly in terms of money, prestige, time, and so forth to go on with a decision as it is to reverse it. Shakespeare's Macbeth (*Macbeth* Act 3, Scene 4) arrives at a similar policy analysis after he has embarked on his series of murders:

I am in blood
Stepp'd in so far that, should I wade no more,
Returning were as tedious as go o'er.

point target A target so small that it requires especially accurate placement of BOMBS or FIRE. Compare to an AREA TARGET, which does not require especially accurate fire and is typically as large as a city. See also PINPOINT TARGET.

polar coordinates 1. In ARTILLERY and NAVAL GUNFIRE SUPPORT, the direction, distance, and vertical correction (shift) from the observer/SPOTTER position to the target. 2. The location of a point in a plane by the length of a radius vector, from a fixed origin in the plane, and the angle the radius vector makes with a fixed line in the plane. See also COORDINATES.

Polaris 1. The first submarine-launched ballistic missile (SLBM) to be deployed, originally put to sea by the U.S. Navy in 1960 aboard the U.S.S. *George Washington*. The range of the Polaris missile is only 4,600 kilometers, thus restricting the total sea area in which the submarines carrying these missiles can operate, and increasing the chances of detection. The TRIDENT C4 class of submarine has an operating range of 7,400 kilometers, and the Trident D5 one of 9,700 kilometers, which greatly improves the safety of the submarines against ANTISUBMARINE WARFARE. 2. The north star, which is used for navigation. See J. J. DiCerto, *Missile Beneath the Sea: The Story of Polaris* (New York: St. Martin's Press, 1967).

polar plot The method of locating a target or point on a map by means of polar coordinates.

pole charge A number of blocks of explosive tied together, capped, fuzed, and mounted on the end of a pole. The minimum weight of the total charge is usually about 15 pounds. A pole charge may be placed in a position out of hand-reach.

police To clear up something; to get something into proper shape. To police an area is to make it clean and ready for inspection.

police, kitchen See KITCHEN POLICE.

police, military See MILITARY POLICE.

policy, national See NATIONAL POLICY.

political intelligence INTELLIGENCE concerning foreign and domestic policies of governments and the activities of political movements.

political objectives The goal of war; the reason why one nation undertakes hostilities against another. It was Machiavelli who said: "Success in war is determined by the political advantages gained, not victorious battles." CLAUSEWITZ concurred: "The political design is the object, while war is the means."

political warfare The aggressive use of political means to achieve national OBJECTIVES; a constant policy of taking political action against the interests of another power in the expectation that the opposing nation will be worn down by the continuous demands of internal and international political problems. Political warfare is a major weapon in the COLD WAR arsenal. See also PSYCHOLOGICAL WARFARE. See Joseph Miranda, "Political Warfare: Can the West Survive?" *Journal of Social, Political and Economic Studies* Vol. 10, No. 1 (Spring 1985); S. Steven Powell, "Deterrence and the Political-Psychological Conflict," *Strategic Review* 14, 1 (Winter 1986).

politico-military gaming The simulation of situations involving the interaction of political, military, sociological, psychological, economic, scientific, and other appropriate factors.

POM A program objective memorandum; a planning document.

POMCUS (Pre-positioning of Material Configured to Unit Sets) The United States military's name for the equipment stored in Europe to be used by reinforcements flow in during a NORTH ATLANTIC TREATY ORGANIZATION (NATO) mobilization. These are not only personal weapons and general stores, but also tanks, guns, and vehicles for entire DIVISIONS, which are pre-positioned and ready for use. This preparation would allow the reinforcement of forward-based troops to proceed much faster than if the equipment, as well as the personnel, required sea-lifting to Europe. It is, of course, an enormously expensive process, not least because divisions based in the United States need separate sets of equipment there for training purposes. There are serious drawbacks to the policy, the main one being that the POMCUS sites, well known to the WARSAW PACT forces, make obvious targets. If either a conventional or nuclear attack could destroy these sites, the incoming divisions would be rendered virtually useless. Furthermore, since the divisions, wherever they enter Europe, have to travel to these few fixed sites, they will themselves become obvious targets. The U.S. Marine Corps tries to operate a similar system by having advance equipment stores on ships deployed in various theaters, particularly the Mediterranean, where they may be required to effect a rapid landing. See also INFRASTRUCTURE and LOGISTICS.

pom-pom guns Antiaircraft cannons usually mounted in groups and which fire alternately, so that when rapidly fired, they sound like the beating of a drum.

pool 1. To maintain and control a supply of resources or personnel upon which other activities may draw. The primary purpose of a pool is to promote maximum efficiency of use of the pooled resources or personnel, e.g., a petroleum pool or a labor and equipment pool. 2. Any combination of resources that serves a common purpose.

poop 1. Information (official or unofficial) or gossip. The "straight poop" is supposed to be unusually accurate. 2. To exhaust, as in "the troops were pooped." 3. To FIRE on a TARGET.

pop-up point (PUP) The location at which fighter aircraft for CLOSE AIR SUPPORT quickly gain altitude for TARGET ACQUISITION and ENGAGEMENT. This point occurs at the end of low-level terrain flight to avoid detection or prevent effective engagement by the enemy. See also TARGET BOX.

port 1. The left hand side of a ship or aircraft when facing forward. 2. A harbor for ships. 3. A place of initial entry into a state even if it is inland. 4. A hole or opening from which WEAPONS may be fired.

port, aerial See AERIAL PORT.

port arms 1. A position in the MANUALS of the RIFLE and CARBINE, in which the weapon is held with the barrel up, diagonally across the body, along a line from the left shoulder to the right hip. 2. The command to take this position.

Poseidon The MISSILE that replaced the POLARIS missile, the first American SUBMARINE-LAUNCHED BALLISTIC MISSILE (SLBM), from 1970 onwards. Poseidon differed from Polaris mainly in that it employed MULTIPLE INDEPENDENTLY TARGETABLE RE-ENTRY VEHICLE (MIRV) technology. Polaris A3, the last model, although it carried three WARHEADS, was only of multiple re-entry vehicle capacity, meaning that the separate warheads all had to be fired at the same target. So although Poseidon was not superior in any other way—its range was much the same as that of Polaris at slightly under 5,000 kilometers—it effectively multiplied the effect of the fleet of 30 or so SSBNs many times over. The United States began to emplace the next generation missile, the TRIDENT I, in the same vessels that had carried both Polaris and Poseidon, in the early 1980s.

position 1. A location or area occupied by a military unit. 2. The location of a weapon, unit, or individual from which FIRE is delivered upon a TARGET. Positions may be classified as primary positions, alternate positions, and supplementary positions. 3. The manner in which a WEAPON is held, as prescribed in the MANUAL of arms. 4. Any of the standard postures taken by a soldier when firing a rifle or other weapon. 5. A civilian position.

position area survey The determination, by surveying, of the relative horizontal and vertical locations of artillery BATTERIES; the establishment of the ORIENTING LINE or lines on the ground, and the determination of their direction.

position correction A correction applied to FIRING DATA to compensate for the difference in location of individual PIECES in a BATTERY.

position defense The type of defense in which the bulk of a defending force is disposed in selected tactical localities where a decisive battle is to be fought. Principal reliance is put on the ability of the forces in the defended localities to maintain their positions and to control the terrain between them. The RESERVE is used to add depth, to block, or to restore the battle position by COUNTERATTACK. A position defense is normally conducted by infantry forces.

positioning band A metal band on some recoilless ammunition, placed to insure the proper positioning of the ROUND inside the CHAMBER and TUBE of the weapon.

positive control A method of airspace control that relies on the positive identification, tracking, and direction of aircraft within an airspace, conducted with electronic means by an agency having the authority and responsibility therein.

post 1. An assigned place of duty. 2. A military installation; usually an army base.

postattack period In nuclear warfare, that period which extends from the termination of the final attack until political authorities agree to terminate hostilities. See also TRANSATTACK PERIOD.

post-boost phase The phase of a MISSILE trajectory, after the BOOSTER's stages have finished firing, in which the various RE-ENTRY VEHICLES are independently placed on BALLISTIC TRAJECTORIES toward their targets. In addition, PENETRATION AIDS are dispensed from the POST-BOOST VEHICLE. The length of this phase is typically 3 to 5 minutes.

post-boost vehicle The portion of a missile PAYLOAD that carries the multiple WARHEADS and has maneuvering capability to place each warhead on its final trajectory to a target; also referred to as a BUS.

post exchange (PX) The exchange service; the food and department stores operated for U.S. Army personnel and their dependents. The U. S. Air Force equivalent is BX, for base exchange. The Navy has a ship's service store.

poststrike damage estimation A revised TARGET ANALYSIS based on new data such as actual weapon YIELD, BURST height, and GROUND ZERO obtained by means other than direct assessment.

poststrike reconnaissance Missions undertaken for the purpose of gathering information used to measure the results of a STRIKE.

post system, fixed See FIXED-POST SYSTEM.

post-traumatic stress disorder A psychiatric disorder defined by the American Psychiatric Association as "the development of characteristic symptoms following a psychologically distressing event that is outside the range of usual human experience." This was traditionally called shell shock or battle fatigue until research during the last two decades on Vietnam veterans identified battle-related problems that continued to show up years after any physical scars might have healed.

POW See PRISONER OF WAR.

powder, black See BLACK POWDER.

powder train 1. An explosive train, usually of compressed BLACK POWDER, used to obtain time action in older FUZE types. 2. A train of EXPLOSIVES laid out for destruction by burning.

power approach A type of descent used to land light aircraft in short fields and over obstacles, whereby the aircraft is slowed to a speed slightly above stalling, and with the application of power is descended to the desired point of landing.

power traverse The turning of a GUN to change the direction of FIRE by means of a power-driven mechanism, as in a tank, aircraft, or ship turret.

power turret A TURRET or enclosed gun mount that is turned by a power driven mechanism, especially in tanks, aircraft, and ships. The guns in the power turret move with it.

powers, emergency See EMERGENCY POWERS.

PPBS See PLANNING PROGRAM BUDGETING SYSTEMS.

practice ammunition Ammunition used for target practice; ammunition with a PROPELLING CHARGE, but with either an inert filler or a LOW-EXPLOSIVE filler to serve as a SPOTTING CHARGE.

prearranged fire FIRE that is formally planned and executed against PLANNED TARGETS or target areas of known location. Such fire is usually planned well in advance, and is executed at a predetermined time or during a predetermined period of time. See also ON CALL; SCHEDULED FIRE.

preassault operation An OPERATION conducted in the OBJECTIVE AREA prior to an ASSAULT. It includes RECONNAISSANCE, MINESWEEPING, bombardment, bombing, UNDERWATER DEMOLITION, and destruction of beach OBSTACLES.

precautionary launch The launching of aircraft loaded with NUCLEAR WEAPONS under conditions of an imminent nuclear attack, so as to preclude friendly aircraft destruction and loss of weapons on the ground or in carriers.

precedence 1. A designation assigned to a message by the originator to indicate to COMMUNICATIONS personnel the relative order of handling of the message and to the addressee the order in which the message is to be noted. For example, an EMERGENCY PRIORITY request would take precedence over URGENT PRIORITY, which in turn would take precedence over a SEARCH-AND-ATTACK PRIORITY. 2. A letter designation, assigned by a unit requesting several RECONNAISSANCE missions, to indicate the relative order of importance, within an established priority, of a particular mission. 3. The act or state of going before; adjustment of place. Precedence is based especially on military grade, and also on position, date of appointment, and other factors. 4. The prescribed order in which medals and SERVICE RIBBONS are worn.

precision adjustment A deliberate adjustment of the FIRE of a WEAPON for the purpose of placing the mean POINT OF IMPACT accurately on the TARGET.

precision bombing Bombing directed at a specific POINT TARGET as opposed to AREA BOMBING.

precision fire FIRE in which the CENTER OF IMPACT is accurately placed on a limited target; fire based on precision adjustment. Precision fire differs from AREA FIRE, which is directed against a general area rather than against a given objective in the area.

precision-guided munitions (PGMs) The heart of the development of EMERGING TECHNOLOGY in conventional warfare. The first PGMs were probably the SMART BOMBS used by the U.S. Air Force in Vietnam. These were LASER-GUIDED WEAPONS which, when dropped from an aircraft onto a target that has been lit up by a laser beam, steered themselves toward the objective while falling. Similar bombs were used by the British in the Falkland Islands campaign and the Americans when they bombed Tripoli in retaliation for Libyan-inspired terrorism. The technology of PGMs is now highly developed and complex, but the enormous change which they are likely to bring to warfare is simple to comprehend. Previously, the accuracy of all weapons was so low that a target, even if clearly identified and within range of a gun, cannon, rocket, or bomb, had a good chance of surviving FIRE directed on it. The chance of hitting the target fired at is known as the single-shot kill probability (SSKP). With PGMs, which can be ARTILLERY SHELLS, multiple-launch rocket systems, SHORT-RANGE BALLISTIC MISSILES, or ANTITANK ROUNDS, as well as bombs and air-launched missiles, the SSKP can easily be over 80 percent—that is, only one or two rounds in 10 would ever miss. The impact of PGM technology is furthered by the way that the techniques are being applied to SUBMUNITIONS, where a single missile WARHEAD may carry dozens of individually precision-guided "mini-bombs." A major task in emerging-technology development, for example, is to design a missile warhead that would explode over an enemy tank squadron, scattering dozens of PGM anti tank bombs that would home in on the heat emitted by the tanks. Thus, according to PGM enthusiasts, a single conventional missile could do the work of a low-yield nuclear warhead. Although the technological problems of PGM development are great, they are probably soluble with the provision of enough money. PGMs are very important to NATO's doctrine of FOLLOW ON FORCES ATTACK.

precision sweep In RADAR reconaissance, a small portion of a normal sweep, usually 2,000 meters, selected and expanded over the entire radar screen in order to permit precise RANGE measurements for weapons.

preclusion-oriented analysis The analysis of initial weapon sizes selected for those AIMING POINTS chosen to maximize lethal coverage or probable enemy locations within SAFETY and COLLATERAL DAMAGE REQUIREMENTS. See also LEAST SEPARATION DISTANCE.

preclusive buying The purchase in a neutral market, regardless of price, of vitally important materials to prevent them from falling into the hands of the enemy.

precursor front An air pressure wave that moves ahead of the main BLAST WAVE of a nuclear explosion for some distance. The pressure at the precursor front increases more gradually than in a true (or ideal) SHOCK WAVE.

precursor sweeping The MINE SWEEPING of an area by relatively safe means in order to reduce the risk to mine-countermeasures vessels in subsequent operations.

predicted fire FIRE that is delivered without adjustment. The term is used to describe the ultimate technique for applying accurately computed corrections (not corrections determined by firing) to standard FIRING DATA for all nonstandard conditions of the weapon-weather-ammunition combination and for rotation of the Earth. It implies the capability of delivering accurate surprise non-nuclear or nuclear fire on a target of known location in any direction from the weapon position and limited in range only by the characteristics of the weapon and ammunition employed.

predicted firing Firing at the point at which a moving TARGET is expected to be when the PROJECTILE that is fired reaches it, according to predictions based on observation.

predicted point The position at which it is expected a moving TARGET will arrive at the instant of firing of a weapon; the point that a moving target is expected to reach at the end of the DEAD TIME between the last observation and the moment of firing. It should not be confused with the set forward point, the predicted position of the target at the moment of impact of a projectile.

predicted position That position at which it is expected a moving TARGET will arrive at the end of the time of flight of a PROJECTILE.

predicted position device A scale ruler, chart, or predictor used in rapidly calculating the probable position of a moving TARGET at a future instant.

predicting dead time The time allowed for calculating and applying FIRING DATA, from the time of observation of a TARGET to the instant of firing.

predicting interval The time interval between successive predictions of future positions of a TARGET.

prediction Determining what the probable future position of a moving TARGET will be at a given time.

prediction angle The angle from the present LINE OF SIGHT to the GUNBORE LINE when properly pointed for a hit; i.e., the angle by which the gunbore line must be offset to account for lead angle, gravity drop, and VELOCITY JUMP in order for a projectile to arrive at its predicted POINT OF IMPACT.

prediction scale An accurately graduated scale or rule used to measure the actual speed of a moving TARGET. A prediciton scale is used together with a SET-FORWARD rule or chart to locate, on a PLOTTING BOARD, the point at which a target will be when a gun is fired at it.

predictor 1. An instrument used in connection with a PLOTTING BOARD to determine the probable future location of a moving TARGET in terms of direction and elevation from a given position. Also called a prediction mechanism. 2. A device used with a plotting board to tell just when a controlled underwater MINE ought to be fired. Such a predictor indicates the exact moment when a target is over the mine.

pre-emptive attack 1. An ATTACK initiated on the basis of supposedly incontrovertible evidence that an enemy attack is imminent. 2. In terms of nuclear strategy, an attack on the nuclear weapons or related assets of an enemy which were highly likely to be used against one's own country. The main point about preemption in nuclear conflict is that the opposite strategy, to "ride out" a nuclear attack and then retaliate, is a much harder option to accept than the general case of waiting for an enemy's attack. On the other hand, the near impossibility of successfully pre-empting such an attack, especially if the enemy has a secure SECOND-STRIKE capacity, persuades most analysts that it is a suicidal option. 3. An attack on a potential opponent just as it begins to develop a nuclear force. Thus, some American planners urged a nuclear strike against the Soviet Union's nuclear weapons stockpile and industry in the early 1950s. There has, for years, been a strong rumor that the Soviet Union contemplated, and invited the United States to join them, in an attack on the nuclear research institutions of the People's Republic of China, in order to prevent that nation from developing a nuclear-challenge capacity. Similarly, the 1981 Israeli attack on the Iraqi Osirak nuclear reactor has been interpreted as an attempt to pre-empt Arab nuclear armament.

pre-emptive war A war in which the country initiating hostilities does so not for an inherently aggressive motive, but because it is certain that it is about to be attacked. Those who try to defend such a stragegy see a pre-emptive war essentially as launching a legitimate defense ahead of the attack. There obviously can be situations in which a country is so vulnerable to an attack that it cannot hope to defend itself successfully if an aggressor is allowed to strike first. Similarly, there can, in principle, be situations in which an opponent's absolutely firm intention to attack can be known ahead of time. However, the situations under which these two necessary conditions would apply will be very rare indeed. More usually, a country that attacked first as a form of defense would risk causing a war that might not otherwise have happened had it simply alerted its forces and mobilized to demonstrate its preparedness, thus adopting a posture of deterrence.

preferential defense The concentration of (usually limited) defensive assets such as tanks or machine guns on locations selected to assure the survival of some of them.

preferential offense The concentration of offensive assets such as tanks or machine guns on a subset of all possible targets.

preflight 1. The totality of procedures undertaken to prepare an aircraft for flight. 2. A briefing prior to a flight. 3. A pilot's preflight check (review) of the aircraft before takeoff.

preflight reliability The percentage of TACTICAL MISSILES that are assembled, prepared for launching, pass all checkout and prefiring tests, and are fired within the required time limits for a missile attack.

pre-H-hour transfer The transfer of CONTROL and TACTICAL LOGISTICS parties for a military operation from their parent ships to assigned control ships in a landing, and the transfer of the necessary troops and accompanying equipment from transports to LANDING SHIPS and transports in preparation for the ship-to-shore movement.

preinitiation The premature detonation of a NUCLEAR WEAPON, resulting in a significantly reduced YIELD.

pre-launch survivability The probability that a delivery or LAUNCH VEHICLE will survive an enemy attack under an established condition of warning.

preliminary bombardment Fire directed toward an attack site in anticipation of a ground assault to disrupt enemy communications and inflict as many casualties as possible. Such fire can come from offshore naval guns, artillery, or conventional bombing.

preliminary demolition A target prepared for demolition preliminary to a withdrawal of military forces, and whose demolition can be executed as soon after preparation as convenient.

preliminary demolition target A target, other than a RESERVED DEMOLITION TARGET, that is earmarked for demolition and which can be demolished immediately after preparation, provided that prior authority has been granted.

preliminary firing Training and practice in firing a GUN, often for the purpose of finding out which individuals have the greatest skills in shooting. Preliminary firing takes place before RECORD FIRING, in which the selected individuals are given additional training and in which a record is kept of their performance.

preload loading The loading of selected items aboard ship at one port prior to the main loading of the ship at another.

premature A malfunction in which a MUNITION functions (for example, explodes) before the expected time or proper circumstance.

preparation fire An intense volume of PREARRANGED FIRE delivered on a target preparatory to an ASSAULT.

preparatory command Part of a DRILL command which states the movement or formation that is to be carried out. A preparatory command is followed by the command of execution, which orders the movement to be carried out. In the command "Forward, March," "Forward" is the preparatory command and "March" is the command of execution.

prepare for action 1. To put a GUN into position for firing. 2. The command to put a gun in position for firing. 3. To put an armored vehicle in readiness for action. 4. Command to put an ARMORED VEHICLE in readiness for action.

preplanned mission request A request for a mission, whether an AIR STRIKE or RECONNAISSANCE, on a target or in support of a maneuver which can be anticipated sufficiently in advance to allow detailed mission coordination and planning. Compare to IMMEDIATE MISSION REQUEST.

preplanned nuclear support NUCLEAR SUPPORT planned in advance of operations.

pre-position To place military units, equipment, or supplies at or near the point of planned use or at a designated location, so as to reduce REACTION TIME, and to insure timely support of a specific force during the initial phases of an operation. Compare to FORWARD BASED.

pre-positioned supplies Supplies located at or near the point of planned use or at other designated locations in order to reduce reaction time and insure resupply.

pre-positioning See POMCUS.

prescribed load That quantity of combat-essential supplies and repair parts (other than ammunition) authorized by major commanders (and technical manuals) to be on hand in units, and which is carried by individuals or on unit vehicles to enable a unit to sustain itself until resupply can be effected. It is normally a 15-day quantity. The prescribed load is continuously reconstituted as it is used.

prescribed nuclear load A specified quantity of NUCLEAR WEAPONS to be carried by a delivery unit. The establishment and replenishment of this load after each expenditure is a COMMAND DECISION, and depends upon the tactical situation, the nuclear logistic situation, and the capability of the delivery unit to transport and utilize the load. See also NUCLEAR WEAPON PACKAGE, NUCLEAR WEAPON SUBPACKAGE.

prescribed nuclear stockage A specified quantity of NUCLEAR WEAPONS, components of nuclear weapons, and WARHEAD test equipment to be stocked in special AMMUNITION SUPPLY POINTS or other logistical installations.

preset guidance A technique of MISSILE control in which a predetermined flight path is set into the control mechanism of a missile and cannot be adjusted after launching. See also GUIDANCE.

presidential call An official order of the President of the United States bringing all or a part of the NATIONAL GUARD into the service of the United States, in time of war or national emergency.

Presidential Directive 59 (PD-59) The declaratory policy of the United States issued by President Jimmy Carter in 1980, which adopted a version of the LIMITED NUCLEAR OPTIONS strategy originally developed under President Nixon's Secretary of Defense, James Schlesinger. As with all United States declaratory policy, it involved the gradual adjustment of a targeting philosophy that had always contained more flexibility than critics had assumed. If PD-59 was any different from its predecessors, it was in stressing the need to be able to fight a prolonged but limited nuclear war. This required both a secure SECOND-STRIKE capacity, also known as a "secure strategic reserve," and enhanced COMMAND, CONTROL, COMMUNICATIONS, AND INTELLIGENCE (C^3I) abilities. PD-59 made public a preference for targeting the Soviet leadership itself, as opposed to only its nuclear forces, for nuclear attack in the event of war. It was argued that a highly centralized state, such as the Soviet Union, might well be paralyzed by hitting a few command bunkers, whereas the U.S. National Command Authorities structure does not display such vulnerability. Compare to: NSC-68. See Jeffery Richelson, "PD-59, NSDD-13 and the Reagan Strategic Modernization Program," *Journal of Strategic Studies* (June 1983); Louis René Beres, "Presidential Directive 59: A Critical Assessment," *Parameters* (March 1981); Milton Leitenberg, "Presidential Directive (P.D.) 59: United States Nuclear Weapon Targeting Policy," *Journal of Peace Research* 18, 4 (1981).

presidential salute A twenty-one gun SALUTE given a president or ex-president of the United States when he visits a military establishment.

pressure-firing device A FIRING DEVICE used to initiate land MINES, BOOBY TRAPS, and other explosive devices by pressure or weight exerted on a contrivance.

pressure force, direct See DIRECT PRESSURE FORCE.

pressure mine 1. In land MINE warfare, a mine whose FUZE responds to the direct pressure of a target. 2. In naval mine warfare, a mine whose circuit responds to the hydrodynamic pressure field of a target.

prestrike reconnaissance MISSIONS undertaken for the purpose of obtaining complete information about known targets for use by a STRIKE FORCE.

preventive war A war initiated in the belief that military conflict, while not imminent, is inevitable, and that to delay would involve greater risk; and therefore that a first strike is necessary. See also PRE-EMPTIVE ATTACK.

prewithdrawal demolition target A target prepared for demolition preliminary to a withdrawal of military forces, the demolition of which can be executed as soon after preparation as convenient.

primacord A trade name for a type of detonating cord.

primary fire sector The principal area to be covered by the gunfire of an individual or unit.

primary gun The principal or main GUN, especially of a TANK or other ARMORED VEHICLE.

primary interest Principal, although not exclusive, interest and responsibility for the accomplishment of a given MISSION, including responsibility for reconciling the activities of other agencies that have collateral interest in the mission.

primary position A place for a weapon, unit, or individual to fight which provides the best means to accomplish an assigned mission. See also ALTERNATE POSITION, BATTLE POSITION, POSITION, SUCCESSIVE POSITIONS, SUPPLEMENTARY POSITIONS.

primary weapon A WEAPON that is the principal arm of a COMBAT unit. The RIFLE is the primary or basic weapon for an INFANTRY rifle company, as compared with GRENADES or CHEMICAL PROJECTILES.

prime mover A vehicle, including heavy construction equipment, designed primarily for towing heavy, wheeled weapons.

primed charge A CHARGE ready in all aspects for ignition.

primer 1. A device used to initiate the functioning of an EXPLOSIVE or IGNITER

TRAIN. It may be actuated by friction, a blow, pressure, or electricity. See also ANVIL. 2. A first coat of paint used on some military equipment.

primer, electric See ELECTRIC PRIMER.

primer, igniting See IGNITING PRIMER.

primer detonator An assembly consisting of a PRIMER and DETONATOR. It may also include a delay element.

primer seat A chamber in the BREECH mechanism of a gun that used SEPARATE-LOADING AMMUNITION into which the PRIMER is set.

primer setback A defect in the firing of a ROUND of FIXED AMMUNITION, in which the explosion of the PROPELLING CHARGE forces the primer against the face of the BOLT. Primer setback is due to a faulty bolt or a defective CARTRIDGE, or to excessive pressure.

priming charge An initial charge that transmits the detonation wave to the whole of a CHARGE.

principal direction of fire The direction of FIRE assigned or designated as the main direction in which a weapon will be fired. It is selected on the basis of the enemy, mission, terrain, and WEAPONS capability of a military unit.

principles of war A list of the most critical considerations in the general conduct of military operations. These change over time in response to new fashions and new technology; but the basic principles were laid out by ancient writers such as SUN TZU and VEGETIUS, and have changed in little other than detail during the last two millennia. Most modern armies have their own official sets of principles of war that tend to be revised every few decades. The "Principles of War" adopted by the U.S. Army in 1978 are:

1. *Objective*—Every military operation should be directed toward a clearly defined, decisive, and attainable objective. The ultimate military objective of war is the defeat of the enemy's armed forces. Correspondingly, each operation must contribute to the ultimate objective. Intermediate objectives must directly, quickly, and economically contribute to the purpose of the ultimate objective. The selection of objectives is based on consideration of the mission, the means and time available, the enemy, and the operational area. Every commander must understand and clearly define his objective and consider each contemplated action in light thereof.
2. *Offensive*—Offensive action is necessary to achieve decisive results and to maintain freedom of action. It permits the commander to exercise initiative and impose his will on the enemy, to set the terms and select the place of battle, to exploit enemy weaknesses and rapidly changing situations, and to react to unexpected developments. The defensive may be forced on the commander as a termporary expedient while awaiting an opportunity for offensive action, or may be adopted deliberately for the purpose of economizing forces on a front where a decision is not sought. Even on the defensive, the commander seeks opportunities to seize the initiative and achieve decisive results by offensive action.
3. *Mass*—Superior combat power must be concentrated at the critical time and place for decisive results. Superiority results from the proper combination of the elements of COMBAT POWER. Proper application of this principle, in conjunction with other principles of war, may permit numerically inferior forces to achieve decisive combat superiority at the point of decision.
4. *Economy of force*—This principle is the reciprocal of the principle of mass. Minimum essential means must be employed at points other than that of the main effort. Economy of force requires the acceptance of prudent risks in selected areas to achieve su-

periority at the point of decision. Economy-of-force missions may require limited attack, defense, cover and deception, or retrograde action.

5. *Maneuver*—Maneuver is an essential ingredient of combat power. It contributes materially in exploiting success, in preserving freedom of action, and reducing vulnerability. The object of maneuver is to concentrate (or disperse) forces in such a manner as to place the enemy in a position of disadvantage and thus achieve results that would otherwise be more costly in men and materiel.
6. *Unity of command*—The decisive application of full combat power requires unity of command. Unity of command results in unity of effort by coordinating the action of all forces toward a common goal. While coordination may be achieved by cooperation, it is best achieved by vesting a single commander with the requisite authority.
7. *Security*—Security is essential to the preservation of combat power. Security results from the measures taken by a command to protect itself from espionage, observation, sabotage, annoyance, or surprise. It is a condition that results from the establishment and maintenance of protective measures against hostile acts or influences. Since risk is inherent in war, application of the prinicple of security does not imply undue caution and the avoidance of calculated risk.
8. *Surprise*—Surprise can decisively shift the balance of combat power. With surprise, success out of proportion to the effort expended may be obtained. Surprise results from striking the enemy at a time, place or both, and in a manner for which he is unprepared. It is not essential that the enemy be taken unaware, but only that he become aware too late to react effectively. Factors contributing to surprise include speed, cover and deception, the application of unexpected combat power, effective intelligence, variations of tactics and methods of operation, and operations security (OPSEC). Operations security consists of signals and electronic security, physical security, and counterintelligence to deny enemy forces knowledge or forewarning of intent.
9. *Simplicity*—Simplicity contributes to successful operations. Direct, simple plans and clear, concise orders reduce misunderstanding and confusion. Other factors being equal, the simplest plan is preferred. See John I. Alger, *The Quest for Victory: The History of the Principles of War* (Westport, Conneticut: Greenwood Press, 1982). Generations of officers have learned to remember these principles by using the mnemonic device MOOSEMUSS, which stands for Maneuver, Objective, Offensive, Surprise, Economy, Mass, Unity, Simplicity, Security.

priority With references to war plans, an indication of the relative importance rather than a exclusive and final designation of the order of their accomplishment.

priority, emergency See EMERGENCY PRIORITY.

priority intelligence requirements See INFORMATION REQUIREMENTS.

priority message A category of precedence reserved for messages that require expeditious action by the addressee(s), or which furnish essential information for the conduct of operations in progress when routine precedence will not suffice. See also PRECEDENCE.

priority of fire A direction to a FIRE-SUPPORT planner to organize and employ support means in accordance with the relative importance of the supported unit's missions.

priority of support Priorities set by a commander in his concept of an operation and during its execution, to insure that COMBAT SUPPORT and COMBAT SER-

VICE SUPPORT are provided to subordinate elements in accordance with their relative importance to accomplishing the operation.

priority target A target on which the delivery of FIRE takes precedence over all the fire for the designated firing unit or element. The firing unit or element will prepare, to the extent possible, for the engagement of such targets. A firing unit or element may be assigned only one priority target on FINAL PROTECTIVE FIRES at a time.

prior permission Permission granted by the appropriate authority of a particular nation for the commencement of a flight or a series of flights landing in or flying over the territory.

prismatic compass A magnetic compass combined with a sighting device, used to measure angles. It is equipped with a prism to assist in reading the scale while sighting. It may be equipped with a clinometer for measuring vertical angles.

prisoners' dilemma See GAME THEORY.

prisoner of war 1. A soldier captured by the enemy. Napoleon wrote that: "There is but one honorable way of being made a prisoner of war; that is by being taken separately and when you can no longer make use of your arms. Then there are no conditions—for there can be none, consistent with honor—but you are compelled to surrender by absolute necessity." 2. A detained person as defined in Articles 4 and 5 of the GENEVA CONVENTION Relative to the Treatment of Prisoners of War of August 12, 1949. In particular, one who, while engaged in combat under orders of his government, is captured by the armed forces of the enemy. As such, he is entitled to the combatant's privilege of immunity from the municipal law of the capturing state for warlike acts that do not amount to breaches of the law of armed conflict. For example, a prisoner of war may be, but is not limited to, any person belonging to one of the following categories of a nation's armed forces who has fallen into the power of the enemy: a member of the armed forces, organized militia or volunteer corps; a person who accompanies the armed forces without actually being a member thereof; a member of a merchant marine or civilian aircraft crew not qualifying for more favorable treatment; or individuals who, on the approach of the enemy, spontaneously take up arms to resist the invading forces. See Michael Walzer, "Prisoners of War: Does the Fight Continue after the Battle?" *American Political Science Review* LXIII, 3 (September 1969); Richard A. Falk, "International Law Aspects of Repatriation of Prisoners of War During Hostilities," *American Journal of International Law* 67, 3 (July 1973).

private The lowest enlisted grade in the army; pay grade E-1.

probable error See HORIZONTAL ERROR.

probable error circular See CIRCULAR ERROR PROBABLE.

probable error deflection An ERROR in DEFLECTION which is exceeded as often as not.

probable error height of burst An ERROR in the height of burst at which a projectile or missile FUZE may be expected to exceed as often as not.

probable error in height That vertical distance above and below the desired HEIGHT OF BURST of a projectile within which there is a 50 percent probability that a WEAPON will detonate.

probable error range That ERROR in RANGE which is exceeded as often as not.

probable line of deployment A line previously selected on the ground where attacking UNITS deploy prior to beginning an ASSAULT; it is generally used under conditions of limited visibility.

probably destroyed A damage assessment on an enemy aircraft seen to break-off combat in circumstances that lead to the conclusion that it must have been downed although it is not actually seen to crash. Examples of a "probably destroyed" are an aircraft on fire or a smoking helicopter.

procedure sign (prosign) One or more letters or characters, or combinations thereof, used to facilitate COMMUNICATIONS by conveying in condensed standard form certain frequently used orders, instructions, requests, and information related to communications.

procedure word (proword) A word or phrase limited to radiotelephone procedure and used in lieu of a PROCEDURE SIGN.

proclamation A document published to the inhabitants of an area under military occupation or control, which sets forth the basis of authority and scope of activities of a commander in a given area and which defines the obligations, liabilities, duties, and rights of the population affected.

proconsul 1. A governor of an ancient Roman province. 2. The highest administrator of a colonial or occupying power. See Robert Wolfe, editor, *Americans as Proconsuls: United States Military Government in Germany and Japan, 1944–1952* (Carbondale, IL: Southern Illinois University Press, 1984).

procurement The process of obtaining personnel, services, supplies, and equipment.

procurement lead time The interval in months between the initiation of procurement action and receipt into the supply system of the production model (excludes prototypes) purchased as the result of such actions, and is composed of two elements, PRODUCTION LEAD TIME and ADMINISTRATIVE LEAD TIME.

production base The total national industrial production capacity available for the manufacture of items to meet materiel requirements.

production base temperature Classification for a production base ("hot or cold") according to its rate of production on D-DAY. A production base operating at its maximum sustained production rate would be termed a hot-base facility. A production facility available but not producing is termed a cold-base facility. If a facility is producing at least a minimum sustaining rate of supplies, but less than maximum sustaining rate, it is generally referred to as a warm base.

production lead time The time interval between the placement of a contract and receipt into the supply system of materiel purchased.

production missile A complete MISSILE of an operational type accepted by the military and allocated for tactical or nontactical use, but not including prototype missiles.

profession of arms 1. The practice of the art and science of war. 2. The profession of a career military officer. But does military service qualify as a true profession? Consider that a professional is a member of an occupation requiring specialized knowledge that can be gained only after intensive preparation. Professional occupations tend to have three features: 1. a body of erudite knowledge that is applied to the service of society; 2. a standard of success measured by accomplishments in serving the needs of society rather than purely serving personal gain; and 3. a system of control over a professional practice that regulates the education of its new members and maintains both a code of ethics and appropriate sanctions. The primary characteristic that differentiates it from a vocation is its theoretical commitment to rendering a public service. By meeting these criteria, the profession of arms is seen to be a profession with underlying characteristics similar to those of the more obvious professions of law and medicine. See Morris Janowitz, *The Professional Soldier* (New York: Free Press, 1971); Sam C. Sarkesian, *The Professional Army Soldier in a Changing Society* (Chicago: Nelson-Hall 1975).

proficiency flying That amount of flying required of each military aviator to maintain a safe minimum level of pilot skill. This amount is usually designated as so many hours a year or month. Also known as combat readiness proficiency flying.

proficiency pay Additional pay to an officer or ENLISTED PERSON designated as having special proficiency in a military skill. Examples include parachute pay and flight pay.

proficiency rating A classification that denotes a specific monthly rate of PROFICIENCY PAY.

program of targets A tabulation of PLANNED TARGETS of a similar nature. A program of targets may be initiated on call, at a specific time, or when a particular event occurs. The targets are fired upon in a predetermined sequence.

projectile An object projected by an applied exterior force and continuing in motion by virtue of its own inertia, such as a BULLET, SHELL, or GRENADE. Also applied to ROCKETS and to GUIDED MISSILES.

projectile, chemical See CHEMICAL PROJECTILE.

projectile, high-explosive See HIGH-EXPLOSIVE PROJECTILE.

projectile, leaflet See LEAFLET PROJECTILE.

projectile, smoke See SMOKE PROJECTILE.

projectile velocity The velocity of a PROJECTILE along the LINE OF DEPARTURE.

proliferation The spread of NUCLEAR WEAPONS and the technology to build them to countries that did not previously have them. The SUPERPOWERS have a self-evident mutual interest in keeping to the minimum the number of countries with nuclear armaments. This is one area of arms control in which there has been considerable cooperation. See also NUCLEAR NONPROLIFERATION TREATY. See Bruce Bueno de Mesquita and William H. Riker, "An Assessment of the Merits of Selective Nuclear Proliferation," *Journal of Conflict Resolution* Vol. 25, No. 2 (June 1982); Bruce D. Berkowitz, "Proliferation, Deterrence, and the Likelihood of Nuclear War," *Journal of Conflict Resolution* Vol. 29, No. 1 (March 1985).

promotion list A list, provided by statute, of officers of the armed forces below the permanent grade of BRIGADIER GENERAL in the order of their standing for promotion. This does not necessarily indicate the precise order of promotion.

promotion zone In many services, a period of several years during which a career officer can be promoted to the next rank. If an officer is not promoted during that period, he or she is "passed-over," and may be forced to retire without further promotion.

prompt radiation The gamma rays produced in NUCLEAR FISSION in weapon materials which appear within a second or less after a nuclear explosion. The radiations from such sources are known either as prompt or instantaneous gamma rays. See also INDUCED RADIATION; INITIAL RADIATION; RESIDUAL RADIATION.

prone position A body posture for firing a SMALL ARMS weapon from the ground. The stomach is flat against the ground, the legs are spread, and the insides of the feet are flat on the ground, but the head and shoulders are raised and supported by the elbows, thus leaving the hands free to operate the gun.

prone shelter An open TRENCH that is deep enough to protect a man lying flat (normally 2 feet by 2 feet by the length of a man) from SMALL-ARMS fire and from the GROUND BURSTS of BOMBS and artillery SHELLS. It gives little or no protection against AIRBURST projectiles or against the crushing action of TANKS.

propaganda 1. A government's mass dissemination of truthful information about its policies and the policies of its adversaries. 2. Similar dissemination that is untruthful. The concept was introduced into American political science after World War I when British news reports of German atrocities (both real and imagined) were indicted as having influenced American attitudes toward entry into the war. This fostered Harold D. Lasswell's landmark analysis *Propaganda Technique in World War I* (Cambridge, MA: MIT Press, 1927, 1971). Ever since World War II, when the German Ministry of Propaganda broadcast one lie after another, the term has taken on a sinister connotation. 3. The manipulation of people's beliefs, values and behavior by using symbols (such as flags, music, oratory, etc.) and other psychological tools. 4. According to Sir Ian Hamilton, *The Soul and Body of an Army* (London: Edward Arnold, 1921), making the "enemy appear so great a monster that he forfeits the rights of a human being." This makes it emotionally easier to kill him. Hamilton further observes that since the enemy cannot bring a libel action, "there is no need to stick at trifles." See Jacques Ellul, *Propaganda: The Formation of Men's Attitudes* (New York: Knopf, 1965); Jon T. Powell, "Towards a Negotiable Definition of Propaganda for International Agreements Related to Direct Broadcast Satellites," *Law and Contemporary Problems* Vol. 45, No. 1 (Winter 1982); Paul A. Smith, Jr., "Propaganda: A Modernized Soviet Weapons System," *Strategic Review* 11, 3 (Summer 1983).

propaganda, black See BLACK PROPAGANDA.

propaganda, grey See GREY PROPAGANDA.

propaganda, white See WHITE PROPAGANDA.

propellant A propelling agent, such as a gas or chemical compound, that when charged will force an object, such as a bullet or rocket, forward.

propellent An adjective which describes anything that will force an object forward.

propelling charge A powder CHARGE that is set off in a weapon to propel a PROJECTILE from it; a propellant. Burning of the confined propelling charge produces gases that force the projectile out of the weapon.

proponent 1. An army organization or staff that has been assigned primary responsiblity for material or subject matter in its area of interest; i.e., a proponent school, proponent staff agency, or proponent center. 2. To be charged with the accomplishment of a task.

proportional deterrence The French nuclear strategic doctrine originally promulgated by Charles de Gaulle, and subsequently developed by French strategic thinkers such as Andre Beaufre. It holds that a country can be deterred from acting against another by the threat of incurring damage proportional to the benefit it would have gained by attacking. Put very simply, France's nuclear force might not do very much damage to the Soviet Union as compared to the impact of an American nuclear strike, but the damage would still be enough to ensure that the Soviet Union was, overall, a "loser" in the exchange even if it went on to conquer and occupy France. This theory was developed further, largely along the lines of "the deterrence of the strong by the weak" to justify, by making credible, the threat from a small nuclear force. Proportional deterrence is heard of less often today, as the French have shifted more to a doctrine of nuclear SANCTUARIZATION, and as the modernization of their forces increases the level of damage that they could inflict. See also FORCE DE FRAPPE.

proportionality A doctrine in the JUST-WAR THEORY which holds that the damage that any military action does can only be justified in terms of its proportionality to both the immediate military accomplishment it brings and the original justification for going to war. For example, the use by the American military of FREE-FIRE AREAS in Vietnam may be judged as disproportionate because, while they prevented Vietcong activity, they did so at the cost of innocent lives when other tactics could have achieved the same result. More dramatically, it can be argued that a nuclear strike on an enemy would be so appalling that, even if it were the only possible response to a clear-cut aggression, the right of self-defense could never be seen as proportional to the evil inflicted.

prosign See PROCEDURE SIGN.

protecting power A neutral nation entrusted by a belligerent with the protection, in territory of or occupied by the enemy, of the interests of the belligerent and its nationals.

protective clothing, impermeable See IMPERMEABLE PROTECTIVE CLOTHING.

protective cover An item or object that may be placed between an individual and a toxic CHEMICAL or BIOLOGICAL AGENT spray source to prevent individual contact with the spray.

protective fire FIRE delivered by supporting guns and directed against the enemy to hinder his fire or movement against friendly forces.

protective fire, final See FINAL PROTECTIVE FIRE.

protective line, final See FINAL PROTECTIVE LINE.

protective mask Individual protective equipment consisting of a face piece with integral filter elements or an attached canister and carrier. A mask should protect the wearer against inhaling toxic CHEMICAL AGENTS, SCREENING SMOKES, BIOLOGICAL AGENTS, and radioactive dust particles. Formerly called a gas mask.

protective minefield 1. In land MINE warfare, a minefield employed to assist a unit in its local, close-in protection. 2. In naval mine warfare, a minefield laid in friendly territorial waters to protect ports, harbors, anchorages, coasts, and coastal routes. See also MINEFIELD.

protective security The defensive measures instituted and maintained at all levels of a military command.

protocol 1. The generally accepted practices of international courtesy that have evolved over the centuries. 2. A supplementary international agreement or an annex to a TREATY. 3. The original or preliminary draft of a treaty. 4. Codes prescribing strict adherence to set etiquette, precedence, and procedure between diplomats and among military services. 5. The records or minutes of a diplomatic conference. 6. The plan of a scientific experiment. 7. The rules for the format of messages to be exchanged between computers.

protractor, declination See DECLINATION PROTRACTOR.

provisional map A hastily made line map based on aerial photographs, used as a map substitute.

provisional unit An assemblage of personnel and equipment temporarily organized for a limited period of time for the accomplishment of a specific MISSION.

provost court A military tribunal of limited jurisdiction convened in occupied territory under military government and usually composed of one officer. It usually tries a civilian who commits offenses against occupying troops.

provost marshal A staff officer who supervises the activities of MILITARY POLICE.

proword See PROCEDURE WORD.

proximity fuze A FUZE whose initiation primarily occurs by remote sensing of the presence, distance, or direction of a target or its associated environment, by means of a signal generated by the fuze or emitted by the target, or by detecting a disturbance of a natural field surrounding the target.

proximity scorer A hit-miss target device triggered by the entry of a projectile into a sphere with the scorer at its center. This scorer (in effect, the target's bulls eye) will indicate only that the munition came within its "sphere of influence." It will not differentiate within this sphere between near misses and far misses.

proxy 1. In voter registration, a person authorized to request or complete registration forms or to obtain an absentee ballot on behalf of another person. A proxy may not cast a ballot for another person. 2. Any person who acts for another in a formal proceeding.

proxy war Any war in which SUPERPOWERS are involved on opposite sides as suppliers, supporters, and advisors to smaller states that are the actual combatants. Many of the Arab-Israeli wars of the last several decades have been proxy wars because the United States was supporting and supplying Israel while the Soviet Union was supporting and supplying the other side. See Janice Gross Stein, "Proxy Wars—How Superpowers End Them: The Diplomacy of War Termination in the Middle East," *International Journal* Vol. 35, No. 3 (Summer 1980).

prudent limit of endurance The time during which an aircraft can remain airborne and still retain a given safety margin of fuel.

prudent limit of patrol The time at which an aircraft must depart from its operational area in order to return to its base and arrive there with a given safety margin (usually 20 percent) of fuel reserve for bad-weather diversions.

psychological activities, strategic See STRATEGIC PSYCHOLOGICAL ACTIVITIES.

psychological operation, consolidation See CONSOLIDATION PSYCHOLOGICAL OPERATION.

psychological operations (PSYOPS) Planned psychological activities in peace and war directed to enemy, friendly, and neutral audiences in order to influence their attitudes and behavior favorably to the achievement of political and military objectives. See also PSYCHOLOGICAL WARFARE.

psychological situation The current emotional state, mental disposition, or other behavioral motivation of a target audience, basically founded on its national political, social, economic, and psychological peculiarities, but also subject to the influence of circumstances and events.

psychological warfare The planned use of PROPAGANDA and other psychological actions having the primary purpose of influencing the opinions, emotions, attitudes, and behavior of hostile foreign groups in such a way as to induce them to support the achievement of national objectives. There has probably never been a time when military commanders have not put effort into psychological warfare, because it is directly concerned with morale, the importance of which has always been obvious. Certainly propaganda efforts in World War I were early attempts to organize psychological warfare, or "PsyOps" as the approach now referred to. Such early efforts however, were dedicated to boosting the morale of one's own army. The early days of the World War II saw a deliberate attempt by Britain to carry out psychological warfare on the enemy by mounting bombing raids over Germany in which only propaganda leaflets were dropped, in an attempt to destroy civilian backing for the war. Later, most forces in most theaters attempted to weaken the morale of the fighting troops by dropping leaflets guaranteeing good treatment and safety if a surrendering enemy soldier presented the leaflet to the opposing troops. The range of techniques available for psychological warfare is enormous, but they fall into two tactical categories: (1) increasing the terror felt by the enemy, or (2) trying to demoralize the enemy by removing such psychological props as faith in the cause being fought for, which all soldiers need to face the ordeal of battle. All armies of any importance have separate psychological warfare departments, but there is very little clear evidence that any of them have achieved much more than earlier primitive techniques such as training soldiers to utter terrifying battle calls when charging. See William E. Daughterty and Morris Janowitz, *A Psychological Warfare Casebook* (Baltimore: Johns Hopkins University Press, 1958); Thomas T. Winant, "A History of Psychological Warfare," *Orbis* Vol. XIII, No. 3 (Fall 1969); Michael T. McEwen, "Psychological Operations Against Terrorism: The Unused Weapon," *Military Review* 66, 1 (January 1986); Thomas R. Hammett, "The Soviet Psychological Threat," *Military Review* 66, 11 (November 1986).

psychological warfare consolidation PSYCHOLOGICAL WARFARE directed toward populations in friendly rear areas or in territory occupied by friendly military forces with the objective of facilitating military operations and promoting maximum cooperation among the civil populace. See also MILITARY CIVIL ACTION; CIVIL AFFAIRS.

public affairs officer The staff officer responsible for the overall conduct of the public affairs activities of a command, to include public information, troop information, and community relations. Often, public affairs officers have the frustration of serving under commanders who take the attitude toward public information that Admiral Ernest J. King (1878–1956) expressed as head of the U.S. Navy in World War II: "Don't tell them anything. When it's over, tell them who won."

pucker factor A slang term for the stress soldiers feel when they are on full alert; too much stress and they "pucker up" and become ineffective.

punishment, company See COMPANY PUNISHMENT.

punishment, corporal See CORPORAL PUNISHMENT.

punishment, nonjudicial See NONJUDICIAL PUNISHMENT.

punitive articles Those articles of the UNIFORM CODE OF MILITARY JUSTICE (Sections 33 to 116) in which military crimes and offenses are enumerated.

puppet state 1. A country whose basic policies are controlled by another power. The modern use of the term dates from World War II, when the Germans established what the Allies called "puppet governments" in the countries they conquered. The Soviet-dominated satellite states of Eastern Europe, while responsive to Soviet policies, have a level of independence far greater than the "puppets" of World War II. 2. Any small nation aligned with an opposing superpower.

Purple Heart An American military DECORATION for wounds sustained in action against an enemy or as a direct result of an act of the enemy. The oldest U.S. military decoration, the Purple Heart was first authorized as an award for heroism or meritorious service by George Washington in 1782; but while never discontinued, it wasn't awarded again until 1932, when it was designated a decoration for those injured in action while serving in any capacity with any of the armed services of the United States.

pursuit 1. An OFFENSIVE operation designed to catch or cut off a hostile force attempting to escape, with the aim of destroying it. It follows a successful ATTACK or EXPLOITATION, and is ordered when the enemy cannot conduct an organized defense and attempts to disengage. See also DIRECT-PRESSURE FORCE, ENCIRCLING FORCE, EXPLOITATION. 2. A tactical aircraft such as the World War II P (for pursuit) 38.

pursuit course interception A course in which an attacker must maintain a lead angle over the velocity vector of a target in order to predict the point in space at which gun or rocket fire would intercept the target.

pursuit navigation A method of HOMING GUIDANCE in which a MISSILE is directed toward the position of a target at a given instant.

puzzle palace 1. An army HEADQUARTERS. 2. The PENTAGON. 3. The NATIONAL SECURITY AGENCY, because of its concern wtih codebreaking.

pylon 1. An externally mounted suspension component on an aircraft, used to support external stores. 2. A projection (post tower) marking a prescribed course of flight or training maneuver ("pylon eight") for an aircraft. 3. The integral part of a helicopter fuselage in which the rotors are mounted.

pyrotechnic 1. Any mixture of chemicals which when ignited reacts to produce light, heat, smoke, sound or gas; may also be used to introduce a delay into an EXPLOSIVE TRAIN because of a known burning time. The term excludes propellants and explosives. 2. Ammunition containing chemicals that produce a smoke or brilliant light in burning, used for signaling or for lighting up an area at night.

pyrotechnic delay A PYROTECHNIC device added to a FIRING SYSTEM which transmits the ignition flame after a predetermined delay.

Pyrrhic victory A victory that is won at great cost. In 280 B.C., Pyrrhus, King of Epirus, lost 15,000 men gaining victory over the Romans. Acording to Plutarch, as Pyrrhus was being congratulated, he responded: "If we are victorious in one more battle with the Romans, we shall be utterly ruined."

Q

q-message A classified message relating to navigational dangers, navigational aids, mined areas, and channels searched or swept for mines.

Q-ship See DECOY SHIP.

quadrant 1. An instrument with a graduated scale used in laying (positioning) an artillery piece for elevation. It measures altitudes by degrees and fixes vertical and horizontal directions. Also called a gunner's quadrant. 2. Any of the four quarters into which an area is divided.

quadrant elevation The angle between the level base of the trajectory/horizontal and the axis of the bore when laid. It is the algebraic sum of the elevation, ANGLE OF SITE, and COMPLEMENTARY ANGLE OF SITE.

quadrant mount A device on a gun that holds the gunner's quadrant while the gun is being laid in elevation; the quadrant seat.

quadrant sight A sighting instrument on a gun that is used in laying the gun in elevation.

qualification in arms The degree of skill that an individual shows in the completion of approved marksmanship or gunnery tests.

quantity/distance tables Regulations pertaining to the amounts and kinds of EXPLOSIVES that can be stored in a given location and the proximity of such storage to buildings, highways, MAGAZINES, or other installations.

quartering Providing shelter for troops, headquarters, establishments, and supplies.

quartering party A group of military unit representatives dispatched to a probable new site of operations in advance of the main body of the unit for the purpose of securing, reconnoitering, and organizing an area prior to the main body's arrival and occupation.

quartering wind Wind other than wind parallel to, or perpendicular to, the OBSERVER TARGET LINE; sometimes called oblique wind.

quartermaster 1. The main supply officer in a large military unit responsible for providing housing (quarters), food, fuel, etc. The Quartermaster Corps in the U.S. Army is the equivalent of the Supply Corps in the U.S. Navy. 2. A petty officer in a navy wiith navigational and signals responsibilities; so called because he was historically located on the quarterdeck of a ship.

quarters 1. A place or structure in which military PERSONNEL are housed. 2. A building or structure used for the housing of military personnel and family members.

quarters, general See GENERAL QUARTERS.

quarters in kind Lodging provided by a goverment without cost to military PERSONNEL.

quibbling The creation of a false impression in the mind of the listener by cleverly wording what is said, omitting relevant facts, or telling a partial truth when one does so with the intent to deceive or mislead. This is a serious honor code violation at military academies.

quick fire SMALL-ARMS fire delivered as soon as possible after the appearance of a fleeting target; it may be single ROUND or several rounds fired at a very fast rate.

quick (hasty) fire plan In FIRE-SUPPORT operations, a fire plan that is prepared quickly at a lower ECHELON level in support of a tactical operation, and which contains the necessary elements of a FIRE PLAN.

quick time A marching rate of 120 steps, each 30 inches in length, per minute. It is a normal cadence for DRILLS and ceremonies.

quorum The number of members who must be in attendance to make valid the votes and other actions of a formal group.

quota post An international post for a military or civilian officer which a particular nation has accepted to fill indefinitely.

R

rad See RADIATION ABSORBED DOSE.

radar A radio-detection device that provides information on the RANGE, AZIMUTH and ELEVATION of prospective targets. It operates, as do all such systems, by emitting electronic or acoustic beams which are reflected by such objects, producing an "echo" in a receiver. The reflected waves (the "returns" or "echoes") provide information on the distance to the target and the velocity of the target, and may also provide information about the shape of the target. (Originally an acronym for "Radio Detection And Ranging.")

radar, bistatic See BISTATIC RADAR.

radar, illuminator See ILLUMINATOR RADAR.

radar, imaging See IMAGING RADAR.

radar, long-range See LONG-RANGE RADAR.

radar, medium-range See MEDIUM-RANGE RADAR.

radar, multistatic See MULTISTATIC RADAR.

radar, phased-array See PHASED-ARRAY RADAR.

radar camouflage The use of RADAR-absorbing or reflecting materials to change the radar echoing properties of a surface or an object. This technology lies at the core of the U.S. STEALTH BOMBER and related projects.

radar clutter Unwanted signals, echoes, or images on the face of a display tube, which interfere with the observation of desired signals. Clutter, which is worst when a radar beam is close to the ground, allows low-flying aircraft to penetrate "under" an enemy radar cover. Also called noise.

radar countermeasures See ELECTRONIC WARFARE; CHAFF.

radar discrimination The ability to distinguish separately, on a RADAR scope, several objects that are in close proximity to each other.

radar echo 1. The electromagnetic energy received after reflection of an electromagnetic signal by an object. 2. The deflection or change of intensity on a cathode-ray tube display produced by a radar echo.

radar fire Gunfire aimed at a target that is tracked by RADAR.

radar horizon The locus of points at which the rays from a RADAR antenna become tangential to the Earth's surface. On the open sea this locus is horizontal, but on land it varies according to the topographical features of the terrain.

radar intelligence (RADINT) INTELLIGENCE information derived from data collected by RADAR.

radar locating Finding the position of the BURST of a projectile or of an enemy gun by RADAR.

radar netting The linking of several RADARS to a single center to provide integrated target information.

radar picket Any ship, aircraft, or vehicle stationed at a distance from the force it is protecting, for the purpose of increasing the RADAR detection range.

radar picket CAP Radar picket combat air patrol.

radar range calibration The adjustment of RADAR set so that when it is "on target," the radar set will indicate the correct range of the target.

radar ranging The use of RADAR transmission to determine the range to a target.

radar reconnaissance RECONNAISSANCE by means of RADAR to obtain information on enemy activity and to determine the nature of terrain.

radar scan The movement of a radio-frequency beam through space in searching for an echo.

radar signature The characteristic reflected signal of an object that can be detected by RADAR (primarily from metallic or otherwise highly reflecting surfaces). Considerable effort is taken by INTELLIGENCE services in countries with advanced defense systems to obtain clear radar records of known enemy ships, submarines, aircraft, and weapons systems so that, in times of crisis or war, these signatures can be identified. In this way defending forces can recognize the type and number of an approaching vehicle or missile. Radar signatures are largely determined by the shape of the reflected object. The more care that goes into designing a very smooth shape for an object, the smaller its radar echo will be. It is this design criterion that lies at the heart of STEALTH BOMBERS—aircraft designed to be as close to invisible as possible to defensive radar.

radar silence An imposed discipline prohibiting the transmission by RADAR of electromagnetic signals on some or all frequencies.

radar tracking station A RADAR facility that has the capability of tracking moving targets.

radian A unit of angular measure. One radian is about 57.3 degrees. One microradian (0.000001 radian) is the angle subtended by an object 1 meter across at a distance of 1,000 kilometers.

radiation That portion, upwards of 50 percent, of the energy released by a NUCLEAR DETONATION that is in the form of "invisible" radiation, ranging from X-rays, neutrons, and GAMMA RAYS to alpha rays and low-energy beta rays. Even a short exposure to a low dose of radiation will do some physical damage to the human body, and as is well known, the exposure from a major nuclear explosion far exceeds such levels. There are two distinct ways in which radiation damage can occur. The more energetic particles, X-rays, neutrons, and gamma rays, can directly penetrate the body and are the source of most short-term radiation illness. Other particles can cause harm only when taken into the body, through breathing contaminated dust particles or drinking and eating contaminated foodstuffs.

radiation, electromagnetic See ELECTROMAGNETIC RADIATION.

radiation, initial See INITIAL RADIATION.

radiation, prompt See PROMPT RADIATION.

radiation, residual See RESIDUAL RADIATION.

radiation, thermal See THERMAL RADIATION.

radiation absorbed dose (RAD) A measure of any ionizing RADIATION in which energy is imparted to any matter: 1 rad = 100 ergs of energy/grams of absorber. For most military operations, this formula is the standard unit of measurement of RAD.

radiation dose The total amount of ionizing radiation absorbed by material or tissues, expressed in centigrades. The term radiation dose is often used in the sense of the exposure dose expressed in roentgens, which is a measure of the total amount of ionization that the quantity of radiation could produce in air. This could be distinguished from the RADIATION ABSORBED DOSE, also given in rads, which represents the energy absorbed from radiation per gram of specified body tissue. Further, the biological dose, in rems, is a measure of the biological effectiveness of a radiation exposure.

radiation dose, acute See ACUTE RADIATION DOSE.

radiation dose, chronic See CHRONIC RADIATION DOSE.

radiation dose rate The RADIATION DOSE (dosage) absorbed per unit of time. A radiation dose rate can be set at some particular unit of time (e.g., H + 1 hour would be called H + 1 radiation dose rate).

radiation intensity The RADIATION DOSE RATE at a given time and place. It may be used, coupled with a figure, to denote the radiation intensity measured at a given number of hours after a NUCLEAR BURST, e.g., RI-3 is a radiation intensity 3 hours after the time of a burst.

radiation scattering The diversion of RADIATION (thermal, electromagnetic, or nuclear) from its original path as a result of interaction or collisions with atoms, molecules, or larger particles in the atmosphere or other media between the source of the radiation (e.g., a nuclear explosion) and a point some distance away. As a result of scattering, radiation (especially GAMMA RAYS and neutrons) will be received at such a point from many directions instead of only from the direction of the source.

radiation sickness An illness resulting from excessive exposure to ionizing RADIATION. The earliest symptoms are nausea, vomiting, and diarrhea, which may be followed by loss of hair, hemorrhage, inflammation of the mouth and throat, and general loss of energy. There is little direct evidence of the amount of radiation illness nuclear detonations of varying sizes will cause. Hiroshima and Nagasaki are inadequate guides because both explosions were of relatively low YIELD compared with that of modern weapons. Even in these cases some degree of radiation damage was found in about half of the survivors. It was estimated that 30 percent of short-term deaths among those who survived for longer than 24 hours at Hiroshima were caused by radiation. The long-term effects of radiation include most forms of cancer.

radiation status A criterion to assist the commander of a military unit in measuring a military unit's exposure to radiation based on total past cumulative dose in rads. Categories are as follows:

1. Radiation status—O (RS-0)—No previous exposure history.
2. Radiation status—1 (RS-1)—Negligible radiation-exposure history (more than 0, but less than 71 rads).
3. Radiation status—2 (RS-2)—Significant but not a dangerous dose of radiation (more than 70, but less than 151 rads).
4. Radiation status—3 (RS-3)—Unit has already received a dose of radiation that makes a further exposure dangerous (more than 150 rads).

radiation weapons, enhanced See ENHANCED RADIATION WEAPONS.

RADINT See RADAR INTELLIGENCE.

radioactivity The spontaneous emission of radiation, generally alpha or beta particles, often accompanied by GAMMA RAYS, from the nuclei of an unstable isotope.

radio control The control of a mechanism or other apparatus by radio waves.

radio day The 24-hour period from midnight to midnight covered by a complete set of radio station logs.

radio deception The employment of radio to deceive the enemy. Radio deception includes sending false dispatches, using deceptive headings, and employing enemy call signs. See also ELECTRONIC WARFARE.

radio detection The detection of the presence of an object by RADIOLOCATION without precise determination of its position.

radio direction finding Radio-location in which only the direction of a station is determined by means of its emissions.

radio fix 1. The locating of a radio transmitter by bearings taken from two or more direction-finding stations, the site of the transmitter being at the point of intersection of the bearing lines. 2. The location of a ship or aircraft by determining the direction of radio signals coming to the ship or aircraft from two or more sending stations.

radiolocation The determination of the relative direction, position, or motion of an object, or its detection, by means of the rectilinear propagation characteristics of constant-velocity radio waves; in short, finding something with radar.

radiological agent Any of a family of substances that produce CASUALTIES by emitting RADIATION.

radiological operation The employment of radioactive materials or RA-

DIATION-producing devices to cause casualties or restrict the use of terrain. It includes the intentional employment of fallout from nuclear weapons.

radiological survey The directed effort to determine the RADIATION DOSE RATES and distribution in an area.

radio navigation guidance A technique of missile trajectory control in which the predetermined path of a missile can be adjusted laterally and in terms of range by a device in the missile that uses radio signals from one or more external transmitters to navigate the missile along the desired path.

radio position finding The process of locating a radio transmitter by plotting the intersection of its AZIMUTHS as determined by two or more radio direction finders; also used for aircraft guidance.

radio procedure A standardized method of radio transmission used by operators to save time and prevent confusion. By insuring uniformity, radio procedure increases military SECURITY.

radio relay system A signal communications system using very high frequency radio waves and line-of-sight radio transmitters and receivers in lieu of trunk wire circuits. While widely used during World War II, today communications satellites are more likely to perform this function.

radio silence A condition in which all or certain radio equipment is kept inoperative to ensure that an enemy does not listen to messages.

radius, combat See COMBAT RADIUS.

radius of action The maximum distance a ship, aircraft, or vehicle can travel away from its base along a given course with a normal combat load and return without refueling, allowing for all safety and operating factors.

radius of damage The distance from GROUND ZERO at which there is a 0.5 probability of achieving a desired degree of damage.

radius of integration The distance from GROUND ZERO that indicates the area within which the effects of both a NUCLEAR DETONATION and conventional weapons are to be integrated.

radius of safety The horizontal distance from GROUND ZERO beyond which the effects of a NUCLEAR WEAPON on friendly troops are acceptable.

radius of target The radius of a circular target area or an AREA TARGET that is equated to a circle.

radius of vulnerability The radius of a circle within which friendly troops may become casualties or equipment may be damaged.

radome The covering (housing) for a RADAR antenna; it is essentially transparent to electromagnetic energy.

RAF The Royal Air Force of the United Kingdom.

raid 1. An offensive tactical operation, usually of small scale, involving a swift penetration of hostile territory to secure information, confuse the enemy, or destroy his installations. It ends with a planned withdrawal upon completion of the assigned mission. 2. As used by European air forces, any attack, even of major scale, by strategic bombers. 3. To conduct or participate in a RAID. See Joshua Shani, "Airborne Raids," *Air University Review* 35, 3 (March–April 1984).

raid, air See AIR RAID.

raid, amphibious See AMPHIBIOUS RAID.

raid, artillery See ARTILLERY RAID.

railhead 1. A railroad depot from which personnel and supplies are further distributed by other means. 2. The end of track; the farthest point of the rails of a railroad in any given direction.

railing In electronics, RADAR pulse JAMMING at high frequencies (50 to 150 kilocycles), resulting in an image on a radar indicator resembling fence railings.

rain, black See BLACK RAIN.

rainfall, nuclear The water that is precipitated from the BASE SURGE clouds after an underwater burst of a NUCLEAR WEAPON. This rain is radioactive, and presents an important secondary effect of such a burst.

rainout 1. A canceled event. 2. The bringing down of radioactive material in the atmosphere by precipitation.

raise pistol 1. A prescribed movement in the MANUAL of the pistol that includes taking a pistol out of its holster and raising it as high as, and six inches in front of, the right shoulder. 2. To perform this movement. 3. The command to carry out this movement.

raison d'état An overwhelmingly important general social or state motive for an action. There may be, it is argued, problems of such utter importance to the entire well-being of a state, or interests so vital to the entire population taken as a whole, that all ordinary moral or political restrictions on government actions must be dropped. The "state" itself is very much to the forefront when one invokes a raison d'état argument—it is the continued existence of the very basic structure of authority and legitimacy that has to be at stake.

raison de guerre A DOCTRINE, akin to the concept of RAISON D'ETAT, justifying certain actions in war. While raison d'état argues that at times when the whole safety of a state is at stake, the state can do anything necessary to protect itself, regardless of ordinary moral or legal restrictions; raison de guerre, by analogy, is the doctrine that the vital necessity of winning a battle may, at times, justify behavior that would not normally be acceptable even in the context of the already less-constrained standards of morality that exist in war. While raison d'état has long been recognized in INTERNATIONAL LAW, raison de guerre is much less well established.

rallying point An area (usually predesignated) where members of a military UNIT who have become separated or dispersed reassemble.

ramp sight A type of metallic sight in which the aperture is raised or lowered by moving it forward or backward on an inclined ramp.

range 1. The distance between any given point and an object or TARGET. 2. The extent or distance limiting the operation or action of something, such as the range of an aircraft, ship, or gun. The range of ships and aircraft can be extended with refueling at sea or in the air. The range of a gun can be extended by adjusting the elevation or the amount of PROPELLANT used. 3. The distance that can be covered over a hard surface by a ground vehicle, with its rated PAYLOAD, using the fuel in its tank and in the fuel cans normally carried as a part of ground-vehicle equipment. 4. An area equipped for practice in shooting at targets. In this meaning, also called a target range or firing range.

range, actual See ACTUAL RANGE.

range, adjusted See ADJUSTED RANGE.

range adjustment Successive changes in FIRING DATA so that the impact or burst of an artillery projectile will be on target with respect to RANGE.

range angle The angle between the aircraft target and the vertical line from the aircraft to the ground at the instant a BOMB is released. Also called the DROPPING ANGLE.

range card 1. A small chart on which RANGES and directions to various TARGETS and other important points in an area under FIRE are recorded. 2. A small chart showing the proper amount of CHARGE to be used for various ranges within the limits of a weapon (typically artillery).

range correction Changes in FIRING DATA necessary to allow for deviations of RANGE, due to weather, ammunition, or other nonstandard conditions.

range deviation The distance by which a PROJECTILE strikes beyond, or short of, its TARGET.

range difference The difference between the RANGES from any two points to a third point; especially the difference between the ranges of a TARGET from two different GUNS.

range disk A graduated disk, used for RANGE setting, connected mechanically with the elevating mechanism of a GUN. A range disk is usually graduated in meters of range and degrees of ELEVATION.

range dispersion diagram A chart indicating the expected percentage of shots fired with the same FIRING DATA that will fall into each of eight areas within the DISPERSION PATTERN for a given RANGE.

range drum A graduated, cylinder-type indicator, connected mechanically with the elevating mechanism of a gun and used for RANGE setting.

range error The difference between the RANGE to the point at which a particular PROJECTILE bursts and the range to the MEAN POINT OF IMPACT of a group of shots fired with the same FIRING DATA.

range flag A red flag displayed on or near a TARGET during firing practice as a warning that firing is being conducted.

range house Buildings with a storeroom, and sometimes offices, on a firing range.

range indicator A card showing the distance in meters from the FIRING POINT to the TARGET, used in target practice. Range indicators are attached to the parts of a landscape target to show the distances of points from the firer.

range K In ARTILLERY fire, a correction expressed in meters/1000 meters of range to correct for nonstandard conditions.

range ladder A naval term used to describe a method of adjusting gunfire by firing successive VOLLEYS, starting with a RANGE that is assuredly over or short of the TARGET and applying small uniform range corrections to the successive volleys until the target is crossed.

range markers Two upright markers, which may be lighted at night, placed so that when aligned, the direction they indicate assists in piloting. They may be used in AMPHIBIOUS OPERATIONS to aid in beaching LANDING SHIPS or LANDING CRAFT.

range officer The officer in charge of, and responsible for the safety of, a FIRING RANGE.

range rake A T-shaped device with pegs set in the cross. The distance between pegs subtends a definite angle at the base of the T. By sighting with a range rake, an observer can get a quick angular measurement of RANGE DEVIATION.

range resolution The ability of RADAR equipment to separate two reflecting objects on a similar bearing, but at different ranges from the antenna.

range scale 1. A scale on the arm of a plotting board where the observer range of a moving target is recorded. 2. A graduated scale on the sight or mount of a gun used to show the ELEVATION of a gun. 3. A table of FIRING DATA giving elevation settings corresponding to various ranges for standard CHARGES.

range table A prepared table that gives ELEVATIONS corresponding to RANGES for a gun under various conditions. A range table is part of a FIRING TABLE.

range wind The horizontal component of true wind in the vertical plane through the LINE OF FIRE of a gun.

rangefinder Any of a variety of optical instruments used for determining the distance between a gun and a target (or between a camera lens and an object to be photographed).

rangefinder, coincidence See COINCIDENCE RANGEFINDER.

rangefinder, laser See LASER RANGEFINDER.

ranger 1. Soldiers specially trained and organized for raiding operations. 2. Soldiers specially trained in guerrilla warfare. An American ranger unit, Roger's Rangers, was first created in the eighteenth century. The name was revived for American COMMANDO units in World War II. The rangers were deactivated after the KOREAN WAR but during the VIETNAM WAR, ranger units were once again created and saw considerable action. Compare to SPECIAL FORCES.

ranging 1. Wide scale scouting, especially by aircraft, designed to search an area systematically. 2. Locating an enemy gun by watching its flash, listening to its report, or by other similar means.

rank 1. One's place in a hierarchical ordering of positions. 2. One's military grade level. 3. The condition of being of a higher grade than another (i.e. the colonel ranked the captain). 4. A row of soldiers arranged in close order side by side. 5. The plural, "the ranks," refers to enlisted personnel in general.

rank and file 1. The masses. 2. Those members of an organization who are not part of its management. 3. The enlisted soldiers in a traditional army who had to line up in ranks, side by side, and files, one behind the other. Officers, being gentlemen, were spared such indignities.

ranks, close See CLOSE RANKS.

rapid deployment force (RDF) 1. Any military force that is maintained so as to be almost immediately prepared to be deployed against an enemy. CENCOM was the "central command" set up by President of the United States Jimmy Carter as a response to America's potential need to intervene at some distance from the American mainland, such as in the Persian Gulf. Rapid deployment forces need not be heavily armed, but must be capable of fighting for at least a brief period before reinforcements and extra equipment can be delivered. CENCOM does not, in fact, consist of permanently prepared troops at all. It consists mainly of a headquarters and planning staff which would, in a time of emergency, be able to deploy combat forces that usually have a more general role, such as the 82nd Airborne Division and various Marine units. It is generally thought that, at its conception, the RDF was meant to be a permanent force, stationed somewhere in the Mediterranean area—but no country was prepared to host the base. See Victor H. Krulak, "The Rapid Deployment Force: Criteria and Imperatives," *Strategic Review* 8, 2 (Spring 1980); Robert P. Haffa, Jr., *The Half War: Planning U.S. Rapid Deployment Forces to Meet a Limited Contingency, 1960–1983* (Boulder, CO: Westview Press, 1984).

rate of fire The number of ROUNDS fired per WEAPON per minute.

rate of march The average number of miles or kilometers to be travelled by a military force in a given period of time, including all ordered halts. It is expressed in miles or kilometers in the hour. See also PACE.

rating 1. British naval terminology for an ENLISTED PERSON. 2. Enlisted personnel grade levels in the U.S. Navy.

ration 1. The allowance of food for the subsistence of one person for one day. 2. Any food. 3. To limit the use of a particular supply because of a shortage. See Elizabeth Sutphen and Victoria Dibbern, "Combat Rations of Warsaw Pact Countries," *Army Logistician* 17, 3 (May–June 1985); John R. Bussert and Bent Ramskov, "Unitized T-Rations," *Army Logistician* 18, 4 (July–August 1986).

ration, field See FIELD RATION.

ration basis of issue Subsistence items issued to a military unit for a given day, which contain the quantity of food necessary for an equal number of breakfasts, lunches, and dinners for each member of the unit.

ration cycle The time covering one day's ration or three meals. It may begin with any meal.

ration dense Foods which, through processing, have been reduced in volume and quantity to a small, compact package without appreciable loss of food value, quality, or acceptance, and with a high yield of nutrition in relation to the space they occupy, such as dehydrates and concentrates.

rationalization 1. Any action that increases the effectiveness of allied forces through more efficient or effective use of defense resources committed to an alliance. Rationalization includes standardization, interoperability, greater cooperation, etc. 2. The mandated merger of industrial plants during wartime to achieve greater efficiency and lower costs.

rations in kind Actual food items issued for consumption; the items may be either cooked or uncooked. In general, opposed to subsistence payment in money for a given period.

ration strength For the FIELD RATION, the actual number of persons present for meals.

ratline 1. An organized effort for moving PERSONNEL or MATERIEL by clandestine means across a denied area or border. 2. The ladder-like rigging on the masts of ships, that allow a sailor to climb to the mast top.

reaction decoy A DECOY deployed only upon warning or suspicion of imminent nuclear attack.

reaction engine A jet or rocket engine; one that produces power in response to the momentum created by gases being ejected.

reaction propulsion A propulsion system in which a forward motion or thrust is produced by the forcing of propellant gases through nozzles, generally longitudinally opposed to the intended line of travel of a rocket, missile or jet aircraft.

reaction time 1. The elapsed time between the initiation of an action and the required response. 2. The time required between the receipt of an order directing a military OPERATION and the arrival of the initial element of the force concerned in the designated area.

reactive armor Thin slabs of explosives fitted to the outside of a tank. If the tank is hit by an enemy round, the reactive armor explodes. This has the effect of disrupting the highly concentrated stream of molten metal and superheated gases that result from the explosion of a High Explosive Anti-Tank (HEAT) projectile. Shaped HEAT rounds burn through a tank's armor, destroying the crew and internal equipment. If the secondary explosion of reactive armor disperses these forces, the effect of such a hit on a tank is no more than the effect of an ordinary high explosive shell, from which tanks are usually immune. Reactive armor was pioneered by the Israeli Army and saw its first use in the 1982 Lebanon War. It has become important because the Soviet Army has adopted it and is fitting it to their most modern tanks. No one knows for sure how effective this form of armor is, but some estimates suggest that all current generation antitank weapons in NATO except for those fired by tank guns themselves will become useless. Tank guns fire KINETIC ENERGY projectiles, which cannot be disrupted in this way. Work is well advanced on a new generation of HEAT projectiles that can make reactive armor useless, usually by firing a small precursor charge on impact which will trigger the reactive armor before the main shaped explosive charge is ignited.

reactor, nuclear See NUCLEAR REACTOR.

readiness A general measure of how well prepared military UNITS are for COMBAT. Readiness covers all the relevant elements, including ammunition and fuel stocks, training levels, MOBILIZATION PLANS and, sometimes, the availability of necessary RESERVES in a hurry. It has come to be exceptionally important because of the likely nature of any CENTRAL FRONT war between NORTH ATLANTIC TREATY ORGANIZATION (NATO) and the WARSAW PACT forces. Such a war will almost inevitably be short, whether or not it "goes nuclear." This means that no inadequacies in equipment, war stocks, or training can be made up for during the course of the war. Unfortunately, it is precisely in the area of readiness that NATO is least well prepared. The stress on weapons acquisition, always more glamorous and professionally interesting for military officers, combined with restricted defense budgets, has led to underinvestment in all aspects of readiness. Air forces, for example, would always prefer to buy more aircraft than to allow their pilots more flying time to train on the aircraft which they already have; armies let their ammunition stocks run down, and restrict the number of training ROUNDS tank crews can fire in order to procure more tanks. Concerns for readiness go back to the dawn of recorded warfare. SUN TZU (400–320 B.C.) wrote in *The Art of War:* "It is a doctrine of war not to assume the enemy will not come, but rather to rely on one's readiness to meet him; not to presume that he will not attack, but rather to make one's self invincible."

readiness, force See FORCE READINESS.

readiness, ground See GROUND READINESS.

readiness, materiel See MATERIEL READINESS.

readiness, operational See OPERATIONAL READINESS.

readiness condition, defense See DEFENSE READINESS CONDITION.

readiness evaluation, operational See OPERATIONAL READINESS EVALUATION.

ready The term used to indicate that a WEAPON(s) is loaded, aimed, and prepared to FIRE.

ready reserve American military units or personnel that can be ordered to active duty by the president in a war or a national emergency.

Reagan Doctrine What the media have come to call the Reagan Administration's policy (in conjunction with that of the U.S. Congress) of militarily supporting guerrilla insurgencies against Communist governments in THIRD WORLD countries such as Afghanistan, Angola, Cambodia, and Nicaragua. See Stephen Rosenfeld, "The Reagan Doctrine: The Guns of July," *Foreign Affairs* Vol. 64, No. 4 (Spring 1986); William R. Bode, "The Reagan Doctrine," *Strategic Review* 14, 1 (Winter 1986).

realpolitik A German word, now absorbed into English, meaning "realist politics." The term is applied to politics—whether of the organizational or societal variety—that are premised upon material or practical factors rather than theoretical or ethical considerations. It is the politics of realism; an injunction not to allow wishful thinking or sentimentality to cloud one's judgment. It has taken on more sinister overtones, particularly in modern usage. At its most moderate, "realpolitik" is used to describe an overly cynical approach, one that allows little room for human altruism, that always seeks an ulterior motive behind another individual's or government's statements or justifications. At its strongest it suggests that no moral values should be allowed to affect the single-minded pursuit of one's own, or one's nation's self-interest, and an absolute assumption that any opponent will certainly behave in this way. See Louis Rene Beres, *Reason and Realpolitik: U.S. Foreign Policy and World Order* (Lexington, Mass.: Lexington Books, 1984).

real time The absence of delay, except for the time required for the transmission by electromagnetic energy, between the occurrence of an event or the transmission of data, and the knowledge of the event, or reception of the data at some other location. See also NEAR REAL TIME.

rear area The area to the rear of the COMBAT and FORWARD AREAS in which supply, maintenance support, communications centers, and administrative ECHELONS are located.

rear-area combat operations Operations undertaken in the REAR AREA to protect military UNITS, LINES OF COMMUNICATIONS, INSTALLATIONS, and facilities from enemy ATTACK or SABOTAGE, or from natural disaster; to limit damage and to re-establish SUPPORT capabilities.

rear echelon 1. Elements of a FORCE which are not required in a COMBAT AREA. This is a generic term used to describe all elements normally located in the REAR AREA. 2. Those units or elements thereof that are not required in an AIRHEAD or BEACHHEAD; They normally remain in the departure area.

rear guard 1. The rearmost elements of an advancing or withdrawing force, which protect a COLUMN from hostile forces. During a withdrawal, the rear guard delays the enemy by armed resistance, destroying bridges, and blocking roads. During an advance, it keeps supply routes open. 2. Any SECURITY detachment that a moving ground force details to the rear to keep it informed and covered.

rearming 1. Replenishing the prescribed stores of ammunition, bombs, and other armament items for an aircraft, ship, tank, etc., to make it ready for COMBAT. 2. Resetting the FUZE on a bomb or other PROJECTILE, so that it will detonate at the desired time. 3. The development of plans of procurement to re-establish a previous military capacity for a nation that has been, or allowed itself to become, disarmed.

rear obstacles Obstacles used by an army corps or division in the REAR AREA to limit deep penetration or to protect critical terrain. Compare to FORWARD OBSTACLES; INTERMEDIATE OBSTACLES.

rear point The group of soldiers in a REAR GUARD that is farthest to the rear, and which observes enemy movements and discourages enemy pursuit by using HARASSING FIRE against the enemy.

rebellion See REVOLUTION.

receding leg That portion of a target's course line in which the SLANT RANGE increases for successive target positions.

receiver The part of a gun that takes the CHARGE from the MAGAZINE and holds it until it is seated in the BREECH.

receptacle box A central electrical distribution box mounted on a gun carriage. A receptacle box serves as a distributor of fire CONTROL DATA from a director to the AZIMUTH, ELEVATION, and FUZE SETTER.

reciprocal laying A method of making the PLANES OF FIRE of two guns parallel by pointing the guns in a parallel direction. In reciprocal laying, the two guns sight on each other, then swing out through supplementary angles to produce equal deflections from the base line connecting the two guns.

reckoning, dead See DEAD RECKONING.

reclama A request to duly constituted authority to reconsider its decision or its proposed action.

recognition signal Any prearranged signal by which individuals or units may identify each other.

recoil 1. The rearward kick of a gun when it is fired. 2. The distance that a gun moves backward when it is fired.

recoil cylinder A fixed cylinder through which a piston attached to a gun is forced by the backward motion or RECOIL of the gun on firing. The recoil is cushioned by springs or by the slow passage of air or a fluid through holes in the piston.

recoilless A term applied to certain weapons that employ high-velocity gas ports (jets) to counteract RECOIL.

recoilless rifle A light CANNON capable of being fired from either a ground mount or from a vehicle, and capable of destroying tanks. By ejecting the explosive force of its ammunition behind it, the weapon itself is not displaced by RECOIL.

recoil mechanism A mechanism designed to absorb the energy of RECOIL gradually and to avoid violent movement of the GUN CARRIAGE. The recoil mechanism is usually a hydraulic, pneumatic, or spring-type shock absorber that permits the BARREL assembly to move to the rear while resistance is progressively built up.

recoil pit A pit dug near the BREECH of a gun to provide space for the breech when it moves backward during RECOIL.

reconfiguration, dynamic See DYNAMIC RECONFIGURATION.

reconnaissance (RECON) A mission undertaken to obtain, by visual observation or other detection methods, information about the activities and resources of an enemy or potential enemy; or to secure data concerning the meteorological, hydrographic, or geographic characteristics of a particular area. See George E. Daniels, "An Approach to Reconnaissance Doctrine," *Air University Review* 33, 3 (March–April 1982).

reconnaissance, aerial See AERIAL RECONNAISSANCE.

reconnaissance, amphibious See AMPHIBIOUS RECONNAISSANCE.

reconnaissance, area See AREA RECONNAISSANCE.

reconnaissance, armed See ARMED RECONNAISSANCE.

reconnaissance, combat See COMBAT RECONNAISSANCE.

reconnaissance, contact See CONTACT RECONNAISSANCE.

reconnaissance, counter See COUNTERRECONNAISSANCE.

reconnaissance, map See MAP RECONNAISSANCE.

reconnaissance, photographic See PHOTOGRAPHIC RECONNAISSANCE.

reconnaissance, poststrike See POSTSTRIKE RECONNAISSANCE.

reconnaissance, prestrike See PRESTRIKE RECONNAISSANCE.

reconnaissance, radar See RADAR RECONNAISSANCE.

reconnaissance, route See ROUTE RECONNAISSANCE.

reconnaissance, strategic See STRATEGIC RECONNAISSANCE.

reconnaissance, zone See ZONE RECONNAISSANCE.

reconnaissance by fire A method of RECONNAISSANCE in which FIRE is placed on a suspected enemy position to cause the enemy to disclose his presence by movement or return of fire.

reconnaissance in force A limited-objective OFFENSIVE operation designed to discover or test an enemy's strength or to obtain other information. A commander ordering such a mission should be prepared to extricate his force or exploit its success.

reconnaissance of position A detailed examination of terrain as a basis for the selection of advantageous locations for guns and troops.

reconnaissance patrol For ground forces, a PATROL used to gain tactical information, preferably without the knowledge of the enemy.

reconnaissance strip A series of overlapping aerial photographs which, when joined together, will provide a rough continuous picture of the area photographed. A reconnaissance strip is generally used in studying a long, narrow piece of terrain, such as a river or a road.

reconstitution site A location selected by the surviving COMMAND authority as the site at which a damaged or destroyed HEADQUARTERS can be reformed from survivors of an attack or personnel from other sources, predesignated as REPLACEMENTS.

record as target In ARTILLERY AND NAVAL GUNFIRE SUPPORT, the order used to denote that a target is to be recorded for future ENGAGEMENT or reference.

record firing Target practice in which a record is kept. For SMALL ARMS, this record is the basis of a soldier's classification in marksmanship.

recover 1. To go back to a position just held in DRILL or practice. 2. The command to go back to any such position. 3. To solve or reconstruct CRYPTOGRAPHIC data or plain text. 4. The return of damaged, unserviceable, or abandoned MATERIEL to supply or maintenance channels; usually only applicable to COMBAT ZONES.

recovery 1. In air operations, that phase of a mission which involves the return of an aircraft to a base. 2. In naval MINE warfare, the salvage of a mine as nearly intact as possible, to permit its further investigation for intelligence or evaluation purposes. 3. In AMPHIBIOUS RECONNAISSANCE, the physical withdrawal to safety of landed froces or their link-up with friendly forces. 4. Extricating damaged or disabled equipment and moving it to locations where repairs can be made on it.

recovery airfield Any airfield, military or civil, at which aircraft might land post-H-hour. It is not expected that combat missions would be conducted from a recovery airfield.

recovery and reconstitution Those actions taken by one nation prior to, during, and following an attack by an enemy nation in order to minimize the effects of the attack, rehabilitate the national economy, provide for the welfare of the populace, and maximize the combat potential of remaining forces and supporting activities.

recovery site 1. In EVASION AND ESCAPE usage, an area from which an evader or an escapee can be evacuated. 2. An area of the ocean where objects from space may descend to be brought aboard ship.

recruit 1. A new entrant into military service. 2. The act of seeking new military enlistees through advertising or visits to high schools and colleges.

recruiting district A geographical subdivision of a military area established for the administration of recruiting activities. The U.S. is divided into recruiting districts by the various services.

recruiting station An INSTALLATION at which prospective RECRUITS are contacted and preliminarily screened for enlistment.

rectangle of dispersion An area, assumed to be rectangular, in which the PROJECTILES of a PIECE will fall when the piece is fired with the same FIRING DATA under apparently identical conditions.

recurring issue SUPPLIES issued on a cyclic basis to replenish material consumed or worn out through wear and tear in operations.

red 1. A Communist, because the red flag has long been the international symbol of Communism. 2. The color used to indicate danger or stop. 3. The concept that the electrical and electronic circuits, components, equipments, and systems that handle classified plain-language information in electrical signal form (red) be separated from those that handle encrypted or unclassified information (black). Under this concept, red and black terminology is used to clarify specific criteria relating to, and to differentiate between such circuits, components, equipments, and systems. See also ENCRYPTION.

redeployment The transfer of a unit, an individual, or supplies deployed in one area to another area, or to another location within the area, or to the ZONE OF INTERIOR for the purpose of further employment.

red legs Slang for artillery. During the American Civil War, Union artillerymen had red stripes on their uniform trousers.

redoubt 1. Any FIELD WORK that provides minimum protection from all directions of attack. 2. An EARTHWORK outside a main fortification, placed at likely attack routes.

redout The blinding or dazzling of INFRARED DETECTORS due to high levels of infrared radiation produced in the upper atmosphere by a nuclear explosion.

reduced charge. 1. The smaller of the two PROPELLING CHARGES available for naval guns. 2. A charge employing a reduced amount of propellant to fire a gun at short ranges, as compared to a normal charge. 3. A plea bargain accepted in a court martial.

reduced-strength unit A unit organized at the minimum organizational strength, consistent with the demands of a long noncombatant period and a limited period of combat.

reduction coefficient The ratio of the observer–target distance to the gun–target distance. The reduction coefficient is the number by which an observed deviation is multiplied in order to correct FIRING DATA at the gun.

reefer 1. A refrigerator. 2. A motor vehicle, railroad freight car, ship, aircraft, or other conveyance so constructed and insulated as to protect commodities from either heat or cold. 3. A marijuana cigarette; something than cannot be legally smoked by American military personnel. 4. A thick woolen sweater popular in the British Navy.

re-entry vehicle (RV) In the context of NUCLEAR WEAPONS, a WARHEAD carried by a BALLISTIC MISSILE. During the first stage, the BOOST PHASE, of the missile's flight, power is provided by ROCKET motors, which drop away when exhausted. The front end or BUS then enters an exoatmospheric suborbital path during what is usually called the mid-course phase. The warheads of the missile are ejected from the bus during this phase, and re-enter the atmosphere, hence the name. Because of the high kinetic energy and the atmospheric heat ablation that RVs encounter, they have to be specially armored against heat damage, while the bus, if it does not remain in orbit, burns up on re-entry.

refer 1. To bring the sights of an artillery PIECE that has been laid for direction to bear on a chosen AIMING POINT without moving the piece itself. 2. Command for this adjustment.

reference line A convenient and readily identifiable line used by the OBSERVER or SPOTTER of a gun crew as the line to which spottings will be related. One of three types of spotting lines. See also SPOTTING LINE.

reference piece One gun of a BATTERY selected as the standard with which to compare the firing of the other guns. Each of the other guns is called a test piece.

reference point A prominent, easily located point in the terrain.

referring point A new AIMING POINT on which gunners are to refer an artillery PIECE that has been laid for direction.

reflected shock wave The wave produced by a SHOCK WAVE traveling in one medium which then strikes an interface between this medium and a denser medium. The reflected shock wave travels back through the less dense medium.

reflector, confusion See CONFUSION REFLECTOR.

reform caucus, military See MILITARY REFORM CAUCUS.

refugees Persons who, because of real or imagined danger, move of their own volition, spontaneously or in violation of established policy, irrespective of whether they move within their own country (national refugees) or across international boundaries (international refugees).

refuse a flank To position a force's FLANK against a natural obstruction, such as a wide river, which makes it impossible for the enemy to attack that point.

regime 1. A form of government (republican, totalitarian, etc.). 2. A particular government that is in power; the group of individuals that constitutes the administration. 3. A system of governance (as opposed to anarchy).

regime, military See MILITARY REGIME.

regiment An administrative and tactical army unit, on a command level below a DIVISION or BRIGADE and above a BATTALION, the entire organization of which is prescribed by a TABLE OF ORGANIZATION. The commanding officer of a regiment is usually a COLONEL. In the U.S. Army regiments are now used just for armored cavalry units. But a regiment consisting of three infantry battalions is still the basic tactical unit of the U.S. Marine Corps. In the British and some other European armies, a regiment may in fact have only one battalion (especially in the armored branches). The usual equation is that an American regiment has three battalions, while three battalions, possibly each being a regiment, make up a European brigade.

regimental combat team A traditional infantry REGIMENT that has been reinforced by attached artillery, engineer, and other appropriate units.

regimental landing team A TASK ORGANIZATION for landing, comprising an infantry REGIMENT reinforced by those elements required for the initiation of its combat functions ashore.

registration The adjustment of FIRE to determine FIRING DATA corrections.

registration, offset See OFFSET REGISTRATION.

registration fire FIRE delivered to obtain accurate data for subsequent effective ENGAGEMENT of targets.

registration point A terrain feature or other designated point on which FIRE is adjusted for the purpose of obtaining corrections to FIRING DATA.

regular army A permanent army maintained in peace as well as in war; a STANDING ARMY; one of the major components of the U.S. Army. Regular military forces do not include reservists, who may be called up in time of war or other natural emergency.

regulating unit A unit within a marching COLUMN that sets the pace for the rest of the column.

rehabilitation 1. The processing, usually in a relatively quiet area, of military units or personnel recently withdrawn from combat or arduous duty, during which units recondition equipment and are rested, furnished special facilities, filled with replacements, issued replacement supplies and equipment, given training, and generally made ready for employment in future operations. 2. The action performed in restoring a military installation to authorized design standards.

reinforce To strengthen a force by committing additional forces, supporting elements, or SUPPORTING FIRE.

reinforcing 1. To strengthen a unit or position by adding additional troops, equipment, or other combat support services. 2. In ARTILLERY usage, a tactical mission in which one artillery unit augments the FIRE of another artillery unit.

reinforcing force, external See EXTERNAL REINFORCING FORCE.

related tell The relay of information between facilities through the use of a third facility.

relative error See ABSOLUTE ERROR.

release line, bomb See BOMB RELEASE LINE.

release point 1. In road movements, a well-defined point on a route at which the ELEMENTS composing a COLUMN return under the authority of their respective commanders, each one of these elements continuing its movement toward its own appropriate destination. 2. In air transport, a point on the ground directly above which the first paratroop or cargo item is airdropped. 3. In DISMOUNTED attacks, especially at night, that point at which a commander releases control of subordinate units to their commanders or leaders.

release point, bomb See BOMB RELEASE POINT.

releasing commander A commander who has been delegated authority to approve the use of NUCLEAR WEAPONS within prescribed limits.

releasing officer A designated individual who may authorize the sending of a message for and in the name of the originator.

relief 1. One's replacement after one's TOUR OF DUTY. 2. The removal of a commander because of inappropriate conduct. 3. The rescue of, or help for, a besieged force by friendly forces. 4. The differences in height and slope of the earth that are represented on maps by contours or shadings.

relief commander A NONCOMMISSIONED OFFICER of the GUARD who instructs and posts SENTRIES, changes RELIEFS, and is in charge of one of the reliefs.

relief in place An operation in which all or part of a unit is replaced by an incoming unit. The responsibilities of the replaced elements for the mission and the assigned zone of operations are transferred to the incoming unit.

relocation 1. Determining the RANGE and AZIMUTH of a target from a given station when the range and azimuth from another station are known. 2. Determining the range and azimuth of a future position of a moving target.

relocation clock A circular diagram used in the ADJUSTMENT OF FIRE to accurately show the positions of a moving target and the deviations of shots as reported by observers.

rem (roentgen equivalent mammal) The quantity of ionizing radiation of any type which, when absorbed by a human or other mammal, produces a physiological effect equivalent to that produced by the absorption of 1 roentgen (a unit of exposure) of X-ray or gamma radiation.

remaining forces The total surviving effective forces at any given stage of combat operations.

remaining velocity The speed of a PROJECTILE at any point along its TRAJECTORY. The remaining velocity is usually measured in feet per second.

remote sensors Remotely monitored devices implanted in an area to monitor personnel or vehicle activity. A SENSOR system consists of remote sensors, sensor relays, and sensor monitoring equipment. Formerly called "unattended ground sensors."

remotely piloted vehicle (RPV) A remotely piloted airborne RECONNAISSANCE, SURVEILLANCE, and TARGET ACQUISITION and target designation device. Such vehicles provide timely and accurate intelligence, and locate targets behind enemy lines.

rendezvous 1. A prearranged meeting at a given time and location from which to begin an action or phase of an OPERATION, or to which to return after an operation. See also JOIN-UP. 2. In an AMPHIBIOUS OPERATION, the area in which the LANDING CRAFT and AMPHIBIOUS VEHICLES rendezvous to form waves after being loaded, and prior to movement to the LINE OF DEPARTURE.

rendezvous, force See FORCE RENDEZVOUS.

rendezvous area, boat See BOAT RENDEZVOUS AREA.

reorganize 1. To restore order in a unit after COMBAT, by replacing CASUALTIES, reassigning PERSONNEL, if necessary, replenishing the AMMUNITION supply, and performing whatever other actions are necessary or possible in order to prepare the unit for further attack or pursuit of the enemy. 2. To change from one type of unit to another within an arm or service, or to change personnel and equipment within a unit in accordance with newly published or revised TABLES OF ORGANIZATION.

repatriate A person who returns to his country or citizenship, having left his native country, either against his will or as one of a group who left for reason of politics, religion, or other pertinent reasons.

repeat In ARTILLERY AND NAVAL GUNFIRE SUPPORT, an order or request to repeat the firing of a given number of ROUNDS with the same method of fire.

replacement company A COMPANY in which personnel are received, administered, and provided appropriate training before assignment to other units that have suffered combat losses.

replacement factor The estimated percentage of equipment or repair parts that will require replacement during a given period due to wear, enemy action, abandonment, pilferage, and other causes except catastrophes.

replacements Personnel required to take the place of others who depart from a unit.

replenishment cycle quantity The quantity of MATERIEL required to sustain normal operations during the interval between successive replenishments. Under normal conditions it is equal to the OPERATING LEVEL OF SUPPLY.

report line A line at which troops, after having reached it, must report to their command ECHELON.

reported unit A unit designation for an enemy force that has been mentioned in an agent report, captured document, or interrogation report, but for which available information is insufficient to include the unit in accepted enemy ORDER OF BATTLE estimates.

reporting time interval 1. In SURVEILLANCE, the time interval between the detection of an event and the receipt of a report about it by the intended user of the information. 2. In COMMUNICATIONS, the time for transmission of data or a report from the originating terminal to the end receiver.

request modify In ARTILLERY and NAVAL GUNFIRE SUPPORT, a request by any person, other than the person authorized to make modifications to a FIRE PLAN, for a modification.

required delivery date 1. The date on which a unit is required to arrive at the MAIN BATTLE AREA in support of a specific operations plan.

required military force The armed forces necessary to carry out a military mission over a specified period of time.

required supply rate The amount of ammunition expressed in ROUNDS per weapon per day for those items fired by weapons, and of all other items of supply expressed in terms of appropriate unit of measure per day, estimated to sustain the operations of any designated force without restriction for a specified period.

rescue stop A piece of rescue equipment that is put around a person's chest to secure that person to a rescue line or helicopter hoist cable. Also called a horse collar.

resection A method of locating a point by computation or by plotting the intersection of rays obtained from sightings taken to three or more points whose locations are already known.

reservation, airspace See AIRSPACE RESERVATION.

reserve 1. A portion of a body of TROOPS that is deep to the rear or withheld from action at the beginning of an ENGAGEMENT, in order to be available for a decisive movement. Any force not engaged or lightly engaged may also be designated or employed as a reserve. The timely use of one's reserves is one of the most important aspects to the art of command. CLAUSEWITZ wrote in *On War:* "Fatigue the opponent, if possible, with few forces and conserve a decisive mass for the critical moment. Once this decisive mass has been thrown in, it must be used with the greatest audacity." 2. Members of military services who are not in active service but who are subject to a call to active duty once a war begins. 3. That portion of an appropriation or contract authorization held or set aside for future operations or contingencies, and in respect to which administrative authorization to incur commitments or obligations has been withheld. See also GENERAL RESERVE; OPERATIONAL RESERVE; RESERVE SUPPLIES.

reserve, general See GENERAL RESERVE.

reserve, ready See READY RESERVE.

reserve, rolling See ROLLING RESERVE.

reserve, selected See SELECTED RESERVE.

reserve, strategic See STRATEGIC RESERVE.

reserve, war See WAR RESERVE.

reserve officer A duly COMMISSIONED OFFICER or WARRANT OFFICER of a RESERVE component.

Reserve Officers' Training Corps (ROTC) The military training organization established at civilian educational institutions; created by the Morrill Act of 1862. It provides military training for and commissions in the U.S. Armed Forces to selected students upon graduation. Few other Western nations operate this system, though the United Kingdom has a similar version. See William C. Stancik and R. Cargill Hall, "Air Force ROTC: Its Origins and Early Years," *Air University Review* 35, 5 (July–August 1984); Robert F. Collins, "ROTC Today and Tomorrow," *Military Review* 66, 5 (May 1986).

reserve supplies SUPPLIES accumulated in excess of immediate needs for the purpose of insuring continuity of an adequate supply. Also called reserves.

reserved demolition target A target for demolition, the destruction of which must be controlled at a specific level of command because it plays a vital part in the TACTICAL or STRATEGIC PLAN, or because of the importance of the structure itself, or because the demolition may be executed in the face of the enemy. See also DEMOLITION TARGET.

reserves, initial See INITIAL RESERVES.

reserves, mobile See MOBILE RESERVES.

residual contamination Radioactive contamination that remains after steps have been taken to remove it. These steps may consist of nothing more than allowing the contamination to decay normally.

residual forces Unexpended portions of reserve forces that have an immediate COMBAT potential but have been deliberately withheld from utilization.

residual radiation Nuclear radiation caused by FALLOUT, radioactive material dispersed artificially, or irradiation that results from a nuclear explosion and persists longer than one minute after burst. See also CONTAMINATION; INDUCED RADIATION; INITIAL RADIATION.

resistance force In UNCONVENTIONAL WARFARE, that portion of the population of a country who is engaged in the resistance movement, i.e., guerrillas, AUXILIARIES, and members of the UNDERGROUND.

resistance movement An organized effort by some portion of the civil population of a country to resist the legally established government or an occupying power, and to disrupt civil order and stability.

resolution in azimuth The angle by which two TARGETS must be separated in order to be distinguished by a RADAR set when the targets are at the same RANGE.

resolution in range The distance by which two TARGETS must be separated in order to be distinguished by a RADAR set when the targets are on the same AZIMUTH line.

response, controlled See CONTROLLED RESPONSE.

response time In INTELLIGENCE usage, the time lapse between the initiation of a request for information and the receipt of that information.

responsible officer A military officer answerable by law or regulations for the discharge of a duty.

responsibility 1. The obligation to carry forward an assigned task to a successful conclusion. With responsibility goes the authority to direct and take the necessary action to insure success of the task. 2. The obligation for the proper custody, care, and safekeeping of property or funds entrusted to the possession or supervision of an individual.

responsor An electronic device used to receive an electronic challenge and display a reply thereto.

rest 1. In ARTILLERY, a command that indicates that the UNIT(s) or GUN(s) to which it is addressed shall not follow up FIRE orders during the time that the command is in force. 2. A mechanical support for a gun in aiming and firing. 3. Limited freedom to move, talk, or smoke while in RANKS. 4. A command allowing military personnel to move, talk, or smoke in ranks, but requiring them to keep one foot in place.

rest and recuperation (R&R) 1. The withdrawal of personnel from combat or duty in a combat area for short periods of rest and recuperation. This is commonly referred to as R&R. See also REHABILITATION. 2. Rest and recreation. This often refers to the American policy of allowing servicemen during the VIETNAM WAR to have a respite from combat in major recreation areas in Vietnam and other parts of the Pacific (such as Hong Kong, Hawaii, Australia and Bangkok).

restart at . . . In ARTILLERY, a term used to restart at a given time a FIRE PLAN after "dwell at . . ." or "check firing," or "cease loading" has been ordered.

restricted area 1. An airspace of defined dimensions, above the land areas or territorial waters of a state, within which the flight of aircraft is restricted in accordance with certain specified conditions. 2. An area in which there are special restrictive measures employed to prevent or minimize interference between friendly forces. 3. An area under military jurisdiction in which special security measures are employed to prevent unauthorized entry.

restriction Punitive restraint imposed on a person subject to MILITARY LAW by NONJUDICIAL PUNISHMENT or a military court which restricts the person to limits, such as a specific place or area, prescribed in the order.

restrictive fire area An area in which specific restrictions are imposed and into which FIRE in excess of those restrictions will not be delivered without prior coordination with the headquarters that establishes the area. It is generally located on identifiable terrain to facilitate recognition from the air.

restrictive fire line A line established to coordinate FIRE between helicopter-borne or airborne forces and LINKUP forces, or between any converging friendly forces (one or both of which may be moving). It prohibits fire or the effects of fire across the line without prior coordination with the affected force. It is established on identifiable terrain by the common commander of the converging forces. Formally called the "fire coordination line."

restrictive fire plan A safety measure for friendly aircraft which establishes an airspace that is reasonably safe from friendly, surface-delivered, non-nuclear FIRE.

resupply, early See EARLY RESUPPLY.

retain A mission requiring a unit to specifically prevent the enemy from occupying a position, terrain feature, or man-made object.

retained enemy personnel Certain protected enemy personnel, such as medical personnel and chaplains and, under certain circumstances, members of the staffs of national Red Cross societies and other recognized volunteer aid societies, retained in PRISONER-OF-WAR camps to assist prisoners-of-war, and who while so retained are not deemed prisoners-of-war but are afforded at least the same protection afforded them.

retaliation, massive See MASSIVE RETALIATION.

retalitory capability, strategic See STRATEGIC RETALIATORY CAPABILITY.

retentions 1. Personnel who elect to remain in a military service after an initial period of enlistment has expired. 2. Ships assigned to the control of a THEATER OF WAR commander for the movement of cargo from one point to another within the theater or between theaters.

retired list A list of OFFICERS and ENLISTED PERSONNEL who have been released from ACTIVE duty because of age, disability, or some other cause, and who have qualified to receive retired pay and benefits.

retirement 1. An operation in which a force out of contact moves away from the enemy. 2. The orderly withdrawal of troops according to their own plan and without pressure by the enemy. 3. Release from active military service because of age, length of service, disability, or some other reason.

retirement route The track or series of tracks along which helicopters move from a specific LANDING SITE or LANDING ZONE back to a rear area.

retraining brigade A correctional treatment facility for the confinement, retraining, and restoration of prisoners as better motivated soldiers.

retransmission The rebroadcasting of a message on a different frequency simultaneously with an original broadcast, by means of an electrically operated linkage device between a radio communication receiver and transmitter of the same set.

retreat 1. At its best, a tactical move away from the enemy; at its worst a mass fleeing from the enemy. While a retreat is a time-honored tactic used effectively in both traditional and unconventional warfare, Western military culture gives it an undeservedly negative connotation. One example of this is Napoleon's attitude that: "An ordinary general occupying a bad position, if surprised by a superior force, seeks safety in retreat; but a great captain displays the utmost determination and advances to meet the enemy." 2. A formal ceremony held at sunset when flags are lowered at most bases and posts.

retrograde 1. Contrary to the normal flow or direction of a process. Thus in an attack during combat a retrograde movement might see troops heading toward the rear. This could be good if the movement were an orderly tactical withdrawal to set a trap for the enemy; or bad if they were simply fleeing in panic. 2. A condition designation for MATERIEL, such as ammunition, earmarked for movement to a rear depot or offshore facility.

retrograde movement Any movement of a command to the rear, or away from the enemy. It may be forced by the enemy or may be made voluntarily. Such movements may be classified as withdrawals, retirements, or delaying actions.

retrograde personnel Personnel evacuated from a THEATER OF WAR; they may include noncombatants and civilians.

return 1. To put a weapon back into a holder. 2. The echoed signal from an object located by radar.

return pistol The command to put a pistol back in its holster. "Return pistol" is a prescribed command in the MANUAL of the pistol.

return program A planned operation to deal with the final disposition of remains of personnel who have died while in military service.

reveille (pronounced ré-ve-lē) 1. A signal, such as a bugle call, for the start of the day. 2. Morning ceremonies with the troops assembled in formal order as the flag is raised.

reverse slope Any slope that descends away from the enemy.

reverse-slope defense A DEFENSE AREA organized on any slope that descends away from the enemy.

reversed standard alphabet In CRYPTOGRAPHY, an alphabet in which the cipher component is in the normal sequence but reversed in direction from the plain component, which is also in the normal sequence.

revetment 1. A protective well (dirt, sandbags, etc.) for parked aircraft, gun emplacements, and other equipment or personnel. 2. Any EARTHWORK that affords protection against explosives.

review 1. The formal inspection of an organization. 2. A ceremony to honor some official or dignitary, or to present DECORATIONS to military personnel.

revolution 1. Any social, economic, agricultural, political, or intellectual change involving major transformations of fundamental institutions. 2. The overthrow by the citizens of a society of bad, incompetent, or unjust rulers, sometimes by violence, in order to establish a better government. The right of revolution serves as a continuous check on potential tyrants. This right of revolution has often been turned into a religious obligation. Both the American Thomas Jefferson (1743–1826) and the Iranian Ayatollah Khomeini (1901–) have preached that resistance to tyrants is obedience to God. Franz Kafka (1883–1924) is alleged to have warned: "Every revolution evaporates, leaving behind only the slime of a new bureaucracy." 3. The goal of an INSURGENCY, victory over and destruction of a present government. Such an insurgent revolution must be contrasted with rebellion. Theoretically, those in rebellion seek power for its own sake. In seeking domination, they violate the structures of a civil society. In this context, the worst rebels are tyrannical rulers who have violated both their personal honor and their political mandates and thus deserve to be overthrown. The ability of those fighting to overthrow a government to sustain their cause ultimately affects whether history calls them revolutionaries or rebels—as the Confederate states discovered when they lost the American Civil War. See Clarence Crane Brinton, *The Anatomy of Revolution,* revised edition (Englewood Cliffs, NJ: Prentice-Hall, 1938, 1952); Hannah Arendt, *On Revolution* (New York: Viking, 1963); D. H. Close and C. R. Bridge, *Revolution: The History of an Idea* (Totowa, NJ: Rowman & Allenheld, 1985).

rhumb line A line on the surface of the Earth that intersects successive meridians at the same oblique angle.

rib rifling Rifling of the BORE of a gun in which the LANDS and GROOVES are of equal width.

ricochet burst The near-surface burst of a high-explosive PROJECTILE after the projectile strikes a surface obliquely and is deflected at an angle.

ricochet fire FIRE in which the projectile glances from a surface after impact with it; it is sometimes used in artillery fire to obtain AIRBURSTS after initial impact.

rifle 1. A shoulder-held small arm with spiral grooves cut inside a long barrel to force its bullet into a rotating, more precise trajectory. 2. The grooves cut inside the barrel. 3. A cannon that has similar grooves cut inside its barrel.

rifle, recoilless See RECOILLESS RIFLE.

rifle company An element of a traditional infantry battalion in a triangular DIVISION. Such a battalion would typically have three rifle companies and one heavy weapons company. Each rifle company would, in turn, have three rifle platoons and a weapons platoon.

rifle grenade A GRENADE or small BOMB, designed to be projected from a special launching device attached to the MUZZLE of a RIFLE or CARBINE. It is propelled by a special blank cartridge fired in the rifle or carbine. There are essentially three types of rifle grenades: fragmentation, antitank, and smoke.

rifles A 19th century term for a military unit primarily equipped with rifles.

rifle salute A salute defined in the MANUAL of arms, in which the rifle is held at RIGHT SHOULDER ARMS or ORDER ARMS position, and the left hand is carried smartly across the body to the rifle, forearms horizontal, palm down, fingers together and extended.

rifling Helical grooves cut into the BORE of a gun that give spin to the PROJECTILE fired from it; this increases its RANGE and accuracy as compared to smooth-bore guns.

right 1. See LEFT (or RIGHT). 2. See STARBOARD.

right bank See LEFT (or RIGHT) BANK.

right face 1. In CLOSE-ORDER DRILL, the movement from the halted position of ATTENTION in which the soldier turns on the heel of the right foot and the ball of the left so as to face 90 degrees to the right of the original position. 2. The command to execute this movement.

right flank 1. The entire right side of a COMMAND, from the leading element to the rearmost element, as it faces the enemy. 2. By the "right flank" is a preparatory command to have every soldier in a formation change direction by 90 degrees to the right of the original direction of march. All of the soldiers in the formation turn at the same time.

right (left) flank, MARCH A two-part command to have every soldier in a formation change his or her direction of march 90 degrees to the right (left) of the original direction of march.

right shoulder arms 1. A movement in the MANUAL of the rifle in which the rifle is placed on the right shoulder, barrel up, and inclined at an angle of 45 degrees. 2. The command to perform the same movement. In the British drill the equivalent order would be "shoulder arms."

ring and bead sight A type of GUNSIGHT in which the front sight is a bead or post and the rear sight a ring.

ring sight 1. Any GUNSIGHT having a ring through which one looks. Ring sights are usually used as rear sights. 2. A gunsight consisting of concentric rings by means of which RANGE is estimated.

riot control agent A substance that produces temporary irritating or disabling physical effects that disappear within minutes after exposure to it has ceased. There is no significant risk of permanent injury, and medical treatment is rarely required.

riot gun Any shotgun with a short barrel, especially a short-barreled shotgun used in guard duty or to scatter rioters. A riot gun usually has a 20-inch, cylindrical barrel.

ripe In MINE WARFARE, a word sometimes used to mean "armed."

risers 1. That part of a personnel parachute harness that extends between the shoulder adapters and the connector links where the suspension lines of the parachute canopy are attached to the harness. 2. That part of a cargo parachute harness which extends between the snap fasteners or point of attachment to the load and the point of attachment of the parachute canopy suspension lines.

risk (nuclear), negligible See NEGLIGIBLE RISK (NUCLEAR).

river bank In river-crossing operations, river banks are referred to as the entry bank and the exit bank. The entry bank is that side of a water obstacle first encountered by friendly troops; the exit bank is opposite the entry bank.

river line 1. The water's edge on the defender's side of a stream. 2. Any tactical line marked by a stream.

R method Method of transmitting a message in which the receiving station is required to give a receipt.

road capacity The maximum traffic flow obtainable on a given roadway, using all available lanes; usually expressed in vehicles per hour or vehicles per day.

road clearance time The total time a COLUMN requires to travel over and clear a section of road.

road discipline The orderly, systematic movement of troops, vehicles, and gun mounts using a road. Road discipline prevents confusion and delay.

road net The system of roads available within a particular locality or area.

road screen Anything that is used to conceal movement along a road from enemy observation, especially artificial concealment or CAMOUFLAGE.

road space The length of roadway allocated to or actually occupied by a COLUMN on a route, expressed in miles or kilometers.

robust 1. A characteristic of a weapons system of the STRATEGIC DEFENSE INITIATIVE (SDI) with an ability to endure and perform its mission against a reactive adversary. 2. An ability to survive under direct attack.

rocket A self-propelled airborne vehicle whose TRAJECTORY or course, while in flight, cannot be controlled. Compare to BALLISTIC MISSILE.

rocket, free See FREE ROCKET.

rocket bomb A bomb with attached rockets to give it great downward speed as it is dropped from an airplane.

rocket motor 1. A chamber in which the propellant for a rocket is burned to provide propelling force. 2. A propulsion device that consists essentially of a thrust chamber(s) and exhaust nozzle(s), and which carries its own fuel combination from which hot gases are generated by combustion and expelled through a nozzle(s). See also REACTION PROPULSION.

roll back 1. The process of progressive destruction or neutralization of enemy defenses, starting at the periphery and working inward, to permit deeper penetration of succeeding defense positions. 2. A phrase used to describe the policy that the United States should seek to push Soviet influence out of Eastern Europe, rather than just containing the Soviet Union inside its current area of influence. 3. A bombing term meaning the gradual elimination of opposition to air superiority over a given time and area.

roll call 1. The process of checking the attendance of a military unit; so called because the names of all members were once inscribed on a roll. 2. A signal to assemble for a roll call.

roll-in-point The point at which aircraft enter the final leg of an attack (e.g., a dive or glide).

roll-on-roll-off 1. A term referring to cargo checked at a point of origin and loaded aboard a trailer-type conveyance and transported to a vessel at a port of loading, rolled onto the vessel, stowed, and rolled off at a port of discharge. 2. A land and water express service comprising a through movement of cargo between domestic depots and overseas depots, and also military intra- and intertheater depots.

roll out That movement of an aircraft from the moment when the wheels make impact with a landing surface until the aircraft leaves the runway at the end of the landing run.

roll-up 1. The process for orderly dismantling of facilities no longer required in support of operations and available for transfer to other areas. 2. A MANEUVER following a successful penetration of an enemy's defenses in which a force turns against the enemy positions from the flanks, as opposed to rupturing successive defense lines.

rolling barrage A BARRAGE in which the fire of units or subunits progresses by leapfrogging; the fire advances as each battery fires further in front of the others until the final objective is reached. 2. A preplanned OFFENSIVE in which a curtain of artillery fire moves toward the enemy at the same speed as attacking troops move forward from behind the line of fire.

rolling reserve RESERVE SUPPLIES held close to troop units. When equipment is available, these supplies are kept stored in railroad cars or in trucks ready for immediate transportation.

room circuit In CRYPTOGRAPHIC operations, a circuit that has no connection with outside stations, and which is used for encipherment and decipherment in off-line operation.

rope An element of CHAFF consisting of a long roll of metallic foil or wire which is designed to interfere with broad, low-frequency responses of radar.

rope-chaff CHAFF that contains one or more ROPE elements.

rosette A distinctive ribbon, folded and fanned into approximately a circular shape and inserted into a fabric-covered metal form with a prong and clutch, for attachment to the lapels of civilian clothing to denote the receipt of a specific, military DECORATION. The only American military decoration that includes a rosette for civilian wear is the Medal of Honor.

rotating band A soft metal band around a PROJECTILE near its base. The rotating band makes the projectile fit tightly in the BORE of a gun by centering the projectile, thus preventing the escape of gas and giving the projectile its spin.

round A ROUND OF AMMUNITION.

round, complete See COMPLETE ROUND.

round clearance distance The total distance over which the head of a motor column must travel for the entire column to clear a given section of a road.

round of ammunition All the components necessary to fire a weapon once. In general, these components are a PRIMER, PROPELLANT, container or holder for the propellant (CARTRIDGE CASE or bag), and PROJECTILE—with a FUZE and BOOSTER if necessary—for the proper functioning of the projectile.

rounds, (number of) See NUMBER OF ROUNDS.

rounds complete In ARTILLERY AND NAVAL GUNFIRE SUPPORT, the term used to report that the NUMBER OF ROUNDS specified for a given effect have been fired.

route 1. A road such as Route 66. 2. The prescribed course to be traveled from a specific point of origin to a specific destination.

route, approach See APPROACH ROUTE.

route capacity 1. The maximum traffic flow of vehicles in one direction at the most restricted point on a route. 2. The maximum number of metric tons that can be moved in one direction over a particular route in one hour. It is the product of the maximum traffic flow and the average PAYLOAD of the vehicles using the route.

route classification A classification assigned to a route indicating the heaviest military vehicle that it can accept. It is based on the weakest bridge or portion of the route.

route column 1. A CLOSE-ORDER formation of troops, suitable for marching. 2. A flexible formation adopted during the early phase of a movement to contact. During this phase, troops need not be tactically grouped, and may move by various means of transportation and by different routes.

route march A march in which the troops are allowed to break step, talk, or sing. More generally, a lengthy and demanding march.

route network, basic military See BASIC MILITARY ROUTE NETWORK.

route order The manner in which a ROUTE MARCH is made with horseback-mounted troops or troops traveling in vehicles. Talking, smoking, and relaxing are permitted, provided that there is no straggling or loss of relative positions.

route reconnaissance RECONNAISSANCE along a specific line of communications, such as a road, railway, or waterway, to provide new or updated information on route conditions and activities along the route.

route step 1. A way of marching in which troops are allowed to break step, talk, or sing, and carry their guns as they please. 2. A preparatory command to march in this manner.

routes of communication A network of roads or other routes over which supplies are carried and combat movements are made. Routes of communication include navigable waters, aircraft landing sites, and rail facilities.

routine ammunition maintenance Maintenance operations not involving the disassembly of AMMUNITION or replacement of its components, and comprising chiefly the cleaning and protecting of the exterior surfaces of individual packages of ammunition, ammunition components, and explosives.

routine message A category of precedence to be used for all types of messages that justify transmission by rapid means unless of sufficient urgency to require a higher precedence.

roving field artillery FIELD ARTILLERY withdrawn from its regular position and assigned to special missions. Roving field artillery is usually moved about and fired from different positions to deceive the enemy about the position and strength of a military force.

roving gun A GUN that is moved about and fired from different postions to mislead or harass the enemy. It is generally used for REGISTRATION when the location of a BATTERY position must remain secret.

row marker In LAND MINE WARFARE, a natural, artificial, or specially installed marker, located at the start and finish of a mine row, where mines are laid by individual rows.

RPV See REMOTELY PILOTED VEHICLE.

rules of engagement 1. Military or paramilitary directives that delineate the circumstances and limitations under which force can be used. For example, soldiers might be told to shoot only if they are fired upon first, or police might be told only to use deadly force when lives (as opposed to property) are in immediate danger. 2. Directives from a military headquarters that delineate the circumstances and limitations under which troops will initiate or continue combat ENGAGEMENT with the enemy. 3. In AIR DEFENSE, directives that delineate the circumstances under which weapons can fire at an aircraft. The right of self-defense is always preserved. See also AIR DEFENSE WEAPONS CONTROL STATUS.

running key system 1. A CIPHER SYSTEM in which a previously agreed upon PLAIN TEXT or book serves as the source of successive key letters for encipherment. 2. A cipher system using a long keying sequence which is not repeated with a given message.

running spare A repair part that is packaged and shipped with an operable set of equipment in addition to the parts needed for initial operation of the equipment, in order to provide replacements as they become necessary; for example, vacuum tubes and dry batteries.

rupture 1. To quickly create a gap in enemy defensive positions. 2. A complete or partial circular break in the metal of a fired CARTRIDGE CASE. A rupture causes loss of power and difficult extraction or jamming of a cartridge case. 3. The breaking of Earth or another substance by the explosion of a PROJECTILE or other CHARGE below the surface.

rupture force Any force that penetrates enemy forces and opens a gap for the remainder of a TASK FORCE to pass through.

rupture zone The region immediately adjacent to the boundary of the crater produced by an explosion in which the stresses produced by the explosion have exceeded the ultimate strength of the medium. It is characterized by the appearance of numerous radial cracks of various sizes. See also PLASTIC ZONE.

ruse A trick, or *ruse de guerre,* designed to deceive the enemy and thereby obtain an advantage. It is characterized by deliberately exposing false information to the collection means of the enemy. For example, prior to the World War II D-Day landings on the beaches of Normandy, the Allies took elaborate measures to "allow" German agents to "discover" evidence that the landings would be elsewhere. SUN TZU advised in *The Art of War:* "When capable, feign incapacity; when active, inactivity. When near, make it appear that you are far away; when far away, that you are near. Offer the enemy a bait to lure him; feign disorder and strike him. When he concentrates, prepare against him; where he is strong, avoid him. Anger his general and confuse him. Pretend inferiority and encourage his arrogance. Keep him under strain and wear him down. When he is united, divide him. Attack where he is unprepared; sally out when he does not expect you. These are the strategist's keys to victory." See also DEMONSTRATION, DISPLAY, FEINT, INDIRECT APPROACH.

S

sabot A lightweight, thrust-transmitting carrier in which a SUBCALIBER projectile is centered to permit firing the projectile in a larger caliber weapon. The carrier fills the BORE of the weapon from which the projectile is fired; it is normally discarded a short distance from the MUZZLE. One example of sabot rounds is the kinetic energy projectiles fired from modern tank cannons that are relatively small and narrow. They are too small for the bore of the tank guns, which have to be able at the same time to take the much larger propellants necessary to give the projectiles the muzzle velocity on which their kinetic energy depends. The projectiles themselves are therefore fitted inside steel cases, the sabots (from the French for boot), to fill up the bore.

sabotage 1. The deliberate destruction of property. 2. The slowing down of work in order to damage a business. During a 1910 railway strike in France, strikers destroyed some of the wooden shoes (sabots) that held the rails in place. "Sabotage" was soon adopted into English usage, but it wasn't until World War II that the word gained widespread popularity as a description of the efforts of secret agents to hinder an enemy's industrial military capabilities.

SACEUR (Supreme Allied Commander Europe) The senior military commander of the entire NORTH ATLANTIC TREATY ORGANIZATION (NATO) land and air contingent on the European continent. His authority extends over six subordinate commands: Northern Europe, Central Europe, Southern Europe, the Allied Command Europe Mobile Force, the United Kingdom Air Forces, and the special NATO Airborne Early Warning Force. His main responsibility, however, is to prepare for air and land strategy on the CENTRAL FRONT where, in case of war, SACEUR would have direct final command authority over the air forces and army corps of all the NATO allies. The SACEUR is always a U. S. Army general, partly because of the need to seek authority for the use of NUCLEAR WEAPONS from the president of the United States, but also in token of the predominant American contribution to NATO. The second-in-command can come from any of the other member states, but is often a West German or British army general because these two countries provide the bulk of the air and land power not provided by the United States. Additionally, NATO has two other command positions at the SACEUR level: SACLANT and CINCHAN. SACLANT, the Supreme Allied Commander Atlantic, is an American admiral based at Norfolk, Virginia, and CINCHAN, the Commander-in-Chief Channel, is a British admiral operating from Northwood, London. See Robert S. Jordon, *Generals in International Politics: NATO's Supreme Allied Commander, Europe* (Lexington, KY: University Press of Kentucky, 1987).

SACLANT See SACEUR

safe 1. As applied to explosives, a condition indicating that an explosive device is in an unarmed position; or a condition during which detonation cannot occur by FUZE action. 2. Constituted and set so as not to detonate accidentally. 3. The opposite of "armed."

safe area A designated area in hostile territory that offers the evader or escapee a reasonable chance of avoiding capture and of surviving until he can be evacuated.

safe burst height The HEIGHT OF BURST of a nuclear weapon at or above which the level of FALLOUT or damage to ground installations is predetermined by a military commander as acceptable.

safe-conduct A document similar to a passport, issued by a military authority, which a person must have if he wishes to enter or remain in an occupied zone or RESTRICTED AREA. A safe conduct may also enable the holder to move goods to or from places within the area, and to engage in trade, which would otherwise be forbidden, and to have military protection.

safe distance, minimum See MINIMUM SAFE DISTANCE.

safeguard 1. To protect. 2. A written order issued by a commander for the protection of persons or property from molestation by troops. A safeguard may be issued for reasons of military discipline, personal consideration, public policy, humanity, or other reasons. "Forging a safeguard" means the disregarding and violation of such an order, and in time of war may be punishable by death. 3. A soldier or detachment that is detailed to protect people, places, or property. 4. A lock on a door or gate for the protection of people, places, or property.

safe house An innocent-appearing house or premises established by an organization for the purpose of conducting CLANDESTINE or COVERT OPERATIONS in relative security.

safe separation distance With regard to a NUCLEAR WEAPON, the minimum distance between the weapon delivery system and the weapon beyond which the hazards associated with detonation are acceptable.

safety 1. A locking or cutoff device that prevents a WEAPON or any MISSILE from being fired accidentally. 2. Freedom from or protection against those hazardous conditions that have the potential to cause injury, illness, or death to personnel, or damage to or loss of equipment or property.

safety, nuclear See NUCLEAR SAFETY.

safety card A card issued for a particular BATTERY position for a particular time, prescribing the area into which FIRE may safely be placed both laterally and in depth.

safety diagram A geographic representation, usually an overlay, depicting the areas into which FIRE may safely be placed.

safety factor 1. An increase in RANGE or ELEVATION that must be set on a gun so that friendly troops, over whose heads FIRE is to be delivered, will not be endangered. 2. An overload factor in the design of a piece of equipment to insure its safe operation.

safety fork 1. A metal clip that fits over the collar of the FUZE in a MINE and prevents the mine from being set off accidentally. Its function is the same as that of a SAFETY PIN or SAFETY LEVER in a GRENADE, BOMB, or PROJECTILE. 2. A metal clip that is part of the quick parachute release assembly, which prevents accidental release of the parachute harness.

safety fuze A PYROTECHNIC device contained in a flexible and weatherproof sheath and which burns at a timed and constant rate; used to transmit a flame to the DETONATOR of an explosive device.

safety lanes Specified sea lanes designated for use in transit by submarines and surface ships to prevent attack by friendly forces.

safety lever 1. A lever that prevents the accidental firing of a GRENADE so long as it remains locked in position. Its function is the same as that of the SAFETY FORK in projectiles, bombs, and mines. 2. A lever that sets the safety mechanism on certain types of AUTOMATIC WEAPONS.

safety limit 1. A line marking off a zone or area in front of friendly troops, over whose heads gunfire is delivered. Shots must clear this zone if the troops are not to be endangered. 2. A boundary set around a target area on a firing range, within which there is a danger from shell fragments or ricocheting shells.

safety lock A locking device that prevents a gun from being fired accidentally.

safety officer 1. An officer who supervises field practice in gunnery to make sure that persons and property will not be endangered; often the assistant to the officer in charge of firing. 2. An officer who administers and directs organizational safety program activities.

safety pin See ARMING WIRE.

safety shoes 1. Special shoes designed to prevent foot injuries. 2. Special shoes (with conductive or nonconductive soles) to prevent sparks when working with explosives or other hazardous material.

safety stake One of the stakes set in the ground to mark the right or left limit of safe fire for a gun.

safety wire See ARMING WIRE.

safety zone An area of land, sea, or air reserved for noncombat operations of friendly aircraft, surface ships, submarines, or ground forces.

safing As applied to weapons and ammunition, the changing from a state of readiness for initiation to a SAFE condition.

salient 1. A bend projecting outward in the front line of friendly troops; one that bulges into the enemy line. 2. The outward projecting part of a FORTIFICATION or TRENCH system. 3. Projecting outward from any line or surface. 4. Something important.

sally An attack outward from forces defending a FORTIFICATION, especially a rushing SORTIE of troops against besieging forces.

sally port A large gate or passage in some fortified places for use by troops making a SORTIE.

SALT Strategic Arms Limitation Talks; the extensive negotiating sessions between the United States and the Soviet

Union to promote balanced and verifiable limitations on strategic NUCLEAR WEAPONS. The first such session, SALT I, started in 1969; after two and one-half years of negotiating, the parties signed the anti-ballistic missile (ABM) Treaty and an Interim Agreement which presumably froze offensive weapons at existing levels for five years. The second session, SALT II, which began in 1972, sought to achieve a comprehensive agreement to replace the Interim Agreement. A SALT II Treaty was signed by the United States and the Soviet Union in 1979; but U.S. Senate ratification of the treaty was indefinitely postponed on January 3, 1980, in response to the December 27, 1979, Soviet invasion of Afghanistan. Nevertheless, the United States abided by the terms of the treaty until 1986, when President Reagan asserted that Soviet violations made it impossible for the United States to continue doing so. Compare to START. See William C. Potter, editor, *Verification and SALT: The Challenge of Strategic Deception* (Boulder, CO: Westview Press, 1980); Gerard Smith, *Doubletalk: The Story of the First Strategic Arms Limitation Talks* (New York: Doubleday and Company, 1980); Samuel B. Payne, *The Soviet Union and SALT* (Cambridge: MIT Press, 1980).

salted weapon A NUCLEAR WEAPON which has, in addition to its usual components, certain elements or isotopes that capture neutrons at the time of the explosion and produce radioactive products over and above the usual radioactive weapon debris.

salute 1. A hand SALUTE; a signal of respect among military personnel. 2. A friendly military greeting between forces; for example, the firing of a cannon or the dropping of a flag.

salute, color See COLOR SALUTE.

saluting distance A distance, generally between 6 and 30 paces, at which SALUTES are given. A recognition of INSIGNIA is not difficult at a distance up to 30 paces, therefore 30 paces is set as the maximum distance.

saluting gun A cannon used for firing SALUTES.

salvage-fused A term that refers to a WARHEAD (usually nuclear) that is set to detonate when it is attacked.

salvo 1. In naval gunfire support, a method of FIRE in which a number of weapons are fired at the same target simultaneously. 2. In CLOSE AIR SUPPORT/AIR-INTERDICTION operations, a method of ORDNANCE delivery in which the release mechanisms are operated to release or fire all ordnance of a specific type simultaneously. 3. In army ARTILLERY, a simultaneous firing of guns.

salvo, bracketing See BRACKETING SALVO.

SAM See SURFACE-TO-AIR MISSILE.

sanction 1. A penalty attached to a law to encourage people to obey it. 2. Ratification by a higher (or another) authority. 3. A foreign policy ranging in a continuum from the suspension of diplomatic or economic relations to outright military intervention, designed to force another nation to change its behavior.

sanctuarization The description of a possible nuclear strategy which is only overtly admitted to by one member of the NUCLEAR CLUB, France, but may well underlie the thinking of all of them. President Charles de Gaulle developed France's independent nuclear force, the FORCE DE FRAPPE or, as it is now sometimes called, the "Force de Dissuasion" because he believed that no nuclear-armed nation would ever use its forces except in its own interests, and that therefore the American guarantee of extended deterrence to Western Europe was invalid. For de Gaulle and subsequent French thinkers, the point of owning a nuclear force was to make one's own country a sanctuary, an area against which no other power would dare use nuclear weapons. It was mainly to express this purpose of creating a sanctuary for France in the midst of a future war that de Gaulle stressed the idea that French missiles point "à tous azimuts" (in all directions), rather than only at the Soviet Union. The doctrine of sanctuarization, therefore, holds that nuclear weapons are not usable against a nuclear-armed power, because even a small nuclear force can present a retaliatory threat too awful to risk. The United States cannot admit to sanctuarization as the basis of its nuclear policy because the NORTH ATLANTIC TREATY ORGANIZATION'S (NATO) FLEXIBLE RESPONSE strategy depends on the American nuclear guarantee. However, even in the United States there are respected analysts and leading politicians who would argue that the arrival of NUCLEAR PARITY, combined with the dangers of uncontrolled ESCALATION from any nuclear use, in reality mean that the only function nuclear weapons fulfil is to deter the Soviet Union from attacking the United States directly. Compare to SECOND DECISION CENTER THESIS.

sanctuary 1. A safe place. 2. A sacred or religious place. 3. The protection once offered by churches to those fleeing the secular law because of political or other crimes. 4. The 1980s movement to help illegal immigrants from Central America find refuge in the United States. 5. A nation or area near or contiguous to a COMBAT AREA that by tacit agreement between the warring powers is exempt from attack and therefore serves as a refuge for staging, logistic, or other activities of the combatant powers.

sanitize To revise a report or other document in order to prevent the identification of sources, the persons and places concerned, or the means by which information was acquired.

sap 1. A narrow TRENCH as an extension of an existing trench, dug toward an enemy position. 2. A tunnel dug to place explosives under enemy positions.

sapper 1. A specialist in field fortifications, especially a builder of SAPS. 2. In the United Kingdom, a military engineer. 3. "Expendable" Viet Cong guerrillas who mined roads and waterways, emplaced booby traps, and led enemy attacks by clearing paths through minefields and other defensive obstacles.

SAS See SPECIAL AIR SERVICE.

satchel charge A number of blocks of explosive taped to a board fitted with a rope or wire loop for carrying and attachment. The minimum weight of the charge is usually about 15 pounds. Often incorrectly called a "satchel bomb."

satellite 1. A country that is informally dominated by another. For example, the Communist states of Eastern Europe are generally considered to be satellites of the Soviet Union. However, no self-respecting country would ever formally admit to being a satellite because that would imply that it is less than a sovereign nation. 2. A unit or activity not located within an installation but dependent upon it for specific support. 3. A device in Earth orbit which is capable, through cameras and electronic eavesdropping equipment, of collecting and transmitting back INTELLIGENCE and RECONNAISSANCE data. See Curtis Peebles, *Guardians: Strategic Reconnaissance Satellites* (Novato, CA: Presidio Press, 1987); Bhupendra Jasani and Toshibomi Sakata, *Satellites for Arms Control and Crisis Monitoring* (New York: Oxford University Press, 1987).

satellite defense All measures designed to identify, nullify, or destroy terrestrial orbiting vehicles, manned and unmanned, in order to prevent or contain the accomplishment of the vehicles' established mission. See also ACTIVE SATELLITE DEFENSE, PASSIVE SATELLITE DEFENSE.

satellite defense, active See ACTIVE SATELLITE DEFENSE.

satellite defense, passive See PASSIVE SATELLITE DEFENSE.

satellite weapon, defensive See DEFENSIVE SATELLITE WEAPON.

scabbing 1. The breaking off of fragments in the inside of a wall of hard material due to the impact or explosion of a projectile on the outside. 2. Dunnage used to fill excess space along the length of railway freight cars or ammunition cars.

scale factor A value by which distance on the ground is multiplied in order to compensate for map distortion when determining the ground distance as represented on a map.

scaling law A mathematical relationship that permits the effects of a nuclear explosion of given energy yield to be determined as a function of distance from the explosion (or from GROUND ZERO), provided the corresponding effect is known as a function of distance for a reference explosion, e.g., a 1-kiloton energy yield.

scan 1. In electromagnetic or acoustic searching, one complete rotation of an antenna. 2. The motion through space of an electronic beam searching for a target. 3. A search through radio channels.

scan period The period taken by a RADAR, sonar, or other detection device to complete a SCAN pattern and return to a starting point.

scarp Abbreviation for ESCARP.

scatter bomb A bomb containing SUBMUNITIONS that scatter over a wide area.

scatterable mine A MINE laid without regard to classical pattern and which is designed to be delivered by aircraft, artillery, missile, a ground dispenser, or thrown by hand.

scenario In strategic thinking, an imaginative account of a series of events, or of some crisis or problem, that might occur in the future. Because there is so little past experience of conflict approaching what a NUCLEAR WAR would be like, the use of historical analogy is rare in modern strategic theory. Instead, analysts concentrate on relatively simple imaginary situations to work out the implications of factors such as doctrine, weaponry, declaratory policy, and so forth. One scenario, for example, might depict a fully-fledged conventional onslaught by WARSAW PACT forces on West Germany, as featured in novels like Tom Clancy's *Red Star Rising,* or a highly technical, precise phase of SUPERPOWER interaction over a crisis in the Middle East. Essentially, scenario planning emulates the traditional WAR-GAME techniques used by all military forces over the last century. Wargaming, if carried out with great care and attention to detail, as well as with imagination, can be a vital tool. It is often said that the U. S. Navy fought no actions in the Pacific during World War II that its admirals had not practiced in war games during the 1930s. It is less clear whether the conditions of CENTRAL STRATEGIC WARFARE can be modelled in this way, and nuclear-strategy scenarios almost inevitably involve a greater degree of cool rationality in their development than anyone thinks would be possible in a real situation. Nevertheless, scenario-writing remains the nearest thing strategic thinkers have to a laboratory. See also GAME THEORY.

schedule control system System of traffic control in which truck columns and troops are dispatched over fixed routes at given rates of speed according to a time schedule.

schedule of fire Groups or series of FIRE planned to take place in a definite sequence and according to a definite program.

schedule of targets In ARTILLERY and NAVAL GUNFIRE SUPPORT, individual targets or groups or series of targets to be fired upon in a definite sequence and according to a definite program.

scheduled fire A type of PREARRANGED FIRE executed at a predetermined time.

scheduled supply A system whereby any UNIT (user or supplier) is furnished some or all of its supply requirements on a previously planned schedule that specifies items, quantities, and time and place of delivery.

scheduled target 1. In ARTILLERY and NAVAL GUNFIRE SUPPORT, a PLANNED TARGET on which FIRE is to be delivered at a specific time. Compare to ON-CALL TARGET. 2. A planned target on which a NUCLEAR WEAPON is to be delivered at a time specified in terms of minutes before or after a designated time, or in terms of the accomplishment of a predetermined movement or TASK.

scheme of command A plan for the control of all ELEMENTS of a COMMAND during a military operation, including provision for communication, observation, and the location of the COMMAND POST.

scheme of maneuver The TACTICAL PLAN to be executed by a FORCE in order to seize assigned objectives or hold its assigned area.

Schlesinger Doctrine The doctrine of LIMITED NUCLEAR OPTIONS, which set in train the design of an American nuclear arsenal that would be able to be used much more flexibly than the force structure created for the assured destruction (see MAD) policy. The details of the doctrine revolved around the need to have a range of options with which to respond to a nuclear attack on the United States, and above all, featured a shift of emphasis away from COUNTERVALUE strikes in favor of COUNTERFORCE targeting. This doctrine was first developed by James Schlesinger (1929–) while he was U. S. Secretary of Defense under President Nixon in 1973. See also COUNTERVAILING STRATEGY. See Douglas Kinnard, "James R. Schlesinger as Secretary of Defense," *Naval War College Review* 32, 6 (November–December 1979).

School of the Americas The United States Army's Spanish-language school, located in Georgia, which was founded in 1946 to train Latin American military commissioned and noncommissioned officers. Prior to 1984 it was located in Panama.

school solution An approach to a battlefield problem that employs conventional wisdom, so called because it comprises the answers expected by instructors at military academies, command and staff colleges, and other military training facilities.

scorched earth defense The systematic destruction, usually by burning, of anything that could be of use to an advancing enemy force. This includes the destroying of all domestic animals, civilian housing, crops, food stores, etc.

scout 1. An individual or military unit whose primary function is RECONNAISSANCE. 2. A VANGUARD force; a unit assigned to make contact with enemy lines.

scramble 1. An order directing that aircraft take off as quickly as possible, usually followed by mission instructions. 2. In telephony, to make unintelligible to casual interception. 3. In CRYPTOGRAPHY, to mix in random or quasirandom fashion.

screen 1. An arrangement of ships, aircraft, or submarines to protect a MAIN BODY of ground forces or a CONVOY. 2. In SURVEILLANCE, CAMOUFLAGE, and CONCEALMENT, any natural or artificial material, opaque to surveillance SENSORS(s), interposed between the sensor(s) and the object to be camouflaged or concealed. 3. A SECURITY element whose primary task is to observe, identify, and report information, and which only fights in self-protection.

screening fire Fire using smoke or other obscurants employed on the battlefield between enemy observation points and friendly units to mask friendly maneuvers or to deceive and confuse the enemy as to the nature of friendly operations.

screening smoke SMOKE employed in areas of friendly operation or in areas between friendly and enemy forces to conceal ground maneuvers, breaching and recovery operations, key assembly areas, supply routes, and logistic facilities, and to degrade enemy ground and aerial observation.

screw picket A metal post with a spiral point resembling a corkscrew. It is used as a support for a wire entanglement or as an anchor for a cable.

sea-air-land team (SEAL) A U.S. Navy Commando-type force especially trained and equipped for conducting unconventional and paramilitary operations and to train personnel of allied nations in such operations, including SURVEILLANCE and RECONNAISSANCE in and from restricted waters, rivers, and coastal areas. Commonly referred to as a SEAL team. Such forces are especially skilled at infiltrating enemy territory by sea in small boats or as frogmen, by air from helicopters or by parachute, and by land in ranger patrols. They also gather intelligence and breech (or emplace) underwater and beach obstacles.

Sea-Air-Land Team Six The United States Navy's elite counterterrorism unit.

sea echelon A portion of ASSAULT shipping that withdraws from, or remains out of, the transport area during an amphibious landing, and operates in designated areas to seaward in an ON-CALL, unscheduled status.

sea level, mean See MEAN SEA LEVEL.

sea-launched ballistic missile Any BALLISTIC MISSILE launched from a submarine or surface ship. See SLBM.

sea-launched cruise missile (SLCM) The most numerous of all CRUISE MISSILES, which may be fired from both submarines and surface ships. The Soviet navy has been arming its ships with cruise missiles of varying degrees of technical sophistication since the late 1950s, and the U. S. Navy has made them very general equipment since the early 1980s. Most of the SLCM (pronounced "slickum") inventories carry only conventional warheads, intended for ship-to-ship combat, but an undeterminable portion are nuclear-armed and capable of striking targets over 2,000 kilometers inland.

sealift The capacity to move troops, equipment, and supplies rapidly to a conflict zone by ship. While the limitations of American AIRLIFT potential are often discussed, traditional sealift capacity faces equally serious problems. A major reason for lack of full sealift capacity is that navies are prone to concentrate their attention on the more "glamorous" business of acquiring combat ships, and to focus on what they see as their primary single-service tasks, rather than supporting army and air force needs, which they regard as marginal (see INTERSERVICE RIVALRY). Sealift inadequacies are common, in fact, to all Western nations with a need or desire for FORCE PROJECTION. The old-fashioned troop carriers of World Wars I and II have not been replaced in naval inventories, and the numbers of supply and amphibious-warfare craft are relatively small in all navies. Of even more pressing concern is the fact that countries such as the United States and the United Kingdom have experienced a very serious shrinkage in the size of their merchant navies. Thus, in any sustained war, the ability of either government to rely on conscripting civilian ships and crews, as has been the pattern in the past, will be much reduced. See Peter J. Luciano, "Sealift Capability: A Dwindling Defense Resource," *Defense Management* 18, 3 (Third Quarter 1982); Andrew S. Prince, "Sealift Is a Critical Component of Maritime Superiority," *Defense Management* 19, 2 (Second Quarter 1983).

sealed orders Secret or confidential orders in a sealed envelope, given to a commander of troops or of a ship with instructions not to open them until a given time or upon arrival at a specified destination.

seaman The enlisted rank in the navy equivalent to a private in the army and an airman in the air force.

search 1. An OPERATION to locate an enemy force known or believed to be at sea. 2. A systematic RECONNAISSANCE of a defined area, so that all parts of the area have passed within visibility. 3. To distribute gunfire over an area in depth by successive changes in gun ELEVATION.

search-and-attack priority The lowest category of IMMEDIATE MISSION REQUEST, involving suspected targets related to an enemy's tactical or logistical capabilities, e.g., those capabilities that are not inhibiting a unit's advance, but which by their fleeting nature and tactical importance should be located and destroyed.

search-and-destroy mission An operation which first required that a unit extensively search for the enemy before engaging in an action designed to destroy them, their base areas, and their supplies. Such missions were very common during the VIETNAM WAR.

search and rescue (SAR) To seek and rescue personnel in distress on land or at sea, through the use of aircraft, surface craft, submarines, and specialized rescue teams and equipment.

search attack unit The designation given to one or more ships separately organized or detached from a formation as a TACTICAL UNIT to search for and destroy submarines.

searching fire FIRE distributed in depth by successive changes in the ELEVATION of a gun.

search radius In SEARCH AND RESCUE operations, a radius centered on a datum point and having a length equal to the maximum range of the lost object plus an additional safety distance to insure a greater than 50 percent probability that a TARGET is in the search area.

sea skimmer A MISSILE designed to transit at less than 50 feet (or 15 meters) above the surface of the sea.

season cracking Cracking in AMMUNITION caused by its aging.

sea superiority That degree of dominance in a sea battle of one FORCE over another; it permits the conduct of operations by the former and its related land, sea, and air forces at a given time and place without prohibitive interference by the opposing force.

sea supremacy That degree of SEA SUPERIORITY wherein an opposing force is incapable of effective interference with a seaborne operation.

sea tail That part of an AIRBORNE or air-transported unit that is not committed to combat by air and will join an armed force by sea travel.

second-decision-center thesis One of the major justifications provided by British governments for the United Kingdom's independent nuclear deterrent capability since the early 1960s. The argument is that the Soviet Union might not believe in the firmness of the American nuclear umbrella covering Western Europe, and might conclude that it could safely use its NUCLEAR WEAPONS to intimidate or destroy European targets. But if Britain, and also possibly France, were known to have nuclear weapons and to be prepared to use them in the case of a Soviet attack on NORTH ATLANTIC TREATY ORGANIZATION (NATO) nations, the uncertainty of what response it might face would deter it from attacking. The argument is not based so much on a guarantee that British nuclear weapons would be used in any particular circumstance as on the impossibility of the Soviet Union's ever being able to calculate when they might be used. The whole argument is an example of deterrence through uncertainty. The Soviet Union can never be sure, no matter what it decides about American policy, that it faces no nuclear threat. Unlike France, where the nuclear deterrent is justified in part on the argument that no European country can trust the United States to risk CENTRAL STRATEGIC WARFARE, British governments have always been careful to insist that they do not doubt America, but that the Soviet Union might, and that a second center of nuclear decision-making is therefore vital to deterrence.

second line 1. Not the best; not first class. 2. Equipment that can be used if its limitations are taken into account. 3. Those combat forces that support the FIRST-LINE troops facing an enemy.

second strike The first counterblow of a war. A second strike in STRATEGIC language is a response to an enemy's nuclear attack, where that attack was the first use of nuclear weapons.

second-strike capability The ability to survive a first military strike with sufficient resources to deliver an effective counterblow. This concept is usually associated with NUCLEAR WEAPONS. Compare to FIRST STRIKE.

Second World The Socialist countries of Eastern Europe plus the Soviet Union, all of which have centrally planned economies.

secondary armament In ships with multiple-size GUNS, that BATTERY consisting of guns next largest to those of the main battery.

secondary fire sector An area not properly in a designation fire sector, but which can be swept by fire if necessary. Such areas are located close in on the FLANKS in the fire sectors of adjacent units.

secondary landing A landing usually made outside a designated LANDING AREA, for the purpose of supporting the main landing directly or indirectly.

secondary road A road supplementing a main road, usually wide enough and suitable for two-way, all-weather traffic at moderate or slow speeds.

secondary source of supply Any source of supply upon which a requisition, shipping order, or purchase request is placed by an INITIAL SOURCE OF SUPPLY.

secondary station 1. An OBSERVATION POST for ARTILLERY fire, at the end of a base line farthest from the gun or DIRECTING POINT. 2. Any station in a radio NET other than the net control station.

secondary target A target against which FIRE is directed when the main fire mission has been accomplished, or when it has become impossible or impractical for a gun or battery to carry out its main fire mission.

secondary weapon A supporting or auxiliary WEAPON of a unit, vehicle, position, or aircraft. It is generally a GUN of smaller CALIBER than the PRIMARY WEAPON, and its purpose is to protect or supplement the FIRE of the primary weapon.

Secretary of the Air Force The head of the Department of the Air Force, who administers that department within the Department of Defense.

Secretary of the Army The head of the Department of the Army, who administers that department within the Department of Defense.

Secretary of the Navy The head of the Department of the Navy, who administers that department within the Department of Defense.

section 1. As applied to ships or naval aircraft, a tactical subdivision of a division. It is normally one-half of a division in the case of ships, and two aircraft in the case of aircraft. 2. A subdivision of an office, installation, territory, works, or organization; especially a major subdivision of a staff. 3. A tactical unit of the Army and Marine Corps. A section is smaller than a PLATOON and larger than a SQUAD. In some organizations the section, rather than the squad, is the basic tactical unit.

sector 1. An area designated by boundaries within which a UNIT operates, and for which it is responsible. The story is often told that when General Dwight D. Eisenhower first arrived in France during World War II, an enemy plane suddenly began STRAFING; the general and his party ran for cover. As soon as the attack was over, a very solicitous officer came running to see that Eisenhower was all right. The general then thanked him for his great concern. "Oh," said the officer, "my concern was just that nothing should happen to you in my sector." 2. One of the subdivisions of a coastal frontier. See also AREA OF INFLUENCE; ZONE OF ACTION.

sector of fire An area that is required to be covered by FIRE by an individual soldier, a weapon, or a military unit.

sector scan A radar scan in which the antenna oscillates through a selected angle.

secure In an operational context, to gain possession of a position or terrain feature, with or without force, and to make such disposition as will prevent, as far as possible, its destruction or loss by enemy action. See also DENIAL MEASURE.

security 1. Measures taken by a military UNIT, ACTIVITY, or INSTALLATION to protect itself against all acts designed to, or which may, impair its effectiveness. 2. A condition that results from the establishment and maintenance of protective measures that insure a state of inviolability from hostile acts or influences. 3. With respect to classified matter, the condition that prevents unauthorized persons from having access to official information that is safeguarded in the interests of NATIONAL SECURITY.

security, communications See COMMUNICATIONS SECURITY.

security industrial See INDUSTRIAL SECURITY.

security assistance The programs of various agencies of the United States government relating to international defense cooperation. United States security assistance (sometimes called military assistance) has five components:

1. *The Military Assistance Program,* in which defense articles and defense services are provided to eligible foreign governments on a grant basis.
2. *International Military Education and Training,* which provides grant military training in the United States and United States territories to foreign military and civilian personnel.
3. *Foreign Military Sales,* which provides credits and loan-repayment guarantees to enable eligible foreign governments to purchase defense articles and defense services.
4. *Security Supporting Assistance,* which promotes economic and political stability in areas where the United States has special foreign policy security interests.
5. *Peacekeeping Operations Programs,* which fund the Sinai Support Mission and the American contribution to the UNITED NATIONS (UN) forces in Cyprus. See Harry J. Shaw, "U. S. Security Assistance; Debts and Dependency," *Foreign Policy* No. 50 (Spring 1983); Francis J. West, Jr., "The U.S. Security Assistance Program: Giveaway or Bargain?" *Strategic Review* 11, 1 (Winter 1983); Henry M. Lewandowski, "Security Assistance Guidelines," *Naval War College Review* 39, 2 (March–April 1986).

security certification The formal indication that a person has been investigated and is eligible for access to CLASSIFIED INFORMATION to the extent stated in the certification.

security classification A category to which NATIONAL SECURITY information and material is assigned to denote the degree of damage that unauthorized disclosure would cause. There are three major categories: top secret, secret and confidential.

security clearance An administrative determination that an individual is eligible for access to CLASSIFIED INFORMATION.

Security Council The most powerful of the elements created by the UNITED NATIONS (UN) Charter, for dealing with questions of international peace and security. The Security Council has five permanent members (China, France, the Soviet Union, the United Kingdom, and the United States) and representatives of ten other nations, five of which are chosen each year for two-year terms. Each of the permanent members has a veto over any decisions of substance. This veto provides the major powers with protection against majority decisions in the larger UN General Assembly, where each member nation (with minor exceptions) has only one vote. The Security Council is the only organ of the United Nations capable of deploying military forces whether to oppose aggression (as during the KOREAN WAR) or to maintain peace by separating hostile parties (such as in Cyprus or Lebanon). See PEACEKEEPING FORCE. See Louis B. Sohn, "The Security Council's Role in the Settlement of International Disputes," *American Journal of International Law* Vol. 78, No. 2 (April 1984).

security intelligence INTELLIGENCE on the identity, capabilities, and intentions of hostile organizations or individuals who are or may be engaged in ESPIONAGE, SABOTAGE, SUBVERSION, or TERRORISM.

security operations Those operations designed to provide reaction time, maneuver space, and information about an enemy, and protection to the MAIN BODY of a military force. While primarily CAVALRY-type operations, other combat units (e.g., ARMOR, MECHANIZED INFANTRY) may be used in security operations.

security risk 1. An employee of a public agency or of a private government contractor determined to be so susceptible to the influence of foreign AGENTS that he or she cannot be trusted with continued employment or continued access to sensitive information. 2. Any disloyal or generally untrustworthy citizen.

seen fire FIRE that is continuously aimed at the future position of an aircraft, the aim being derived from visual observation.

seizure As an operational purpose, the capture of a restricted portion of an enemy-controlled territory. The capture of an isolated landmass such as an island is also usually categorized as a seizure.

selected reserve American military reservists, including all NATIONAL GUARD members, who complete a specified amount of inactive duty training each year and may be called up for active duty by the president of the United States at any time. Compare to READY RESERVE.

selective jamming See SPOT JAMMING.

selective loading The arrangement and stowage of military equipment and supplies aboard ship in a manner designed to facilitate its issue to UNITS.

selective service The name given to the American system of CONSCRIPTION, operated by the U.S. Selected Service System. The system is designed to cope with the fact that the United States could never use, let alone handle, all the young men available for military service, and that therefore some way must be found of making a selection that seems to be socially just. This has never been achieved, and the "draft" has sometimes occasioned bitterness, especially during the VIETNAM WAR era. After the Vietnam War the draft was suspended, and the United States has operated with ALL-VOLUNTEER FORCES, but all men still have to be registered in case there arises a future need to conscript. See James B. Jacobs and Dennis McNamara, "Selective Service Without a Draft," *Armed Forces and Society* 10, 3 (Spring 1984).

Selective Service System The federal agency that administers the law requiring all male citizens and all other males who are in the United States and are between the ages of 18 and 26 years to register for possible induction into the Armed Forces of the United States. Subsequent induction can only come about by future action of the U. S. Congress. Exempted from the requirement to register (at U. S. Post Offices) are foreign diplomatic and consular personnel, member of the active armed forces, and nonimmigrant aliens.

selective strikes Attacks on precise and carefully delimited targets, chosen so as so maximize the military value of attacking them or to send a very clear signal to an enemy while minimizing the risks of ESCALATION. As such, selective strikes are at the heart of the more general concept of LIMITED NUCLEAR OPTIONS. The concept is not limited to NUCLEAR WARFARE, however. In any OFFENSIVE it is possible to distinguish between general attacks against any targets of military value that are within range, and carefully defined targets that have been singled out. The essential point is that a selective strike is as much a matter of signalling to the enemy as of achieving purely military goals. The trouble is that the friction of war usually makes strikes much less selective than intended. Some military thinkers like to talk about SURGICAL STRIKES, with the implication that such a use of violence is analogous to removing a tumor for the benefit of the whole body. Yet even with the TARGET-ACQUISITION capacity of modern weapons, this is seldom achieved, and COLLATERAL DAMAGE occurs; for example, a hospital might be bombed in a raid meant only to destroy a TERRORIST headquarters.

selective unloading In an AMPHIBIOUS OPERATION, the controlled unloading from assault shipping, and movement ashore, of specific items of cargo at the request of the LANDING-FORCE commander. See also COMBAT LOADING; SELECTIVE LOADING.

self-destroying fuze A FUZE designed to detonate a PROJECTILE before the end of its flight.

self-propelled 1. Able to go on its own power. 2. A gun or rocket that has a motorized vehicle as a base. 3. A military unit that possesses self-propelled guns.

semiautomatic weapon A WEAPON whose trigger must be pulled for each shot fired in contrast to a fully automatic weapon which fires continuously as long as the trigger is held.

semi-fixed ammunition AMMUNITION in which the CARTRIDGE CASE is not permanently attached to the PROJECTILE.

semimobile unit A military UNIT with an insufficient number of ORGANIC vehicles to transport all of assigned personnel and equipment from one point to another in one trip.

sending state A nation, party to an international agreement, which pursuant thereto orders its military personnel to enter and remain in the territory of another party to the agreement.

senior service A country's oldest armed service. In the United States the senior service is the Army; in the United Kingdom it is the Royal Navy.

sensor 1. Any technical means to extend natural human senses. 2. A device that detects and indicates the configuration of terrain, the presence of military targets, and other natural and man-made objects and activities by means of energy emitted or reflected by such targets or objects. The energy may be nuclear, electromagnetic (including the visible and invisible portions of the spectrum), chemical, biological, thermal, or mechanical (including sound, atmospheric compression from a blast, and earth vibration). See John Rhea, "Sensors Across the Spectrum," *Air Force,* 70, 11 (November 1987).

sensor, active See ACTIVE SENSOR.

sentinel 1. A guard posted to protect a particular thing, person, or place; he is obligated to challenge all who would pass his post to determine if they have authorization and to warn against a surprise attack. Also called a SENTRY. 2. A now obsolete U.S. missile.

sentry A soldier assigned to duty as a member of a GUARD, to keep watch, maintain order, protect persons or places against surprise, or warn of enemy attack; a sentinel.

sentry box A small building, only large enough to hold one person, provided to protect sentries from bad weather.

separate-loading ammunition AMMUNITION in which the PROJECTILE and CHARGE are loaded into a gun separately.

separated ammunition A PROJECTILE and its PROPELLING CHARGE, which is sealed into a CARTRIDGE CASE. Both are loaded into a weapon in one operation.

separation distance, safe See SAFE SEPARATION DISTANCE.

sergeant 1. The NONCOMMISSIONED OFFICER above CORPORAL. There are many gradations of sergeant, including sergeant, staff sergeant, technical sergeant, master sergeant, sergeant major and gunnery sergeant.

sergeant, drill See DRILL SERGEANT.

sergeant, first See FIRST SERGEANT.

sergeant, master See MASTER SERGEANT.

sergeant of the guard The title given to the senior NONCOMMISSIONED OFFICER of an INTERIOR GUARD, no matter what his grade may be.

sergeant's call, first See FIRST SERGEANT'S CALL.

serial 1. The numerical or alphabetical designation given to an ELEMENT or a group of elements within a SERIES, for convenience in planning, scheduling, and control. 2. A compact FORMATION of aircraft, under the control of a formation commander, and separated from other formations by time and space. 3. A troop UNIT or integral part thereof, with accompanying equipment, which is assigned a numerical designation for convenience of scheduling or for the control of its ship-to-shore movement in AMPHIBIOUS OPERATIONS.

serial assignment table A table that is used in AMPHIBIOUS OPERATIONS, and which shows the SERIAL NUMBER and the title of a UNIT, the approximate number of PERSONNEL, MATERIEL, vehicles, or equipment in the serial; the number and type of LANDING CRAFT or AMPHIBIOUS VEHICLES required to boat the serial; and the ship on which the serial is embarked.

serial number 1. A number assigned to a document by the originating office, for the purpose of counting the copies prepared and of controlling their distribution. 2. The number of a message in a series. 3. The specific number or symbol assigned for identification to an individual item of MATERIEL in a series of items. 4. An arbitrary number, assigned to a UNIT or grouping, and including its equipment, that is: embarked entirely in one ship; to be landed as a unit on one beach or helicopter LANDING ZONE; and to be landed at approximately the same time. 5. The identification number once assigned to each member of the U.S. Armed Forces; today it has been replaced by the Social Security number.

series of targets In ARTILLERY AND NAVAL GUNFIRE SUPPORT, a number of targets or group(s) of targets planned for destruction in support of the MANEUVER phase of an operation. A series of targets may be indicated by a nickname.

service 1. The armed forces of a nation in general. 2. A military UNIT primarily concerned with providing COMBAT SERVICE SUPPORT or ADMINISTRATION as a whole; for example, the finance corps or quartermaster corps. 3. A branch of an army whose primary function is to render noncombatant support to other organizations rather than to engage in combat. 4. All activities of a unit or COMMAND other than combat activities. 5. One of the components of the armed forces of a nation (army, navy, etc.). 6. In COMMUNICATIONS, the notes covering routing instructions, time of delivery or receipt of messages, the radio frequency used, the operator's identifying sign, or similar information written on a message blank by the sending and receiving operators. 7. A career employment or a civil agency, such as the U. S. Foreign Service.

service ammunition AMMUNITION intended for combat rather than training purposes.

service calls Regular calls given by bugle, drum, or other means, to assemble personnel for FORMATIONS or routine duties. It is the largest classification of bugle calls, including calls not classified as warning calls, formation calls, or alarm calls.

service cap A uniform cap made of authorized material and having a visor.

service club A recreation and social-activities facility for ENLISTED PERSONS and their families at a military installation.

service echelon That subdivision of a military command that is responsible for the supply, evacuation, maintenance, and administration of the command.

service force A naval TASK ORGANIZATION that performs missions for the LOGISTIC SUPPORT of military operations.

service group A major naval administrative or tactical organization designed to exercise OPERATIONAL CONTROL and administrative command of assigned squadrons and units in providing logistic support to FLEET operations.

service number A combination of numbers or letters and numbers assigned to each individual in the military service as a means of positive personal identification.

service obligation An obligation to perform military service for a period of time, incurred by law or by some voluntary undertaking such as training, the acceptance of a promotion, a change of station, or by express agreement.

service of the piece The operation and maintenance of a gun or other military equipment by its crew.

service practice Part of the training program for ARTILLERY units, consisting primarily of practical problems in the preparation, execution, and conduct of FIRE with SERVICE or target practice AMMUNITION.

service ribbon The small, narrow, variously colored ribbons worn on a military uniform often in place of the much larger medals that the ribbons represent.

service stripe A stripe worn on the left sleeve of the shirt or coat of an ENLISTED PERSON, one for each three-year period of military service.

service test A test of an item, system of MATERIEL, or technique conducted under simulated or actual operational conditions to determine whether it satisfies specified military requirements.

service troops Those UNITS designed to render supply, maintenance, transportation, evacuation, hospitalization, and other services required by air and ground combat units.

service uniform A uniform prescribed by regulations for wear by military personnel on routine duty, as distinguished from DRESS, full dress, or work UNIFORMS.

service unit A unit or organization designed primarily to render noncombatant support to other units or for an entire theater, in order to insure the continuity of operations.

setback The rearward movement caused by inertia, of the free moving parts in a PROJECTILE when it is fired. This movement is used to push back a spring or plunger in a TIME FUZE and start the operation of the fuze.

set forward The forward movement of the component parts within a PROJECTILE, missile, or bomb when impact occurs.

set-piece battle A carefully planned and rehearsed battle; a battle rehearsed by its initiators.

setting rounds ROUNDS OF AMMUNITION fired at varying angles of elevation to seat the SPADE and BASE PLATE of a gun mount firmly in the ground.

shallow fording capability The ability of a SELF-PROPELLED gun or ground vehicle equipped with built-in weatherproofing, with its wheels or tracks in contact with the ground, to negotiate a water obstacle without the use of a special waterproofing.

SHAPE Supreme Headquarters Allied Powers Europe; the headquarters of the entire European NORTH ATLANTIC TREATY ORGANIZATION (NATO) military command under the direct authority of SACEUR. Situated near Mons, in Belgium, it is essentially an administrative headquarters rather than a command location. Were NATO ever to go to war, SACEUR would change from being a largely political functionary to a direct operational commander and, together with SHAPE, would move to some secret and highly-fortified COMMAND, CONTROL, COMMUNICATIONS AND INTELLIGENCE (C^3I) bunker.

shaped charge A CHARGE shaped so as to concentrate its explosive force in a particular direction.

sharpshooter 1. Any soldier who is extraordinarily accurate with a rifle or pistol. 2. The second highest grade of MARKSMAN.

sheaf In ARTILLERY AND NAVAL GUNFIRE SUPPORT, a planned plane of fire that produces a desired pattern of BURSTS with ROUNDS fired by two or more weapons. Compare to CONVERGED SHEAF, OPEN SHEAF, PARALLEL SHEAF, SALVO.

sheet explosive A PLASTIC EXPLOSIVE provided in a sheet form.

shelf life The length of time during which an item of supply, subject to deterioration or having a limited life and which cannot be renewed, is considered serviceable while stored.

shell 1. A hollow PROJECTILE filled with an explosive, chemical, or other material, as opposed to a SHOT, which is a solid projectile. 2. A shotgun CARTRIDGE. 3. A command or request indicating the type of projectile to be used in the firing of a gun.

shell-destroying tracer An ignited element, placed between the EXPLOSIVE in an air-defense projectile and the TRACER element, that is designed to permit activation of the explosive by the tracer after the projectile has passed the target point but is still high enough to be harmless to ground troops.

shellproof shelter, heavy See HEAVY SHELLPROOF SHELTER.

shell wave An audible disturbance or wave emitted from SHELLS moving at supersonic speeds. See also BALLISTIC WAVE.

shield, ablative See ABLATIVE SHIELD.

shielding 1. Material of suitable thickness and physical characteristics to protect personnel from RADIATION during the manufacture, handling, and transportation of fissionable and radioactive materials. 2. Obstructions which tend to protect personnel or materials from the effects of a nuclear explosion.

shift 1. The transfer of FIRE from one target to another. 2. To transfer fire from one target to another. 3. The DEFLECTION DIFFERENCE from one designated point to another, used when opening or closing the SHEAF of FIELD ARTILLERY or MORTAR units.

shifting fire FIRE delivered at constant RANGE but at varying DEFLECTIONS; used to cover the width of a target that is too great to be covered by an OPEN SHEAF.

ship, landing See LANDING SHIP.

ship, marker See MARKER SHIP.

ship will adjust In naval gunfire support, a method of control in which the ship that is providing gunfire can see the target and, with the concurrence of the SPOTTER, will adjust its fire as needed.

shipment, block See BLOCK SHIPMENT.

shipping, administrative See ADMINISTRATIVE SHIPPING.

shock, ablative See ABLATIVE SHOCK.

shock action The combined destructive physical and psychological effect on the enemy produced by the FIREPOWER of ARTILLERY, tanks and other mobile ARMOR, and supporting TROOPS.

shock front The boundary between the pressure disturbance created by an explosion (in air, water, or earth) and the surrounding atmosphere, water, or earth.

shock troops 1. ELITE TROOPS. 2. Troops trained for offensive action or for SHOCK ACTION.

shock wave The continuously propagated pressure pulse formed by the blast from an explosion in air, under water or under ground. See also BLAST WAVE.

shore bombardment line A ground line established to delimit bombardment by friendly surface ships.

shore fire control party A unit for controlling naval gunfire in support of troops ashore, consisting of a spotting team to adjust fire and a NAVAL GUNFIRE LIAISON TEAM to perform liaison functions for the supported battalion commander.

shore party A TASK ORGANIZATION of a LANDING FORCE, formed for facilitating the landing and movement onto the beach of troops, equipment, and supplies; for the evacuation from the beach of CASUALTIES and PRISONERS OF WAR; and for facilitating the beaching, retraction, and salvaging of LANDING SHIPS and LANDING CRAFT.

shore-to-shore movement An ASSAULT movement directly from a shore STAGING AREA to an OBJECTIVE, involving no transfer of troops or materiel between the shore and landing craft or ships.

short In ARTILLERY AND NAVAL GUNFIRE SUPPORT, a term used to indicate that a BURST(s) has occurred short of the target in relation to the SPOTTING LINE.

shortfall That lack of forces, equipment, personnel, materiel, or capability, apportioned to and identified as the requirement for a planned mission, that would adversely affect a command's ability to accomplish the mission.

short guard The prescribed guard position in bayonet drill in which the point of the bayonet is directed at the opponent's stomach, and the right hand holds the small of the stock of the rifle on which the bayonet is mounted near the right hip. 2. The command to take this position.

short-range attack missile An air-to-surface missile armed with a NUCLEAR WARHEAD; the missile range, speed, and accuracy allow a carrier or strategic bomber force to "stand off" from its intended targets and launch missiles away from enemy anti-aircraft defenses. Also called a standoff bomb.

short-range ballistic missile A BALLISTIC MISSILE with a range capability up to about 600 nautical miles.

short round 1. The unintentional or inadvertent delivery of ORDNANCE on friendly troops, installations, or civilians by a friendly WEAPON SYSTEM. 2. A defective CARTRIDGE in which the PROJECTILE has been seated too deeply.

short take-off and landing (STOL) The ability of an aircraft to clear a 50-foot (15 meters) obstacle within 1,500 feet (500 meters) of commencing take-off or to stop within 1,500 feet (500 meters) after passing over a 50-foot (15 meters) obstacle in landing. Compare to VERTICAL TAKE-OFF AND LANDING.

short thrust A BAYONET thrust in which the arm is extended and the body thrown forward, with the weight shifted to the leading foot; a short lunge. It is delivered at a distance of about three feet.

shot 1. The discharge of a projectile toward a target. 2. The distance traveled by a projectile. 3. Small balls or pellets of lead that are combined in a cartridge to be fired by a shotgun. 4. A collective term for all projectiles fired from guns. 5. A solid projectile as opposed to a SHELL.

shot across the bow A phrase from traditional naval warfare. A ship on blockade duty would often fire a cannon shot just in front of a BLOCKADE RUNNER to make it clear that if it did not halt it would be sunk. In a modern context it refers to one possible strategy for an early use of NUCLEAR WEAPONS by the NORTH ATLANTIC TREATY ORGANIZATION (NATO). The idea is that a very small and selective strike, possibly using only one WARHEAD, would be made at that point in a central-front battle in which NATO's tactics demanded the use of nuclear weapons. There would be a pause following this "shot across the bow" to give the WARSAW PACT the chance to halt its advance, and perhaps a chance for INTRA-WAR BARGAINING to take place. If the Warsaw Pact chose, as it were, to ignore the warning, a serious use of THEATER NUCLEAR FORCES would commence. (See also FIRE-BREAK.)

shoulder patch The cloth INSIGNIA that indicates a soldier's UNIT; worn on the left outer sleeve, one half inch below the shoulder seam of a shirt or jacket.

showdown inspection An inspection of all individual clothing and equipment or organizational equipment to insure its complements (full numbers) and serviceability.

show of force The deploying of military forces to deter a potential aggressor.

shrapnel 1. Any fragment from an exploded MUNITION. The term is derived from Sir Henry Shrapnell (1761–1842), the nineteenth-century British general who invented artillery ammunition designed to explode in midair and send a multitude of projectiles toward the enemy.

shuttered fuze A FUZE in which inadvertent initiation of the DETONATOR will not initiate either the BOOSTER or the BURSTER CHARGE.

shuttle bombing The bombing of objectives, utilizing two bases. By this method, a bomber formation bombs its target, flies on to its second base, reloads, and returns to its home base, again bombing a target if required.

shuttle marching The alternate riding and marching of a unit in a troop movement, usually because of insufficient vehicles to carry the entire unit.

sick call 1. A daily assembly when all sick and injured personnel, other than those in the hospital, report to a medical officer for examination or treatment. 2. The bugle call or signal for this assembly.

sickness, radiation See RADIATION SICKNESS.

side arm Any weapon that can be carried attached to a soldier's side or waist (i.e., a sword, pistol, etc.).

side spray Fragments of a bursting SHELL that are thrown sideways from the line of flight of the shell. See also BASE SPRAY; NOSE SPRAY.

side step A single step 12 inches to the right or left of a given standing position.

siege A prolonged attack on a fortress or town. Traditional sieges were designed to starve out any enemy. Attacking forces would "lay siege to" a place by surrounding it and cutting all its outside communications. Compare to BESIEGE.

siege works Offensive FIELDWORKS surviving a besieged position.

sight 1. A GUNSIGHT. 2. Visual contact as opposed to radar or other electronic means.

sight base The mount for a gunsight.

sight blade A thin, flat metal post used as the front sight on some firearms.

sight bracket A clamp used to hold a detachable GUNSIGHT in position when mounted on a gun.

sight cover A protective metallic cover fastened about a GUNSIGHT to guard it from being moved out of adjustment by jars or blows.

sighting Direct visual contact with personnel or objects.

sighting angle In bombing, the angle between the line of sight to the AIMING POINT and the vertical.

sighting bar A wooden device with enlarged front and rear sights, an eyepiece, and a movable target. It is used to train men in the proper method of aiming a SMALL ARMS weapon. The eye piece forces the student to hold his eye in proper position.

sighting disk A cardboard or metal disk with a small bullseye painted on it, and an attached handle. With a gun in fixed position, a sighting disk is used in aiming practice. The instructor moves the disk across a sheet of paper as the student directs, until the student believes it is in line with the sights of the gun. He then marks the position. Three such marks make up a shot group or aiming group.

sighting shot A trial shot fired from a gun to find out whether the SIGHTS are properly adjusted.

sight leaf A movable, hinged part of the rear SIGHT of a gun that can be raised and set to a desired RANGE or snapped down when not in use.

sign, call See CALL SIGN.

sign off A PROSIGN denoting the termination of a transmission.

signal 1. As applied to electronics, any transmitted electrical impulse. 2. Operationally, a type of message, the text of which consists of one or more letters, words, characters, signal flags, visual displays, or special sounds with prearranged meaning, and which is conveyed or transmitted by visual, acoustical, or electronic means. 3. A PYROTECHNIC item designed to produce a sign by means of illumination, smoke, sound, or a combination of these effects in order to provide identification, location, or a warning.

signal axis 1. The line or route on which lie the starting position and probable future locations of the COMMAND POST of a unit during a troop movement. 2. The main route along which messages are relayed or sent to and from combat units in the field.

signal corps A branch of an army responsible for communications.

signaling panel A strip of cloth used in sending CODE signals between ground and aircraft in flight.

signals intelligence (SIGINT) Intelligence information concerned with SIGNAL characteristics of communication emissions, electronics, and telemetry. It is not always necessary to be able to decode the enemy's radio communications to gain useful information. Changes in the volume, pattern, and location of radio messages by themselves can give vital clues. See also COMMUNICATIONS INTELLIGENCE, ELECTRONIC INTELLIGENCE, RADAR SIGNATURE.

signature 1. The characteristic pattern of a target as displayed by detection and identification equipment. 2. In naval MINE warfare, the variation in the influence field produced by the passage of a ship or sweep. 3. The visible or audible effects produced when a WEAPON is fired or a piece of equipment operated, such as noise, smoke, flame, heat, or debris; also, an electronic emission subject to detection and traceable to the equipment producing it.

signature, radar See RADAR SIGNATURE.

signature, weapon See WEAPON SIGNATURE.

signature equipment Any item of equipment that reveals the type and nature of the unit or formation to which it belongs.

silhouette target 1. A target whose shape is outlined against a light background although its body features cannot be clearly seen. 2. A practice target consisting of the dark image of a person or object outlined against a light background.

silo The predominant basing mode for INTERCONTINENTAL BALLISTIC MISSILES (ICBMs) among the superpowers, although increasing attention is being given to mobile missile launchers. This is because the orthodox silo is becoming increasingly vulnerable. A silo is simply a hole in the ground walled with a thick concrete and steel shell, inside which a missile sits, sometimes on a sprung floor, with a heavy concrete cover that can be electrically opened. Silos were originally capable of considerable protection, and were built to withstand blast OVERPRESSURES of more than 2,000 pounds per square inch (psi). (An ordinary brick building will collapse at between five and 10 psi.) However, combinations of high nuclear weapons YIELDS and great accuracy (that is, a very low CIRCULAR ERROR PROBABLE) mean that most of America's silo-based missiles now suffer from problems of vulnerability to the main Soviet ICBM force.

simulated mustard A harmless substance composed of molasses residuum and used in training to simulate toxic liquid CHEMICAL AGENTS.

simultaneous engagement The concurrent engagement of hostile targets by a combination of INTERCEPTOR aircraft and SURFACE-TO-AIR-MISSILES.

single envelopment A MANEUVER made against one FLANK, or around one flank of a military force and against the rear of the initial DISPOSITIONS of the enemy.

single integrated operations plan (SIOP) The United States's strategic nuclear war plan, supposed to cover all eventualities and contain options ranging all the way from a demonstration strike, or SHOT ACROSS THE BOW, to an intentional "city-busting" assured destruction bombardment of the Soviet Union. The original SIOP (pronounced sigh-op) was drawn up at the end of President Dwight D. Eisenhower's administration, when the utter confusion and incoherence of the United States's existing nuclear war plans became apparent. Before its introduction, each commander of a force that had any nuclear weapons attached to it drew up a separate target list and strike plan. Since this involved not only the STRATEGIC AIR COMMAND (SAC) but also tactical air commanders in various places around the world, the commanders of all navy fleets that had nuclear weapons on their aircraft carriers, the new ballistic missile submarine force (see SSBN), and even land troop commanders with medium-range rockets, the combined effect was absurd. Investigators found that dozens of vital targets had been put on the attack list by sometimes as many as three different commanders, and that there was no centralized priority list or agreed set of alternatives to suit various contingencies. Eisenhower forced the creation of a single, centralized targeting agency that always has a SAC general as its chief, with a navy officer as deputy. It is responsible to the JOINT CHIEFS OF STAFF committee for maintaining a target list and preparing options from within this list that can be presented for the president to choose from in any given situation. The SIOP is continually updated, and undergoes major revisions from time to time as presidential directives change the emphasis of America's overall nuclear war policy. The SIOP is not one war plan, but many. For example, it is believed to contain more than 40,000 targets, yet the total American WARHEAD inventory is only about 10,000; thus there can be no meaning to the phrase occasionally used of "executing the SIOP." See Peter Pringle and William Arkin, *SIOP: The Secret U.S. Plan for Nuclear War* (New York: W. W. Norton and Company, 1983).

single section charge The PROPELLING CHARGE in SEPARATE-LOADING AMMUNITION, which is loaded into a single bag. A single section charge cannot be reduced or increased for changes of RANGE, as a MULTISECTION CHARGE can be.

single-shot kill probability (SSKP) A measure of the general effectiveness of a WEAPONS SYSTEM. It is essentially a combination of the accuracy of the weapon, whether it be a RIFLE, antitank rocket, or ANTIBALLISTIC MISSILE, and the probability that a target, if hit, will be destroyed. It is the enormous increase in SSKP of modern weapons that has transformed the patterns of warfare today. Techniques of TARGET ACQUISITION and guidance of projectiles have made the accuracy of a single ROUND so high that it is sometimes said that, if one can see the target, one will be able to destroy it. The U. S. Army officially calculates that the SSKP for an antitank missile fired from one of its most modern tanks at an enemy tank 3,000 meters away is now 0.9 (that is, a 90 percent probability, since SSKP figures are rated on a scale from 0 to 1). (See also PRECISION-GUIDED MUNITIONS and SMART BOMBS.)

sinker See BOUQUET MINE.

situation map A map showing the tactical or administrative situation of a military operation at a particular time.

situation report A report giving the situation in the area of the reporting unit or formation.

six-by-four As applied to motor vehicles, a six-wheeled vehicle of which four are driving wheels, dual wheels being considered as one wheel. It is usually written 6×4.

six-by-six As applied to motor vehicles, a six-wheeled vehicle of which six are driving wheels, dual wheels being considered as one wheel. It is usually written 6×6.

skate mount A MOUNT for a MACHINE GUN that permits it to travel on a continuous track extending around the inside of the body of a vehicle. The gun can be locked in any position for use.

skip bombing A method of aerial bombing in which a bomb is released from such a low altitude that it slides or glances along the surface of the water or ground and strikes its target at or above water level or ground level.

skirmish A small-scale fight between hostile forces, lasting a brief time and involving few soldiers.

skirmish line The troops who precede an attacking force or who are deployed in front of a defending force. Their task is to engage small enemy units and to prevent large enemy units from achieving surprise.

skirmishers Soldiers who precede the MAIN BODY of a military force to make contact with and disrupt or delay the enemy.

skirting armor The outermost plate or piece of a SPACED ARMOR structure.

sky glow The illumination caused by the firing of WEAPONS from a defiladed position.

slant range 1. The distance in a straight line from the center of BURST of a projectile at the instant of its detonation (ZERO POINT) to a target. 2. The distance in a straight line from a gun, point of observation or radar set to a target, especially an air target.

SLBM A submarine-launched ballistic missile; the naval version of an INTERCONTINENTAL BALLISTIC MISSILE (ICBM). Carried on nuclear-powered submarines known as SSBNs, SLBMs have ranges varying from 2,000 kilometers for the first version of the U. S. POLARIS missile to between 9,000 and 11,000 kilometers for the most advanced Soviet and American weapons. For much of their brief history, submarine-carried missiles were much less accurate than land-based ICBMs, and could only be used for COUNTERVALUE strikes against urban-industrial targets. The great value of this leg of the triad was that submarines on patrol were virtually undetectable, so that the SLBM part of a nation's nuclear deterrent could ensure its SECOND STRIKE CAPABILITY.

SLCM See SEA-LAUNCHED CRUISE MISSILE.

sleigh Part of a gun CARRIAGE which supports the RECOIL MECHANISM and BARREL of the gun, and slides with the gun on recoil, guiding it in runways in the CRADLE.

slew time The time needed for a weapon to re-aim at a new target after having just fired at a previous one.

slice An average logistic planning factor used to obtain estimates of requirements for PERSONNEL and MATERIEL. A personnel slice, for example, generally consists of the total strength of the stated basic combatant ELEMENTS, plus its proportionate share of all support and higher HEADQUARTERS personnel.

slightly wounded A term referring to a CASUALTY that is a sitting or a walking case.

sling arms To place a rifle or other weapon in position with its sling over the shoulder.

slit trench 1. A narrow TRENCH especially useful for protection during air attack. 2. A straddle trench field latrine.

slow fire A type of FIRE used in instructing beginning gunners and in record firing, in which no time limit for completing a score is set.

small arms All arms, including AUTOMATIC WEAPONS, up to and including a BORE of 20 millimeters (.787 inches).

small-bore practice Practice in firing with SMALL ARMS using .22-caliber ammunition instead of the standard service rounds.

small of the stock Part of the STOCK of a SMALL ARMS weapon, ordinarily gripped by the right hand; part of the stock immediately behind the RECEIVER and TRIGGER assembly; a pistol grip in some styles of stock.

smart bombs Bombs that have some capacity to steer themselves directly onto targets. Thus their accuracy is not entirely dependent on the initial accuracy of aiming, which can be affected by such variables as wind speed. There are essentially two basic technologies for smart bombs. The first generation of such weapons depended on the target being illuminated by LASER beams from an aircraft or ground observer. The reflections from the laser were picked up by sensors on the bomb, which then controlled fins to angle the path the bomb took. The next generation of smart bombs, which is currently coming into operation, is self-contained. With these weapons a sensing device detects heat or other emanations from the target directly. The most common form of this new generation relies on an INFRARED DETECTOR which allows the bombs to home in on sources of heat. For this reason most second-generation smart bombs are primarily antitank weapons, the heat of a tank engine making the vehicle a perfect target. Compare to PRECISION-GUIDED MUNITIONS. See Don Wylie, "The Smart Bomb," *Airman* (October 1972); Paul F. Walker, "Smart Weapons in Naval Warfare," *Scientific American* 248, 9 (May 1983).

smart rocks See KINETIC ENERGY AMMUNITION.

smoke An artificially produced aerosol of solid, liquid, or vapor in the atmosphere, which attenuates the passage of visible light or other forms of ELECTROMAGNETIC RADIATION. Smoke is classified in three general categories: IDENTIFICATION SMOKE, OBSCURATION SMOKE, SCREENING SMOKE.

smoke-and-flash defilade 1. A condition in which the smoke and flash of a gun are concealed from enemy observation in an intervening OBSTACLE, such as a hill or ravine. 2. The vertical distance by which the smoke and flash of fire from a gun are concealed from enemy observation.

smoke haze A light concentration of SMOKE placed over friendly installations to restrict accurate enemy observation and fire, but not dense enough to hamper friendly operations. The density of the haze is equivalent to that of light fog.

smoke projectile Any PROJECTILE containing a SMOKE-producing agent that is released on impact or burst. Also called a smoke shell.

snake A specially constructed EXPLOSIVE CHARGE used for clearing paths through MINEFIELDS or for the denudation of wooded areas. It is so constructed that it may be pulled near an OBSTACLE and finally pushed into place by a tank.

sniper A concealed, skilled rifleman who shoots at exposed enemy soldiers.

sniperscope A device combining a SNOOPERSCOPE and a CARBINE or other firearm, which enables the operator to see and shoot at targets in the dark.

snooperscope A hand-carried device combining a source of infrared rays with a viewer to enable the operator to see in the dark.

social defense A phrase used to cover not so much one policy as a collection of attitudes toward defense. The idea comes from the left of the British political spectrum, and particularly from those opposed to NUCLEAR WEAPONS. It refers to any or all of the forms of defense that operate without nuclear weapons, and as much as possible without any weapons that could be used for large-scale offensive warfare. A mainstay of social defense philosophy is the argument that the best way to defend a country is to make it extremely difficult to occupy. Social defense can range all the way from passive civil disobedience through total non-cooperation with occupying authorities, to large-scale GUERRILLA WARFARE. It is argued that no occupying power can afford enough troops to effectively subdue a population that thoroughly resists it and, at the same time, that a country which completely lacks the ability to attack another nation is much less likely to become an object of attack itself. The obvious problem with social defense is that the historical record, especially during World War II, shows that collaboration with an occupying enemy is at least as common as partisan warfare against it.

soft missile base Missile-launching base that is not protected against a nuclear explosion. Compare to HARD TARGET.

soldier 1. Any person in military service. 2. Any ground combat participant, as opposed to those who fight in the air or on the sea. 3. An ENLISTED PERSON as opposed to an OFFICER. 4. To serve as a soldier. 5. In the industrial world, historically, to malinger, to shirk one's duty, to feign illness, or to make a pretense of working. The usage comes from naval history. In earlier centuries, soldiers aboard ship did not have duties as arduous as those of the regular ship's company, so the sailors construed soldiering to be synonymous with loafing and other nonproductive activities.

soldier's qualification card The basic classification record of every enlisted person in the U. S. Army. It contains a summary of personal history, schooling, occupational and military experience, test scores, and other information.

sonar An acronym for sound, navigation, and radar; a general label for the sound-acquiring detection systems used in underwater naval combat. There are two general types of sonar, which are similar to those found with RADAR. One is passive: a submarine, or submarine-hunting ship or helicopter, simply listens for sounds transmitted through the water from a target submarine or ship. The alternative, active sonar, involves sending out noise "blips" which create echoes. As the echoes return to the transmitting ship or submarine, a picture of the underwater environment can be built up which looks much like the picture on a radar screen.

sonne photography Continuous strip photography; photography of a strip of terrain, in which the image remains unbroken throughout its entire length, being produced by a moving film passing an aperture of extremely narrow width and synchronized with the speed of the aircraft that is taking the photograph.

SOP See STANDARD OPERATING PROCEDURE.

sortie 1. A sudden attack made from a defensive position. In this meaning, it is sometimes called a SALLY. 2. One operational flight by one aircraft. Thus five sorties might be made up of five trips by one aircraft or one trip by five aircraft. 3. To depart from a port or anchorage, with an implication of departure for OPERATIONS or MANEUVERS. 4. A set of photographs obtained on a photograph mission.

sound off 1. A command to count cadence while marching. 2. A command for a unit to count off while at attention. 3. A command given at PARADE and GUARD MOUNT when the band is to play a short series of chords before beginning to play the march.

sound surveillance system (SOSUS) A major American asset in antisubmarine warfare (ASW) that would be crucial in a war between the NORTH ATLANTIC TREATY ORGANIZATION (NATO) and the WARSAW PACT. It is a chain of underwater listening devices planted on the seabed at various strategic locations around the world, most notably across the approaches to the Atlantic Ocean, such as the Greenland-Iceland-United Kingdom Gap, through which Soviet submarines would have to pass. With these listening posts strung across such seaways and, continuously monitored in land-based stations, the U. S. Navy is able to follow, even during peacetime, a large part of all Soviet submarine movements. Thus, in the event of a "hot" war, American naval intelligence would have a fairly accurate idea of the size and deployment of Soviet submarines outside their territorial waters, and would be able to assess the main areas of Soviet threat.

source 1. A person, thing, or activity from which INTELLIGENCE information is obtained. 2. In CLANDESTINE operations, a person (AGENT), normally a foreign national, in the employ of an intelligence activity for intelligence purposes. 3. In interrogation activities, any person who furnishes intelligence information, either with or without the knowledge that the information is being used for intelligence purposes. In this context, a controlled source is in the employment or under the control of the intelligence activity, and knows that the information it delivers is to be used for intelligence purposes. An uncontrolled source is a voluntary contributor of information and may or may not know that the information is to be used for intelligence purposes.

source of supply, secondary See SECONDARY SOURCE OF SUPPLY.

Soviet bloc 1. The WARSAW PACT. 2. Those Eastern European states under more or less firm control by Moscow, and governed by Communist parties. It includes the major East European Powers of East Germany, Hungary, Poland, Czechoslovakia, Romania, Bulgaria. It can also cover Albania, and Yugoslavia.

Soviet Doctrine The military policies of the Soviet Union in the event of war. This is only known in part to Western analysts, and most of it must be deduced from Soviet and Russian traditions. A few points are fairly clear. It is axiomatic to top Soviet military thinking that their own strategy must be based on observing the strengths and weaknesses of the opponent: it is in this way a reactive strategy. The way in which the Soviets analyze the NORTH ATLANTIC TREATY ORGANIZATION (NATO) leads them to the conclusion that surprise is all important in order to take advantage of NATOs greatest weakness, which is its mobilization time. This accords with long-term Soviet, and indeed Russian, preference for an offensive strategy. Second, speed is of the essence if Soviet objectives are to be achieved before any nuclear exchange can take place, and if the early stages of a war are to be conducted in such a way as to make NATO's use of NUCLEAR WEAPONS as difficult as possible. Thus it is likely that the Soviet Union would attach great importance to the use of OPERATIONAL MANEUVER GROUPS, to try to prevent the forming of a rigid "front line" with Western forces. If they can move very fast at a stage when an unmobilized NATO has not yet firmly established its defensive position on the CENTRAL FRONT, and succeed in inserting their armies inside the area covered by the mobilizing NATO units, it will be very difficult indeed for NATO to use BATTLEFIELD NUCLEAR WEAPONS. Finally, the Soviet Union fears that NATO, given any time at all, will gain air superiority, and thus the doctrine stresses the need to open the offensive with an immediate attack on NATO airbases and command, control centers (see COMMAND, CONTROL, COMMUNICATIONS, AND INTELLIGENCE ASSETS), possibly using CHEMICAL WARFARE techniques, against which they need at the moment fear little retaliation. A further reason for the emphasis on speed and surprise in Soviet doctrine is that a long, drawn-out war, even if it did not go nuclear, would pit the greater Western economic-military potential against the WARSAW PACT. See John Baylis and Gerald Segal, *Soviet Strategy* (London: C. Helm; Montclair,

NJ: Allanheld, Osmun, 1981); Derek Leebaert, ed., *Soviet Military Thinking* (London: George Allen and Unwin, 1981).

space, deep See DEEP SPACE.

spaced armor A protective ARMOR covering consisting of two or more plates or pieces (of the same material or of different material) with intervening unfilled space or spaces.

spade 1. A tool for digging in the earth; a variety of shovel. 2. The very end of the tail of an artillery piece that is forced into the ground by or previous to the first discharge of the gun in its present position; this prevents movement of the gun chassis during RECOIL.

spall Fragments torn from either the outer or inner surface of ARMOR plate as a result of a complete or partial penetration of the armor, or by dynamic effects of an EXPLOSIVE CHARGE. See also SCABBING.

span of detonation That total period of time, resulting from a timer error, between the earliest and the latest possible detonation of an ATOMIC DEMOLITION MUNITION.

spanning tray A removable hollowed tray on which the elements of SEPARATE-LOADING AMMUNITION slide when being inserted in the BREECH of a CANNON.

spasm war Retaliation against an aggressor with all possible NUCLEAR WEAPONS without any concern for military targets or war aims, but rather destruction. The spasm war theory held that the outbreak of a nuclear war would involve almost instantaneous launching of every missile that the United States and the Soviet Union possessed. It stresses the uncontrollability of nuclear weapons interchange, and contradicts all theories of carefully calculated escalation. While many analysts still believe that nuclear war would have this uncontrolled spasm character, more fashionable nuclear strategic thinking stresses control and rational targeting, as with LIMITED NUCLEAR OPTIONS or, at the opposite extreme to spasm war, the doctrine of nuclear WAR-FIGHTING.

spearhead The leading elements of an attack.

special air operation An operation, conducted at any level of conflict, in support of UNCONVENTIONAL WARFARE and CLANDESTINE, COVERT, and PSYCHOLOGICAL WARFARE.

Special Air Service (SAS) The regiment of the British Army whose wartime mission closely parallels that of the Soviet SPETSNAZ troops, to destroy rear headquarters and vital logistics points of the enemy. In peacetime they have a major antiterrorist mission. For example, they have been used extensively, but very secretly, against the IRA in Northern Ireland. The men and officers of the SAS are already in the army, with long-term careers in their own regiments. They are seconded to the SAS for periods of a few years, and then return to regimental service. They are organized into four-man "bricks," making up 20-man troops, and the whole regiment consists of no more than a few hundred men in three "squadrons."

special ammunition supply point A MOBILE SUPPLY POINT where special ammunition is stored and issued to combat units.

special atomic-demolition munition A very low-YIELD, man-portable, ATOMIC DEMOLITION MUNITION that is detonated by a timer device.

special court-martial In the United States, a COURT-MARTIAL consisting of at least three members, having jurisdiction to try any person subject to MILITARY LAW for any crime or offense not capital and made punishable by the UNIFORM CODE OF MILITARY JUSTICE. A special court-martial may impose sentences not in excess of 6 months' confinement at hard labor and forfeitures of two-thirds pay per month for a like period, except that a BAD-CONDUCT DISCHARGE may be imposed in addition to the authorized punishments.

special forces 1. Military units trained for unconventional operations; especially COUNTERINSURGENCY warfare. The U.S. Army Special Forces, often called the Green Berets because of their hats, was formed in 1952. Originally they were designed to organize guerrilla bands behind enemy lines. During the early 1960s, they began to evolve into a major counterinsurgency force; in this capacity they provided significant service during the Vietnam War. 2. Any specially trained military unit. While the Green Berets tend to be *the* special forces, others include the RANGERS and DELTA FORCE, and TASK FORCE 160.

special guard A soldier or group of soldiers detailed for various GUARD duties that have not been assigned to the MAIN GUARD. Guards of honor and other guards posted to protect personnel, and members of the INTERIOR GUARD of a command are assigned to the special guard.

specialist An enlisted person or warrant officer with technical or administrative duties whose pay grade corresponds to those of corporal through sergeant but whose duties do not require the exercise of command.

specialization An arrangement within an alliance wherein a member or group of members most suited by virtue of technical skills, location, or other qualifications assume(s) greater responsibility for a specific task or significant portion thereof than other members.

special operations 1. Operations conducted by specially trained forces against strategic or tactical targets in pursuit of national military, political, economic, or psychological objectives. 2. Secondary or supporting operations that may be adjuncts to various other operations and for which no one service is assigned primary responsibility. See Frank R. Barnett, B. Hugh Tovar, and Richard H. Shultz, *Special Operations in U.S. Strategy* (Washington, D.C.: National Defense University Press in cooperation with National Strategy Information Center, Inc., 1984); Rudolph C. Barnes Jr., "Special Operations and the Law," *Military Review* 66, 1 (January 1986).

special sheaf In ARTILLERY and NAVAL GUNFIRE SUPPORT, any SHEAF other than a PARALLEL, CONVERGED, or OPEN SHEAF.

special staff All staff officers assigned to a HEADQUARTERS but not included in the GENERAL STAFF group or personal staff group. Examples include a CHAPLAIN and a JUDGE ADVOCATE GENERAL.

special troops Troops attached or assigned to the headquarters of a DIVISION or larger unit.

specific intelligence collection requirement An identified gap in INTELLIGENCE information that may be satisfied only by collection action.

specific search RECONNAISSANCE of a limited number of points for specific information.

specified command A COMMAND that has a broad continuing MISSION and is normally composed of forces from one SERVICE under the Department of Defense; the STRATEGIC AIR COMMAND, for example.

specified task Those tasks delineated in a particular MISSION by higher headquarters. See also: IMPLIED TASKS, CONCEPT OF OPERATIONS.

spectrum of war A term that encompasses the full range of conflict; from COLD, through LIMITED, to GENERAL WAR.

speed of sound Generally 1,088 feet per second at sea level. It varies with temperature and in different media. For example, sound in the air that travels a mile in five seconds will travel the same distance under water in one second. See also MACH NUMBER.

speed ring sight A reticular sight, which may be metallic or an optiprismatic apparatus, with concentric ring elements by which the valves of the lead angles required for certain target speeds can be determined.

SPETSNAZ Russian special forces trained by the KGB to land by parachute or submarine behind NATO lines and attack Western military sites at the onset of war. See Viktor Suvorov, "Spetsnaz: The Soviet Union's Special Forces," *Military Review* (March 1984).

sphere of influence A geographic area in which one major power is dominant and there exists a tacit or formal agreement that other powers will not intrude. For example, the Soviet Union considers Eastern Europe to be within its sphere of influence. See Paul Keal, "Contemporary Understandings about Spheres of Influence," *Review of International Studies* 9, 3 (July 1983).

spherical coordinates A system for locating a point in space by the length of a radius vector from a fixed origin, the angle this vector makes with a reference plane through the origin, and the angle the projection of the radius vector on the reference plane makes with a fixed line in the reference plane.

spider-wire entanglement An irregular crisscross grouping of barbed wire fences.

spike To make an enemy gun unusable; or to make one's own gun unusable if it is likely that the next person to use the gun will be the enemy.

spit and polish 1. Excessive attention to superficial appearances, such as shined shoes. 2. Well-warranted attention to COMBAT and ENGAGEMENT efficiency. 3. A method of shining shoes to achieve a mirror-like finish.

splash 1. In ARTILLERY AND NAVAL GUNFIRE SUPPORT, word transmitted to an observer five seconds before the estimated time of the impact of a SALVO or ROUND. 2. In AIR INTERCEPTION, target destruction verified by visual or radar means.

splinterproof shelter A shelter that protects against RIFLE and MACHINE GUN fire, splinters of HIGH-EXPLOSIVE projectiles, and GRENADES, but not against direct hits by 3-inch shells or larger.

split unit An element of a UNIT which is stationed at a different location than that of the MAIN BODY of the unit.

spoiling attack A tactical maneuver employed to seriously impair a hostile attack while the enemy is in the process of forming or assembling for an attack. Usually employed defensively by armored units through an attack on enemy assembly positions in front of a main LINE OF RESISTANCE or BATTLE POSITION.

spoils 1. Plunder or loot taken from an enemy during war. 2. Historically used to denote rape, pillage, and plunder; whatever commanders allow troops to do in a conquered area.

sponson A hollow enlargement on the side of the hull of a tank, used for storing ammunition or as a space for radio equipment or guns.

spook 1. Any SPY. 2. An employee of the CENTRAL INTELLIGENCE AGENCY. 3. A ghost writer.

spot 1. To determine, by observation, deviations of ORDNANCE from the target for the purpose of supplying necessary information for the ADJUSTMENT OF FIRE. 2. To place in a proper location.

spot elevation A point on a map or chart whose ELEVATION is noted.

spot jamming The JAMMING of a specific telecommunications or electronic detection channel or frequency. See also BARRAGE JAMMING; ELECTRONIC WARFARE.

spot net A radio communication NET used by a SPOTTER in calling FIRE.

spot requisition Items of civilian supply, essential or supplemental, but not included in the planned supplies requisitioned by theater commanders.

spotter 1. An observer stationed for the purpose of observing and reporting results of naval gunfire; a spotter may be employed in designating targets. 2. A small, black metal disk attached to a long wooden pole which is used by a target operator in practice shooting to show the marksman exactly where the target was hit.

spotting A process of determining, by visual or electronic observation, deviations of artillery or naval gunfire from a target in relation to a SPOTTING LINE, for the purpose of supplying necessary information for the ADJUSTMENT OF FIRE.

spotting board A device for determining the direction and size of deviations from a target. It converts the readings of SPOTTERS into usable form for FIRING DATA.

spotting charge A charge of LOW EXPLOSIVE used in PRACTICE AMMUNITION to show the striking point of a PROJECTILE or BOMB.

spotting line A reference line used by an observer in making spotting corrections.

spray dome The mound of water spray thrown up into the air when the SHOCK WAVE from the underwater detonation of a NUCLEAR WEAPON reaches the surface.

spreading fire A notification by a SPOTTER or naval gunfire ship, depending on who is controlling the FIRE, to indicate that fire is about to be distributed over an area.

springing charge A small explosive charge used to enlarge (spring) the diameter of a BOREHOLE (a hole drilled in the earth), or to form a chamber at the bottom of a borehole in which a larger charge may be placed.

spy An INTELLIGENCE agent who operates in a foreign or enemy-occupied country.

squad 1. The smallest formal TACTICAL UNIT; usually from 8 to 12 men. The squad is the basic fighting element in almost all military organizations. It is normally commanded by a sergeant and has two fire teams of four men each. In an armor unit, the tank and its crew and, in an artillery unit, the gun and its crew are the equivalent of a squad. Four squads usually make up a PLATOON. 2. Any small military group assigned a specific function.

squad, firing See FIRING SQUAD.

squad column A formation in which the personnel in a SQUAD are arranged in an irregular column behind the leader, usually about five paces apart.

squadron 1. A battalion-sized unit of cavalry commanded by a lieutenant colonel normally consisting of three or more troops. Some squadrons are an integral part of an infantry division; others are independent. The strength of an independent squadron is approximately 1,000 officers and men; divisional squadrons have fewer men. 2. An organization consisting of two or more divisions of ships. It is normally, but not necessarily, composed of ships of the same type and commanded by a flag officer. 3. The basic administrative aviation unit, consisting of several flights of five aircraft each, and commanded by a lieutenant colonel (or commander in a Navy). In the U.S. Air Force, the squadron is the main administrative and operational unit consisting of three or four elements; two or three squadrons make up a group. See Napoleon B. Byars, "When a Squadron Deploys," *Air Force* 70, 10 (October 1987).

squadron, amphibious See AMPHIBIOUS SQUADRON.

square base The rear end of a PROJECTILE, which is cylinder-shaped and does not taper off from the ROTATING BAND to the end, as in a projectile having a BOATTAIL.

squib A small PYROTECHNIC that burns with a hissing sound. It may be used to fire an igniter. While a squib itself does not explode, it may be the first phase of an EXPLOSIVE CHAIN.

SS 1. A United States Navy's letter designation for submarines. 2. Supersonic as in the SST (supersonic transport). 3. A steam ship. 4. *Schutzsaffel;* the World War II military unit of Nazi Germany noted for its war crimes such as the shooting of American prisoners during the Battle of the Bulge and its operations of concentration camps, where it killed millions of innocent civilians.

SS-18 The largest of all the Soviet Union's INTERCONTINENTAL BALLISTIC MISSILES (ICBMS) and, given that the Soviet Union has always built much heavier missiles than anyone else, the largest in the world. The second STRATEGIC ARMS LIMITATION TALKS (SALT II) agreement placed a limit of 308 SS-18s on the Soviet Union, and deployment to this level has taken place in SILOS in the western Soviet Union. There are several varieties of the SS-18, but the main one deployed, called the Mode 4 by the NORTH ATLANTIC TREATY ORGANIZATION (NATO), has a THROW-WEIGHT of eight tons and a RANGE of at least 10,500 kilometers. This allows it to carry 10 or more MULTIPLE INDEPENDENTLY TARGETABLE RE-ENTRY VEHICLE (MIRV) warheads, each of 500 kilotons; the Mode 5 version may carry the same number of even more powerful 750-kiloton warheads. Compared with most Soviet missiles, it is especially accurate, with a CIRCULAR ERROR PROBABLE estimated to be as low as 250 meters.

SSBN The technical acronym standing for submarine, ballistic, nuclear, used in the American and British navies for submarines, such as the British Polaris or American TRIDENT and POSEIDON, which carry part of the Western nations' nuclear deterrent force. The missiles they carry are known as submarine-launched ballistic missiles (SLBMS). The Soviet Union has a variety of similar submarines, the largest class of which, the Typhoon, is considerably larger than anything in a Western navy. The French also have a squadron of such submarines, which is in the process of being upgraded and may number as many as seven by the end of the century. It is thought that the fifth acknowledged nuclear power, the People's Republic of China, does not as yet have an operating SSBN fleet. Submarines of the SSBN type have the major advantage of being, given the current state of the art in ANTISUBMARINE WARFARE (ASW) almost undetectable. They leave their home ports on long, secret patrols of three months or longer, during which they never broadcast, so that an opponent has no idea of their position. Being nuclear-powered they need never surface, and the range of their missiles is such that they have a huge area of sea in which to hide. The modern U. S. Trident II submarine, due to come into service in the late 1980s, will be able to carry up to 24 missiles with MULTIPLE INDEPENDENTLY TARGETABLE RE-ENTRY VEHICLES (MIRVS) each with a range of up to 10,000 kilometers. These submarines are likely to remain the most secure leg of the nuclear TRIAD, unless some as yet unforeseeable breakthrough occurs in ASW capacity.

SSN The standard naval acronym for a nuclear-powered submarine that is not a ballistic missile submarine (SSBN). They are variously described as attack submarines, fleet submarines and HUNTER-KILLER SUBMARINES. Their general purposes are to protect the surface fleet against enemy submarine attack, to seek out the enemy's SSBNs, and to prey on shipping. The Soviet Union, United States, France, and the United Kingdom all deploy SSNs, although with the exception of the United States, traditional diesel-powered submarines are also used.

SS-20 The first really successful Soviet weapon for THEATER NUCLEAR FORCES. It is solid-fuelled, and so can be kept at one hour's readiness. It is mobile, and so can be moved and dispersed, making it much less vulnerable to attack. It has a range of at least 5,000 kilometers, and so can be based as far east as the Ural Mountains and still hit targets in Western Europe. It has a front end based on MULTIPLE INDEPENDENTLY TARGETABLE REENTRY VEHICLE (MIRV), with three WARHEADS of about 150 kilotons rather than one hugh MEGATON-level warhead such as in the SS-4. Smaller warheads are far more suitable for theater use. The SS-4, if used, might have carried hugh amounts of FALLOUT over WARSAW PACT troops. Finally, the SS-20 is vastly more accurate, nearly six times more, than the SS-4. All this marked a quite dramatic improvement in the lethality of Soviet theater weapons, particularly when it is remembered that doubling the accuracy of a weapon is equivalent in effect to an eight-fold increase in megatonnage. The INF TREATY calls for the removal and destruction of all of the Soviet Union's SS-4 and SS-20 missiles. See John Cartwright and Julian Critchley, *Cruise, Pershing and SS-20: The Search for Consensus: Nuclear Weapons in Europe* (Washington, D.C.: Brassey's Defense Publishers, 1985).

stabilizer In AMMUNITION, a material added to a propellant, chemical, incendiary, or smoke composition to inhibit or reduce deterioration or change by the chemical decomposition of explosive materials while the ammunition is in storage.

stabilizing fin A fin on the tail of some PROJECTILES and BOMBS that helps to maintain balance during their flight, so that the projectile or bomb strikes nose-first.

stabilizing sleeve The cloth tube attached to an aircraft FLARE to hold it in proper position while it is descending.

stack arms 1. To put a number of rifles in a group, upright with their butts on the ground. Three of them are linked together with the stacking swivels. Additional rifles are stacked leaning against this group. 2. The command to do this.

stacking swivel A hinged hook near the muzzle of a rifle that allows a group of rifles to be fastened together to form a standing stack.

staff 1. Specialists who assist line officers in carrying out their duties. Generally, staff units do not have the power of decision, command, or control of operations. Rather, they make recommendations (which may or may not be adopted) to the line personnel. 2. OFFICERS who perform staff functions (such as planning and administration) for a commander.

staff, allied A staff composed of OFFICERS from two or more allied nations.

Staff, Chief of See CHIEF OF STAFF.

staff, combined See COMBINED STAFF.

staff, general See GENERAL STAFF.

staff, parallel See PARALLEL STAFF.

staff, personal Those members of an organization who report directly to an executive rather than through an intermediary such as a CHIEF OF STAFF. Personal staff members could be relatively low level, such as secretaries or chauffeurs, or high-level technical experts.

staff estimate A staff officer's expert evaluation of how factors in his particular field of interest will influence the COURSE OF ACTION under consideration by the commander. See also COMMANDER'S ESTIMATE of the situation.

staff organization Those segments of a larger organization that provide support services and have no direct responsibilities for line operations or production. Personnel administration has traditionally been a staff function. See Ernest Dale and Lyndall F. Urwick, *Staff in Organization* (New York: McGraw-Hill, 1960).

staff out The process that involves soliciting a variety of views or recommendations on an issue so that a decision-maker will be aware of all reasonable options.

staff planning factor A properly selected multiplier, based on experience, used in planning to estimate the amount and type of effort involved in a contemplated military operation.

staff principle The principle of military ADMINISTRATION which states that the executive should be assisted by officers who are not in the line of operations but are essentially extensions of the personality of the executive, and whose duties consist primarily of assisting the executive in controlling and coordinating the organization and of offering advice.

staffing plan A planning document that minimally lists a military organization's projected PERSONNEL needs by occupation and grade level and identifies how these needs will be met.

stage field A predetermined area where aircraft assemble prior to conducting an AIRMOBILE OPERATION.

staged crew An aircrew pre-positioned at a specific point along an air route to allow the continuous operation of an aircraft.

staggered column, double See DOUBLE STAGGERED COLUMN.

staging area A locality between the MOUNTING AREA and the OBJECTIVE of an AMPHIBIOUS or AIRBORNE EXPEDITION, through which the expedition, after mounting, passes for refueling, regrouping, inspection, and redistribution of troops. 2. A locality established for the concentration of troop units and transient personnel between movements over lines of COMMUNICATION.

staging base 1. An advanced naval base for the anchoring, fueling, and refitting of transports and cargo ships, and for replenishing combat service squadrons. 2. A landing and takeoff area with minimum servicing, supply, and shelter, provided for the temporary occupancy of military aircraft during the course of movement from one location to another.

stalemate, nuclear See NUCLEAR STALEMATE.

standard ballistic conditions A set of BALLISTIC CONDITIONS arbitrarily assumed as standard for the computation of FIRING TABLES and RANGE TABLES.

standard day of supply The total amount of SUPPLIES required for an average day. See also ONE-DAY'S SUPPLY.

standardization The process (within military services and between allies) of developing concepts, doctrines, procedures, and designs to achieve and maintain the most effective levels of compatibility, interoperability, interchangeability and commonality in the fields of OPERATIONS, ADMINISTRATION, and MATERIEL. See Phillip Taylor, "Weapons Standardization in NATO: Collaborative Security or Economic Competition?" *International Organization* 36, 1 (Winter 1982).

standardization agreement An agreement among several or all of a group of allied nations to adopt like or similar military equipment, ammunition, supplies, and stores, as well as operational, logistic, and administrative procedures.

standard muzzle velocity The speed at which a given PROJECTILE is supposed to leave the MUZZLE of a gun. The speed is calculated on the basis of the particular gun, the PROPELLING CHARGE used, and the type of projectile fired from the gun. FIRING TABLES are based on standard muzzle velocity.

standard nomenclature A system of uniform designation of items of U.S. Army MATERIEL, in which the designating noun or phrase is given first, followed by the modifiers in reverse of the normal conversational order. "Tank, Medium, M46," is an example of standard nomenclature.

standard operating procedure See STANDING OPERATING PROCEDURE.

standard pattern In LAND MINE WARFARE, the agreed-upon pattern in which mines are normally laid.

standard trajectory A path through the air that it is calculated a PROJECTILE will follow under given conditions of weather, position, and materiel, including the particular FUZE, projectile, and PROPELLING CHARGE that are used. FIRING TABLES are based on standard trajectories.

standard trench A TRENCH of uniform cross section that can be used either as a FIRE or COMMUNICATIONS trench.

standby 1. A command to troops to take posts without delay and be ready for action. 2. A condition in which electronic equipment, such as radios, radars, and computers, is kept in readiness for instantaneous action to avoid the warm-up time required by most vacuum tubes. 3. A WARNING ORDER given to an observer five seconds before the expected time of a BURST.

standby reserve United States military reservists who may be ordered to ACTIVE DUTY if Congress declares that war or a national emergency exists. Compare to: READY RESERVE, RESERVE, and SELECTED RESERVE.

stand down 1. To descend to a lower level of alert or combat readiness. 2. To relax; to be off duty.

stand fast 1. In ARTILLERY, the order at which all action against a position ceases immediately. 2. An order to hold a position. 3. A command given to prevent the movement of indicated units while others move.

standing army A peacetime military establishment; one that is maintained in readiness even when there is no immediate threat of war. At the American Constitutional Convention of 1787, Elbridge Gerry is supposed to have said: "A standing army may be likened to a standing member—an excellent assurance of domestic tranquility, but a dangerous temptation to foreign adventure."

standing operating procedure (SOP) A set of instructions covering those features of OPERATIONS that lend themselves to a definite or standardized procedure without loss of effectiveness. Standard operating procedure is applicable unless ordered otherwise. Also called standard operating procedure.

standing order A promulgated ORDER that remains in force until amended or cancelled.

standing patrol A PATROL which, having taken up its allotted position, is not free to maneuver in the performance of its task without permission.

standoff 1. A tactical stalemate wherein neither side can gain an advantage. 2. The desirable characteristic of an aerial WEAPON SYSTEM that permits attacking aircraft to launch an attack on a target at a safe distance from it, usually outside the range of COUNTERFIRE.

standoff bomb See SHORT-RANGE ATTACK MISSILE.

standup assault A direct frontal dismounted infantry assault on an enemy position. Compare to FIGHTING THROUGH.

starboard The right side of a ship or aircraft when facing forward.

star gage An instrument for measuring the diameter of the BORE of a gun.

star shell A projectile that contains a chemical that is ignited when the projectile bursts. The chemical burns with a brilliant flame, and is used to illuminate targets at night.

START Strategic Arms Reduction Talks (START), the title chosen by President Ronald Reagan of the United States for the round of negotiations that were to follow the second round of STRATEGIC ARMS LIMITATION TALKS (SALT II) with the Soviet Union. Thus they were a continuation of Strategic Arms Limitation Talks that had begun under President Richard Nixon in the early 1970s. However, the word "reduction" was included in the title to suggest a major development from its predecessors, which had been merely strategic arms limitation talks.

start point (SP) A clearly defined initial control point on a route, at which specified elements of a column of ground vehicles or flight of aircraft come under the control of the commander having responsibility for the movement.

star wars See STRATEGIC DEFENSE INITIATIVE.

state of alert 1. Degree of readiness. 2. As used in AIR DEFENSE, the combat readiness maintained by a MISSILE unit, as expressed in terms of the period of time within which the unit must be capable of launching at least one missile. States of alert are: BATTLE STATIONS (fire within 30 seconds), 5-minute, 15-minute, 30-minute, 1-hour and 3-hour.

static defense See MAGINOT MENTALITY.

static line A line attached to a PARACHUTE pack and to a strop or anchor cable in an aircraft, so that when a load is dropped, its parachute is deployed automatically.

station 1. Any military or naval activity at a fixed land location. 2. A particular kind of activity to which other activities or individuals may come for a specific service, often of a technical nature, e.g., an aid station. 3. An assigned or prescribed position in a naval FORMATION or cruising DISPOSITION; or an assigned area in an approach, contact, or battle disposition. 4. Any place of duty or post or position in the field to which an individual, group of individuals, or a unit may be assigned. 5. Transmitters or receivers.

station, action See ACTION STATION.

station, base end See BASE END STATION.

station authentication A SECURITY measure designed to establish the authenticity of a transmitting or receiving station.

station patrol, fixed See FIXED-STATION PATROL.

station time In air transport operations, the time at which crews, passengers, and cargo are to be on board and ready for flight.

staybehind An AGENT or agent organization established in a given country to be activated in the event of its being overrun by hostile forces or in other circumstances under which normal access to the country would be denied.

stay-behind force A military FORCE that is left in position to conduct a specified mission when the remainder of the force withdraws or retires from the area.

steady on An element of a tank FIRE command, until "ON," which is a command to stop TRAVERSE of the tank gun, is given.

steal a march A military MANEUVER that gains a tactical advantage by stealth.

Stealth Bomber The next planned generation of American long-range heavy bombers, after the B1-B, which will come into service in the early 1990s. The Stealth bomber was originally commissioned by the Carter Administration as a way of retaining the viability of the piloted bomber leg of the nuclear TRIAD, which was rapidly becoming too vulnerable to Soviet antiaircraft defenses. Stealth technology involves a variety of methods to make aircraft as near as possible invisible to RADAR and other, principally infrared, forms of SURVEILLANCE. Little is known of the ultra-secret program, but at its core appear to be design techniques that avoid straight or sharp edges, protruding control surfaces, and other features that produce clear radar images. In addition, effort is taken to reduce heat emission, and there are some suggestions that special paints have been developed which absorb rather than reflect radar beams. A really successful stealth technology would, at least for a time, radically alter strategic calculations. A bomber force that could penetrate Soviet airspace without being detected would offer a much more flexible and controllable deterrent than missiles, which can never be very accurate (see BIAS and CIRCULAR-ERROR PROBABLE), cannot be recalled, and are not really testable in warlike conditions. On the other hand, many analysts believe that the response to stealth aircraft would be a scientific ARMS RACE to develop more powerful detection methods, and the initial advantage would not last for long.

Stealth Fighter The F-117A, the United States Air Force's strike and attack plane that incorporates stealth technology to allow it to evade radar and penetrate enemy territory.

stellar guidance A system wherein a GUIDED MISSILE may follow a predetermined course, with reference primarily to the relative position of the missile and certain preselected celestial bodies.

step 1. A pace in walking or marching. A full step is 30 inches, a half step is 15 inches in QUICK-TIME marching. See also PACE. 2. To move a CIPHER element (e.g., a rotor or key tape) from one enciphering position to another.

sterilize 1. In naval MINE warfare, to permanently render a mine incapable of firing by means of a device (a "sterilizer") within the mine. 2. To remove from material to be used in COVERT and CLANDESTINE OPERATIONS all marks or devices that could identify it as emanating from a particular nation or organization.

stern The back part of a ship, boat, or aircraft.

stick 1. The number of paratroopers who jump from one door of an aircraft during one run over a DROP ZONE. 2. A succession of MISSILES fired or released separately at predetermined intervals from a single aircraft.

stick commander The individual who controls parachutists from the time they enter an aircraft until they exit it. See also JUMPMASTER.

sticky charge An improvised EXPLOSIVE CHARGE, covered with heavy grease, tar, or another adhesive material, and thrown against or stuck on an object by hand. Also incorrectly called a sticky bomb.

Stinger A lightweight, man-portable, shoulder-fired, air defense artillery missile weapon for low altitude air defense of forward area combat troops. The Stinger is that rare example of a tactical weapon that has had strategic significance. Supplied to the Afghan resistance fighters indirectly by the United States after the 1979 Soviet invasion of Afghanistan, this weapon allowed the Afghans, without any air force of their own, to gain the local air superiority necessary for success on the ground.

stock The rigid wood, metal or plastic part of a rifle (or other shoulder-held small arm) to which the barrel is attached;

that part of a rifle that is held against the shoulder when firing.

stockade 1. A wooden fort. 2. A military prison.

stockpile 1. A stock of materials (strategic or critical) stored and maintained for use in times of emergency. 2. Quantities of SUPPLIES and equipment authorized to be procured for current operations. Stockpiles are established in lieu of or in addition to normal levels of supply, usually because procurement economies, procurement difficulties, or unpredictable issue demands such action. 3. Stores of special AMMUNITION, such as major assemblies (the large parts) of NUCLEAR WEAPONS.

stockpile-to-target sequence The order of events involved in removing a NUCLEAR WEAPON from storage and assembling, testing, transporting, and delivering it on a target.

stoppage A failure of an AUTOMATIC or SEMIAUTOMATIC weapon to extract or effect a spent CARTRIDGE CASE, or to load or fire a new ROUND.

stopping power, basic See BASIC STOPPING POWER.

stopping power, general See GENERAL STOPPING POWER.

storage life The length of time for which an item of SUPPLY, including explosives, may be expected to remain serviceable and, if relevant, safe, given specific storage conditions. See also SHELF LIFE.

storm A very forceful ATTACK against a fortified place.

storm boat A small, light, very rugged ASSAULT CRAFT equipped with a high-powered outboard motor, designed to transport and beach personnel in forced crossings of wide streams under conditions in which secrecy can be sacrificed for speed.

storm flag A national flag used at posts and national cemeteries and flown in lieu of the post flag in inclement weather. The storm flag is 9 1/2 feet long (fly) by 5 feet wide (hoist); it may be used in lieu of the interment flag to drape the casket of the honored dead in a military funeral.

stowage loading, block See BLOCK STOWAGE LOADING.

straddle trench A TRENCH used as a latrine during field operations and combat. Compare to SLIT TRENCH.

strafing The delivery of AUTOMATIC WEAPONS fire by aircraft on ground targets.

straggler 1. Any personnel, vehicles, ships, or aircraft which, without apparent purpose or assigned mission, become separated from their UNIT, COLUMN, or FORMATION. 2. A ship separated from its CONVOY.

straggler line A MILITARY POLICE control line that may be manned or unmanned, extending across the ZONE OF ACTION or a sector of a defense, usually in the rear of division medium artillery positions, and designated by a commander for the apprehension of stragglers, line-crossers, and infiltrators. It may consist of fixed posts, patrols, or both when manned.

straight leg infantry Traditional soldiers that travel by foot as opposed to airborne paratroopers (who land with bent legs). In conversation "straight leg" is often shortened to "leg."

strategic 1. Of or relating to STRATEGY. 2. Necessary to or important to initiate, conduct or complete a war strategy. 3. Required for a war effort, but not available domestically. 4. Of great importance to an integrated whole plan. 5. The combination of "long-range" and "powerful" in the context of military planning; designed to strike the enemy at the sources of political, military, or economic power. Thus, a strategic MISSILE is one that can reach transoceanic targets, and has a high-yield nuclear WARHEAD. The U.S. STRATEGIC AIR COMMAND (SAC) is a "strategic" air force because it would be used to destroy vital targets very far away from its bases. A strategic weapon, in a parallel shade of meaning, can be directly used to achieve the war aims that lie behind strategy, rather than simply to carry out one of the thousands of moves (classed as TACTICS) that collectively add up to a strategy. A tank, for example, cannot be regarded as a strategic weapon, because the aim of American strategy, to prevent war and if necessary to defeat the Soviet Union for example, cannot directly be achieved by any number of tanks by themselves. See Robert P. Berman and John C. Baker, *Soviet Strategic Forces: Requirements and Responses* (Washington: The Brookings Institution, 1982); Dale Herspring and Robin Laird, *The Soviet Union and Strategic Arms* (Boulder, Colorado: Westview Press, 1984); Robert H. Kupperman and William J. Taylor, *Strategic Requirements for the Army to the Year 2000* (Lexington, Mass.: Lexington Books, 1984).

strategic advantage The overall relative power relationship of opponents that enables one nation or group of nations to effectively control the course of a military or political situation.

Strategic Air Command (SAC) That part of the U. S. Air Force which operates part of America's nuclear deterrent, both with its bomber squadrons and the land-based INTERCONTINENTAL BALLISTIC MISSILE (ICBM) force. When the U.S. Air Force was set up as a separate service in 1947, SAC took over the role and traditions of the U. S. Army's air forces that had played so major a role in the bomber offensives against Germany and Japan in World War II. As the EISENHOWER DOCTRINE of massive retaliation became the key to American defense policy, SAC, the only service that could deliver NUCLEAR WEAPONS over the Soviet Union, was paramount. As missiles were developed with the potential to replace the piloted bomber, the SAC ensured that they came under its control. The dominance of the SAC was finally challenged by the U. S. Navy when SUBMARINE-LAUNCHED BALLISTIC MISSILES (SLBMS) arrived on board the POLARIS submarines. From that point on, the development of the nuclear war plan was centralized in the SINGLE INTEGRATED OPERATIONS PLAN (SIOP), although the senior officer of the group that draws up the SIOP is always an SAC general. Despite the advent of the ICBM, and the development of very powerful antiaircraft defenses in the Soviet Union, the SAC has managed to retain, and is now entirely re-equipping, its piloted bomber force. It currently has over 300 long-range bombers in some 16 squadrons, as well as five squadrons with some 60 medium-range bombers. The bombers, particularly the B-52, which was designed in the mid-1950s, are being equipped to carry air-launched CRUISE MISSILES. They are to be replaced by the new B-1B bomber, and ultimately by the highly secret STEALTH BOMBER. See Michael E. Brown, "The Strategic Bomber Debate Today," *Orbis* 28, 2 (Summer 1984); Tim Wrixon, "Rejuvenation of the Manned Bomber," *Jane's Defense Weekly* (August 25, 1984); Bill Yenne, *SAC: A Primer of Modern Strategic Air Power* (Novato, CA: Presidio Press, 1985).

strategic air transport The movement of PERSONNEL and MATERIEL by air in accordance with a STRATEGIC PLAN. Compare to AIRLIFT and TACTICAL AIRLIFT. See William M. Leary, "Strategic Airlift: Past, Present, and Future," *Air University Review* 37, 6 (September–October 1986).

strategic air warfare Air operations designed to effect the progressive destruction and disintegration of the enemy's war-making capacity.

Strategic Arms Limitation Talks See SALT.

Strategic Arms Reduction Talks See START.

strategic balance The comparative destructive power of the military forces of two rivals (such as the United States and the Soviet Union) or rival alliances (such as the NORTH ATLANTIC TREATY ORGANIZATION (NATO) and the WARSAW PACT). Compare to BALANCE OF POWER.

strategic bombardment Bombing undertaken to destroy the enemy's industry, especially its war-related industrial plant, and thereby to shorten or even directly end a war. A strategic bombardment may also be aimed directly at killing the enemy population, in order to destroy civilian morale and therefore the will to fight. The destruction of anything that will help this end, for example by hampering food supplies, is a legitimate strategic aim.

strategic concentration The assembly of designated forces in areas from which it is intended that OPERATIONS of the assembled force shall begin so that they are best disposed to initiate an ATTACK.

strategic concept The COURSE OF ACTION accepted as the result of the estimate of the strategic situation. It is a statement of what is to be done in broad terms, and sufficiently flexible to permit its use in framing the military, diplomatic, economic, psychological, and other measures that stem from it. See also BASIC UNDERTAKINGS; ESTIMATE OF THE SITUATION.

strategic deception 1. The hiding of strategic assets (bombers, missiles, etc.) from a potential enemy. The main problem with hiding such items during peacetime is that, if they are not known to a potential enemy, they lose their deterrent effect. 2. Cheating on ARMS CONTROL agreements in order to maintain greater strategic forces than formally allowed or agreed to. See also DECEPTION and MILITARY DECEPTION. See Donald C. Daniel and Katherine L. Herbig, *Strategic Military Deception*, (New York; Oxford: Pergamon Press, 1982); Brian D. Dailey and Patrick J. Parker, *Soviet Strategic Deception*, (Stanford, Calif.: Hoover Institution Press; Lexington, Mass.: Lexington Books, 1987).

strategic defense A military posture that protects a nation's people, productive capacity, and armed forces against potential military aggression.

Strategic Defense Initiative (SDI) The Reagan Administration's effort to create a space-stationed American defense against enemy MISSILES. Popularly known as "Star Wars" because the purported capabilities of the final defense system sound (to its critics) so much like a high-technology fantasy, SDI signalled a U-turn in American policy, which had previously rejected the strategy of building a BALLISTIC MISSILE DEFENSE (BMD) when the United States and the Soviet Union signed the Anti-Ballistic Missile Treaty as part of the first round of STRATEGIC ARMS LIMITATION (SALT I) talks in 1972. The United States had given up BMD because it was felt to be technically and economically impossible. The belief then was that the exchange ratio, which means the relative cost of building the capacity to destroy an incoming missile compared with the cost of adding to the attacking missile force, would always be unfavorable to the former. But the SDI initiative is based on the assumption that modern technologies can invert this exchange ratio, and this is at the core of the disputes about SDI's viability. The range of defensive systems advocated for SDI is large, and many depend on technology that has not yet been devised, let alone tested. Whether the system is designed for POINT or AREA DEFENSE, it is clear that it will be a "layered" or "tiered" system (see LAYERED DEFENSE). This means that successive belts of weapons will attempt to hit incoming missiles at successive stages in their flight path. The most "glamorous" aspect of SDI technology is undoubtedly the idea of using space-based weapons, probably LASER WEAPONS of some sort, in orbiting SATELLITE battle stations, which would seek to hit Soviet missiles in either their BOOST PHASE or MIDCOURSE PHASE. The more likely system will simply be a modernization of the ground-based interceptor rocket system on which the earlier American and Soviet BMD projects were based. Versions of these, including the use of kinetic-energy weapons in the form of "smart rocks," that is, self-guiding solid projectiles, and possibly ground-based lasers, are within current technological capacity, although at a cost that may be prohibitive. The single most important criticism about the technical feasibility of SDI relates not to the weapons themselves, but to the computing support they will need. The "Battle Management" program, as it is called, will have to be both enormous, and of extreme sophistication. The problems of TARGET ACQUISITION and tracking (following), of allotting each target to a specificed weapon, of dealing with system failures and endless forms of electromagnetic interference, will make the computing aspect paramount. Many leading computer scientists have pointed out that it is absolutely impossible to write a program of this magnitude and be sure that it will work the first time, yet by the very nature of the activity there can be no rehearsals. According to McGeorge Bundy, George F. Kennan, Robert S. McNamara, and Gerard Smith, in "The President's Choice: Star Wars or Arms Control," *Foreign Affairs* 63, 2 (Winter 1984–5): "The inescapable reality is that there is literally no hope that Star Wars can make nuclear weapons obsolete . . . [But] as long as the American People believe that Star Wars offers real hope of reaching the President's asserted goal, it will have a level of political support unrelated to reality." Nevertheless, other technical experts of similar stature say without qualification that SDI is both technologically feasible and strategically desirable. See Alvin M. Weinberg and Jack N. Barkenbus, "Stabilizing Star Wars," *Foreign Policy* No. 54 (Spring 1984); Sidney D. Drell, Philip J. Farley, and David Holloway, *The Reagan Strategic Defense Initiative: A Technical, Political and Arms Control Assessment* (Palo Alto: Stanford University Press, 1984); James R. Schlesinger, "Rhetoric and Realities in the Star Wars Debate," *International Security* Vol. 10, No. 1 (Summer 1985).

strategic defensive A large-scale defensive action of a nation at war, as opposed to tactical defensive, which refers to a particular operation.

strategic envelopment See INDIRECT APPROACH.

strategic intelligence Information gathered by INTELLIGENCE agencies that can be used to formulate policy and military plans at national and international levels: it is long ranged and widely focused as opposed to TACTICAL INTELLIGENCE, which is short ranged and narrowly focused. See Harry Howe Ransom, "Strategic Intelligence," *Proceedings of the Academy of Political Science* Vol. 34, No. 4 (1982); Loch Johnson, "Seven Sins of Strategic Intelligence," *World Affairs* Vol. 146, No. 2 (Fall 1983).

strategic logistics All military action concerned with the provision of LOGISTIC support to a THEATER OF WAR.

strategic material A material required for essential use in a war emergency, the procurement of which in adequate quantity, quality, or time is sufficiently uncertain, for any reason, as to require prior provision for its supply. See John D. Morgan, "Past Is Prologue: Strategic Materials and the Defense Industrial Base," *Defense Management,* 18, 1 (First Quarter 1982).

strategic mission A MISSION directed against one or more of a selected series of enemy TARGETS with the purpose of progressive destruction and disintegration of the enemy's war-making capacity and his will to make war. Targets include key manufacturing systems, sources of raw material, critical material, stockpiles, power systems, transporting systems, communication facilities, and others. As opposed to tactical operations, strategic operations are designed to have a long-range, rather than immediate, effect on the enemy and its military forces.

strategic mobility The capability to deploy and sustain military forces worldwide in support of NATIONAL STRATEGY. See also MOBILITY.

strategic offensive A large-scale offensive action of a nation at war, as opposed to a tactical offensive, which refers to a particular operation.

strategic plan A plan for the overall conduct of a war.

strategic planning See PLANNING, STRATEGIC.

strategic psychological activities Planned PSYCHOLOGICAL OPERATIONS designed to gain the support and cooperation of friendly and neutral countries and to reduce the will and the capacity of hostile or potentially hostile countries to wage war. See PSYCHOLOGICAL WARFARE.

strategic reconnaissance The searching of wide areas, usually by air, to gain information about enemy force concentrations or movements that should aid in making strategic or large-scale decisions.

strategic reserve 1. An external reinforcing military force that is not committed in advance to a specific combat operation, but which can be deployed later in response to circumstances. 2. A quantity of material put in a particular geographic location due to strategic considerations or in anticipation of major interruptions in the supply distribution system.

strategic retaliatory capability The capacity to carry out a SECOND STRIKE.

Strategic Submarine Ballistic Nuclear See SSBN.

strategic superiority The position of strategic military forces, usually nuclear, in such numbers and in such placement as to be able to overwhelm a potential enemy in the event of war. Strategic superiority is one of the more elusive concepts in modern strategic thought. Unsurprisingly it refers to a situation where one SUPERPOWER has a significantly greater number of powerful strategic missiles than the other. However, it is not just any imbalance that constitutes strategic superiority, because there are natural limits to the numbers of missiles that can be provided with suitable targets. Thus, the much greater number of submarine-launched ballistic missile (SLBM) warheads possessed by the United States as compared with the Soviet Union is not thought to confer strategic superiority. Such missiles are generally usable only against cities, rather than missile fields. Thus the Soviet Union only needs the ability to destroy most American cities, not most American missiles, to preserve parity. There are two powerful arguments which suggest to some analysts that the concept of strategic superiority is largely meaningless. Any FIRST STRIKE would require a guarantee of near perfection in execution before it could be risked. If such a strike went even slightly wrong in timing, accuracy, or effectiveness, it would immediately engender retaliation in kind by the surviving part of the defender's force. Since it is impossible ever to be sure that a high-technology weapons system will operate effectively the first time, and since nothing approaching a first strike can ever be rehearsed, it would be extraordinarily difficult to be certain that any superiority on paper could be realized in practice. Second, and perhaps more importantly, the idea of a SURGICAL STRIKE so perfectly executed that minimum COLLATERAL DAMAGE occurs is impossible. Most analyses suggest that even a strike aimed only at America's missile fields in the sparsely-populated Midwest would kill many millions of American civilians. Faced with that, the Soviet Union could not possibly be sure that the president of the United States would not feel obliged to retaliate in kind, that is, with an SLBM strike against Soviet cities. Nevertheless, there are significant elements in both the Soviet and American defense communities that believe that strategic superiority is a very meaningful concept that can be attained; and that their side should attain it. See Barry M. Blechman and Robert Powell, "What in the Name of God is Strategic Superiority?" *Political Science Quarterly* Vol. 97, No. 4 (Winter 1982–83).

strategic vulnerability The susceptibility of vital elements of national power to being seriously decreased or adversely changed by the application of actions within the capability of another nation to impose. Strategic vulnerability may pertain to political, geographic, economic, scientific, sociological, or military factors.

strategic warfare, central See CENTRAL STRATEGIC WARFARE.

strategic warning A notification that enemy-initiated hostilities may be imminent.

strategic withdrawal Withdrawal for causes of strategic importance; withdrawal designed to improve the strategic situation.

strategies, competitive See COMPETITIVE STRATEGIES.

strategist, military See MILITARY STRATEGIST.

strategy The art and science of developing and using political, economic, psychological, and military forces as necessary during peace and war, to afford the maximum support to national policies, in order to increase the possibility of their favorable outcome. Strategy is the science of directing the overall use of a nation's power to achieve major long-term ends. In its specifically military context it refers to large-scale planning of broadly-defined force structures with which to pursue a nation's war aims. Thus, for example, it is possible to talk about a country adopting a maritime strategy, a strategy of horizontal escalation or one of FLEXIBLE RESPONSE. While strategy is an overall plan that links a nation's resources, DOCTRINE, and goals, it deals hardly at all with details. For example, the NORTH ATLANTIC TREATY ORGANIZATION (NATO) strategy for defeating the WARSAW PACT in a third world war might be to hold the invading armies as long as possible, and as far east in Germany as possible, to give American reinforcements time to cross the Atlantic Ocean. Meanwhile, American naval and marine forces will "take the war to the enemy" by attacking the Soviet Union, while preventing the Soviet use of THEATER NUCLEAR FORCES (TNFs) by the threat of escalation. Once reinforcements were in place, an attempt to force the Warsaw Pact back over the East German borders would be made, but they would not be pursued far into East Germany, because then the war-aim—to return matters to the pre-war status—would have been achieved. The easiest way to grasp why the above is a description of a strategy, rather than something else such as a doctrine of operational tactics, is that the obvious question to ask at the end of the description is "How?" The "how" is the domain of detailed plans, of weapons design and acquisition, of training doctrine, of logistics, of war-gaming and staff courses and so on. Although strategy is grander than these more mundane matters (and indeed the phrase "grand strategy" is often encountered), it is in fact dependent on them. A single example will illustrate this. Forward defense, doing everything humanly possible to prevent the Warsaw Pact armies from penetrating anywhere on the front, is without doubt a major part of NATO's strategy, because of the need to avoid turning the whole of West Germany into a battlefield. Forward defense implies a particular tactical doctrine, fixed defense, which is today both unfashionable and arguably impossible. The fixed-defense tactics themselves require the development of particular weapons, for example cheap, infantry-borne antitank missiles, and also involve training and morale-building plans. It just may not be possible to come up with a way of persuading infantry soldiers to stand up in the middle of a Soviet tank attack and fire a modernized bazooka. It might also not be possible to build such weapons which are accurate enough even if the soldiers can be motivated to fire them, and the logistics needed to rapidly supply enough reloads over the entire front may be inaccurate. Strategy is therefore inevitably tied to much smaller-scale problems, or at least it should be. See Julian Lider, "Towards a Modern Concept of Strategy," *Cooperation and Conflict* Vol. 16, No. 4 (December 1981); Peter Paret, editor, *Makers of Modern Strategy from Machiavelli to the Nuclear Age* (Princeton, NJ: Princeton University Press, 1986); Donald R. Baucom, "A Historical Framework for the Concept of Strategy," *Military Review* 67, 3 (March 1987); Edward N. Luttwak, *Strategy: The Logic of War & Peace* (Cambridge, MA: Harvard University Press, 1987).

strategy, Fabian See FABIAN STRATEGY.

strategy, grand See GRAND STRATEGY.

strategy, maritime See MARITIME STRATEGY.

strategy, national See NATIONAL STRATEGY.

streamer 1. A malfunctioning parachute that does not open into a canopy which allows for a safe descent but into a long stream instead. 2. A colored strip of cloth attached to the top of a military unit's flag with the name of a battle or campaign in which the unit participated; collectively these are called HONORS.

stream takeoff Aircraft taking off in TRAIL or column FORMATION, one behind the other.

strength, accountable See ACCOUNTABLE STRENGTH.

strength, attached See ATTACHED STRENGTH.

strength, authorized See AUTHORIZED STRENGTH.

strength, bargaining See BARGAINING STRENGTH.

strength, end See END STRENGTH.

strength, initial See INITIAL STRENGTH.

strength, intransit See INTRANSIT STRENGTH.

strength, military See MILITARY STRENGTH.

strength, operating See OPERATING STRENGTH.

strength, organized See ORGANIZED STRENGTH

strength, peak See PEAK STRENGTH.

strength, ration See RATION STRENGTH.

stretch out The military procurement practice of buying fewer large price items each year (such as tanks or fighter aircraft) than originally planned. This stretches out production, which means a lower cost per procurement year, but also may cause production to lose the advantages of large order economies of scale; thus making prices higher for each item.

strike 1. An ATTACK intended to inflict damage on, seize, or destroy an objective. 2. To take down, remove, or prepare for transfer; especially a flag, tent, camp, etc.

strike, deep See DEEP STRIKE.

strike, surgical See SURGICAL STRIKE.

strike capability, first See FIRST-STRIKE CAPABILITY.

strike force 1. A military force organized to undertake an OFFENSIVE mission. 2. By analogy, any formally structured government effort to "attack" a problem.

strike photography Aerial photographs taken during an AIR STRIKE.

striker 1. A kind of FIRING PIN that is designed to hit the primer in a GRENADE fuze. 2. An enlisted person who is paid to do extra-duty work for an officer. 3. An enlisted person in the United States Navy who is training for a technical rating.

striking force, carrier See CARRIER STRIKING FORCE.

striking force, mobile See MOBILE STRIKING FORCE.

striking forces, surface See SURFACE STRIKING FORCES.

string 1. A series of radio messages sent from one station to another. The receiving station does not signal receipt of each message individually, but waits until the whole series is given to acknowledge receipt. 2. A given number of shots fired within a certain time interval.

strip area A built-up area of interconnecting villages and towns along roads or valleys.

strip marker In land MINE warfare, a marker, natural, artificial, or specially installed, located at the start and finish of a MINE STRIP.

strip search RECONNAISSANCE along a straight line between two given reference points.

strong point 1. A key point in a DEFENSIVE POSITION, usually strongly fortified and heavily armed with AUTOMATIC WEAPONS, around which other positions are grouped for its protection. 2. Any defensive position, fortified as extensively as time and materials permit, which is essentially an antitank nest that cannot be quickly overrun or bypassed by tanks, and which can by reduced by enemy infantry only with the expenditure of much time and overwhelming forces. A strong point is located on a terrain feature critical to the defense, or one which should be denied to the enemy. All weapons should be dug in with overhead cover in primary and alternate positions. Adequate time to construct a strong point must be given to the force assigned to establish it.

structured attack The arrival of a sequence of warheads at their targets in a manner timed to create the maximum destructive effect.

strut 1. Part of the firing mechanism that puts pressure on the hammer in automatic pistols and revolvers. 2. A brace or supporting piece, especially in an aircraft or an artillery weapon.

subcaliber ammunition PRACTICE AMMUNITION of a CALIBER smaller than standard for the gun on which practice is being given. Subcaliber ammunition is economical and may be fired in relatively crowded areas. It is therefore used with special subcaliber equipment to simulate firing conditions with standard ammunition.

subkiloton weapon A NUCLEAR WEAPON producing a YIELD below one kiloton. See also KILOTON WEAPON; MEGATON WEAPON; NOMINAL WEAPON.

submachine gun Any hand-held, lightweight automatic weapon designed to be fired from the shoulder or hip.

submarine, ballistic, nuclear See SSBN.

submarine, fleet ballistic missile See FLEET BALLISTIC MISSILE SUBMARINE.

submunition Any item, device or munition dispensed from or carried in PROJECTILES, dispensers, or CLUSTER BOMB

UNITS and intended for employment therefrom. ROCKETS are not considered submunitions.

subpackage A grouping of division-based and lower-echelon NUCLEAR WEAPONS.

subsequent operations phase The final phase of an AIRBORNE, AIRMOBILE, or AMPHIBIOUS OPERATION conducted after the ASSAULT PHASE. Operations in the objective area may consist of offense, defense, linkup, or withdrawal.

subsidiary landing In an AMPHIBIOUS OPERATION, a landing usually made outside the designated landing area, the purpose of which is to support the main landing.

subversion Secret actions designed to weaken the military, economic, or political strength of a nation from within by undermining the morale, loyalty, or reliability of its citizens. See Paul W. Blackstock, *The Strategy of Subversion; Manipulating the Politics of Other Nations* (Chicago: Quadrangle Books, 1964).

subversive activity An action that lends aid, comfort, and moral support to individuals, groups, or organizations that advocate the overthrow of incumbent governments by force and violence. All willful acts that are intended to be detrimental to the best interests of the government and which do not fall into the categories of TREASON, sedition, SABOTAGE, or ESPIONAGE can be considered subversive.

subversive political action A planned series of activities designed to accomplish political objectives by influencing, dominating, or displacing individuals or groups who are so placed as to affect the decisions and actions of another government.

successive formation A FORMATION in which the various UNITS move into their positions one after another.

successive level training A training concept under which a low-skill MILITARY OCCUPATIONAL SPECIALTY is established as a base from which the highest-caliber personnel are selected and trained into a military occupational specialty of higher skill without interruption.

successive objectives OBJECTIVES in sequence, where one objective is initially assaulted by a portion of a main force, supported by the remainder. As soon as the commander is assured the assaulting force can mop up the initial objective, other portions of the command attack the next objective. This process can be continued until the final objective is reached. The process is usually carried out by armored units.

successive positions Defensive fighting positions located one after another on the battlefield. A force can conduct a delaying action from successive DELAY POSITIONS.

summary court-martial A COURT-MARTIAL composed of one officer dealing with offenses by enlisted personnel. It may impose punishment of confinement for up to 1 month, hard labor without confinement for up to 45 days, restriction for up to 2 months, and forfeitures for up to two-thirds of 1 month's pay. It is analogous to a civil trial before a justice of the peace.

summit 1. The highest altitude above mean sea level that a PROJECTILE reaches in its flight from a gun to a target; the algebraic sum of the maximum ordinate and the altitude of the gun. 2. A meeting between the highest level executives of independent organizations.

Sun-Tzu (4th century B. C.) The ancient Chinese writer whose essays, under the title *The Art of War,* have influenced all Western military analysts ever since they were first published in the West in Paris in the late eighteenth century. Sun-Tzu was the first writer to formulate a rational basis for the conduct and planning of military operations. He believed that rulers and generals needed a systematic thesis to guide and direct them in war. He believed that skillful strategists should be able to beat an adversary without engaging him, to take cities without destroying them, and to overthrow states without bloodshed. He advocated the use of secret agents—spies and intelligence—to keep his leaders informed and to make them able to plan better. His book concentrates on how to conduct a war of maneuver. See Tau Hanzhang, *Sun Tzu's Art of War: The Modern Chinese Interpretation* (New York, NY: Sterling Publishing Company, 1987).

supercharge, weapons and ammunition A PROPELLING CHARGE intended to give the highest standard MUZZLE VELOCITY authorized for a PROJECTILE in the weapon for which the projectile is intended. Sometimes used as an identifying designation when more than one type of CHARGE is available for a weapon.

supercritical The condition of fissionable material in which a chain reaction will multiply with such speed as to cause an explosive energy release. See NUCLEAR FISSION.

superelevation An added positive angle in AIR DEFENSE gunnery that compensates for the fall of a PROJECTILE during its time of flight due to the pull of gravity.

superencryption A further ENCRYPTION of encrypted text for privacy or increased security.

superiority, air See AIR SUPERIORITY.

superiority, sea See SEA SUPERIORITY.

superiority, strategic See STRATEGIC SUPERIORITY.

superpower The most militarily powerful of nations. Today only the United States and the Soviet Union have this status; thus the term "superpowers" always refers to them.

superquick fuze A FUZE that functions immediately upon the impact of a PROJECTILE with its target; the FIRING PIN is driven into the PRIMER immediately upon first contact of the projectile with the target. Also called an instantaneous fuze.

supersensitive fuze A FUZE that will set off a PROJECTILE quickly when it strikes even a very light target, such as an airplane wing.

supplementary position A place to fight that provides the best means to accomplish a task that cannot be accomplished from the PRIMARY or ALTERNATE POSITIONS.

supplementary target A target other than the original target assigned to a GUN or BATTERY. It is a target on which FIRE is delivered when the original targets have been destroyed, or when it is impossible to deliver effective fire on them.

supplies All items necessary for the equipment, maintenance, and operation of military forces.

supplies, accompanying See ACCOMPANYING SUPPLIES.

supplies, housekeeping See HOUSEKEEPING SUPPLIES.

supplies, reserve See RESERVE SUPPLIES.

supply control The process by which an item of supply is controlled within the supply system, including requisitioning, receipt, storage, stock control, shipment, disposition, identification, and accounting.

supply, credit system of See CREDIT SYSTEM OF SUPPLY.

supply-point distribution A method of distributing supplies in which the receiving unit obtains supplies at a supply point, railhead, or truckhead and moves the supplies to its own area using its own transportation.

supply rate, controlled See CONTROLLED SUPPLY RATE.

support 1. The action of a force which aids, protects, complements, or sustains another force in accordance with a directive requiring such action. 2. A unit that helps another unit in battle. Aviation, artillery, or naval gunfire may be used as a support for infantry. 3. A part of any unit held back at the beginning of an attack as a RESERVE. 4. An element of a command that assists, protects, or supplies other forces in combat. See Edward A. Corcoran, "Support Troops in Combat Operations in Europe," *Army Logistician* 10, 1 (January–February 1978).

support, direct See DIRECT SUPPORT.

support, nuclear See NUCLEAR SUPPORT.

support agreement, interservice See INTERSERVICE SUPPORT AGREEMENT.

support ammunition service, direct See DIRECT SUPPORT AMMUNITION SERVICE.

support area A designated area in which COMBAT SERVICE SUPPORT elements, some STAFF elements, and other elements locate to support a unit.

support area, beach See BEACH SUPPORT AREA.

support area, brigade See BRIGADE SUPPORT AREA.

support artillery, direct See DIRECT SUPPORT ARTILLERY.

support artillery, general See GENERAL SUPPORT ARTILLERY.

support command, division See DIVISION SUPPORT COMMAND.

support coordination center, fire See FIRE-SUPPORT COORDINATION CENTER.

support coordination line, fire See FIRE SUPPORT COORDINATION LINE.

support coordinator, fire See FIRE SUPPORT COORDINATOR.

support craft Naval craft designed for the employment of ROCKETS, MORTARS, and AUTOMATIC WEAPONS at close range to a landing site from seaward, both in support of an assault against enemy held beaches and in the continuation of the attack.

support echelon 1. Those ELEMENTS of a military force that furnish logistical assistance to COMBAT UNITS. 2. Those units that support by FIRE the commander's plan of maneuver.

support equipment, ground See GROUND SUPPORT EQUIPMENT.

support mission, close See CLOSE SUPPORT MISSION.

support officer, fire See FIRE-SUPPORT OFFICER.

support officer, morale See MORALE-SUPPORT OFFICER.

support-reinforcing, general See GENERAL SUPPORT-REINFORCING.

support unit A unit that acts with and assists or protects another unit, but that does not act under the orders of the commander of the protected unit, of which it is not an ORGANIC part.

supporting arms Air, sea, and land weapons of all types employed to support ground units.

supporting artillery ARTILLERY that executes FIRE missions in support of a specific unit, usually INFANTRY, but which remains under the command of the artillery commander.

supporting attack An OFFENSIVE operation carried out in conjunction with a MAIN ATTACK and designed to deceive the enemy; destroy or pin down enemy forces that could interfere with the main attack; control ground whose occupation by the enemy will hinder the main attack; prevent the enemy from reinforcing the elements opposing the main attack; or force the enemy to commit RESERVES prematurely or in an indecisive area. Also called a HOLDING ATTACK.

supporting distance The distance between two units that can be traveled in time for one to come to the aid of the other. Also, for small units, the distance between two units that can be covered effectively by their FIRES. See also MUTUAL SUPPORT.

supporting fire FIRE delivered by supporting units to assist or protect a unit in combat. See also CLOSE SUPPORTING FIRE; DEEP SUPPORTING FIRE; DIRECT SUPPORTING FIRE.

supporting fire, close See CLOSE SUPPORTING FIRE.

supporting fire, deep See DEEP SUPPORTING FIRE.

supporting fire, direct See DIRECT SUPPORTING FIRE.

supporting force Forces stationed in, or to be deployed to, an AREA OF OPERATIONS to provide support for the execution of an OPERATION ORDER. The operational command of supporting forces is not passed to the supported commander.

supporting operations In AMPHIBIOUS OPERATIONS, those operations conducted by forces other than those assigned to the AMPHIBIOUS TASK FORCE. They are ordered by higher authority at the request of the amphibious task force commander, and are normally conducted outside the area for which the amphibious task force commander is responsible at the time of their execution.

supporting range That distance within which effective FIRE can be delivered by available WEAPONS.

supporting weapon Any WEAPON used to assist or protect a UNIT of which it is not an organic part.

suppression DIRECT and INDIRECT FIRE, ELECTRONIC COUNTERMEASURES, or SMOKE brought to bear on enemy personnel, weapons, or equipment to prevent effective fire on friendly forces. When suppressive measures are lifted, the enemy may once again be fully effective.

suppression mission A mission to suppress an actual or suspected WEAPONS SYSTEM for the purpose of degrading its performance below the level needed to fulfill its mission objectives at a specific time for a specific duration.

suppressive fires, immediate See IMMEDIATE SUPPRESSIVE FIRE.

supremacy, air See AIR SUPREMACY.

supremacy, sea See SEA SUPREMACY.

Supreme Allied Commander Europe See SACEUR.

surface burst See NUCLEAR SURFACE BURST. Compare to BURST.

surface burst, nuclear See NUCLEAR SURFACE BURST.

surface line A telephone or telegraph line that is laid on the ground hastily during the early stages of an attack or defense. In an organized area, surface lines are replaced by more permanent installations.

surface of impact The plane tangent to the ground or coinciding with the surface of the target at the point of impact of a projectile.

surface of rupture The area on the surface of the ground that is broken up by the explosion of an underground CHARGE.

surface striking forces Naval forces that are organized primarily to do battle with enemy forces or to conduct shore bombardment.

surface-to-air-missile (SAM) A surface-launched MISSILE designed to operate against a target above the ground or sea surface. Although antiaircraft guns still exist and have great importance in special roles, defense against enemy aircraft is now very largely the business of specialized surface-to-air missiles. These can vary from very small, short-range weapons fired from a portable disposable rocket launcher to highly complex, radar-guided missile systems. Missiles exist for differing needs, and the entire family of SAMs in a modern arsenal can cope with aircraft at all altitudes from a few hundred meters to 25 or more kilometers. Experience in modern wars, especially in Vietnam and the Middle East, but also the Falkland Islands conflict, has shown that well-trained troops can put up a very effective defense against attacking aircraft, and modern SAM designs can be highly mobile but still based on very sophisticated radar guidance. As with antitank missiles, it is not yet clear whether these relatively cheap weapons will make a massive inroad on the threat posed by the enormously complex and expensive weapons systems they are designed to counter.

surface-to-air missile envelope That airspace within the kill capabilities of a specific SURFACE-TO-AIR MISSILE system.

surface-to-air missile site A plot of ground prepared in such a manner that it will readily accept the hardware used in a SURFACE-TO-AIR MISSILE system.

surface-to-surface missile A surface-launched MISSILE designed to operate against a target on the surface.

surface zero See GROUND ZERO.

surgeon, flight See FLIGHT SURGEON.

surgical strike A euphemistic term that describes plans to destroy vital targets by means of carefully controlled force, so as to minimize COLLATERAL DAMAGE. It can refer to the use of conventional force, such as a precisely planned bombing raid on a terrorist headquarters intended to spare all innocent civilians nearby. It can also be used in nuclear strategy, where a surgical strike might be considered as a LIMITED NUCLEAR OPTION against a COMMAND, CONTROL, COMMUNICATIONS, AND INTELLIGENCE (C^3I) bunker or a set of missile SILOS. However, surgical strikes can rarely be as clean and discriminating in their destruction as planned, and the idea of such an attack using NUCLEAR WEAPONS is patently absurd by any normal standards. The lowest-YIELD strategic nuclear WARHEAD in any country's arsenal is of about 150 kilotons—at least 10 times the power of the Nagasaki bomb in 1945. Unless the target was in a most extremely deserted area, hundreds, and probably thousands, of "innocent" people would surely become casualties.

surprise attack An attack at the enemy when and where he is least prepared. Surprise in combat can bring success out of proportion to the numbers of forces involved. Surprise does not necessarily mean that the enemy is taken unaware, but simply that he is unable to react effectively. According to SUN TZU in *The Art of War:* "The enemy must not know where I intend to give battle. For if he does not know where I intend to give battle, he must prepare in a great many places. . . . If he prepares to the front his rear will be weak and if to the rear, his front will be fragile. If he prepares to the left, his right will be vulnerable and if to the right, there will be few on his left. And when he prepares everywhere he will be weak everywhere." See Julian Critchley, *Warning and Response: A Study of Surprise Attack in the 20th Century and an Analysis of Its Lessons for the Future,* (New York: Crane, Russak, 1978); Richard K. Betts, *Surprise Attack: Lessons for Defense Planning* (Washington, D.C.: Brookings Institution, 1982); Richard W. Bloom, "Military Surprise: Why We Need a Scientific Approach," *Air University Review* 35, 5 (July–August 1984).

surprise dosage attack A chemical operation which establishes on target a dosage OF CHEMICAL AGENT sufficient to produce the desired casualties before the enemy's troops can mask or otherwise protect themselves.

surrender, unconditional See UNCONDITIONAL SURRENDER.

surveillance The systematic observation of airspace or surface areas by visual, aural, electronic, photographic or other means.

surveillance, acoustical See ACOUSTICAL SURVEILLANCE.

surveillance, air See AIR SURVEILLANCE.

survey meter A portable instrument, such as a Geiger counter or ionization chamber, used to detect nuclear RADIATION and to measure the RADIATION DOSE RATE.

survivability In the context of NUCLEAR WARFARE, an aspect of weapons and other military assets as important as such traditional qualities as accuracy or troop morale. Except for static assets like MISSILES, which can be placed in superhardened SILOS, survivability is seldom a long-term aim in weapons design because it simply cannot be achieved. Aircraft and tanks can be "hardened" up to a point, so that they can survive a near miss by a NUCLEAR WEAPON long enough to carry out one mission, although their crews are unlikely to live much longer than that because it is so difficult to protect them against RADIATION SICKNESS. In the earlier scenarios and planning for nuclear war, survivability was restricted to the ability of a nuclear retaliatory force to survive a first strike by the enemy—hence the positioning of INTERCONTINENTAL BALLISTIC MISSILES (ICBMS) in silos and the deep-sea deployment of SUBMARINE-LAUNCHED BALLISTIC MISSILES (SLBMs), which were expected to be entirely used up in an immediate SECOND STRIKE. What has become clear since the development of war-fighting scenarios is the vital importance of ensuring the survivability of COMMAND, CONTROL, COMMUNICATIONS, AND INTELLIGENCE (C^3I) assets, and the personnel to staff them, as well as the NATIONAL COMMAND AUTHORITIES for whose benefit they exist. However, estimates for the survivability of such assets are not optimistic, even over the short period of a few weeks that the most prolonged of nuclear war scenarios require.

survivability operations The development and construction of protective positions such as earth berms (walls), dug-in positions, overhead protection (such as an ELEPHANT STEEL SHELTER), and countersurveillance measures to reduce the effectiveness of enemy WEAPONS SYSTEMS.

survival, combat See COMBAT SURVIVAL.

suspect battery 1. A hostile BATTERY whose existence is known, but whose location is uncertain. 2. An accurately located position about which there is doubt as to whether it is occupied or unoccupied. 3. A dummy position.

suspend 1. To stop something. 2. To deprive an OFFICER of some of the privileges of his rank, such as sitting as a member of a COURT-MARTIAL, selecting QUARTERS, or exercising COMMAND, as a punishment for some offense.

suspension equipment All aircraft devices such as racks, adapters, missile launchers, and pylons used for the carriage, employment, and jettisoning of aircraft stores.

suspension of arms A short truce arranged by local commanders for a special purpose, such as to collect the wounded, to bury the dead, or arrange for an exchange of prisoners.

suspension strop A length of webbing or wire rope between a helicopter and a cargo sling.

sustainability The ability to maintain the necessary level and duration of combat activity to achieve objectives. Sustainability is a function of providing and maintaining those levels of force, materiel, and consumables necessary to support a military effort.

sustained rate of fire The rate of FIRE that a WEAPON can continue to deliver for an indefinite length of time without seriously overheating.

sweep 1. To employ technical means to uncover planted microphones or other SURVEILLANCE devices. 2. A swift flight of a FORMATION of combat aircraft over enemy territory. 3. To cover a wide area by fire by successive changes in DEFLECTION line. 4. A trace produced on the screen of a cathode ray tube by linear deflection of the electron beam. 5. To drag a body of water to find and remove or explode MINES. 6. To pass a MINE DETECTOR over an area to detect any mines that may be contained therein.

sweep, influence See INFLUENCE SWEEP.

sweeping fire FIRE, especially from AUTOMATIC WEAPONS, that shifts gradually in elevation or direction.

swinging traverse A type of machine-gun FIRE used against dense troop formations moving toward a machine-gun position or against rapidly moving targets; the gunner makes continuous movements, back and forth, so that an arc of fire is created in front of him.

switch position A defense position diagonal to and connecting successive defensive positions that are parallel to the FRONT.

switch trench A TRENCH diagonal to and connecting successive trenches that are parallel to the FRONT.

syllabary In a CODE BOOK, a list of individual letters or a combination of letters or syllables accompanied by their equivalent CODE GROUPS, to be used for spelling out words or proper names not present in the vocabulary of a code. Also known as a spelling table.

sympathetic detonation The detonation of a CHARGE by exploding another charge adjacent to it.

syndrome, China See CHINA SYNDROME.

synthesis In INTELLIGENCE usage, the examining and combing of process information and intelligence for final interpretation.

synthetic exercise A training exercise in which enemy or friendly forces are generated, displayed, and moved by electronic or other means on simulators, radar scopes, or other training devices.

system 1. Any organized collection of parts that is united by prescribed interactions and designed for the accomplishment of a specific goal or general purpose. 2. The political process in general. 3. The establishment; the powers that be; the governance; the domain of a ruling elite. 4. The bureaucracy.

system indicator A symbol or group of symbols that identify a specific CRYPTOSYSTEM.

systems analysis The methodologically rigorous collection, manipulation, and evaluation of data about mechanical or social units to determine the best way to improve their functioning and to aid a decision-maker in selecting a preferred choice among alternatives. See Edward

S. Quade and W. I. Boucher, eds. *Systems Analysis and Policy Planning: Applications in Defense* (New York: American Elsevier, 1968).

T

T 1. A designation for a trainer aircraft that is specially equipped for instructional purposes; for example, the T-33 or T-37. 2. The classic naval maneuver whereby one line of battleships crosses and concentrates its fire on the leading ships of the enemy's line, thus forming a T.

tabard The silk banner attached to a bugle or trumpet.

table of distribution and allowance A table that prescribes the organizational structure, personnel, and equipment authorizations and requirements of a military unit for a specific mission for which there is no appropriate TABLE OF ORGANIZATION AND EQUIPMENT.

table of organization See ESTABLISHMENT.

table of organization and equipment A table that prescribes the normal mission, organizational structure, and personnel and equipment requirements for a military unit.

TACAMO 1. Take Charge and Move Out. 2. A specific type of aircraft exercise. In the U.S. Navy, TACAMO squadrons are responsible for communicating instructions to ballistic missile submarines (SSBNs) in the event of nuclear war. That is why they are called TACAMO, for "take charge and move out." Because radio communication with submerged submarines is extremely difficult, and otherwise depends on huge groundbased aerials, TACAMO aircraft are vital. They follow slow flight paths, trailing a very long aerial of about 9 kilometers, capable of sending commands to submarines. At any time there is at least one TACAMO aircraft patrolling each of the Atlantic and Pacific oceans.

tacit arms control agreement A COURSE OF ACTION in ARMS CONTROL in which two or more nations participate without any formal agreement having been made.

tac-log group Representatives designated by troop commanders in an AMPHIBIOUS OPERATION to assist U.S. Navy control officers aboard CONTROL SHIPS in the ship-to-shore movement of troops, equipment, and supplies.

tactical 1. Pertaining to the employment of military units in combat. 2. Skillful or adroit maneuvering. 3. Pertaining to a plan created to gain an advantage over an enemy.

tactical air command (TAC) 1. An air force organization designed to conduct offensive and defensive air operations in conjunction with land or sea forces. 2. A subordinate command in an air force. The United States Air Force's Tactical Air Command (TAC), unlike the Strategic Air Command, does not keep its units and aircraft under its own operational control but places them under the operating control of the command that actually does the fighting.

tactical air control party A subordinate operational component of a tactical air control system designed to provide air liaison to land forces and to control aircraft.

tactical air control party support team An army team that provides armored combat or special purpose vehicles and crews to tactical air control parties.

tactical air coordinator An officer who coordinates, from an aircraft, the action of combat aircraft engaged in the close support of general or sea forces; effectively functions as a GROUND CONTROL INTERCEPTOR. See also FORWARD OBSERVER.

tactical air doctrine Fundamental principles having to do with air superiority, interdiction, close air support, and isolating the battlefield; designed to provide guidance for the employment of air power in TACTICAL AIR OPERATIONS, in order to attain established OBJECTIVES. See Thomas J. Mayock, "Notes on the Development of AAF Tactical Air Doctrine," *Military Affairs* 14, 4 (Winter 1950); S. J. Deitchman, "The Implications of Modern Technological Developments for Tactical Air Tactics and Doctrine," *Air University Review* 29, 1 (November–December 1977).

tactical air force An air force charged with carrying out TACTICAL AIR OPERATIONS in coordination with ground or naval forces; more generally, to carry out tasks, including air defense and air superiority, in a particular THEATER OF WAR. Thus the air assets under the command of SACEUR in the NORTH ATLANTIC TREATY ORGANIZATION (NATO) system are all tactical air forces, although only the U.S. Air Force makes the distinction overtly, because it is scheduled to operate in the context of immediate warfare needs on the CENTRAL FRONT. In contrast, the U.S. STRATEGIC AIR COMMAND would be used to carry out long-range separate raids, most probably nuclear, as part of a wider war strategy that might have only incidental impact on any immediate battle going on in Europe. See William D. White, *U.S. Tactical Air Power: Missions, Forces, and Costs* (Washington: Brookings, 1974); Basil H. Liddell-Hart, "The Employment of Tactical Air Power: A Study in the Theory of Strategy," *Air University Review* (Sep.–Oct. 1975); James A. Machos, "Tacair Support for Airland Battle," *Air University Review* 35, 4 (May–June 1984).

tactical airlift 1. The movement of PERSONNEL and MATERIEL by TACTICAL AIR FORCES. 2. An airlift that provides an immediate delivery of combat troops and supplies directly into objective areas through airlanding, extraction, AIRDROP, or other delivery techniques. See Paul L. Wilke, "Tactical Airlift Tactics and Doctrine: More Carts, More Horses," *Air University Review* 37, 4 (May–June 1986).

tactical air observer An officer trained as an air observer whose function is to observe from airborne aircraft and report on the movement and disposition of friendly and enemy forces, terrain, weather, hydrography, and other factors.

tactical air operations 1. The employment of air power in coordination with ground or naval forces to attain and maintain air superiority; prevent movement of enemy forces into and within the combat zone and to seek out and destroy these forces and their supporting installations; and assist in attaining ground or naval forces objectives by combined or joint operations. 2. The USAF term for air operations involving the six combat functions of COUNTERAIR: CLOSE AIR SUPPORT, AIR INTERDICTION, tactical air reconnaissance, tactical aircraft operations (including air evacuation), and special operations performed by TACTICAL AIR FORCES.

tactical air operations center A subordinate unit of a tactical air command that directs and controls all en route air traffic and air defense operations, including manned interceptors and surface-to-air weapons, in an assigned sector.

tactical air reconnaissance The use of aircraft to obtain information concerning terrain, weather, and the disposition, composition, movement, installations, lines of communications, electronic and communication emissions of enemy forces. Also included are artillery and naval gunfire adjustment, and systematic and random observation of ground battle areas, targets, or sectors of airspace.

tactical air support Air operations carried out in co-ordination with surface forces and which directly assist land or maritime operations.

tactical area of responsibility (TAOR) A defined area of land for which responsibility is specifically assigned to the commander of the area as a measure for the control of assigned forces and coordination of SUPPORT.

tactical column A phase of a MOVEMENT TO CONTACT when contact is improbable and during which troops are tactically grouped to facilitate the prompt adoption of COMBAT FORMATIONS.

tactical command The authority delegated to a commander to assign tasks to forces under his command for the accomplishment of a mission assigned by higher authority. See Arthur S. Collins, Jr. "Tactical Command," *Parameters: Journal of the US Army War College* 8, 3 (September 1978).

tactical communications Those COMMUNICATIONS provided by, or under the OPERATIONAL CONTROL of, commanders of combat forces, COMBAT TROOPS, COMBAT SUPPORT troops, or forces assigned a COMBAT SERVICE SUPPORT mission.

tactical concept A statement, in broad outline, that provides a common basis for the future development of a tactical doctrine.

tactical control The detailed and usually local direction and control of movements or MANEUVERS necessary to accomplish assigned MISSIONS or TASKS.

tactical counterintelligence Actions designed to thwart the enemy's collection of INTELLIGENCE by denying and shielding friendly intentions and actions.

tactical damage assessment A direct examination of an actual STRIKE area by air observation, aerial photography, or direct ground observation.

tactical diversion See DIVERSION.

tactical inspection An INSPECTION to evaluate the COMBAT EFFICIENCY of a unit.

tactical intelligence INTELLIGENCE required for the planning and conduct of tactical operations. Tactical intelligence and STRATEGIC INTELLIGENCE differ primarily in their level of application, but may also vary in terms of scope and detail.

tactical intelligence zone A geographic guideline for the acquisition and transfer of information about the enemy, weather, and terrain. The tactical intelligence zone is used in planning for the employment of intelligence assets. The depth of the tactical intelligence zone varies with the level of command. A higher commander's tactical intelligence zone includes the zones of subordinate units.

tactical loading See COMBAT LOADING; UNIT LOADING.

tactical locality An area of terrain which, because of its location or features, has a tactical significance in the particular circumstances existing at a particular time.

tactical logistics The provision of LOGISTICS support to combat forces deployed within a theater of operations.

tactical map A large-scale map used for tactical and administrative purposes.

tactical minefield A MINEFIELD that is part of a formation obstacle plan and is laid to delay, channel, or break up advancing enemy formations.

tactical mining In naval MINE warfare, mining designed to influence a special operation or to counter a known or presumed tactical aim of the enemy. Implicit in tactical mining is a limited period of effectiveness of the minefield.

tactical missile A MISSILE produced for COMBAT use in an immediate theatre of operations.

tactical movement A movement of troops and equipment with a tactical mission under combat conditions when not in direct ground contact with the enemy.

tactical nuclear weapons. See BATTLEFIELD NUCLEAR WEAPONS.

tactical operations center (TOC) An ELEMENT within a main COMMAND POST that contains STAFF elements that permit the commander to see the battle, allocate resources, and position COMBAT SERVICE SUPPORT.

tactical operations center, field artillery See FIELD ARTILLERY TACTICAL OPERATIONS CENTER.

tactical organization, basic See BASIC TACTICAL ORGANIZATION.

tactical plan A plan for a particular combat operation, exclusive of arrangements for supply, evacuation, maintenance, or administration.

tactical plan, ground See GROUND TACTICAL PLAN.

tactical planning See PLANNING, TACTICAL.

tactical reserve A part of a force that is held under the control of the commander as a MANEUVERING FORCE to influence future action.

tactical sub-concept A statement, in broad outline, for a specific field of military capability within a TACTICAL CONCEPT that provides a common basis both for equipment- and WEAPONS SYSTEM development and for the future development of tactical DOCTRINE.

tactical training The training of troops in all phases of combat operations, including marching, security, offensive and defensive action, and withdrawals.

tactical troops COMBAT TROOPS, together with any SERVICE TROOPS required for their direct support, who are organized under one commander to operate as a unit and engage the enemy in combat.

tactical unit An organization of troops, aircraft, or ships that is intended to serve as a single unit in combat. It may include SERVICE UNITS required for its direct support.

tactical unit, basic See BASIC TACTICAL UNIT.

tactical warning 1. A notification that the enemy has initiated hostilities in a given sector of a theater of war. 2. In SATELLITE SURVEILLANCE, a notification to operational command centers that a specific threatening event is occurring.

tactical wire Wire entanglements used to break up the attack formations of the enemy and hold the enemy in areas that can be covered by intensive defensive fire; barbed wire.

tactics 1. The science and art of maneuvering troops, ships, or aircraft on a battlefront in preparation for, and in the conduct of, combat. By analogy, a tactical weapon (as opposed to a strategic one) is any device available to a commander during the course of a battle. NUCLEAR WEAPONS, while normally designed for STRATEGIC purposes, have also been designed for tactical use. 2. The methods of employing units in combat. 3. The detailed means for carrying out the directions set for a military force by the strategy to which it is bound. A general's strategy may require a DIVISION to capture a town. The divisional commander makes the tactical decisions, such as which BATTALION to send by which route to take which intermediate point. The appropriate combination of ARMOR and INFANTRY to be used, and the point in the battle plan when ARTILLERY should be employed to suppress enemy troops, are further typical tactical questions.

Tactics, in many ways, are the essence of MILITARY SCIENCE. They deal with the more ascertainable and measurable variables of one's own and the enemy's military hardware and troop capabilities. It is possible, at least in principle, to develop a generally "correct" tactical doctrine to deal with the sort of COMBAT, FIREPOWER, and FORCE ratios expected in any particular context, and to train troops accordingly. Strategy, being both more wide-ranging and less reducible to military technicalities, cannot be turned into doctrine in quite this way. Both levels of analysis, strategy and tactics, are, however, interrelated. It is not meaningless to talk about a division commander's strategy for taking a town; it simply means that tactical considerations are being confined to a much lower level of decision-making.

tactics, barrier See BARRIER TACTICS.

tail 1. COMBAT SUPPORT forces. 2. That part of an AIRBORNE force that does not move by air. 3. The rear of an aircraft.

take charge and move out See TACAMO.

tank An armored combat vehicle that is self-propelled, is usually armed with cannon and machine-guns, and travels on caterpillar tracks. The British invented the prototype of the modern tank during World War I in order to break the stalemate caused by the machine-gun and trench warfare. The British code word for their highly secret weapon, "tank," has stuck in spite of early efforts to formally call them "combat cars" or "assault carriages."

tank, flame thrower See FLAME THROWER.

tank, heavy A vague classification for the heaviest class of tank in general use. During World War II a heavy tank was one that weighed more than 35 tons. Today a heavy tank weighs more than 50 tons.

tank, light A vague classification for the lightest class of tank; one that weighs less than 25 tons.

tank, main battle (MBT) A TRACKED VEHICLE providing mobile FIREPOWER and crew protection for offensive combat. Modern versions of the tank, such as the American Abrams, the M-1 tank, use highly-advanced LASER aiming devices and stabilizing engineering, which allows them to fire with very great accuracy and at long range, even while on the move. They have ARMOR coating impenetrable by anything but extremely powerful antitank missiles and cannons. The M-1's main disadvantages are its expense and the inevitable fragility of its high-technology engines and armaments. There is a school of thought that suggests that even the most advanced main battle tanks will always be vulnerable to relatively cheap antitank weapons, and that their ability to deal with well-equipped infantry and, even more, with specialized antitank helicopters, is low. As a result, increasingly sums of money have to be invested in providing ARMORED FIGHTING VEHICLES so that they can accompany the tank units to protect them from infantry and airborne attack. Despite these doubts, the tank units are central to both WARSAW PACT and NORTH ATLANTIC TREATY ORGANIZATION (NATO) tactics, and it is here that the Warsaw Pact numerical superiority is at its highest. The Warsaw Pact is estimated to have over 25,000 MBTs available for a CENTRAL FRONT war, against a NATO total of about 13,000. The Soviet Union has not attempted to match the very high technology of the American MBT, preferring instead to rely on numbers, but it must be remembered that most of the NATO tanks are of an earlier generation than the U.S. Abrams. See William P. Baxter, "T-72: An Impressive Rival," *Army* 31, 9 (September 1981); Thomas L. McNaugher, *Collaborative Development of Main Battle Tanks: Lessons from the U.S.-German Experience, 1963–1978* (Santa Monica, CA: Rand Corporation, 1981); John T. Revelle,

"Quick-Fixing the M1 Tank," *Army Logistician* 17, 2 (March–April 1985).

tank, medium A vague classification for the tanks that are neither light nor heavy; one that weighs between 25 and 50 tons.

tank destroyer A SELF-PROPELLED antitank gun.

tank landing ship A naval ship designed to transport and land AMPHIBIOUS VEHICLES, TANKS, COMBAT VEHICLES, and equipment in an amphibious assault.

tank recovery vehicle A full-tracked motor vehicle, usually armored, designed to remove disabled or abandoned heavy vehicles from a battlefield to a collection point or maintenance establishment.

tank sweep An offensive operation by ARMOR forces designed to deliver a rapid, violent attack against an enemy force so as to inflict maximum casualties, disrupt control, and destroy equipment. It is normally associated with a SPOILING ATTACK, RECONNAISSANCE IN FORCE, or a COUNTERATTACK.

tank trap An obstruction that will stop a tank from going through it; examples include ditches dug so that tanks cannot cross them, and tetrahedrons, which are pyramid-like steel obstacles imbedded in the ground.

tank vehicle A vehicle that is usually wheeled, incorporating, in lieu of a body, a tank-type container for transporting bulk liquid. It normally includes dispensing valves, and may have pumps, hoses, or devices for segregating water and impurities from fuel.

taps 1. The final bugle call of a day; also played at military funerals.

target 1. A geographical area, complex, or installation planned for capture or destruction by military forces. 2. In INTELLIGENCE usage, a country, area, installation, agency, or person against which intelligence operations are directed. 3. An area designated and numbered for future firing. 4. In gunfire support usage, an impact BURST that hits its target. 5. In RADAR, generally any discrete object that reflects or retransmits energy back to the radar equipment; specifically, any object of radar search or surveillance. 6. The object of a search and rescue operation. 7. A mark to shoot at. 8. A goal to be achieved.

target, area See AREA TARGET.

target, auxiliary See AUXILIARY TARGET.

target, denial See DENIAL TARGET.

target, engageable See ENGAGEABLE TARGET.

target, fleeting See FLEETING TARGET.

target, hard See HARD TARGET.

target, lucrative See LUCRATIVE TARGET.

target, mark See MARK TARGET.

target, on-call See ON-CALL TARGET.

target, pinpoint See PINPOINT TARGET.

target, planned See PLANNED TARGET.

target, point See POINT TARGET.

target, priority See PRIORITY TARGET.

target, scheduled See SCHEDULED TARGET.

target, secondary See SECONDARY TARGET.

target, silhouette See SILHOUETTE TARGET.

target, supplementary See SUPPLEMENTARY TARGET.

target, transient See TRANSIENT TARGET.

target acquisition The detection, identification, and location of a TARGET in sufficient detail to permit the effective employment of weapons against it. Target acquisition is one of the two principal tasks of any long-range WEAPONS SYSTEM, and is most usually used in reference to antiaircraft or ANTIBALLISTIC MISSILE defenses. These defenses usually involve two radar systems, one of which makes a general search for incoming threatening objects, whether they be missiles or aircraft. When such an object is identified it is said to be "acquired," and its coordinates are passed to another radar set that tracks the threat continuously, in turn passing data to the defensive weapon itself in order for it to fire. More generally, target acquisition is the process of spotting and identifying enemy units or hardware which pose a tactical or strategic threat, and communicating this identification either to a decision-maker, to a weapons system, or to another SURVEILLANCE system. Target acquisition at long range is at the heart of Western plans for EMERGING TECHNOLOGY weaponry, and presents the most significant problems in the high-technology arms race, which is focused on COMMAND, CONTROL, COMMUNICATIONS, AND INTELLIGENCE ASSETS (C^3I) capacity.

target allocation In AIR DEFENSE, the process, following weapon assignment, of allocating a particular target or area to a specific SURFACE-TO-AIR MISSILE unit or INTERCEPTOR aircraft.

target analysis An examination of potential targets to determine their military importance, priority of attack, and the weapons required to obtain a desired level of damage or casualties in attacking them.

target approach point In air transport operations, a navigational checkpoint over which the final turn into a DROP ZONE or LANDING ZONE is made.

target area designator grid A grid system employing numbers and letters for the area designation of targets, with the numbers indicating a 1,000-meter square, and the letters indicating a 200-meter square within the numbered square.

target-area survey That portion of a survey concerned principally with the location of TARGETS AND OBSERVATION POSTS.

target array A graphic representation of enemy forces, personnel, and facilities in a specific situation, accompanied by a TARGET ANALYSIS.

target audience An individual or group selected for influence or attack by means of PSYCHOLOGICAL OPERATIONS.

target bearing 1. The true compass bearing of a TARGET from a ship that is firing upon it. 2. The relative bearing of a target measured in the horizontal from the bow of one's own ship clockwise from 0 degrees to 360 degrees, or from the nose of one's own aircraft in hours of the clock.

target box Areas on identifiable terrain in which enemy targets are expected to appear and against which air support will be employed.

target chart A large-scale map or diagram showing the target or targets assigned to bombing aircraft. A target chart is one type of aeronautical chart.

target combat air patrol A PATROL of fighter aircraft maintained over an enemy TARGET AREA to destroy enemy aircraft and to cover friendly shipping in the vicinity of the target area in AMPHIBIOUS OPERATIONS.

target complex A geographically integrated series of TARGET CONCENTRATIONS.

target concentration A grouping of geographically proximate targets.

target date The date on which it is desired that an action be accomplished or initiated.

target designating system A system for transmitting to one instrument the position of a TARGET that has been located by another instrument.

target designating system, laser See LASER TARGET DESIGNATING SYSTEM.

target discrimination The ability of a SURVEILLANCE or GUIDANCE SYSTEM to identify or engage any one target when multiple targets are present.

target dossier A file of assembled TARGET INTELLIGENCE about a specific geographic area.

target echo A radio signal reflected by an air or other target and received by the RADAR station that transmitted the original signal.

target evaluation The review of targets to determine their military importance and their relative priority for attack.

target folder A file folder containing TARGET INTELLIGENCE and related materials prepared for planning and executing action against a specific target.

target-information center An INTELLIGENCE center set up afloat or ashore for the assembly, evaluation, interpretation, dissemination, and coordination of target information for supporting weapons, i.e., artillery, naval gunfire, and airborne weapons. See also COMBAT INFORMATION CENTER.

target intelligence INTELLIGENCE that portrays and locates the components of a TARGET or TARGET COMPLEX, and indicates its vulnerability and relative importance.

target list A tabulation of confirmed or suspected TARGETS maintained by any ECHELON for information and FIRE-SUPPORT planning purposes.

target number The reference number given to a target by a FIRE-CONTROL unit.

target of opportunity 1. A TARGET visible to a surface or air SENSOR or OBSERVER, which is within range of available WEAPONS, and against which FIRE has not been scheduled or requested. 2. A target for a nuclear strike observed or detected after an operation begins, and which has not been previously considered for a nuclear strike.

target oriented analysis An analysis of initial AIMING POINTS chosen to determine the capabilities of available WEAPONS for an attack on a TARGET.

target overlay A transparent sheet which, when superimposed on a particular chart, map, drawing, tracing, or other representation, depicts TARGET locations and designations. The target overlay may also show boundaries between MANEUVER elements, OBJECTIVES, and friendly forward DISPOSITIONS.

target pattern The flight path of an aircraft during the ATTACK phase of its mission. Also called the attack pattern.

target priority The indicated sequence of attack for a grouping of TARGETS.

target reference point An easily recognizable point on the ground (either natural or man-made) used for identifying enemy TARGETS or controlling DIRECT FIRE.

target response The effect on men, material, and equipment of blast, heat, light, and nuclear radiation resulting from the explosion of a nuclear weapon.

target selector An observing instrument not carried on the gun CARRIAGE provided for the purpose of selecting an initial or new target, and which is electrically connected to a gun MOUNT in such a manner as to slew the gun to the approximate AZIMUTH and ELEVATION of a selected target, or give the tracker an indication of the direction of approach of a selected target.

target servicing The capability of a force to acquire, engage, and neutralize or destroy enemy firepower systems (tanks, combat vehicles, etc.) within a central battle. It includes the tasks of employing and coordinating support weapons such as MORTARS, FIELD ARTILLERY, and TACTICAL AIR SUPPORT as well as ELECTRONIC WARFARE assets that enhance the target servicing effort.

target system 1. All of the TARGETS situated in a particular geographic area and functionally related to one another. 2. A group of targets which are so related that their destruction will produce some particular effect desired by the attacker. See also TARGET COMPLEX.

target system component A set of targets belonging to one or more groups of industries and basic utilities required to produce component parts of an end product, such as periscopes, or one type of a series of interrelated commodities, such as aviation gasoline.

target-tracking radar RADAR which, as an integral part of a WEAPONS SYSTEM, is used to track a TARGET. It provides target-position data to a computer when used in a missile COMMAND GUIDANCE system. In a BEAM RIDER system, target-tracking radar provides coordinate information that permits the guidance of a missile to the target by an onboard computer.

targeting The process of selecting targets and matching the appropriate response to them, while taking account of operational requirements and capabilities. This has vast implications for nuclear policy. For example, WARHEADS targeted on non-military targets would not be considered FIRST STRIKE weapons. See Benjamin S. Lambeth and Kevin N. Lewis, "Economic Targeting in Nuclear War: U.S. and Soviet Approaches," *Orbis*, 27, 1 (Spring 1983); Desmond Ball and Jeffrey Richelson, *Strategic Nuclear Targeting* (Ithaca, NY: Cornell University Press, 1986).

targeting, flexible In a nuclear context, weapons that can be rapidly moved from one target to another, such as from military to non-military targets. See Stephen J. Cimbala, "Flexible Targeting, Escalation Control, and War in Europe," *Armed Forces and Society* Vol. 12, No. 3 (Spring 1986).

targets, group of See GROUP OF TARGETS.

targets, series of See SERIES OF TARGETS.

targets, time-sensitive See TIME-SENSITIVE TARGETS.

task fleet A mobile COMMAND consisting of ships and aircraft necessary for the accomplishment of a specific major task or tasks that may be of a continuing nature.

task force 1. A temporary grouping of units, under a single commander, formed for the purpose of carrying out a specific OPERATION or MISSION. 2. A semi-permanent organization of units, under one commander, formed for the purpose of carrying out a continuing specific TASK. 3. A component of a FLEET organized by the commander for the accomplishment of a specific task or tasks. 4. By analogy, a temporary interdisciplinary team, within a larger organization, that is charged with accomplishing a specific goal. Task forces are typically used in government when a problem crosses departmental lines. 5. A temporary government commission charged with investigating and reporting on a problem.

task force, airmobile See AIRMOBILE TASK FORCE.

task force, amphibious See AMPHIBIOUS TASK FORCE.

task force, battalion See BATTALION TASK FORCE.

Task Force 160 The United States Army's transportation and support unit for counterterrorist operations. Its helicopter, equipped for night operations, can move DELTA Force members up to 200 miles in pitch-black conditions. For longer distance counterterrorism operations, the United States Air Force has a Special Operations Wing with similar capabilities.

task group, amphibious See AMPHIBIOUS TASK GROUP.

task organization 1. A temporary grouping of forces designed to accomplish a particular MISSION. 2. A naval organization which assigns to responsible commanders the means with which to accomplish their assigned TASKS in any planned action. 3. A TABLE OF ORGANIZATION that pertains to a specific naval DIRECTIVE.

tasking The process of translating an allocation of personnel, supplies and equipment into ORDERS, and passing these orders to the UNITS involved. Each order normally contains sufficiently detailed instructions to enable the executing agency to accomplish the ordered mission successfully.

tattoo 1. A nighttime signal via bugle or drum that calls enlisted ranks to return to QUARTERS; usually played just before taps. 2. Military music and marching exercises designed as entertainment.

technical characteristics Those desired military characteristics of equipment that pertain primarily to the engineering principles involved in producing it.

technical damage assessment A direct DAMAGE ASSESSMENT conducted by special teams to obtain technical information.

technical escort Individuals technically qualified and properly equipped to accompany designated material requiring a high degree of safety or security during shipment.

technical intelligence INTELLIGENCE concerning foreign technological developments, and the performance and operational capabilities of foreign material, which have or may eventually have a practical application for military purposes.

technical observer A civilian technical expert, representing a commercial firm, who accompanies troops in the field to observe and report on the operation of mechanical equipment or armament under field conditions.

technical proficiency inspection An inspection of a NUCLEAR WEAPONS storage, support, or delivery organization to see that it is adhering to standard procedures in the storage, maintenance, safety testing, handling, and assembly of such weapons.

technical representative A civilian expert, usually working for a contractor to the government, who attends to equipment in the field.

technical survey A complete electronic and physical inspection to ascertain that offices, conference rooms, war rooms, and other similar locations where CLASSIFIED INFORMATION is discussed are free of monitoring systems. See also SWEEP.

technique The method of performance of any act, especially the detailed methods used by troops or commanders in performing assigned tasks. Technique refers to the basic methods of using equipment and personnel. The phrase "tactics and technique" is often used to refer to the general and detailed methods used by commanders and forces in carrying out their assignments.

technology, emerging See EMERGING TECHNOLOGY.

technology transfer 1. The application of technologies developed in one area of research or endeavor to another, frequently involving a concomitant shift in institutional setting (e.g., from one federal agency to another). Examples include the application of space technology developed under the auspices of the National Aeronautics and Space Administration (NASA) to the problems of public transportation or weather prediction. Claims regarding the future possibilities for technology transfer are frequently factors in decisions about continuing financial support for technology development. 2. The movement of new technologies from one nation to another through sales or espionage. The United States government is continually concerned about restricting the flow of new technology to potential enemies, because their technologies often have considerable military application. See Richard Perle, "The Eastward Technology Flow: A Plan of Common Action," *Strategic Review* 12, 2 (Spring 1984); Sumner Benson, "The Impact of Technology Transfer on the Military Balance," *Air University Review* 36, 1 (November–December 1984); Arthur F. Van Cook, "Checks on Technology Transfer: The Defense Stakes Are High," *Defense Management* 21, 1 (First Quarter 1985).

teeth-to-tail ratio The ratio of direct COMBAT forces (the teeth) to SUPPORT forces (the tail). See Robert P. Johnson, "Tooth-to-Tail," *Army Logistician* 16, 5 (September–October 1984).

temperature, ballistic See BALLISTIC TEMPERATURE.

tempest An unclassified, short name referring to investigations and studies of compromising electronic emanations.

templating, doctrinal See DOCTRINAL TEMPLATING.

tenant A unit or activity that occupies facilities on a military installation of another department or command, and receives supplies or other support services from that installation.

terminal ballistics The subdivision within BALLISTICS that deals with the effects of MISSILES at their TARGETS.

terminal clearance capacity The amount of cargo or personnel that can be moved through and out of a terminal on a daily basis.

terminal guidance 1. The guidance applied to a GUIDED MISSILE between MIDCOURSE GUIDANCE and arrival of the missile in the vicinity of its target. 2. Electronic, mechanical, visual, or other assistance given an aircraft pilot to facilitate arrival at, operation within or over, landing upon, or departure from an air landing or AIRDROP facility. See also GUIDANCE.

terminal phase The final phase of the TRAJECTORY of a BALLISTIC MISSILE; that portion of the trajectory of a ballistic missile between its re-entry into the atmosphere or the end of its MIDCOURSE PHASE and its impact or arrival in the vicinity of its target. By this stage the missile WARHEAD is not under motorized propulsion, and follows a predictable ballistic path. Although the terminal phase is short, lasting only a few minutes, the warheads are at their most vulnerable; they are approaching any defense systems deployed around the targets, and they are less easily masked by DECOYS. Furthermore, by the terminal phase it is much clearer to the defenders just what the targets are, and they can concentrate defenses to protect their most vital assets. For this reason, much of the effort in the STRATEGIC DEFENSE INITIATIVE is being put into terminal-phase interception. See also BOOST PHASE; RE-ENTRY VEHICLE.

terminal velocity 1. The hypothetical maximum speed a body could attain along a specified flight path under given conditions of weight and thrust, if diving through an unlimited distance in air of specified uniform density. 2. The remaining speed of a PROJECTILE at the point in its downward path where it is level with the MUZZLE of the weapon that fired it.

terrain, compartment of See COMPARTMENT OF TERRAIN.

terrain, dominant See DOMINANT TERRAIN.

terrain, key See KEY TERRAIN.

territory, liberated See LIBERATED TERRITORY.

territory, occupied See OCCUPIED TERRITORY.

terrain analysis The process of interpreting a geographic area to determine the effect of its natural and man-made features on military operations.

terrain break angle The angle between two adjacent slope facets, which may or may not constitute an OBSTACLE.

terrain evaluation The evaluation and interpretation of an area of probable military operations to determine the effect of the terrain on the lines of action open to opposing forces in this area.

terrain exercise A training exercise in which a stated military situation is solved on the ground, the troops being imaginary and the solution usually being in writing.

terrain factor Any attribute of a given area of terrain that can be adequately described at any point (or instant of time) by a single measurable value; for example, slope or obstacle height.

terrain flight Flight close to the Earth's surface, during which airspeed, height, or altitude are adapted to the contours and cover of the ground in order to avoid enemy detection and fire.

terrain intelligence Processed information on the military significance of the natural and man-made characteristics of an area.

terrain masking The ability of terrain features to deny observation of an object.

terrain reinforcement The development of terrain to degrade enemy mobility (COUNTERMOBILITY OPERATIONS) or to enhance friendly survivability through the construction of fighting positions and cover.

terrain return The reflection of radiation from the ground, and its return as an echo to the RADAR set that sent it.

terrain spotting The positive spotting of a ROUND OF AMMUNITION not on the OBSERVER–TARGET LINE, based on a knowledge of the terrain near the target.

terrain study An analysis and interpretation of the natural and man-made features of an area, their effects on military operations, and the effect of weather and climate on these features.

terrorism 1. Highly visible violence directed against randomly selected civilians in an effort to generate a pervasive sense of fear and thus affect government policies. 2. Violence against representatives (police, politicians, diplomats, etc.) of a state by those who wish to overthrow its government; in this sense terrorism is REVOLUTION, and thus the cliché that one man's terrorist is another man's freedom fighter. 3. Covert warfare by one state against another; in effect, state-sponsored terrorism. 4. The acts of a regime that maintains itself in power by random or calculated abuse of its own citizens; in this sense, all oppression and dictatorial regimes are terrorist. See Donald B. Vought and James H. Fraser Jr., "Terrorism: The Search for Working Definitions," *Military Review* 66, 7 (July 1986); Christopher Hitchens, "Wanton Acts of Usage: Terrorism: A Cliché in Search of a Meaning," *Harper's* Vol. 273, No. 1636 (September 1986).

terrorism, counter See COUNTERTERRORISM.

test-ban treaty The 1963 agreement signed by the United States, the Soviet Union, the United Kingdom, and more than 90 other countries prohibiting the testing of NUCLEAR WEAPONS in the atmosphere, in space, or under water, but allowing underground testing. France and China, the most conspicuous non-signers of the treaty, have continued to test weapons in the atmosphere. See Glenn T. Seaborg, *Kennedy, Khrushchev and the Test Ban* (Berkeley: University of California Press, 1981).

Test Ban Treaty, Limited See LIMITED TEST BAN TREATY.

test piece Any GUN that is compared with another gun in calibration. The gun used as a basis of comparison is called the reference piece; any other gun adjusted accurately with reference to it is a test piece.

theater 1. A field of operations; a place where action is occurring. 2. The geographical area outside the continental United States for which a commander of a unified or specified command has been assigned military responsibility.

theater nuclear forces (TNF) Intermediate nuclear forces (INF); Euromissiles; any of a range of BALLISTIC and CRUISE MISSILES that are deployed by the NORTH ATLANTIC TREATY ORGANIZATION (NATO) and the WARSAW PACT for relatively long-range nuclear strikes inside Europe. No TNF weapon can reach the United States, although NATO missiles could, in fact, carry out attacks on the western Soviet Union as far as Moscow, and the Soviet Union's missiles could destroy almost any target in Western Europe. The somewhat arbitrary upper limit of 5,500 kilometers of range has been used in ARMS CONTROL negotiations to define theater nuclear weapons, and to distinguish them from TACTICAL or BATTLEFIELD NUCLEAR WEAPONS, which are often defined as having a range under 500 kilometers. Some TNF systems (mainly the ground-launched cruise missiles and PERSHING II ballistic missiles under NATO command, and the Soviet SS-20s) are being destroyed under provisions of the INF TREATY. Compare to LONG-RANGE THEATER NUCLEAR FORCES. See T. Wood Parker, "Theater Nuclear Warfare and the U.S. Navy," *Naval War College Review* 35, 1 (January–February 1982).

theater of operations That part of a theater of war that is engaged in military operations as well as their support while being under a single overall command for all land, air, and sea operations.

theater of war A major geographical area of conflict, or potential conflict, that is covered by one unified COMMAND authority. A theater may include more than one theater of operations or they may be coterminous. STRATEGY is generally determined within a theater independently of military considerations applying elsewhere in the world. Thus the NORTH ATLANTIC TREATY ORGANIZATION (NATO) oversees the European theater of any war, but in a third world war, the United States and Soviet Union might well carry on operations quite separately in, for example, a Pacific or even a Latin American theatre (see HORIZONTAL ESCALATION.) Soviet military doctrine recognizes an intermediate level of command organization, known by the initials TVD, whereas NATO tends to make functional rather than geographical divisions within the main theater, for naval, land, or air operations. See Donald R. Cotter, "NATO Theater Forces: An Enveloping Military Concept," *Strategic Review,* 9, 2 (Spring 1981).

thermal energy The energy emitted from a nuclear FIREBALL as thermal radiation.

thermal kill The destruction of a target by heating it, using directed energy, to the degree that its structural components fail.

thermal radiation 1. The heat and light produced by a nuclear explosion. 2. Electromagnetic radiations emitted from a heat or light source as a consequence of its temperature; these consist essentially of ultraviolet, visible, and infrared radiations.

thermite A standard incendiary agent used as filling for incendiary munitions; a mixture of thermite (iron oxide and aluminum) and other oxidizing agents burns at about 4,300 degrees Fahrenheit.

thermometer, maximum See MAXIMUM THERMOMETER.

thermonuclear A term referring to the explosion of NUCLEAR FUSION weapons, more commonly called H-bombs, or HYDROGEN BOMBS. An initial atomic, or NUCLEAR-FISSION explosion is required to generate the enormous temperatures under which nuclear fusion can occur, releasing the energy of a thermonuclear explosion. "Thermonuclear" is frequently found as an adjective attached to, for example, the noun "war" to indicate the fully fledged CENTRAL STRATEGIC WARFARE in which MEGATON-level hydrogen weapons are used, in contrast to the use of a few relatively small weapons, possibly only atomic bombs, in a tactical confrontation.

thickened fuel Gasoline with a thickener (gelling agent) added, used as an INCENDIARY fuel in FLAME THROWERS and FIRE BOMBS. See also NAPALM.

thickening The reinforcing of units in the conduct of an ACTIVE DEFENSE in order to concentrate forces so as to attain a desired COMBAT RATIO. Thickening may also include the adjusting of boundaries in order to concentrate more forces in a smaller area.

think tank A colloquial term that refers to an organization or organizational segment whose sole function is research, usually in the policy and behavioral sciences. The first important think-tank was probably the Research and Development Corporation (RAND), created by the newly-independent U.S. Air Force in 1947 as a civilian research institute. It operated on a contract basis, and studied problems ranging from the highly technological to the social scientific. This wideranging area of expertise has, perhaps, been the hallmark of the most successful think-tanks. As think-tanks became established, they developed a less passive role, and much of the research they now carry out is conceived of "in house" and "sold to the client." Another example is the Institute for Defense Analyses, which conducts scientific and engineering research on the STRATEGIC DEFENSE INITIATIVE, but also studies battlefield nuclear tactics and prepares reports on such diverse topics as British arms-control policy and Soviet military thought. There are perhaps a half dozen major, and dozens of minor think-tanks in the United States involved with strategic policy. See Max Beloff, "The Think Tank and Foreign Affairs," *Public Administration (Great Britain)* Vol. 55 (Winter 1977); Tyrus W. Cobb, "National Security Perspectives of Soviet 'Think Tanks'," *Problems of Communism* Vol. 30, No. 6 (November–December 1981).

thin natural screen Natural growth left in front of ENTRENCHMENTS and EMPLACEMENTS to aid in concealing them.

thinning The removal of forces and intentional weakening of COMBAT POWER in one part of a battle area.

thin red line The front line of INFANTRY, this is how Rudyard Kipling described the British red-coated infantry in his 1892 poem *Tommy*.

Third World Those countries with underdeveloped but growing economies and low per-capita incomes, often with colonial pasts. "Third World" is often used interchangeably with or as a synonym for "LDC"s (less developed countries), "the South," "developing countries," or "underdeveloped countries." India, Nigeria, Ecuador, and Morocco are examples. In the 1970s, a Fourth World was distinguished from the Third World, and included those developing countries with little economic growth, few natural resources, slight financial reserves, and extremely low per-capita incomes. Bangladesh, Ethiopia, and Sudan are examples. Armies tend to be of particular importance in the politically undeveloped countries of the Third World, where military rule is a common feature. Although the explanations for any specific occurrence of military rule vary, what such occurrences usually have in common is that the army has a near-monopoly of bureaucratically efficient and disciplined personnel, often trained in the developed countries. As civil services develop and civilian governments acquire an aura of legitimacy in Third World nations, the fear of military COUPS D'ETAT will diminish and armies will become servants rather than masters of the state. See Mark N. Katz, *The Third World in Soviet Military Thought* (London: Croom Helm, 1982); H. A. Reitsma and J. M. G. Kleinpenning, *The Third World in Perspective* (Totowa, NJ: Rowman & Allanheld, 1985); Robert E. Harkavy and Stephanie G. Neuman, *The Lessons of Recent Wars in the Third World* (Lexington, Mass.: Lexington Books, 1985); Jerry F. Hough, *The Struggle for the Third World: Soviet Debates and American Options* (Washington, D.C.: The Brookings Institution, 1986).

threat 1. The anticipated inventory of an enemy's weapons and capabilities. 2. In the context of the STRATEGIC DEFENSE INITIATIVE (SDI), the inventory is of NUCLEAR WEAPONS and their delivery systems, as well as of DECOYS, PENETRATION AIDS, and other countermeasures for BALLISTIC MISSILE DEFENSE (BMD). See Grayson Kirk and Nils H. Wessel, eds. *The Soviet Threat: Myths and Realities* (New York: The Academy of Political Science, 1978).

threat assessment The MILITARY INTELLIGENCE calculation of the danger presented by another country; or more specifically the threat posed by a particular action of that country. Assessing the threat of a particular action is usually quite unambiguous, because trying to analyze the motivations and intentions of the opponent hardly matters.

General threat assessment, however, ought to, but usually does not, involve a consideration of the reasons behind an opponent's armament programs. Instead the assessment is usually made on a worst-case basis. For example, the threat posed to the West by the Soviet Union is assessed by counting up the hardware and personnel it has, and assuming that whatever this force structure would be used for is what it will be used for. The Soviet Union has amassed a sufficiently large tank army in Europe to make invasion of West Germany at least a conceivably successful strategy. In military threat assessment this information is enough to establish that there is therefore a real threat of invasion. But is the capacity to do something evidence of such an intention? This kind of political calculation is a necessary part of any sound threat assessment on which strategies and procurement decisions can be based. Other elements besides personnel and hardware totals have to be taken into account, the primary one being the opponent's strategic doctrine. How a potential enemy thinks it can, or must, fight a war is a key element in assessing the threat that it poses, because it determines what the potential enemy thinks it can do with its available capacity. For example, had the British and French militaries, in 1940, taken seriously the German doctrine of BLITZKRIEG, they would have assessed the threat Germany posed to France more highly than indicated on the basis of hardware arithmetic, which correctly showed that the number and quality of French tanks was considerably superior to that of the German Wehrmacht. Increasing attention is paid in the West to Soviet Doctrine, although it is less obvious that defense policy is based on a threat assessment derived from these studies. See James John Tritten, "Threat Assessment," *Navy International* (October 1984).

threat clouds Incoming (from space) concentrations of both threatening and nonthreatening (DECOYS) objects. A BALLISTIC MISSILE DEFENSE must distinguish between them.

threat study An INTELLIGENCE assessment of enemy capabilities in terms of combat MATERIEL, employment doctrine, environment, and force structures.

throughput distribution The shipment of SUPPLIES from their points of origin as far forward as possible, bypassing intermediate supply activities.

throw-weight A measure of MISSILE's capacity to carry a military PAY-LOAD into the atmosphere on a BALLISTIC TRAJECTORY capable of reaching a target of a particular, possibly intercontinental, range. The throw-weight which the missile can carry determines the size and nature of its BUS (or front end), and therefore the size and number of the WARHEADS, PENETRATION AIDS, and DECOYS, and the complexity of the guidance mechanism that the bus itself can contain. Throw-weight is usually a very small fraction of the total weight of a missile. For example, the new MIDGETMAN, planned by the United States as a small, mobile INTERCONTINENTAL BALLISTIC MISSILE (ICBM), will probably have a throw-weight of 450 kilograms against a total missile weight of perhaps 14,000 kilograms. The throw-weight calculation is a function of the drive that the missile's engines can produce and the range over which the front end has to be projected. To some extent these are interchangeable: the same missile can "trade" range for a heavier front end, but this is obviously not an alteration that can be done easily. Not surprisingly, the bigger the missile is, the greater is its throw-weight. Land-based ICBMs, which are often three-stage rockets, can have significantly higher throw-weights than submarine-launched ballistic missiles (SLBMs). See James John Tritten, "Throw-Weight and Arms Control," *Air University Review* 34, 1 (November–December 1982).

thrust, long See LONG THRUST.

thrust line The line forming the base of all coordinates in the thrust-line system of locating the position of objects on a map. It is a line designated by the commander of a unit and located on the map by two reference points, or by a reference point and a direction. Somewhere on the thrust line is a base point, designated by the commander, from which all coordinates are measured. Points are located by giving their distance along the thrust line, forward or in back of the base point, and their distance perpendicular to the thrust line.

ticket-punching A career strategy that calls for an OFFICER to get all of the appropriate assignments and training in order to qualify for the next promotion. A ticket-puncher is more interested in being able to say that he commanded something than in what really happens to what he commands. Examples of ticket-punching also include attending a war college, being a general's aide, or working in the Pentagon. Compare to CAREERISM.

tiger suits Camouflage fatigue uniforms; so called because the pattern of the cloth resembles a tiger's stripes.

time, dead See DEAD TIME.

time distance The time required for any one vehicle to travel between two given points at a given rate of speed.

time fire FIRE in which FUZES are set to act after a fixed time interval and before impact.

time fuze A FUZE that contains a graduated time element to regulate the time interval after which the fuze will function.

time of attack The hour at which an ATTACK is to be launched. If a LINE OF DEPARTURE is prescribed, the time of attack is the hour at which the line is to be crossed by the leading elements of the attack.

time of delivery The time at which the addressee or responsible relay agency acknowledges receipt of a message.

time of flight In ARTILLERY AND NAVAL GUNFIRE SUPPORT, the time in seconds from the instant a WEAPON is fired, launched, or released from its delivery vehicles or WEAPONS SYSTEM to the instant it strikes its target or detonates.

time of origin The time at which a message is released for transmission.

time of receipt The time at which a receiving station completes its reception of a message.

time on target 1. The time at which aircraft are scheduled to attack or photograph a target. 2. The actual time at which aircraft attack or photograph a target. 3. The time at which a NUCLEAR DETONATION is planned at a specified GROUND ZERO. 4. A method of firing on a target in which various ARTILLERY units (or naval gunfire support ships) so time their fire that the initial ROUNDS strike the target simultaneously.

time-over-target conflict A situation wherein two or more DELIVERY VEHICLES are scheduled in such a way that their proximity violates the established separation criteria for YIELD, time, distance, or all three.

time-sensitive targets Those targets requiring an immediate response because they pose (or will soon pose) a clear and present danger to friendly forces or are highly lucrative, fleeting TARGETS OF OPPORTUNITY.

time-urgent target A target that must be destroyed very rapidly in any nuclear exchange, and most probably in a FIRST STRIKE or the earliest stages of a SECOND STRIKE. Typically, time-urgent targets are targets such as missile SILOS. If a power is launching a nuclear first strike against an opponent, clearly the primary need is to destroy as many as possible of that opponent's missiles and air and submarine bases as quickly as possible in order to minimize retaliation. If a second strike is being launched in retaliation, much the same applies, but in addition, enemy radar and antiaircraft facilities may become time-urgent targets so that a clear corridor through the enemy's air defenses can be established for piloted bombers carrying CRUISE MISSILES. The speed with which time-urgent targets must be destroyed presents a problem, since it means that only BALLISTIC MISSILES can be used. Cruise missiles would take so long to reach the targets that an enemy would have ample opportunity to retaliate before its own missiles could be destroyed. (This is a major reason why cruise missiles cannot, in general, be regarded as first strike weapons.) A further problem is that many time-urgent targets are also HARD TARGETS, such as missile silos and command bunkers (see COMMAND, CONTROL, COMMUNICATIONS, AND INTELLIGENCE), and even some airbases with super-hardened aircraft shelters. To guarantee the destruction of such targets requires a combination of very high YIELD and extreme accuracy. Consequently, until the advent of the third generation submarine-launched ballistic missiles (SLBMS), such as the American TRIDENT, only land-based INTERCONTINENTAL BALLISTIC MISSILES (ICBMS) have been regarded as suitable for time-urgent targeting.

TNT Trinitrotoluene, a common explosive.

TNT equivalent A measure of the energy released from the detonation of a NUCLEAR WEAPON, or from the explosion of a given quantity of fissionable material, in terms of the amount of TNT (trinitrotoluene) that could release the same amount of energy when exploded.

to the color A bugle call used when the colors are raised or lowered sounded as a SALUTE to the COLOR, to the president of the United States, or to a foreign chief of state.

tolerance dose The amount of RADIATION that may be received by an individual within a specific period with negligible results.

tommy 1. A British soldier. 2. The Thompson .45 caliber submachine gun.

tone down In CAMOUFLAGE and CONCEALMENT, the process of making an object or surface less conspicuous by reducing its contrast to the surroundings or background.

tonne A metric ton; 1000 kilograms.

top carriage The upper, movable part of a gun carriage.

top kick The FIRST SERGEANT of a COMPANY.

topographic map A map that presents the vertical position of terrain features in measurable form, as well as their horizontal positions. See also MAP.

torpedo 1. A self-propelled cigar-shaped underwater PROJECTILE with an explosive charge that can be launched against ships by submarines, other ships, or aircraft. 2. An older term for a naval MINE. Admiral Farragut was referring to such devices in the 1864 Battle of Mobile Bay when he said: "Damn the torpedoes. Full speed ahead."

torpedo, Bangalore See BANGALORE TORPEDO

Torpex A high explosive consisting of TNT, cyclonite, and aluminum powder, used especially in TORPEDOES, MINES, and DEPTH BOMBS.

toss bombing A method of bombing in which an aircraft flies on a line toward a target, pulls up in a vertical plane, and releases its bomb at an angle that will compensate for the effect of gravity drop on the bomb. It is similar to LOFT BOMBING, yet unrestricted as to altitude. See also OVER-THE-SHOULDER BOMBING.

total dosage attack A chemical attack or fire mission used to build up the required dosage of a toxic chemical or of gunfire over an extended period. It is normally employed against troops who have no protection available. See also SURPRISE DOSAGE ATTACK.

total materiel assets The total quantity of an item available on a worldwide basis in a military system.

total materiel requirement The sum of the peacetime materiel requirement and war-reserve materiel requirement of a military service.

total war 1. A conflict that threatens the survival of a nation and in which all weapons are used. 2. A war effort that mobilizes all sectors of a nation's economy. It was CLAUSEWITZ who first developed the modern concept of total war in the early 1800s. Today total war is a way of describing confrontations such as the two World Wars. The expected nature of a CENTRAL FRONT war between the NORTH ATLANTIC TREATY ORGANIZATION (NATO) and the WARSAW PACT would place it in the same class. According to this characterization, such wars are different from most wars in world history in two crucial ways. Wars are "total" when the entire population of each combatant is affected both by being potential targets and by being drawn into the entire economic and social efforts of the nations involved in the conduct of the war. In "ordinary" war it would be possible to raise an army with perhaps only marginal extra taxation. The army could either be entirely made up of volunteers, or by conscripting a small and typical part of the population. Since the weapons used would not be capable of reaching outside the battlefield, the bulk of the population would be entirely safe and might hardly be affected by the war. In contrast to this, World Wars I and II saw bombing of civilian populations, near starvation because of blockade or submarine warfare, wide-ranging conscription of nearly all men under middle age into mass armies, and the need for the whole adult population not in military service, including women, to be available for war-related industrial jobs. Thus warfare had become "total," involving the effort and risk of every aspect of the social structure. A third world war, if such a disaster occurs, may be too short to be "total" in terms of military and industrial mobilization. However, the risk to the whole population from NUCLEAR WEAPONS would obviously qualify it for classification as a total war.

totalitarianism A governing system in which an autocracy holds all power and controls all aspects of society. No opposition is allowed, and power is maintained by internal terror and secret police. Nazi Germany and Stalinist Russia are two examples of totalitarian states. See Waldemar Gurian, "The Totalitarian State," *Review of Politics,* Vol. 40, No. 4 (October 1978).

touch and go A practice landing and takeoff in a fixed-wing aircraft.

touchdown 1. The safe landing of an aircraft. 2. In AMPHIBIOUS OPERATIONS, the initial landing of the first element of assault forces on a hostile beach.

tour of duty 1. A daily work schedule; for example the day shift or the night shift. 2. The length of time for a prescribed duty; for example, two years with an American division in West Germany; one year in Vietnam. 3. The place of a military assignment; for example, the Pentagon.

tous azimuts The French military term used to indicate NUCLEAR WEAPONS pointed "in all directions." This doctrine was first pronounced in 1967, when the FORCE DE FRAPPE was first being deployed, and France was pulling out of the NORTH ATLANTIC TREATY ORGANIZATION's (NATO) integrated military organization. Obviously it has never been seriously contended that there is any power other than the Soviet Union that needs to be deterred from making a nuclear attack on France, but the doctrine is an important part of French nuclear declaratory policy. In effect it strengthens the classic French doctrine that the possession of nuclear weapons can only credibly be intended for self-protection, and must not in any way be restricted by alliance obligations. For the French, the purpose of possessing a nuclear capability is for the SANCTUARIZATION of their own country. The independence which they have retained in their targeting strategy implies that no country can rely absolutely on the French refraining from using (or, indeed, not using) its nuclear force in concert with the rest of NATO.

TOW Missile System The TOW (Tube-Launched, Optically Tracked, Wire Command-Link Guided) missile is the most powerful antitank weapon used by the United States infantry. It is found at battalion level in ground units and is also mounted on the BRADLEY FIGHTING VEHICLE, the HMMWV, and attack helicopters. When the missile is fired, a sensor in the launcher tracks a beacon in the tail of the missile. The gunner need only keep his crosshairs on the target. A computer in the launcher corrects any deviation of the missile from the crosshair aim points and sends corrections to the missile via two extremely thin wires that deploy in flight.

toxic alarm system Any system of alarm used to give warning of a chemical, biological, or radiological attack.

toxin agent A poison formed as a specific secretion product in the metabolism of a vegetable or animal organism, as distinguished from inorganic poisons. Such poisons can also be manufactured by synthetic processes.

tracer AMMUNITION whose flight is made visible by smoke or fire caused by a PYROTECHNIC mixture that is ignited upon firing; this helps trace the bullets so that the aim can be adjusted.

track 1. A series of related contacts displayed on a PLOTTING AND RELOCATING

BOARD. 2. To display or record the successive positions of a moving object. 3. To lock onto a point of RADIATION and obtain guidance therefrom. 4. To keep a GUN properly aimed, or to point a target-locating instrument continuously at a moving target. 5. The actual path of an aircraft above, or a ship on, the surface of the Earth. The course is the path that is planned; the track is the path that is actually taken. 6. One of the two endless belts on which a FULL-TRACK vehicle runs. 7. A metal part forming a path for a moving object, e.g., the track around the inside of a vehicle for moving a mounted machine gun (see SKATE MOUNT). 8. To follow or pursue an enemy. 9. Any vehicle that moves on tracks as opposed to wheels; for example, a tank or an armored personnel carrier.

track file Information stored in a computer memory and containing the position coordinates and velocity components of a target. In the context of the STRATEGIC DEFENSE INITIATIVE (SDI), it refers to such information concerning offensive weapons during their trajectories: e.g., BOOSTERS, RE-ENTRY VEHICLES, and DECOYS.

track off To deliberately underestimate or overestimate the predicted path of a TARGET, in order to bring about an eventual intersection of the LINE OF FIRE with the target.

track telling The path along which information flows to higher, lower or parallel levels of a command. This is the process of communicating AIR SURVEILLANCE and tactical data between COMMAND AND CONTROL systems or between facilities within such systems. Telling may be classified into the following types:

1. *Back tell*—The transfer of information from a higher to a lower echelon of command.
2. *Cross tell*—The transfer of information between facilities at the same operational level. Also called lateral tell.
3. *Forward tell*—The transfer of information to a higher level of command.
4. *Lateral tell*—See cross tell.
5. *Overlap tell*—The transfer of information, to an adjacent facility, about tracks detected in the adjacent facility's area of responsibility.
6. *Relateral tell*—The relay of information between facilities through the use of a third facility. This type of telling is appropriate between automated facilities in a DEGRADED communications environment.

tracked vehicle Any vehicle that travels on two or more endless tracks mounted on each side. A tracked vehicle has high mobility and maneuverability, is usually armed, and is frequently armored; it is intended for tactical use. TANKS are one example.

tracking The monitoring of the course of a moving TARGET. Objects that follow a BALLISTIC TRAJECTORY may have their tracks predicted by a defensive tracking system, using several observations and physical laws.

tracking, birth-to-death See BIRTH-TO-DEATH TRACKING.

traction capacity The ability of a soil to provide sufficient resistance to the tread or track of a vehicle to furnish necessary forward thrust.

tractor group The group of LANDING SHIPS in an AMPHIBIOUS OPERATION that carry the AMPHIBIOUS VEHICLES of the LANDING FORCE.

trail 1. A term applied to the manner in which a BOMB trails behind the aircraft from which it has been released, assuming the aircraft does not change its velocity after the release of the bomb. 2. To TRACK or shadow a person or object. 3. The part of a GUN CARRIAGE that rests on the ground after a PIECE has been unlimbered.

trail formation 1. A FORMATION in which all aircraft are in single file, each directly behind the other. 2. A formation in which vehicles proceed one behind the other at designated intervals. See also COLUMN FORMATION.

train 1. A SERVICE FORCE or group of service elements that provides LOGISTIC SUPPORT to combat units e.g., an organization of naval auxiliary ships or merchant ships or merchant ships attached to a fleet for this purpose; similarly, the vehicles and operating personnel that furnish supply, evacuation, and maintenance services to land units. 2. BOMBS dropped in short intervals or a short sequence. 3. A moving file of persons or vehicles. 4. A connected column of railroad cars. 5. A series of parts that together form a system producing a result, e.g., an IGNITER TRAIN.

train, ammunition See AMMUNITION TRAIN.

train, baggage See BAGGAGE TRAIN.

train, explosive See EXPLOSIVE TRAIN.

train, field See FIELD TRAIN.

train, igniter See IGNITER TRAIN.

training, advanced individual See ADVANCED INDIVIDUAL TRAINING.

training, advanced unit See ADVANCED UNIT TRAINING.

training, basic See BASIC TRAINING.

training, basic combat See BASIC COMBAT TRAINING.

training, basic military See BASIC MILITARY TRAINING.

training, basic unit See BASIC UNIT TRAINING.

training, combined See COMBINED TRAINING.

training, military See MILITARY TRAINING.

training, operational See OPERATIONAL TRAINING.

training, parallel See PARALLEL TRAINING.

training procedures, joint See JOINT TRAINING PROCEDURES.

training, tactical See TACTICAL TRAINING.

training, unit See UNIT TRAINING.

training company, military See MILITARY TRAINING COMPANY.

trains, combat See COMBAT TRAINS.

trajectory The flight path, or curve, of a projectile such as a spear, bullet, or rocket once it has been fired.

trajectory, ballistic See BALLISTIC TRAJECTORY.

trajectory, base of See BASE OF TRAJECTORY.

trajectory, standard See STANDARD TRAJECTORY.

trajectory chart A diagram of a side view of the paths of PROJECTILES fired at various ELEVATIONS under standard conditions. The trajectory chart is different for different GUNS, projectiles, and FUZES.

trajectory shift The degree to which the TRAJECTORY of a projectile, under the action of a thrust mechanism, departs from a purely BALLISTIC TRAJECTORY.

transattack period In NUCLEAR WARFARE, the period from the initiation of an attack to its termination.

transceiver A combined radio transmitter and receiver in which some circuits, other than those of the power supply, are common to both transmitter and receiver, and do not provide for simultaneous transmission and reception.

transfer area In an AMPHIBIOUS OPERATION, the water area in which the transfer of troops and supplies from LANDING CRAFT to AMPHIBIOUS VEHICLES is effected.

transient 1. Personnel, ships, or craft stopping temporarily at a post, station, or port to which they are not assigned or attached, and having a destination elsewhere. 2. An individual awaiting orders or transport at a post or station to which he is not attached or assigned.

transient target A moving TARGET that remains within observing or firing distance for such a short period that it affords little time for deliberate adjustment and FIRE against it. Transient targets may include aircraft, vehicles, ships, and marching troops. Usually called a FLEETING TARGET.

transmission, blind See BLIND TRANSMISSION.

transmission factor The ratio of the RADIATION DOSE inside a SHIELDING material to the outside (ambient) dose. The transmission factor is used to calculate the dose received through the shielding material.

transponder An electronic device that receives radio, radar, or sonar signals and automatically transmits them upon reception of a predetermined incoming signal.

transport area In AMPHIBIOUS OPERATIONS, an area assigned to a transport organization for the purpose of debarking troops and equipment.

transport capacity The capacity of a vehicle, as defined by the number of persons and the tonnage (or volume) of equipment that it can carry under given conditions.

transport lift, highway See HIGHWAY TRANSPORT LIFT.

transport network The complete system of routes pertaining to all means of transport available in a particular area. It is made up of the network particular to each means of transport.

transport stream A group of TRANSPORT VEHICLES proceeding in TRAIL FORMATION.

transport vehicle A motor vehicle designed and used without modification to the chassis, to provide general transport service in the movement of personnel and cargo.

transportation system A CODE system in which the plain text symbols are retained but are rearranged to form a cryptogram.

trans-shipment point A location where material is transferred between vehicles.

transverse Mercator projection A MAP projection that in mathematical principle is identical to the Mercator projection, except that the surface on which the meridians and parallels are developed is rotated (transversed) 90 percent in AZIMUTH. Unlike the Mercator projection, the meridians and parallels in a transverse Mercator projection are curved except for the equator and central meridian.

trap, booby See BOOBY TRAP.

trap mine A mine designed to explode unexpectedly when PERSONNEL attempt to move an object near it.

travel, angular See ANGULAR TRAVEL.

traveling overwatch See MOVEMENT TECHNIQUES.

traverse 1. To turn a WEAPON to the right or left on its MOUNT. 2. A method of surveying in which the lengths and directions of lines between points on the Earth are obtained by or from field measurements, and are used in determining the positions of the points.

traverse level That vertical displacement above low-level AIR DEFENSE systems, expressed both as a height and as an altitude, at which aircraft can cross a defended area.

tray, breechblock See BREECHBLOCK TRAY.

treason Violation of the allegiance owed to one's sovereign or state; betrayal of one's country. Treason, as defined by Article III, Section 3, of the U.S. Constitution, "shall consist only in levying War against them [the United States], or in adhering to their enemies, giving them aid and comfort. No person shall be convicted of treason unless on the testimony of two witnesses to the same overt act, or on confession in open court."

Treason is the only crime defined by the Constitution. The precise description of this offense reflects an awareness that persons holding unpopular views might be branded as traitors. Recent experience in other countries with prosecutions for conduct loosely labeled as "treason" confirms the wisdom of the authors of the Constitution in expressly stating what constitutes this crime and how it shall be proved. See Rebecca West, *The New Meaning of Treason* (New York: Viking, 1964); James Kirby Martin, "Benedict Arnold's Treason as Political Protest," *Parameters: Journal of the US Army War College* Vol. 11, No. 3 (September 1981).

treason, high What the modern world considers TREASON. It was "high" to distinguish it from "petit" or small treason, which was the killing of someone to whom one owed obedience, such as a husband or overlord. Today petit treason is treated in the same way as any other murder.

treaty A formal, international agreement negotiated between two or more sovereign states which establishes rights as well as obligations for each of the parties. In the United States, once treaties have been negotiated with foreign states they become, in effect, proposals by the executive branch of the government, and must be submitted to the Senate for approval by two-thirds of the senators present. After approval by the Senate, they are signed by the president. A ratified treaty binds the states as well as the federal government. Indeed, the Supreme Court held in *Missouri v. Holland* 252 U. S. 416 (1920) that a treaty may interfere with some of the rights reserved to the states by the Tenth Amendment to the Constitution.

trench A long ditch dug into the ground for protection from enemy FIRE.

trench burial A method of burial used when CASUALTIES are heavy. In trench burial, a trench is prepared and the individual remains of the dead are laid side by side in the trench, thus obviating the need for digging and filling in individual graves.

trench foot A disabling of the feet, similar to frostbite, caused by living in cold, wet TRENCHES for prolonged periods.

trench knife A dagger with a short, sharp point and a double-edged blade; it was often used during World War I for HAND-TO-HAND COMBAT in TRENCHES.

trench warfare The World War I tactic in which opposing armies faced each other for an extended period on a wide front, and dug into an elaborate system of TRENCHES. The area in between the trenches was designated "NO-MAN'S LAND." See Tony Ashworth, *Trench Warfare, 1914–1918: The Live and Let Live System* (New York: Holmes & Meier, 1980).

triad The three-part structure of nuclear strategic forces, consisting of the U.S. Air Force's bomber squadrons, the land-based INTERCONTINENTAL BALLISTIC MISSILE (ICBM) force, and the SUBMARINE-LAUNCHED BALLISTIC MISSILE (SLBM) fleet. Three of the five members of the nuclear club, the United States, Soviet Union and France deploy their forces using this triad, though with differing emphases. For example, the Soviet Union, has no truly intercontinental bomber aircraft, although it is sometimes argued that Soviet "Backfire" bombers (TU-22s) might have such range with enough mid-air refuelling. In terms of numbers of WARHEADS, the land-based ICBM is the mainstay of the Soviet triad, while for the United States it is the SLBM fleet, which carries 70 percent of all American strategic warheads. France faces very different problems, being so much smaller a nuclear power, and it might be argued that the land-based leg, and perhaps even the air force leg, of its triad are kept in being for political and symbolic, rather than strictly military, reasons. As originally conceived, the diversification of the triad is intended to insure the strongest possible defense by dispersing an enemy's attention; each leg is designed to be an independent retaliatory force. The least justifiable of the legs of the triad was, for a long time, the bomber force, because of its vulnerability both on the ground and while trying to penetrate the enemy's airspace. The recently developed air-launched CRUISE MISSILE, however, can be fired from a considerable distance outside the range of AIR-DEFENSE facilities. Potential developments in stealth technology (see STEALTH BOMBER) could make further additions to the viability of this element of the triad. See David R. Anderton, *Strategic Air Command: Two-Thirds of the Triad* (New York: Scribner, 1976); Colin S. Gray, "The Strategic Forces Triad: End of the Road?" *Foreign Affairs* (July 1978).

triage The evaluation and classification of casualties for purposes of treatment and evacuation. It consists of the immediate sorting of patients according to type and seriousness of injury and likelihood of survival, and the establishment of priority for treatment and evacuation to assure medical care of the greatest benefit to the largest number of wounded personnel. See also CASUALTY.

trial elevation In artillery, the ELEVATION at which FIRE FOR EFFECT is begun.

triangle exercise A form of RIFLE target practice in which the rifleman fires three shots, making three holes as close together in the target as possible.

triangle of velocities The fundamental triangle associated with DEAD RECKONING. It is composed of the following vectors: heading and true airspeed; track and groundspeed; wind speed and wind direction.

triangular division A DIVISION in which the combat elements are organized into three REGIMENTS, each of which reports directly to the division commander.

triangulation A method of surveying in which the locations of different terrain features are found by a system of triangles, each of whose base lines are established accurately as to location and length. As locations of new points are determined, new base lines are established, and the locations of other points are determined from them.

tribrach A universal mounting device for all engineering surveying instruments and accessories that are mounted on tripods and have a standard base.

trick 1. A graduation mark on a lens or recticle. 2. A tour of duty or period during which one is on duty. Also called a WATCH.

Trident The most recent development in American submarine-launched ballistic missiles (SLBMs), which comes in two forms: The first, known either as Trident I or as Trident C4, is already deployed. This missile was a considerable advance over the previous, second generation SLBM, the POSEIDON, and enormously more powerful than the old POLARIS missiles. However, it differs only in having greater range and a greater throw-weight, reportedly carrying 10 warheads with 10 MULTIPLE INDEPENDENTLY TARGETABLE RE-

ENTRY VEHICLES (MIRVS), with each vehicle having a nuclear force of about 100 kilotons. Despite this greater power, the Trident is much the same size as the Poseidon, and can be carried in the old Poseidon submarines. Trident II, or D5, is really more like a fourth generation of SLBM than a revised version of Trident I. It is much bigger, requiring a new and much larger model of submarine likewise named the Trident class. Its greater range increases the sea area in which its submarines can patrol. Trident II also has the capacity to carry up to 14 MIRVs in the missile front end, each possibly of as much as 400 kilotons. The real significance of Trident II is that it is the first submarine-launched missile that will have the accuracy to strike HARD TARGETS, because it is expected to have a CIRCULAR ERROR PROBABLE of less than 100 meters. See D. Douglas Dalgleish and Larry Schweikart, *Trident* (Carbondale, IL: Southern Illinois University Press, 1984); Jeffrey A. Merkley, *Trident II Missiles: Capability, Costs and Alternatives* (Washington, D.C.: Supt. of Docs., Government Printing Office, 1986).

trigger 1. The small projecting part of a firearm designed to be pressed by a finger to actuate the discharge mechanism. 2. Any stimulus that starts a reaction.

trigger, hair See HAIR TRIGGER.

trigger thesis A justification advanced to defend small strategic forces, such as those of France and the United Kingdom, against the charge that they were so insignificant in comparison with the Soviet armory that they could not have credibility as deterrents. The argument was that these small forces got their power not from what they could do themselves, but by their ability to "trigger" a supporting strike by the United States. It was claimed that the United States might not actually wish to use its NUCLEAR WEAPONS in defense of an ally, and that the NUCLEAR UMBRELLA might not hold. But if the French or British, in their own defense, made a strike on the Soviet Union, the whole world would be plunged into an uncontrollable process of ESCALATION. The Soviet Union might believe it was under attack from the United States. Alternatively, the United States might think that the risk of the Soviet Union attacking it was so great that it would have to launch a pre-emptive attack against the Soviet Union anyway. The trigger thesis was probably never actually held by any government as real policy, but it made some sort of macabre sense in the early days of nuclear strategy. Now that Western strategies are based so firmly on the hope of being able to exercise very precise control over escalation processes, there is even less room for the theory.

tripwire 1. A military force situated on a DEFENSE line that is not expected to be able to hold off a major enemy ASSAULT, but whose function is to buy time so that RESERVES can be brought into the battle or decisions can be made to use TACTICAL or strategic NUCLEAR WEAPONS. 2. The United States forces in the NORTH ATLANTIC TREATY ORGANIZATION (NATO) and Korea. The tripwire thesis was a way of describing the role of conventional troops on NATO's CENTRAL FRONT during the 1950s and early 1960s, when NATO was committed to a strategy of MASSIVE RETALIATION. According to this thesis, any serious incursion by the Soviet Union would have been met not by a prolonged defensive war, but by immediate nuclear retaliation. Because there was a problem of credibility with this theory, some symbol of intent was required. It was argued that a relatively thin screen of troops would serve this purpose. They would put up some real resistance, so that the WARSAW PACT would have to make its intentions and determination quite clear by defeating them. This would signal the appropriate moment to commence nuclear war, and it was hoped that the Warsaw Pact would realize this. At the same time, the presence of the troops, and even more so their dependents, would leave the president of the United States no alternative but to activate the NUCLEAR UMBRELLA to protect American citizens in Europe.

troop 1. A subordinate unit of a CAVALRY SQUADRON. The troop has both administrative and tactical functions; it is equivalent to a COMPANY or BATTERY. 2. To march or walk by or to a place. 3. To ceremonially carry a flag before assembled soldiers.

troop test A test conducted in the field for the purpose of evaluating operational or organizational concepts, doctrine, tactics, and techniques, or to gain further information on MATERIEL. See also SERVICE TEST.

trooper 1. A mounted (on horseback) soldier. 2. A member of an ARMORED CAVALRY force. 3. A member of a state police force.

troops A collective term for uniformed military personnel (usually not applicable to naval personnel afloat).

troops, combat See COMBAT TROOPS.

troops, corps See CORPS TROOPS.

troops, elite See ELITE TROOPS.

troops, expeditionary See EXPEDITIONARY TROOPS.

troops, ordnance See ORDNANCE TROOPS.

troops, service See SERVICE TROOPS.

troops, shock See SHOCK TROOPS.

troops, special See SPECIAL TROOPS.

troops, tactical See TACTICAL TROOPS.

troop the line To ceremonially review assembled soldiers. This is what a head of state might be invited to do during a formal visit to another state.

trophy of war Any item of captured enemy equipment, the retention of which by its captors is not prohibited by INTERNATIONAL LAW, national law, or current army regulations.

truce 1. An ARMISTICE. 2. A suspension of hostilities for a specific limited period by the warring parties; for example, a three-hour truce to remove wounded from the battlefield. See FLAG OF TRUCE.

true altitude The height of an aircraft as measured from mean sea level.

true azimuth An AZIMUTH referenced to true north or true south.

true bearing The direction to an object from a point, expressed as a horizontal angle measured clockwise from true north.

true control A navigation or mapping system of a common control originating from a point whose true coordinates are known and from which a true direction is known.

true course The course of an aircraft, tank, or ship as indicated by the horizontal angle between the true north–south line and the direction of motion that also takes into account the wind and other variables.

true north The direction from an observer's position to the geographic North Pole. The direction of any geographic meridian.

Truman Doctrine The policy of the Truman Administration of the United States's giving military and economic aid to those countries (Greece and Turkey specifically) seeking to resist "totalitarian aggression." This doctrine, which was first presented by President Truman in 1947 in his address to a joint session of Congress in support of the Greek-Turkish aid bill, became a cornerstone of the American policy of containment of communism. It is generally considered the more or less formal acceptance by the

United States of its self-imposed responsibility to protect the "free world" from Communist or other totalitarian takeover. Compare to the NIXON DOCTRINE. See Joseph C. Satterthwaite, "The Truman Doctrine: Turkey," *Annals of the American Academy of Political and Social Science* Vol. 401 (May 1972); John Lewis Gaddis, "Was the Truman Doctrine a Real Turning Point?" *Foreign Affairs* Vol. 52, No. 2 (January 1974).

tube artillery Traditional GUNS or MORTARS as opposed to ROCKETS and GUIDED MISSILES.

tump line A kind of sling formed by a strap slung over the forehead or chest and used by a person carrying a pack on his or her back.

Turks, young See YOUNG TURKS.

turnaround The length of time between arrival at a point and readiness to depart from that point. It is used in this sense for the loading, unloading, refueling, and rearming of military vehicles, aircraft, and ships.

turning error, northerly See NORTHERLY TURNING ERROR.

turning movement A variation of the ENVELOPMENT in which the attacking force passes around or over the enemy's principal defensive positions to secure OBJECTIVES deep in the enemy's rear, in order to force the enemy to abandon his position or divert major forces to meet the threat. See DOUBLE ENVELOPMENT, SINGLE ENVELOPMENT.

turret 1. A tower that is a part of another larger structure. 2. A domelike structure, usually armored and able to revolve horizontally, in which guns are placed. Thus the revolving top of a tank is a turret. 3. A tall wheeled structure used to breach or scale the walls of a fort.

turret, power See POWER TURRET.

TVD An acronym that refers to the main organizing units of Soviet military posture, and is usually translated into English as standing for "Theatre of Military Operations." It highlights a difference between NORTH ATLANTIC TREATY ORGANIZATION (NATO) and WARSAW PACT doctrine, because Soviet doctrine demands a much more centralized control at a high level—the TVD headquarters—than does NATO's more diffuse command structure. The whole of the CENTRAL FRONT will be one Soviet TVD.

twenty-five percent rectangle In gunnery, a rectangle, eight probable DEFLECTION ERRORS wide and one probable error deep, within which will fall 25 percent of a large number of shots fired from guns with the same setting. One range limit of this rectangle is the CENTER OF IMPACT of a large number of shots fired with the same setting. There are two 25 percent rectangles, one on either side of the center of impact. These two rectangles form the 50 percent rectangle.

two-man rule A system designed to prohibit access by an individual to NUCLEAR WEAPONS and certain designated components of such weapons by requiring the presence at all times of at least two authorized persons, each capable of detecting incorrect or unauthorized procedures with respect to the task to be performed. Also referred to as the two-man concept or two-man policy.

two-up A FORMATION with two ELEMENTS disposed abreast and the remaining element(s) in the rear.

U

umbilical cord A cable fitted with a quick-disconnect plug on a MISSILE, through which the missile is controlled and tested while still attached to launching equipment or a parent aircraft.

unaccompanied A description of military personnel sent to a new DUTY location without dependent family members.

uncommitted force A force that is not in contact with an enemy force and is not already deployed on a specific MISSION or COURSE OF ACTION. Uncommitted forces or least committed forces will normally constitute the RESERVE when no formal reserve has been constituted.

unconditional surrender Putting a defeated military force, a military installation, or an entire nation into the hands of an enemy without terms of any kind. The phrase first gained wide currency during the American Civil War, when Union General Ulysses S. Grant responded to the Confederate offer to surrender on terms during the 1862 battle for Fort Donelson in Tennessee by saying: "No terms except an unconditional and immediate surrender can be accepted." During World War II, unconditional surrender of the Axis powers was the immediate war aim of the Allies.

unconventional mine A MINE that is fabricated at or near its point of use from other explosives (e.g., a BOMB, artillery SHELL, TNT blocks).

unconventional warfare A broad spectrum of military and paramilitary operations conducted in enemy-held, enemy-controlled, or politically sensitive territory. Unconventional warfare includes, but is not limited to, the interrelated fields of GUERRILLA WARFARE, EVASION AND ESCAPE, SUBVERSION, SABOTAGE, and other operations of a low-visibility, covert or clandestine nature. See also LOW INTENSITY WARFARE.

uncover 1. To remove one's hat, cap, or helmet. 2. To move certain designated soldiers of a formation to either side in order to get more space between individuals. 3. The command to move in this manner. 4. To expose or leave unprotected by MOVEMENT or MANEUVER.

uncovered movement A MOVEMENT made when security normally provided by friendly forces is lacking.

under arms Bearing ARMS, especially with arms in hand; or with a SMALL ARMS weapon, or with equipment such as a holster attached to the person.

underground A covert, organization for UNCONVENTIONAL WARFARE, established to operate in areas denied to GUERRILLA forces or conduct operations not suitable for guerrilla forces. See Henry C. Hart, "US Employment of Underground Forces," *Military Review* 26, 3 (March 1947).

underwater demolition team A group of officers and men specially trained and equipped for making hydrographic reconnaissance of approaches to prospective landing beaches; for effecting the demolition of OBSTACLES; for locating, and marking of usable channels; etc. See also SEA-AIR-LAND TEAM.

undesirable discharge A form of DISCHARGE given an ENLISTED PERSON under conditions other than honorable; it may be issued for unfitness, misconduct, homosexuality, or for SECURITY reasons. It is worse than a general discharge and not as bad as a bad conduct discharge.

unexploded explosive ordnance Explosive ORDNANCE that has been primed, fuzed, armed, or otherwise prepared for action, and which has been fired, dropped, launched, projected, or placed in such a manner as to constitute a hazard either by malfunction or for any other cause.

unified command A COMMAND with a broad continuing MISSION under a single commander and composed of significant components of two or more SERVICES.

unified logistic support LOGISTIC SUPPORT provided to two or more of the military SERVICES, or their ELEMENTS, by a single agency or service, by joint, common, or cross-servicing, or by any other appropriate method.

uniform Standard prescribed dress for members of military units. According to Major Mark M. Boatner III, *Military Customs and Traditions* (1956), "The first purpose of a military uniform is, of course, to distinguish between friend and foe. A second consideration, in the early days, was strictly one of economics: in foreign countries the colonel was responsible for clothing his regiment and it was cheaper to buy the material in bulk lots."

uniform, dress See DRESS UNIFORM.

Uniform Code of Military Justice The code of laws governing the conduct of all persons in the ARMED FORCES of the United States.

unilateral arms control measure A COURSE OF ACTION taken by a nation for ARMS CONTROL, without any compensating concession being required of other nations.

unit 1. Any military ELEMENT whose structure is formally prescribed by higher authority; specifically, part of a larger organization. 2. An organization title of a subdivision of a group in a TASK FORCE. 3. A standard or basic quantity into which an item of SUPPLY is divided, issued, or used. In this meaning, also called a UNIT OF ISSUE.

unit, administrative See ADMINISTRATIVE UNIT.

unit, beachmaster See BEACHMASTER UNIT.

unit, cellular See CELLULAR UNIT.

unit cohesion The result of controlled, interactive forces that lead to solidarity within military UNITS, directing the soldiers toward common goals with an express commitment to one another and to the unit as a whole. See William Darryl Henderson, *Cohesion: The Human Element in Combat* (Washington, D.C.: National Defense University Press, 1985); Roger Kaplan, "Army Unit Cohesion in Vietnam: A Bum Rap," *Parameters: U.S. Army War College Quarterly* 17, 3 (September 1987).

unit distribution A method of distributing SUPPLIES by which the receiving unit is issued supplies in its own area through transportation furnished by the issuing agency.

unit emplaning officer In air transport, a representative of a transported UNIT responsible for organizing the movement of that unit.

unit in contact A FORCE that is engaged with an enemy force and is normally not available for commitment elsewhere. However, the least committed unit in contact with an enemy may be designated the RESERVE.

unit integrity 1. A characteristic of a military unit that is complete, that has retained all of its normal component elements. 2. The personal or military honor of an entire unit; for example, something might be done for the "honor" of the regiment.

unit journal A logbook or chronological record of events kept by a UNIT or staff section.

unit loading The loading of troop units with their equipment and supplies into vessels, aircraft, or land vehicles designated only for those units, equipment, and supplies.

unit of issue The quantity of an item: a dozen, gallon, pair, pound, ream, set, or yard. Usually termed a "UNIT OF ISSUE" to distinguish it from a "unit of price."

unit pack A single item or a set quantity of a group of items that has a single stock number.

unit pathfinders Selected personnel from ground UNITS or INSTALLATIONS, who are trained in PATHFINDER techniques.

unit price The cost or price of an item of supply based on the UNIT OF ISSUE.

unit reserves Prescribed quantities of SUPPLIES carried by a military UNIT as a reserve to cover emergencies.

unit status report An evaluation of the combat readiness status of a military organization; a measure of the unit's ability to perform an assigned mission.

unit strength A measure of the number of personnel and amount of supplies, armaments, equipment, and vehicles, as well as the overall logistic capabilities, of a friendly or enemy unit.

unit train 1. All transportation and COMBAT SERVICE SUPPORT operating under the immediate orders of UNIT commander, and primarily concerned with supply, evacuation, and maintenance. 2. A unit train below the DIVISION level. In this meaning, armored division unit trains are classified as COMBAT TRAINS (or A trains) and FIELD TRAINS (or B trains), depending on whether or not they are required for the immediate support of combat elements.

unit training A phase of military training in which emphasis is placed upon training individuals to function as members of a team unit. This training, which usually follows individual training (basic, technical, or specialist), is usually conducted in the field under conditions which the unit would be likely to encounter in combat.

unit training, advanced See ADVANCED UNIT TRAINING.

United Nations (UN) The term "United Nations" was devised by President of the United States Franklin D. Roosevelt, and was first used in the Declaration of United Nations on January 1, 1942. The UN Charter, drawn up at a conference in San Francisco, was signed on June 25, 1945 by 50 nations. The United Nations, an international organization devoted to the peaceful resolution of international conflicts, formally came into existence on October 24, 1945, when a majority of the signatory nations had ratified the Charter. By the 1980s it had three times as many members as when it started. The United Nations' business is conducted primarily through its GENERAL ASSEMBLY and SECURITY COUNCIL. See John F. Murphy, *The United Nations and the Control of International Violence: A Legal and Political Analysis* (Totowa, NJ: Allanheld, Osmun Publishers, 1982).

United States Air Force Academy (USAFA) Established in 1955 in Denver; then moved to Colorado Springs, Colorado, this is the United States's educational institution for training new Air Force officers. After four years of instruction a cadet normally earns a bachelor's degree and a COMMISSION as a SECOND LIEUTENANT in the regular U. S. Air Force.

United States Military Academy (USMA) Established in 1802 at West Point, New York, this is the United States's educational institution for training new Army officers. After four years of instruction, cadets normally earn a bachelor's degree and a COMMISSION as a second LIEUTENANT in the regular U. S. Army.

United States Naval Academy (USNA) Established in 1845 at Annapolis, Maryland, this is the United States's educational institution for training new naval and marine officers. After four years of instruction, MIDSHIPMEN normally earn a bachelor's degree and a COMMISSION as an ENSIGN in the regular U.S. Navy or as a SECOND LIEUTENANT in the regular Marine Corps.

unitized load A single item or a number of items packaged or arranged in a manner capable of being handled as a unit. Unitization may be accomplished by placing items in a container or by banding them together.

unity of command 1. The concept that each individual in an organization should be accountable to only a single superior. 2. The concept that a military unit be led by a single commander. This has long been known. In 1513, Machiavelli wrote "It is better to confide any expedition to a single man of ordinary ability, rather than to two, even though they are men of the highest merit, and both having equal ability." NAPOLEON would later agree: "Nothing is more important in war than unity in command. When, therefore, you are carrying on hostilities against a single power only, you should have but one army acting on one line and led by one commander." It was General George C. Marshall who insisted on unity of command in Europe (under Eisenhower) during World War II.

unobserved fire FIRE for which the POINTS OF IMPACT or burst are not observed.

unthickened fuel A blend of gasoline and light fuel oils or lubricating oils used as an incendiary fuel in portable and small mechanized FLAME THROWERS. See also THICKENED FUEL.

up 1. A term used in a CALL FOR FIRE to indicate that the target is higher in altitude than the point that has been used as a reference point for the target location. 2. A correction used by an artillery or naval gunfire observer to indicate that an increase in HEIGHT OF BURST is desired.

up from the ranks Descriptive of an officer who started military service as an enlisted rank. Such an officer may also be called a mustang.

urgent priority A category of IMMEDIATE MISSION REQUEST that is lower than emergency priority but takes precedence over ordinary priority, e.g., enemy artillery fire that is falling on friendly troops, or enemy troops moving up in such force as to threaten a BREAKTHROUGH.

usable rate of fire The normal rate of FIRE of a gun in actual use, measured in units of shots per minute. The usable rate of fire is considerably less than a gun's maximum rate of fire, which is a theoretical value based on the purely mechanical operation of a WEAPON.

USAFA See UNITED STATES AIR FORCE ACADEMY.

use them or lose them An expression that summarizes the predicament in which the NORTH ATLANTIC TREATY ORGANIZATION (NATO) could very well find itself in with regard to its short-range tactical or BATTLEFIELD NUCLEAR WEAPONS in the event of a CENTRAL FRONT war. The problem is that a very large number of these low-YIELD battlefield weapons, particularly the thousands of nuclear artillery shells but also some missiles, have very short ranges. They have to be kept very near the front line if they are to be used at all. Yet a successful attack by the WARSAW PACT, particularly if achieved with surprise, could penetrate this far into NATO territory in a matter of hours, and certainly within a couple of days. Consequently there is a strong risk of these weapons being overrun and captured unless they are used very early in the conflict. Although NATO is committed to the first use of nuclear weapons if necessary, it naturally does not want to be forced into their use prematurely. This has led to the suggestion that these weapons, several thousands of WARHEADS for which are stored in Europe, can serve no valid military purpose at all.

USMA See UNITED STATES MILITARY ACADEMY.

USMC United States Marine Corps (see MARINES).

USNA See UNITED STATES NAVAL ACADEMY.

USO United Service Organizations, which provides social clubs and entertainment for American soldiers. Through the USO, movie and TV stars (such as Bob Hope) often perform for American forces at locations around the world.

Uzi An Israeli manufactured SUBMACHINE GUN, best known for its small size and light weight, compared to traditional submachine guns.

V

vane 1. A small propeller that rotates and arms the FUZE of a bomb when the bomb is dropped. 2. A lengthwise partition in a CHEMICAL PROJECTILE that makes the contained liquid rotate with the casing. This is necessary for accuracy in flight.

vanguard 1. Those forces that are closest to confronting the enemy. 2. The name of a 1960s era American missile.

VC 1. Viet Cong. 2. Victoria Cross, the highest British Military DECORATION. 3. Veterinary Corps.

V-Day The day of victory. V-E (Victory: Europe) Day was May 8, 1945. V-J (Victory: Japan) Day was August 15, 1945. These dates saw the end of World War II in Europe and the Pacific.

vector 1. A course or compass heading especially issued to an aircraft to provide navigational guidance. 2. A guide to a specified place or area. 3. The carrier of a biological agent. In this context "vector control" is considered with unhealthy numbers of insects and rats in a particular area. 4. A line indicating direction and magnitude. 5. To change the course of a rocket or jet aircraft.

vector gunsight A GUNSIGHT that computes the course heading required for a PROJECTILE to strike its target.

vectored attack An attack in which a weapon carrier (air, surface, or subsurface) that is not holding contact on its target is vectored to the weapon delivery point by a unit (air, surface or subsurface) which holds contact on the target.

Vegetius, Renatus The fourth century A. D. Roman writer whose description of the operation of the Roman Army, *De Re Militari* (378) remains the most detailed and influential source on its workings.

vehicle, amphibious See AMPHIBIOUS VEHICLE.

vehicle, armored See ARMORED VEHICLE.

vehicle, armored fighting See ARMORED FIGHTING VEHICLE.

vehicle, combat See COMBAT VEHICLE.

vehicle, half-track See HALF-TRACK VEHICLE.

vehicle, military-designed See MILITARY-DESIGNED VEHICLE.

vehicle, tracked See TRACKED VEHICLE.

vehicle distance The clearance between vehicles in a COLUMN; it is measured from the rear of one vehicle to the front of the following vehicle.

vehicles, delivery See DELIVERY VEHICLES.

velocity, angular See ANGULAR VELOCITY.

velocity, high See HIGH VELOCITY.

velocity jump The angle between the GUNBORE LINE (or launch line) of a projectile and the actual LINE OF DEPARTURE. Also called the angle of jump.

velocity, low See LOW VELOCITY.

velocity, projectile See PROJECTILE VELOCITY.

velocity, terminal See TERMINAL VELOCITY.

verification The essence of an ARMS-CONTROL agreement. It is a two-step process involving: 1. an assessment before a treaty of whether the other side could violate the treaty and evade detection; and 2. the continuous monitoring of compliance after a treaty is signed and ratified. Verification can either be carried out by scientific instruments such as satellites, usually called NATIONAL TECHNICAL MEANS, or by on-site inspection by observers from the countries involved or from neutral states. The INF Treaty of 1988 created the first major instance of on-site inspections when it provided that Soviet weapons experts would be able to inspect specified weapons facilities in the United States and American experts would be able to inspect parallel facilities in the Soviet Union. See Stephen M. Meyer, "Verification and Risk in Arms Control," *International Security* Vol. 8, No. 4 (Spring 1984); Richard A. Scribner and Kenneth N. Luongo, *Strategic Nuclear Arms Control Verification: Terms and Concepts* (Washington D.C.: The American Association for the Advancement of Science, 1985); David Hafemeister, "Advances in Verification Technology," *Bulletin of the Atomic Scientists* 41, 1 (January 1985).

verify 1. To ensure that the meaning and phraseology of a transmitted message conveys the exact intention of the originator. 2. A request from an observer or FIRE-CONTROL agency to re-examine FIRING DATA and report the results.

vertical and/or short takeoff and landing See V/STOL.

vertical envelopment A tactical MANEUVER in which troops, either air-dropped or air-landed, attack the rear and FLANKS of a force, in effect cutting off or encircling the force. Compare to ENVELOPMENT, DOUBLE ENVELOPMENT, SINGLE ENVELOPMENT, TURNING MOVEMENT.

vertical interval The difference in altitude between two specified locations, e.g., a BATTERY and its TARGET; an OBSERVER location and a target; location of a target previously fired on and a new target.

vertical loading A type of loading whereby items of like character are vertically tiered throughout the holds of a ship, so that selected items are available at any stage of the unloading operation.

vertical replenishment The use of a helicopter for the transfer of MATERIEL to or from a ship.

very-long-range radar 1. RADAR equipment whose maximum range on a reflecting target of 1 square meter normal to the signal path exceeds 965 kilometers, provided a line of sight exists between the target and the radar.

Very (or verry) light A colored signal flare from a special pistol; also called a Very signal light. Named for its inventor Edward W. Very (1847–1907) of the United States Navy.

Very pistol A special pistol used to fire colored signal FLARES.

very-short-range radar 1. RADAR equipment whose range on a reflecting target of 1 square meter normal to the signal path is less than 50 miles, provided a line of sight exists between the target and the radar.

veteran 1. An experienced soldier. 2. A former soldier.

Veterans Administration (VA) The federal agency, created in 1930, that administers benefits for veterans and their dependents. These benefits include compensation payments for disabilities or death related to military service; pensions for totally disabled veterans; education and rehabilitation; home loan guaranty; burial, including cemeteries, markers, flags; and a comprehensive medical program involving a widespread system of nursing homes, clinics and more than 170 medical centers. In 1988 legislation was passed by the Congress and signed by President Ronald Reagan making the VA a cabinet-level agency.

veterans' benefits Any government advantages available to those who served in the armed forces of the United States that are not available to citizens who did not serve. Veterans' benefits may include government-supplied health care, advantageous home mortgage terms, and pensions.

Veterans of Foreign Wars A membership organization of American veterans who served overseas.

veterans' preference The concept that dates from 1865, when the Congress, toward the end of the Civil War, affirmed that "persons honorably discharged from the military or naval service by reason of disability resulting from wounds or sickness incurred in the line of duty, shall be preferred for appointments to civil offices, provided they are found to possess the business capacity necessary for the proper discharge of the duties of such offices." The 1865 law was superseded in 1919, when preference was extended to all "honorably discharged" veterans, their widows, and wives of disabled veterans. The Veterans' Preference Act of 1944 expanded the scope of veterans' preference by providing for a five-point bonus on federal examination scores for all honorably separated veterans (except for those with a service-connected disability, who are entitled to a ten-point bonus). Veterans also received other advantages in federal employment (such as protections against arbitrary dismissal and preference in the event of a reduction in force).

All states and many other jurisdictions have veterans' preference laws of varying intensity. New Jersey, an extreme example, offers veterans absolute preference; if a veteran passes an entrance examination, he or she must be hired (no matter what the score) before nonveterans can be hired. Veterans competing with each other are rank-ordered, and all disabled veterans receive preference over other veterans. See Charles E. Davis, "Veterans' Preference and Civil Service Employment: Issues and Policy Implications," *Review of Public Personnel Administration* 3 (Fall 1982); Gregory B. Lewis and Mark A. Emmert, "Who Pays for Veterans' Preference?" *Administration and Society* 16 (November 1984).

victory 1. Defeating an enemy unequivocally in a battle or war. Victory, whether it is limited or total, is the aim of war—the reason for the creation and use of armies. One of the most famous descriptions of victory comes from a Robert Southey (1774–1843) poem, *The Battle of Blenheim* (1798):

> They say it was a shocking sight
> After the field was won;
> For many thousand bodies here
> Lay rotting in the sun;
> But things like that, you know, must be
> After a famous victory.

Victory Medals Medals awarded to all PERSONNEL who served in the U.S. Armed Forces during World War I or II.

Vietnam War The 1956 to 1975 war between the non-Communist Republic of Vietnam (South Vietnam) and the Communist Democratic Republic of Vietnam (North Vietnam), which resulted in the victory of the north over the south and the unification of the two countries into the Communist Socialist Republic of Vietnam on July 2, 1976. The United States first offered major military assistance to South Vietnam during the Kennedy Administration in 1961. By 1963 the United States had 16,000 military "advisors" in South Vietnam. In 1964 the GULF OF TONKIN RESOLUTION allowed the administration of President Lyndon Johnson to expand American involvement in the war. By 1968 the United States had over half a million men engaged in the most unpopular foreign war in American history. As a direct result the Democrats lost control of the White House to Republican Richard M. Nixon. The Nixon Administration's policy of "Vietnamization" called for the South Vietnamese to gradually take over all the fighting from the Americans. The Americans continued to pull out and the south held off the north for a while. As the American forces dwindled, the north got more aggressive and successful. Finally, the north's January 1975 offensive led to the south's unconditional surrender by April. More than 58,000 Americans died in the Vietnam War; another 150,000 were wounded. See Frances Fitzgerald, *Fire in the Lake: The Vietnamese and the Americans in Vietnam* (Boston: Atlantic-Little, Brown, 1972); Stanley Karnow, *Vietnam: A History* (New York: Viking Press, 1983); Harry G. Summers, Jr., *On Strategy: A Critical Analysis of the Vietnam War,* (Novato, CA: Presidio Press, 1984).

visibility 1. A measure of how far one can see in given weather conditions. See also INTERVISIBILITY. 2. The sometimes dangerous tactic of calling attention to oneself in an effort to advance one's reputation or career. This can be achieved by bravery on the field of battle, or bureaucratic bravery by writing a memo or publishing an article critical of some policy or program.

visibility chart A map or photograph showing which areas can be seen, and which cannot be seen, from a given observation point.

visible path segment A portion of the path a moving target takes, over which visibility of the target is continuous to a particular SENSOR.

visible time segment The length of time a TARGET is on a VISIBLE PATH SEGMENT.

vision slit Any narrow opening or slit in ARMOR through which to look, especially in a TANK or other ARMORED VEHICLE.

visit of courtesy A formal visit paid by one OFFICER to another in conformity with military customs.

visual elevation The distance above a TARGET at which the white TRACER streak from a MACHINE GUN must appear in order to allow for the drop in the TRAJECTORY of the bullets.

visual flight A flight in which the FLIGHT PATH and altitude of an aircraft are controlled by visual reference to the ground, water, or clouds; commonly called VFR for visual flight rules, as opposed to IFR for instrument flight rules.

vocoder A type of voice coder, consisting of a speech analyzer and a speech synthesizer, used to reduce the bandwidth requirement of speech signals. The analyzer circuitry converts the incoming speech into digital signals. A synthesizer then converts the encoded, digital output into artificial speech. For communications SECURITY purposes, a vocoder may be used in conjunction with a key generator and a modulator-demodulator device to transmit digitally encrypted speech signals over normal, narrow-band voice communications channels.

Voice of America The U.S. Information Agency's radio broadcasting service, which transmits American news, public affairs, cultural, and musical programs to overseas audiences in 41 different languages.

volley 1. A method of ARTILLERY firing in which each PIECE fires the specified number of rounds without any attempt to synchronize with other pieces. 2. A BURST of fire, especially a SALUTE fired by a detachment of riflemen.

volley bombing The simultaneous or nearly simultaneous release of a number of BOMBS.

V-I, V-II See V WEAPONS.

V-ring The inner circle of the bull's-eye on a target, used to decide a tie score in a rifle match without changing the total score. A hit within the V-ring is designated as a "V fire."

V series A group of toxic CHEMICAL AGENTS first developed in the 1950s that are generally colorless and odorless. In liquid or aerosol form, they affect the body in a manner similar to that of other NERVE AGENTS but were five times more lethal than the G SERIES.

V/STOL 1. Vertical and/or short takeoff and landing; a term designating an aircraft that can take off or land almost vertically, using a very short runway. The British Harrier Jumpjet is an example. Because it is likely that traditional airfield runways would be early targets for enemy weapons in wartime, V/STOL aircraft, which can use easily improvised short runways may have an increasing role as a second strike weapon. 2. A type of aircraft that provides navies lacking conventional aircraft carriers with seaborne fighter support. See Peter P. W. Taylor, "The Impact of V/STOL on Tactical Air Warfare," *Air University Review* 29, 1 (November–December 1977); William D. Siuru, Jr., "V/STOLS—A Myth of a Promise," *Air University Review* 33, 3 (March–April 1982).

vulnerability 1. The susceptibility of a nation or military force to any action by any means through which its war potential or combat effectiveness may be reduced or its will to fight diminished. 2. The characteristics of a system that makes it susceptible to suffer damage or degradation (incapability to perform the designated mission) as a result of having been subjected to a hostile environment or attack. Vulnerability principally refers to the extent to which a MISSILE system is at risk of destruction in a nuclear preemptive strike. It has no specific technical definition, but the vulnerability of a missile BATTERY can be seen as a function of its detectability, the ease with which it can be destroyed by a nuclear explosion of any given yield and accuracy (see BIAS and CIRCULAR ERROR PROBABLE), and the speed with which it can be launched. There are really three approaches to reducing vulnerability. One is to hide a potentially targetable weapons system by deploying it underwater in submarines. BALLISTIC MISSILE submarines are generally regarded as the least vulnerable nuclear delivering systems. (See SUBMARINE LAUNCHED BALLISTIC MISSILE). The traditional alternative has been to reduce vulnerability by making missiles exceptionally HARD TARGETS by putting them into superhardened SILOS capable of withstanding immense BLAST pressures. However, improvements in missile accuracy have made even these silos highly vulnerable. Consequently, new plans are now aimed at reducing vulnerability by making missile launchers mobile. Both the American MIDGETMAN plans and the newest Soviet missile project, the SS-25, rely on mobility rather than silo deployment. See John D. Steinbruner and Thomas M. Garwin, "Strategic Vulnerability: The Balance Between Prudence and Paranoia," *International Security* 1, 1 (Summer 1976); Les AuCoin, "Nailing Shut the Window of Vulnerability," *Arms Control Today* (September 1984); Michael Nacht, *The Age of Vulnerability: Threats to Nuclear Stalemate* (Washington, D.C.: The Brookings Institution, 1985).

vulnerability, strategic See STRATEGIC VULNERABILITY.

vulnerability assessment, nuclear See NUCLEAR VULNERABILITY ASSESSMENT.

vulnerability program A program to determine the degree of, and to remedy insofar as possible, any existing susceptibility of NUCLEAR WEAPON systems to enemy countermeasures, accidental fire, and accidental shock.

vulnerability study An analysis of the capabilities and limitations of a FORCE, in a specific situation, to determine vulnerabilities capable of exploitation by an opposing force.

V weapons The vengeance ROCKET weapons developed by the Germans toward the end of World War II. The V-1, or flying bomb, was sent against the civilian population of London. The V-2, also used against London, was the first effective MISSILE of the modern era. The V-weapons' developers formed the nucleus of both the American and Russian INTERCONTINENTAL BALLISTIC MISSILE (ICBM) programs in the postwar era. See Walter Dornberger, *V-2: The Nazi Rocket Weapon.* Translated by James Cleugh and Geoffrey Halliday (New York: Viking, 1954); Basil Collier, *The Battle of the V-Weapons, 1944–1945* (New York: Morrow, 1965).

W

WAC Woman's Army Corps. Created during World War II, it ceased to exist as a separate, all-female organization in 1978, when its personnel were integrated with the male personnel of the U.S. Army

war 1. A state of violent, armed conflict between organized political entities in their efforts to gain, defend, or add to their national or political sovereignty or power, or to achieve a lesser political goal. 2. Diplomacy by force. As American General Walter Bedell Smith once said: "Diplomacy has rarely been able to gain at the conference table what cannot be gained or held on the battlefield." 3. What ROBERT S. MCNAMARA once called "the high end of the spectrum of conflict." 4. "An organized bore," according to Oliver Wendell Holmes, Jr. (1841–1935), the U.S. Supreme Court Justice who served in the Civil War as a Union officer. He expressed the common frustration that so much of military action was inaction—specifically, waiting. 5. A state of hostility, conflict or antagonism, e.g. a COLD WAR. 6. Competition for a particular end, e.g. a class war.

war, absolute See ABSOLUTE WAR.

war, accidental See ACCIDENTAL WAR.

war, art of See ART OF WAR.

war, articles of See ARTICLES OF WAR.

war, bilateral A war between adjacent nations.

war, broken-back The fighting that might continue after a full nuclear exchange by the United States and the Soviet Union. In this highly theoretical concept, it is assumed that any surviving military forces would be conventional and thoroughly disorganized.

war, brushfire 1. A short, local war that does not involve outside powers. 2. A local war whose opposing sides become surrogates for and clients of outside powers. 3. Minor military campaigns engaged in by major powers. See Michael Dewar, *Brush Fire Wars: Minor Campaigns of the British Army since 1945* (New York: St. Martin's Press, 1984).

war catalytic See CATALYTIC WAR.

war, central See CENTRAL WAR.

war, civil See CIVIL WAR.

war, cold See COLD WAR.

war, come-as-you-are See COME-AS-YOU-ARE WAR.

war, controlled counterforce A war in which one or both sides concentrates on destroying the military forces of the opponent, and makes a special effort to avoid the civilian population.

war, conventional See CONVENTIONAL WAR.

war, council of See COUNCIL OF WAR.

war, declaration of See DECLARATION OF WAR.

war, dogs of See DOGS OF WAR.

war-fighting 1. Actual combat 2. The events of a nuclear war that would not involve an immediate massive exchange, which would be over in hours, but which instead would follow a slow and controlled path of ESCALATION and involve periods of CEASE-FIRE (or FIRE-BREAKS) for INTRAWAR BARGAINING. The doctrine of war-fighting, which has never quite gained official recognition, is deeply unpopular among the more orthodox proponents of assured-destruction deterrence for two reasons. First, it is claimed that rational preparation to fight a nuclear war makes it more probable that such a war will break out, because it will come to be seen as a possible policy option in a way that an all-out spasm war of the orthodox DETERRENCE theories could not. Second, the demands in terms of sophisticated weapons and COMMAND, CONTROL, COMMUNICATIONS, AND INTELLIGENCE (C^3I) are extremely expensive, whereas a simple, secure SECOND STRIKE capacity need not be. Those who are tempted toward developing war-fighting DOCTRINES and WEAPONS SYSTEMS understandably scorn the established wisdom that a nuclear war can have no winners. Their greatest problem is in explaining how the carefully controlled escalation processes of war-fighting can be expected to work in the unprecedented catastrophic conditions of any CENTRAL STRATEGIC WARFARE. See Barry D. Watts and James O. Hale, "Doctrine: Mere Words, or a Key to War-Fighting Competence?" *Air University Review* 35, 6 (September–October 1984); Stephen J. Cimbala, "War-Fighting Deterrence and Alliance Cohesiveness," *Air University Review* 35, 6 (September–October 1984); John Van Oudenaren, *Deterrence, War-Fighting and Soviet Military Doctrine* (London: International Institute for Strategic Studies, 1986).

war, fortunes of See FORTUNES OF WAR.

war, general See GENERAL WAR.

war, Great Patriotic See GREAT PATRIOTIC WAR.

war, holy See HOLY WAR.

war, honors of See HONORS OF WAR.

war, hot A war in which nations fight each other, as opposed to a COLD WAR, in which bluster, rhetoric, and maneuver are the main armaments.

war, inadvertent A war that results from an error of judgement, a miscalculation of the opposition's intentions, a misinterpreted threat, false intelligence, etc. Compare to ACCIDENTAL WAR.

war, just See JUST-WAR THEORY.

War, Korean See KOREAN WAR.

war, law of See LAW OF WAR.

war, limited See LIMITED WAR.

war, local A war that is confined to a specific geographical area, such as Vietnam or Afghanistan, even though some combatants may be from outside the area. Compare to LIMITED WAR.

war, peripheral See PERIPHERAL WAR.

war, pre-emptive See PRE-EMPTIVE WAR.

war, preventive A war initiated in the belief that fighting, while not necessarily imminent, is inevitable, and that delay would involve even greater risk. A preventive war differs from a PRE-EMPTIVE

ATTACK in the matter of time. The latter is undertaken in the belief that an enemy attack is imminent or under way.

war, principles of See PRINCIPLES OF WAR.

war, proxy See PROXY WAR.

war, total See TOTAL WAR.

War, Vietnam See VIETNAM WAR.

war college A training facility for mid-career military officers studying STRATEGY, TACTICS, STAFF planning, etc. The United States operates five major war colleges. The National War College educates officers to serve on the staff of the Joint Chiefs or multiservice field commands. The Industrial College of the Armed Forces prepares officers for assignments in logistics and procurement. Both schools are part of the National Defense University at Fort McNair, in Washington, D.C. The Army War College in Carlisle, Pennsylvania, the Navy War College in Newport, Rhode Island, and the Air War College in Montgomery, Alabama, each educate lieutenant colonels and colonels and Navy commanders and captains for large commands and senior staff positions. Marine Corps officers attend one of these three. The majority of officers who are promoted to general and admiral are graduates of these war colleges where they study the various arts of military operations.

War College, National See NATIONAL WAR COLLEGE.

war correspondent A reporter assigned by his or her news medium to a theater of operations. Modern war correspondents are products of modern newspapers. While they were initially only able to send letters (or correspondence) back to their papers, they have, since the advent of the telegraph, been relatively instant sources of war news. Their main occupational frustration, besides the prospect of being killed by the warring parties either purposely or accidentally, is military censorship. Today's archetypal war correspondent is more likely to be a television news reporter or a member of a camera team than a traditional print journalist. See Philip Knightly, *The First Casualty* (London: Quartet, 1978).

war crime 1. An act that retains its essential criminality even though it is committed during war and under orders. Thus, many German and Japanese officers were convicted of war crimes after World War II because their conduct went beyond what was considered allowable in war—especially when it involved the murder of PRISONERS OF WAR or the systematic killing of whole populations of innocent, non-combatant civilians. 2. An act that the victor determines to be illegal after a war is over. As Winston Churchill once told his generals apropos the Nuremberg trials of Nazi war criminals after World War II: "You and I must take care not to lose the next war." See Kenneth A. Howard, "Command Responsibility for War Crimes," *Journal of Public Law* 21, 1 (1972); Kurt Steiner, "War Crimes and Command Responsibility: From the Bataan Death March to the MyLai Massacre," *Pacific Affairs* Vol. 58, No. 2 (Summer 1985); Jonathan P. Tomes, "Indirect Responsibility for War Crimes," *Military Review* 66, 11 (November 1986).

war game A simulation, by whatever means, of a military OPERATION involving two or more opposing forces, using rules, data, and procedures designed to depict an actual or assumed real-life situation. See Peter P. Perla and Raymond T. Barrett, "What Wargaming is and is Not," *Naval War College Review* 38, 5 (September-October 1985); Thomas B. Allen, *War Games: The Secret World of the Creators, Players, & Policy Makers Rehearsing World War III Today* (New York, NY: McGraw-Hill Book Company, 1987).

war hawk See HAWK.

war of attrition See ATTRITION.

war of national liberation A Communist phrase for an insurgency that attempts to overthrow an established non-Communist regime.

war powers The legal authority to initiate war. The U. S. Constitution gives to the Congress the authority to declare war, but the president, as COMMANDER-IN-CHIEF, has implied powers to commit the United States's military forces to action. In World War II, the last war in which it actually declared war, Congress was called into emergency joint session by President Roosevelt the day after Pearl Harbor (December 8, 1941), and voted to declare war on Japan. More recently, Congress, concerned with presidential military initiatives during the Vietnam War, has sought to place substantial controls on the president's power to commit American troops to combat.

The War Powers Resolution of 1973 clarifies the respective roles of the president and the Congress in cases involving the use of military forces without a declaration of war. According to the Resolution, the president "in every possible instance" shall consult with the Congress before introducing troops and shall report to the Congress within 48 hours. Use of the armed forces is to be terminated within 60 days (with a possible 30-day extension by the president) unless Congress acts during that time to declare war, enacts a specific authorization for use of the Armed Forces, extends the 60 to 90 day period for termination of the war, or is physically unable to meet as a result of an attack on the United States. At any time before the 60-day period expires, Congress may direct by concurrent resolution that American military forces be removed by the president. While many have contested the constitutionality of the War Powers Resolution, it has never been tested in federal court. See Michael J. Glennon, "The War Powers Resolution Ten Years Later: More Politics Than Law," *American Journal of International Law* Vol. 78, No. 3 (July 1984); Cyrus R. Vance, "Striking the Balance: Congress and the President under the War Powers Resolution," *University of Pennsylvania Law Review* Vol. 133, No. 1 (December 1984); Michael Rubner, "The Reagan Administration, the 1973 War Resolution, and the Invasion of Grenada," *Political Science Quarterly* Vol. 100, No. 4 (Winter 1985–86).

war reserve A stock of MATERIEL amassed in peacetime to meet the increase in military requirements consequent upon an outbreak of war. A war reserve is intended to provide the interim support essential to sustain operations until resupply can be effected.

war room A room in a HEADQUARTERS where current information about a war is maintained on situation maps or charts, together with such other pertinent data as may be desired. It is primarily an orientation, briefing, and conference room.

warfare, acoustic See ACOUSTIC WARFARE.

warfare, antiair See ANTIAIR WARFARE.

warfare, antisubmarine See ANTISUBMARINE WARFARE.

warfare, biological See BIOLOGICAL WARFARE.

warfare, central strategic See CENTRAL STRATEGIC WARFARE.

warfare, chemical See CHEMICAL WARFARE.

warfare, counter-guerrilla See COUNTER-GUERRILLA WARFARE.

warfare, economic See ECONOMIC WARFARE.

warfare, electronic See ELECTRONIC WARFARE.

warfare, guerrilla See GUERRILLA WARFARE.

warfare, integrated See INTEGRATED WARFARE.

warfare, mine See MINE WARFARE.

warfare, mobile See MOBILE WARFARE.

warfare, NBC See NBC WARFARE.

warfare, nuclear See NUCLEAR WARFARE.

warfare, political See POLITICAL WARFARE.

warfare, psychological See PSYCHOLOGICAL WARFARE.

warfare, trench See TRENCH WARFARE.

warfare, unconventional See UNCONVENTIONAL WARFARE.

warhead That part of a MISSILE, PROJECTILE, TORPEDO, ROCKET, or other MUNITION which contains either a nuclear or thermonuclear system, high-explosive system, chemical or biological agents, or inert materials intended to inflict damage. A warhead is very simply the "business end" of a missile or other delivery vehicle, whether it be a submarine torpedo or a SMART BOMB. The term is not used very often by professional strategists and scientists, perhaps because it is too blunt, if honest, and lacks the neutrality of terms such as "re-entry vehicle" or "mission package." It is important to distinguish between the complex machinery, possibly in several stages, that propels, guides, and delivers the explosive or nuclear device, and the device itself. In most forms of weapons in which the warhead/delivery vehicle distinction applies, the latter is more complex, much bigger, and even more expensive than the warhead itself. Furthermore, there is an increasing tendency to try to diminish the sheer destructive power of nuclear warheads in order to reduce COLLATERAL DAMAGE, and as a consequence the importance of accuracy and survivability (against counterweapons) has been emphasized, making more demands on the delivery vehicle than on warhead design. Compare to BUS, FRONT END.

warhead mating The act of attaching a WARHEAD section to a ROCKET or MISSILE body, TORPEDO, airframe, motor, or guidance section.

warning, early See EARLY WARNING.

warning, tactical See TACTICAL WARNING.

warning net A COMMUNICATIONS system established for the purpose of disseminating warning information of enemy movement or action to all interested commands.

warning order An advance notice of an action or an ORDER that is to follow. Usually issued as a brief oral or written message, it is intended to give subordinates time to make necessary plans and preparations. It states a MISSION, in general terms, and the time and place at which a complete order will be issued.

warning phenomenology, dual See DUAL WARNING PHENOMENOLOGY.

warning time, minimum See MINIMUM WARNING TIME.

warrant officer 1. An officer appointed, by warrant, by the United States Secretary of the Army. A highly skilled technician who is provided to fill those positions above the enlisted level which are too specialized in scope to permit the effective development and continued utilization of a broadly trained, branch-qualified COMMISSIONED OFFICER. The rank and precedence of a warrant officer are below that of a second LIEUTENANT, but above those of a CADET and of all NONCOMMISSIONED OFFICERS. A warrant officer has all of the social rights of an officer but none of the command rights. 2. In the United States Navy an officer senior to all chief petty officers and junior to all commissioned officers. A Navy warrant officer ranks below an ensign or a second lieutenant in the United States Marine Corps.

Warsaw Pact A multilateral military alliance formed by the Treaty of Warsaw, signed May 14, 1955, by the Soviet Union, Bulgaria, Czechoslovakia, East Germany, Hungary, Poland, and Rumania. This was ostensibly created as a Communist counterpart to the NORTH ATLANTIC TREATY ORGANIZATION (NATO) and to counter the threat of a remilitarized West Germany. In fact, the parties were already integrated into the Soviet military system through standard treaties of alliance concluded between 1945 and 1948. The primary use of the Warsaw Pact forces and structure is to support the various Soviet military deployments, especially the northern group of forces in Poland, the central group in Hungary and, above all else, the group of Soviet Forces in East Germany. The non-Soviet Warsaw Pact (NSWP) armies are attached in one way or another to these groups of forces, and exercise with them. However, unlike NATO, this peacetime organization, with its attempt to appear as an alliance of equal partners, is not intended to be the command structure for a war. If war were to break out, all NSWP forces would come under the direct command of their associated Soviet force groups, and would serve as reserves and lines-of-communication troops to aid what would in name, as well as fact, be the Soviet Union's war. There is considerable doubt about the political reliability of NSWP forces and, with the exception of some East German divisions, little direct use is expected to be made of the East European forces in front-line combat. Each NSWP army has a Soviet military mission attached to it which supervises and, to some extent, controls training and liaison, with the whole of the Warsaw Pact commanded and organized from Moscow. See Richard A. Gabriel, editor, *NATO and the Warsaw Pact: A Combat Assessment* (Westport, CT: Greenwood Press, 1983); David Holloway and Jane M. O. Sharp, editors, *The Warsaw Pact: Alliance in Transition* (Ithaca, NY: Cornell University Press, 1984); David N. Nelson, editor, *Soviet Allies: The Warsaw Pact and the Issue of Reliability* (Boulder, CO: Westview Press, 1984); Jeffrey Simon, *Warsaw Pact Forces: Problems of Command and Control* (Boulder, CO: Westview Press, 1985).

wartime load The maximum quantity of SUPPLIES of all kinds that a ship can carry.

watch 1. A period of responsibility such as the night watch. 2. Soldiers assigned to guard duty. 3. One of the periods of each day which naval units use for assignments. Such a watch is usually four or eight hours. 4. The naval crew assigned to a watch.

wave A formation of forces, landing ships, landing craft, amphibious vehicles, or aircraft, moved into action at about the same time. Compare to ECHELON.

WAVES Women in the U.S. Navy during World War II; acronym for Women Accepted for Voluntary Emergency Service.

weapon Any object that may be employed to damage an enemy in any way. The degree of simplicity ranges from a rock or a fist to a nuclear warhead.

weapon, absolute See ABSOLUTE WEAPON.

weapon, airborne assault See AIRBORNE ASSAULT WEAPON.

weapon, antipersonnel See ANTIPERSONNEL WEAPON.

weapon, antisatellite See ANTISATELLITE WEAPON.

weapon, binary See BINARY WEAPON.

weapon, biological See BIOLOGICAL WEAPON.

weapon, chemical See CHEMICAL WEAPON.

weapon, conventional See CONVENTIONAL WEAPON.

weapon, directed energy See DIRECTED ENERGY WEAPON.

weapon, electronic beam See ELECTRONIC BEAM WEAPON.

weapon, implosion See IMPLOSION WEAPON.

weapon, kiloton See KILOTON WEAPON.

weapon, megaton See MEGATON WEAPON.

weapon, nominal See NOMINAL WEAPON.

weapon, nuclear See NUCLEAR WEAPON.

weapon, primary See PRIMARY WEAPON.

weapon, salted See SALTED WEAPON.

weapon, secondary See SECONDARY WEAPON.

weapon, semiautomatic See SEMIAUTOMATIC WEAPON.

weapon, subkiloton See SUBKILOTON WEAPON.

weapon, supporting See SUPPORTING WEAPON.

weapon degradation, nuclear See NUCLEAR WEAPON DEGRADATION.

weapon delivery The total action required to locate a TARGET, establish the necessary release conditions for a weapon directed against the target, and maintain guidance of the weapon to the target if required. It includes the detection, recognition, and acquisition of the target, the weapon release, and weapon guidance.

weapon employment time, nuclear See NUCLEAR WEAPON EMPLOYMENT TIME.

weapon maneuver, nuclear See NUCLEAR WEAPON MANEUVER.

weapon of last resort The way the British nuclear force is usually described by the United Kingdom's Ministry of Defence. The British independent deterrent has no WAR-FIGHTING role, as it would now be called, and is not to be seen as part of a strategic escalation doctrine. Instead, it exists only for use "in the last resort," which would presumably be to retaliate against a nuclear strike on the United Kingdom. However, since the United Kingdom does not give clear declaratory policy guidelines for the conditions under which Britain would launch its missiles (so as to increase the deterrent effect through uncertainty), it is unclear what the "last resort" is to be taken to mean. Furthermore, the British nuclear force is officially dedicated to the NORTH ATLANTIC TREATY ORGANIZATION (NATO), and is supposed to come under the targeting plans developed by SACEUR, and since these NATO plans are based on the possible early use of nuclear weapons, the term may have no functional meaning.

weapon package, nuclear See NUCLEAR WEAPON PACKAGE.

weapon–target line An imaginary straight line from a WEAPON to a TARGET.

weapons, dual (multi)-capable See DUAL (MULTI)-CAPABLE WEAPONS.

weapons, dual (multi)-purpose See dual (multi)-purpose weapons.

weapons, enhanced radiation See ENHANCED RADIATION WEAPONS.

weapons, laser See LASER WEAPONS.

weapon(s) accident, nuclear See NUCLEAR WEAPON(S) ACCIDENT.

weapons and ammunition supercharge See SUPERCHARGE, WEAPONS, AND AMMUNITION.

weapons free An AIR DEFENSE WEAPONS CONTROL STATUS used to indicate that WEAPONS SYSTEMS may be fired at any TARGET not positively identified as friendly. See also WEAPONS HOLD; WEAPONS TIGHT.

weapons hold An AIR DEFENSE WEAPONS CONTROL STATUS used to indicate that WEAPONS SYSTEMS may be fired only in self-defense or in response to a formal order. See also WEAPONS FREE; WEAPONS TIGHT.

weapon signature Any smoke, vapor trail, noise, heat, flash tracer, or flight characteristic that denotes a specific WEAPONS SYSTEM (i.e., the sharp crack of an AK-47 rifle). See also SIGNATURE.

weapons list A list of WEAPONS authorized and on hand within tactical or other units employed in a combat role. It includes hand-carried weapons, towed ARTILLERY, and weapons mounted on wheeled or TRACKED VEHICLES.

weapons of mass destruction In ARMS-CONTROL usage, WEAPONS that are capable of a high order of destruction or of being used in such a manner as to destroy large numbers of people. They can be nuclear, chemical, biological, or radiological weapons, but the means of transporting or propelling the weapon, where such means is a separable and divisible part of the weapon, is excluded from the definition.

weapons package A number of NUCLEAR WEAPONS by type and YIELD, planned or approved for employment within a specific area and time span.

weapons system A weapon and those components required for its operation. It can be as small as a RIFLE and its ammunition or as large as an aircraft carrier and its support ships. This general concept is made necessary by the complex technology of modern warfare, and has a wide usage internationally. The whole of a modern military aircraft, including its radar and navigational equipment as much as its missiles, bombs, and guns, can be regarded as one weapons system, because on a technical level they are so highly interdependent. Similarly, it would be pointless to treat only the WARHEADS or RE-ENTRY VEHICLES on a long-range missile as the "weapon," given their uselessness unless they can be dropped onto the target. Therefore, strategists as much as engineers or arms controllers increasingly talk about weapons systems in describing the whole of mechanical units. See Barry S. Cossel and Stephen H. Waters, "A Clear, Simple Guide to Weapon System Troubleshooting," *Defense Management* 19, 1 (First Quarter 1983).

weapon-system employment concept A description in broad terms of the application of a particular piece of equipment or WEAPONS SYSTEM within the framework of tactical concept and future doctrines.

weapons tight An AIR DEFENSE WEAPONS CONTROL STATUS used to indicate that WEAPONS SYSTEMS may be fired only at TARGETS identified as hostile. See also WEAPONS FREE; WEAPONS HOLD.

weather, marginal See MARGINAL WEATHER.

weather central An organization that collects, collates, evaluates, and disseminates meteorological information in such a manner that it becomes a principal source of such information for a given area.

weather minimum The worst weather conditions under which aviation operations may be conducted using either visual or instrument flight rules. Usually prescribed by DIRECTIVES and STANDING OPERATING PROCEDURES in terms of minimum ceiling, visibility, or specific hazards to flight.

wedge 1. A simple tool; a triangular-shaped piece of wood or metal. 2. A tactical FORMATION used since ancient times, which calls for a strong unit to spearhead an attack; it is followed by constantly increasing supporting units.

Wehrmacht The entire German military establishment prior to and during World War II. See Roger A. Beaumont, "On the Wehrmacht Mystique," *Military Review* 66, 7 (July 1986).

weight, gross See GROSS WEIGHT.

weight ton As used in U.S. Army marine operations, a long ton (2,240 pounds).

weighting Those actions taken by a commander to increase the capabilities of a UNIT (e.g., allocation of additional organizations, allocation of priorities of FIRE, or reducing the size of the unit's AREA OF RESPONSIBILITY).

West Point See UNITED STATES MILITARY ACADEMY.

West Point honor code See HONOR CODE, WEST POINT.

West Point Protective Association (WPPA) An imaginary organization that works to further the military careers of all WEST POINT graduates, especially, it is assumed, at the expense of officers who have graduated from other institutions. While the WPPA has not been formally incorporated as an ongoing organization, the fact that some West Point graduates act as if such an organization does exist gives many graduates of other fine educational institutions cause for concern.

West Pointer A graduate of WEST POINT.

white paper Any formal statement of an official government policy with its associated background documentation.

white propaganda PROPAGANDA disseminated and acknowledged by the sponsor or by an accredited agency thereof. Compare to black propaganda.

whole range distance The horizontal distance between the RELEASE POINT and the WHOLE RANGE POINT of a bomb.

whole range point The point vertically below an aircraft at the moment of impact of a BOMB released from that aircraft, assuming that the aircraft's velocity has remained unchanged.

width of sheaf The lateral interval between the centers of flank bursts or impacts of artillery fire. The comparable naval gunfire term is deflection pattern.

WILCO A word used to report that a radiotelephone order has been received and will be carried out. It stands for "Your last message (or message indicated) received, understood, and *will* be *co*mplied with."

wild weasel A missile or an aircraft specially modified with ELECTRONIC COUNTERMEASURES (ECM) equipment to identify, locate, and destroy ground-based enemy AIR-DEFENSE systems that employ SENSORS radiating electromagnetic energy.

wind, quartering See QUARTERING WIND.

windage 1. The force of the wind that acts on a PROJECTILE and alters its original course. 2. A gun-sighting correction made to account for wind. 3. The distance between a gun BORE and the size (diameter) of the AMMUNITION used in it.

windage scale A scale for adjusting a gunsight to allow for the effect of the wind on a bullet in flight. Also called a wind gage.

wind-chill The combined cooling effect of wind and air temperature on heated bodies. The wind-chill is expressed in kilogram-calories per square meter per hour.

wind correction Any adjustment that must be made to allow for the effect of wind; especially, the adjustments to correct for the effect on a PROJECTILE in flight, on sound received by sound-ranging instruments, and on an aircraft flown by DEAD-RECKONING navigation.

wind corrector A mechanical device that computes the correction necessary for the effect of wind, used in sound ranging and artillery FIRE CONTROL.

window 1. Metal reflective material used to confuse enemy RADAR. 2. A military, diplomatic, or political term for a time frame in which there is an opportunity for some achievement or an event of some danger. For example, international developments may create a "window" of opportunity for a diplomatic breakthrough; or a new WEAPONS SYSTEM may be deployed to close a "window" of vulnerability. See Robert H. Johnson, "Periods of Peril: The Window of Vulnerability and Other Myths," *Foreign Affairs* Vol. 61, No. 4 (Spring 1983); Richard Ned Lebow, "Windows of Opportunity: Do States Jump Through Them?" *International Security* Vol. 9, No. 1 (Summer 1984).

wind drift 1. A shift in the apparent position of a sound source or target observed by sound apparatus. Wind drift is caused by the effect of wind on sound waves, which changes their direction and increases or decreases sound lag. 2. The amount of drift experienced by parachutists and supplies dropped by parachute.

wing 1. A U.S. Air Force unit normally composed of one primary MISSION group and the necessary supporting organizations, i.e., organizations designed to render supply, maintenance, hospitalization, and other services required by the primary mission group. Primary mission groups may be functional, such as combat, training, airlift, or service. Commanded by a colonel, a typical fighter wing consists of three fighter squadrons of 25 aircraft each. 2. A fleet air wing; the basic organizational unit for naval aviation. A carrier air wing is commanded by a captain, and contains about 75 aircraft. A typical wing consists of two fighter squadrons, four attack squadrons, and reconaissance and early warning detachments. 3. A balanced Marine Corps TASK ORGANIZATION that contains the aviation elements normally required for the air support of a Marine division. Commanded by a major-general, it may include hundreds of aircraft. 4. A flank unit; that part of a military force to the right or left of the MAIN BODY.

wingman 1. An aviator subordinate to and in support of a designated section leader. 2. The aircraft flown in this role. 3. A U.S. Air Force term for an important assistant; one who "covers" for you in a fight either of a combat or administrative nature.

wings The INSIGNIA awarded to qualified military aviators; thus "to get one's wings" is to qualify as a pilot.

wire entanglement An OBSTACLE of barbed wire. A wire entanglement is used to hold the enemy to areas that can be covered by gunfire, and to delay or prevent an ASSAULT.

wire head The forward limit of telephone or telegraph communications in a command.

wire roll A barrier consisting of a roll of steel wire wound in a continuous spiral which becomes entangled in, and jams, the propelling wheels or tracks of a vehicle. A wire roll is similar to a concertina wire.

wisdom, conventional See CONVENTIONAL WISDOM.

withdrawal, strategic See STRATEGIC WITHDRAWAL.

withdrawal operation A planned operation in which a FORCE in contact disengages from an enemy force; a RE-

TROGRADE operation in which a force frees itself for a new mission. The operation may be conducted with or without enemy pressure, and assisted or unassisted by another unit.

withhold The limiting of authority to employ NUCLEAR WEAPONS by denying their use within specified geographical areas or certain countries.

wooden bomb A theoretical concept that pictures a WEAPON as being completely reliable and having an infinite shelf life while at the same time requiring no special handling, storage, or surveillance.

working limit, forward See FORWARD WORKING LIMIT.

works, field See FIELD WORKS.

World Court The International Court of Justice established by the UNTIED NATIONS Charter in 1945. Located in The Hague (Netherlands), the court consists of fifteen judges elected by the United Nations; each from a different country. Because the Court has no powers of enforcement, it usually considers only cases brought before it by the disputing nations themselves. Consequently, the World Court is not really a court, but rather a panel for the arbitration of minor international disputes. How is "minor" defined here? It means that the case is of such significance to a nation's vital interests that the nation is willing to let an "objective" third party resolve it. In 1986 the World Court ruled against the United States in a case concerning the CONTRAS brought before the Court by the government of Nicaragua. But since the United States did not recognize the Court's jurisdiction in this matter, the ruling had no effect—except that it was of considerable PROPAGANDA value to the government of Nicaragua. See David S. Patterson, "The United States and the Origins of the World Court," *Political Science Quarterly* Vol. 91, No. 2 (Spring 1976); Abram Chayes, "Nicaragua, the United States and the World Court," *Columbia Law Review* Vol. 85, No. 7 (November 1985).

World-Wide Military Command and Control System (WWMCCS) A communications network linking American military forces.

wounded in action A battle CASUALTY other than "KILLED IN ACTION." The term encompasses all kinds of wounds and other injuries incurred in action, whether there is a piercing of the body or not.

X

x-axis A horizontal axis in a system of rectangular coordinates; that line on which distances to the right or left (east or west) of the reference line are marked, especially on a map, chart, or graph.

x-site An outside storage site, unbarricaded but with temporary covers, for the temporary storage of AMMUNITION.

Y

yankees 1. The United States in general; all American citizens. 2. The northerners during the American Civil War. 3. All American military personnel. 4. The colonial settlers of New England. There are several explanations of the origin of the word. It was the theory of James Fenimore Cooper (1789–1851), the novelist who wrote *The Last of the Mohicans* (1826) and the *Deerslayer* (1841), that "yankee" is how the Indians pronounced "l'anglais" the French word for "the English." But other authorities such as H. L. Mencken (1880–1956) contend that it is a corruption of "Jan Kees," an unkind nickname for a Dutchman meaning John Cheese. The Dutch who first settled New York applied the name to the English who followed. By the mid-eighteenth century the word was used to refer to any resident of the English colonies.

yaw 1. The rotation of an aircraft, ship, or missile about its vertical axis so as to cause the longitudinal axis of the aircraft, ship, or missile to deviate from the flight line or heading in its horizontal plane. 2. The angle between the longitudinal axis of a PROJECTILE at any moment and the tangent to the TRAJECTORY in the corresponding point of flight of the projectile.

y-axis A vertical axis in a system of rectangular coordinates; that line on which distances above or below (north or south) the reference line are marked, especially on a map, chart, or graph.

yield See NUCLEAR YIELDS.

young Turks 1. Younger Turkish army officers who sought reforms in the Ottoman Empire in the decade prior to World War I. 2. Any newer members of an organization or party that seek to significantly reform it.

Y-site An outside storage site, with earthen barricades on four sides, for the temporary storage of AMMUNITION. Although normally open, a Y-site may be covered with improvised coverings.

Z

zero deflection An adjustment of a GUNSIGHT exactly parallel to the axis of the BORE of the gun to which it is attached.

zeroed out 1. A nonoperating condition in which a military unit has been reduced to zero strength, and in which the unit's equipment has been placed in administrative storage under the care of a designated unit or activity. 2. The total destruction of something.

zeroize 1. To align the variable cryptographic elements (e.g., rotors) of a piece of CRYTOEQUIPMENT to a specified basic setting unrelated to operational settings. 2. To destroy the setting of machine elements automatically upon the occurrence of an untoward event; e.g., a crashing impact, the loss of electrical power, or possible capture by the enemy.

zero-length launching A technique in which the first motion of a missile or aircraft removes it from its LAUNCHER.

zero point The location of the center of the BURST of a NUCLEAR WEAPON at the instant of detonation. The zero point may be in the air, or on or beneath the surface of land or water, depending upon the type of burst, and it is thus to be distinguished from GROUND ZERO.

zero-sum game 1. An international political/military perspective that views potential gains for one side as a loss for the other; for one player to win, another must lose. 2. A game in which the total of the payoffs to the players is zero regardless of the outcome.

zone 1. The area of responsibility assigned to a UNIT by the drawing of boundaries. It generally applies to OFFENSIVE operations. 2. Any tactical area of importance, generally parallel to the front, such as a fortified area, a defensive position, a combat zone, or a traffic control zone. 3. A strip of several bands or belts of wire entanglements placed in depth. 4. An area in which PROJECTILES will fall when a given PROPELLING CHARGE is used and the elevation of the weapon firing the projectiles is varied between the minimum and the maximum. Compare to SECTOR.

zone, beaten See BEATEN ZONE.

zone, killing See KILLING ZONE.

zone, landing See LANDING ZONE.

zone fire ARTILLERY or MORTAR fire that is delivered in a constant direction at several QUADRANT ELEVATIONS.

zone of action A tactical subdivision of a larger area, the responsibility for which is assigned to a TACTICAL UNIT; generally applied to offensive action. See also SECTOR.

zone of attack An area forward of the LINE OF CONTACT assigned to a force having a mission to attack, normally delineated by boundaries extending forward into enemy territory. It delineates an area and direction of movement when close coordination and operation is required between adjacent units. The next higher commander assigns a zone of attack, and subordinate commanders may further subdivide it for their units. When assigned a zone of attack, units should not maneuver into an adjacent zone without coordination with the adjacent unit commander or the next higher commander. They may, however, engage clearly identified enemy in an adjacent zone by DIRECT FIRE so long as no friendly units are endangered.

zone of fire An area within which a designated ground unit or fire support ship delivers, or is prepared to deliver, FIRE SUPPORT. Fire may or may not be observed. Normally, the zone of fire of an ARTILLERY unit coincides with the zone of action of the supported force. See also CONTINGENT ZONE OF FIRE.

zone of fire, contingent See CONTINGENT ZONE OF FIRE.

zone of interior That part of a national territory not included in a THEATER OF OPERATIONS.

zone reconnaissance A directed effort to obtain detailed information about all routes, obstacles (including chemical or radiological contamination), terrain, and enemy forces within a ZONE defined by boundaries. A zone-reconnaissance mission is normally assigned when the enemy situation is in doubt, or when information about cross-country trafficability is desired. Compare to AREA RECONNAISSANCE, ROUTE RECONNAISSANCE.

zulu time See Greenwich Mean Time.